Lecture Notes in Civil Engineering

Volume 356

Lecture Notes in Civil Engineering (LNCE) publishes the latest developments in Civil Engineering—quickly, informally and in top quality. Though original research reported in proceedings and post-proceedings represents the core of LNCE, edited volumes of exceptionally high quality and interest may also be considered for publication. Volumes published in LNCE embrace all aspects and subfields of, as well as new challenges in, Civil Engineering. Topics in the series include:

- Construction and Structural Mechanics
- Building Materials
- Concrete, Steel and Timber Structures
- Geotechnical Engineering
- Earthquake Engineering
- Coastal Engineering
- Ocean and Offshore Engineering; Ships and Floating Structures
- Hydraulics, Hydrology and Water Resources Engineering
- Environmental Engineering and Sustainability
- Structural Health and Monitoring
- Surveying and Geographical Information Systems
- Indoor Environments
- Transportation and Traffic
- Risk Analysis
- Safety and Security

To submit a proposal or request further information, please contact the appropriate Springer Editor:

- Pierpaolo Riva at pierpaolo.riva@springer.com (Europe and Americas);
- Swati Meherishi at swati.meherishi@springer.com (Asia—except China, Australia, and New Zealand);
- Wayne Hu at wayne.hu@springer.com (China).

All books in the series now indexed by Scopus and EI Compendex database!

Wenhui Duan · Lihai Zhang · Surendra P. Shah
Editors

Nanotechnology in Construction for Circular Economy

Proceedings of NICOM7, 31 October–02 November, 2022, Melbourne, Australia

Editors
Wenhui Duan
Department of Civil Engineering
Monash University
Clayton, VIC, Australia

Lihai Zhang
Department of Infrastructure Engineering
University of Melbourne
Parkville, VIC, Australia

Surendra P. Shah
Civil and Environmental Engineering
Northwestern University
Evanston, IL, USA

ISSN 2366-2557 ISSN 2366-2565 (electronic)
Lecture Notes in Civil Engineering
ISBN 978-981-99-3332-7 ISBN 978-981-99-3330-3 (eBook)
https://doi.org/10.1007/978-981-99-3330-3

This Springer imprint is published by the registered company Springer Nature Singapore Pte Ltd.
The registered company address is: 152 Beach Road, #21-01/04 Gateway East, Singapore 189721, Singapore

Organization

Advisory Committee

Scientific Advisory Committee

C. Caprani
H. Huang
D. Wu
B. Sainsbury
D. Robert
T. Yu
T. Ren
A. Remennikov
Y. Zhang
P. D. Silva
W. Gao
G. Li
X. Liu
K. Vessalas
R. Erkmen
R. Shrestha
J. Li
P. Thomas
N. Gowripalan
S. Nejadi
H. Wang
P. Mutton
R. Zou
Y. Huang
V. Tam
Y. Zhuge
M. Ghodrat
J. Zhao
Q. Zhang
M. M. Alam
K. Le

Local Organizing Committee

L. Zhang (Chair)
K. Sagoe-Crentsil (Co-Chair)
B. Chang (Secretary)
S. Miramini (Coordinator)
D. Chen (Coordinator)
S. Zhang
X. Yao
H. Sui
Y. Liu

F. Basquiroto
H. Nguyen
W. Wang

Award Committee

Y. Mai
D. Nethercot
J. Torero
S. T. Quek
R. Amal
M. Bradford
S. Kitipornchai
R. Kell
P. Phillip
A. Paradowska
R. Yeo

NICOM7 Preface

The Seventh International Symposium on Nanotechnology in Construction (NICOM7), with the theme of "Nanotechnology in Construction for Circular Economy", provided a bridge between research advances and industry/commercial opportunities within the construction sector underpinned by nanoscience. To this end, NICOM7 showcased the latest developments across the entire supply chain involving researchers, manufacturers, suppliers, and end users.

NICOM7 was delivered via a hybrid mode (onsite & online) for a total of 8 plenary/keynote speeches and 12 parallel sessions containing 61 presentations and attracting nearly 100 attendees from Australia, China, Germany, Brazil, USA, India, UK, and Singapore. It was jointly hosted by Monash University and the University of Melbourne, Australia, and coordinated by an Advisory Committee (S. P. Shah, USA; K. Wang, USA; K. Sobolev, USA; M. S. Konsta-Gdoutos, Greece; L. Ferrara, Italy), Scientific Advisory Committee (49 members), Local Organising Committee (12 members), and Awards Committee (11 members). The editors acknowledge their excellent contributions that enabled successful delivery of NICOM7, as well as support from all reviewers, helpers, and conference organizers.

This proceeding volume is based on contributions presented at NICOM7. The 56 high-quality papers/extended abstracts cover a wide range of topics across nanotechnologies in concrete structures, structural health monitoring, nanocomposite cement replacements rails, pavements, AI and nanomodification of cementitious materials, the alkali–silica reaction, concrete durability, geopolymer concretes, etc. The focus of the proceedings is next-generation nanotechnologies for the broader construction sector. The editors sincerely thank the authors for their outstanding contributions, and hope the NICOM7 experience will promote innovation to enhance the rapid promotion and integration of nanoscience into mainstream construction practices. The editors also anticipate that the proceedings of NICOM7 will have great value for both the research and engineering communities by advancing nanotechnology initiatives towards a globally sustainable and innovative construction sector.

Clayton, Australia

Surendra P. Shah
Honorary Chair

Parkville, Australia

Wenhui Duan
Conference Chair

Evanston, USA

Lihai Zhang
Conference Co-Chair

Contents

Nonlinear Wind-Induced Vibration Behaviors of Multi-tower Suspension Bridges Under Strong Wind Conditions

R. Zhou, Y. J. Ge, Y. Yang, Y. D. Du, and L. H. Zhang

Abstract The aerodynamic characteristics of a multispan suspension bridge differ from those of a two-span suspension bridge. In this study we investigated the nonlinear aerodynamic characteristics of the Maanshan Bridge under nonstationary flow using combination quasi-3D finite element (FE) bridge models of 2D nonlinear aerodynamic force models and 3D nonlinear FE bridge models. The developed model predictions were validated by wind tunnel tests involving a 2D sectional stiffness model and 3D full-bridge aeroelastic model. Results showed that the developed model could potentially describe the nonlinear and unsteady aerodynamic effects on the bridge. Furthermore, the flutter behavior of the Maanshan Bridge under uniform flow changed from the stable limit cycle of soft flutter to unstable limit cycle with the disconnection of two hangers at the 1/2L of the right main span, while the flutter behavior of the bridge under turbulence flow could be defined as the fracture failure of the hangers from the 1/2L of the left main span.

Keywords Failure mode · Multispan suspension bridges · Nonlinear flutter behavior · Quasi-3D FE model

1 Introduction

Due to the advantages of not sharing anchorage, relatively shorter main span and lower cost, more and more multi-tower suspension bridges are being built worldwide, especially for the bridges with the main span >1000 m [1]. The Taizhou and Maanshan

R. Zhou (✉) · Y. D. Du
College of Civil and Transportation Engineering, Shenzhen University, Shenzhen, China
e-mail: zhourui@szu.edu.cn

Y. J. Ge · Y. Yang
State Key Lab for Disaster Reduction in Civil Engineering, Tongji University, Shanghai, China

L. H. Zhang
Department of Infrastructure Engineering, University of Melbourne, Melbourne, VIC, Australia

W. Duan et al. (eds.), *Nanotechnology in Construction for Circular Economy*, Lecture Notes in Civil Engineering 356,
https://doi.org/10.1007/978-981-99-3330-3_1

Yangtze River bridges with a main span of 1080 m are two excellent examples of super long-span three-tower suspension bridges in China. The structural characteristics and aerodynamic performance of a multi-tower suspension bridge are significantly different from that of a two-tower suspension bridge [2], and they are susceptible to flutter instability under wind loading [3]. Under strong wind loads, the wind-induced vibrations and post-critical flutter behaviors of long-span multi-tower bridges remain challenges for wind engineers.

Because the nonlinear flutter behavior of long-span multi-tower bridges is a complex phenomenon involving aerodynamic nonlinearities and structural nonlinearities due to structural large deformations, which can result in damage of partial components, and ultimately the collapse of the whole bridge structure [4]. In the past decades, most research focused on modeling the self-excited forces using rational or indicial functions in the time domain (e.g., Chen and Kareem [5]; Diana et al. [6]). However, the unsteady and nonlinear effects of the aerodynamic forces produced by the wind–structure interaction were been simultaneously considered in those studies. Recently, Wu and Kareem [7] presented a nonlinear convolution scheme based on Volterra-Wiener theory, and Arena et al. [8] used a nonlinear quasi-steady aerodynamic model for time-periodic oscillations of suspension bridges and conducted global bifurcation analysis of post-critical behaviors. Gao et al. proposed a nonlinear self-excited force model in terms of nonlinear flutter derivatives [9], and Xu et al. studied the flutter performance and hysteresis phenomena of a streamlined bridge deck sectional model using a large-amplitude free vibration test [10]. More importantly, Liu proposed a nonlinear aerodynamic force model (NAFM) based on nonlinear differential equations that could produce aerodynamic hysteresis phenomena [11], and Zhou et al. developed the NAFM by considering the vortex-induced force and then analyzed the nonlinear wind-induced behavior of long-span bridges [12, 13]. Through a series of wind-tunnel tests and three-dimensional nonlinear finite-element (FE) analyses, we further compared the comprehensive wind-resistance performance of a suspension bridge with various slot ratios [12], vertical stabilizers [14, 15], grid porosities [16], guide plates [17, 18], and combination of aerodynamic measures [19–21]. Previous studies have shown that nonlinear FE models of a 3D bridge incorporated with the NAFM is a commendable approach to simulating the nonlinear behavior of wind-induced vibration of bridges under strong wind.

This study aimed to understand the nonlinear dynamic behaviors in flutter and post-flutter of multi-tower suspension bridges under strong wind excitation. Firstly, the 2D displacement responses and the aerodynamic forces in the NAFM of a closed-box girder were calculated based on the computational fluid dynamics (CFD) simulation. Subsequently, an integrated numerical approach for a 3D three-tower suspension bridge under different wind excitations using a combination of a NAFM and nonstationary flows was developed. Finally, the nonlinear displacement responses and flutter collapse of the 3D bridge under uniform and turbulent flow, respectively, were analyzed. The present study could potentially contribute to further understanding of the flutter mechanism of multi-tower suspension bridges.

2 NAFM of Bridges

2.1 Three-Tower Suspension Bridge

A typical three-tower suspension bridge with a span arrangement of 360 + 2 × 1080 + 360 m and a sag-to-span ratio of the main cable of 1/9 was studied. All three towers are 176 m high with the middle tower being 128 m above the deck, and each of the side towers was 143 m above the deck. The distance between the two cables was 35 m, and the spacing between two adjacent hangers was 16 m. As shown in Fig. 1, the deck cross-section of the bridge was a closed streamlined box steel girder of 38.5 m wide × 3.5 m deep, in which the vertical and torsional frequencies were 0.08 ad 0.26 Hz, respectively.

2.2 2D Displacement Responses of the Closed-Box Girder

Based on the fluid–structure interaction in the CFD simulation, an unstructured grid system incorporating large eddy simulation modeling with a Smagorinsky subgrid-scale model was used to discretize the governing Navier–Stokes equations. A steady uniform flow velocity was applied at the inlet boundary, and an opening pressure condition was given at the outlet boundary. The overall mesh division around two closed-box girders in the CFD simulations with the computational domain of a 60B × 40B rectangle is described in Fig. 2a. As illustrated in Fig. 2b, the vertical displacement of the bridge deck remained stable when the wind velocity (U) increased from 0 to ≈78 m/s, and then rapidly increased when the U was >80 m/s. Finally, the maximum vertical and torsional displacement responses were as high as ≈2.5 h/H and 80°, respectively. It can be seen that the torsional displacement reached ≈5° under U = 75.5 m/s as the soft flutter phenomenon. Therefore, the critical flutter wind velocity of the three-tower suspension bridge from the CFD simulation was 147.5 m/s. Moreover, the hysteresis loops of F_D, F_L, and F_M predicted by the NAFM were close to those from the CFD simulation shown in Fig. 2c, d. The parameters of static force, self-excited force, and buffeting force in the NAFM of the closed-box girder were identified.

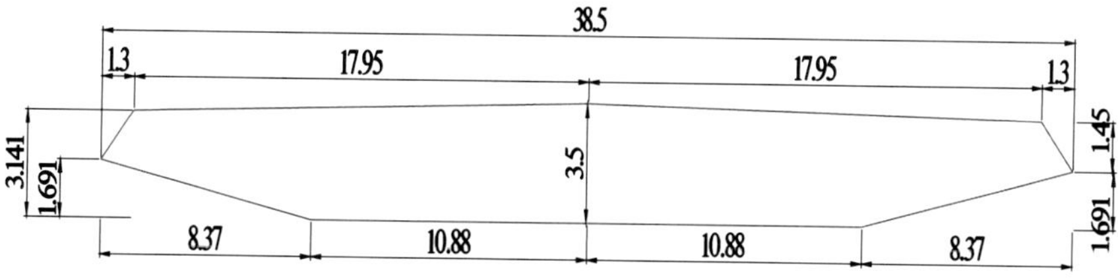

Fig. 1 Cross-section of the closed-box girder (unit: m)

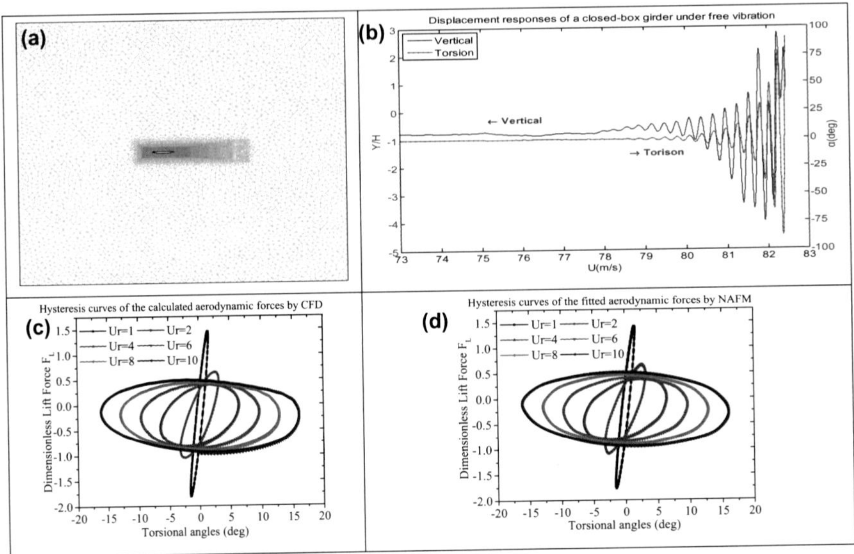

Fig. 2 Parameters in the NAFM: **a** overview of the CFD simulation of the closed-box deck; **b** vertical and torsional displacement responses; **c**, **d** hysteresis curves of lifting force and torsional angle of the CFD and NAFM

3 Nonlinear Flutter and Post-Flutter Behaviors of a Multi-tower Bridge

3.1 Turbulence Flow at Bridge Site

The reference height and the corresponding average wind velocity at the bridge site were $Z_{ref} = 57.83$ m and $U_{ref} = 39.3$ m/s, respectively, and $z_0 = 0.01$ and $\alpha = 0.12$ because the bridge site belongs to the B-type terrain. Based on the combination of weighted amplitude wave superposition and Fast Fourier Transform (FFT) technique, the time histories of the vertical and horizontal (along-bridge) wind velocities at the height of $z = 54.8$ m at the middle point of the two main spans, are shown in Fig. 3.

Nonlinear 3D FE models of the three-tower suspension bridge were established with a total of 1228 elements. The nonlinear governing coupled equations in the integrated FE model were numerically solved using the Newton–Raphson method in combination with the Newmark-β method. Accordingly, the turbulent flow at the bridge site was firstly simulated to reflect the influence of turbulent flow on aerodynamic performance. Then the nonlinear behavior of the displacement responses, structural frequencies, oscillation configurations, and failure modes of the 3D bridge under uniform and turbulent flow, respectively, were obtained.

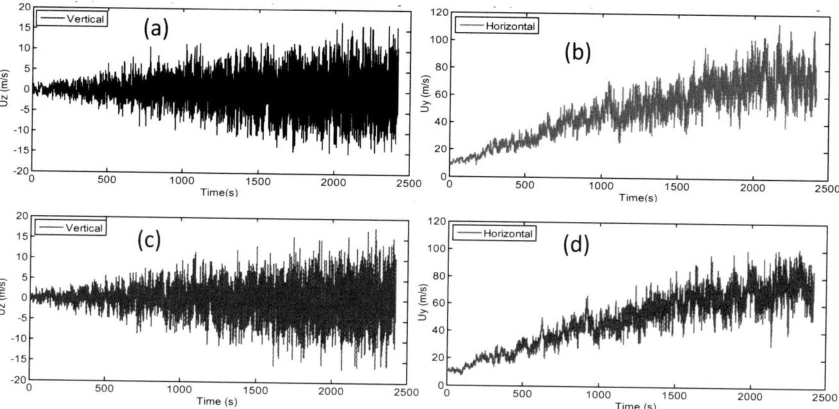

Fig. 3 Time histories of turbulent flow of a multi-tower suspension bridge: **a**, **b** vertical and horizontal wind velocities at the left main span; **c**, **d** vertical and horizontal wind velocities at the right main span

3.2 *Flutter and Post-Flutter Behaviors Under Uniform Flow*

As presented in Fig. 4, the relative vertical displacement (Y/H) and torsional displacement responses (α) of the right main span rapidly decreased and approached a balance location at wind velocity of U = 70 m/s. However, there was an obvious soft flutter phenomenon of the bridge when the wind velocity increased to U = 75 m/s. In particular, the vertical and torsional displacements gradually become larger after 100 s, and then maintained a sinusoidal oscillation after 250 s with the relative value of Y/H ≅ 0.5 and α = 3.5°. Subsequently, the displacement responses presented an ever-increasing trend with increasing wind velocity. Both the vertical and torsional displacements rapidly increased after 60 s under U = 82.5 m/s and divergence finally occurred with the extreme values of Y/H ≅ 5 and α = 30°. As a result, the soft flutter phenomenon of the bridge under U = 75 m/s fell into a stable limit cycle. In addition, Fig. 4 shows that the spatially dependent lateral amplitudes along the bridge span were generally very small in comparison with the vertical or torsional components of oscillation, and the motion configuration of the two main spans was an antisymmetrical vertical and torsional coupled oscillation. The failure of the whole bridge occurred after two hangers were finally damaged at the middle ½ L of the right main span under U = 82.5 m/s. Therefore, the whole flutter collapse of the three-tower suspension bridge under uniform flow can be defined as the change from the stable limit cycle of soft flutter to the unstable limit cycle of bending-torsional coupled divergence.

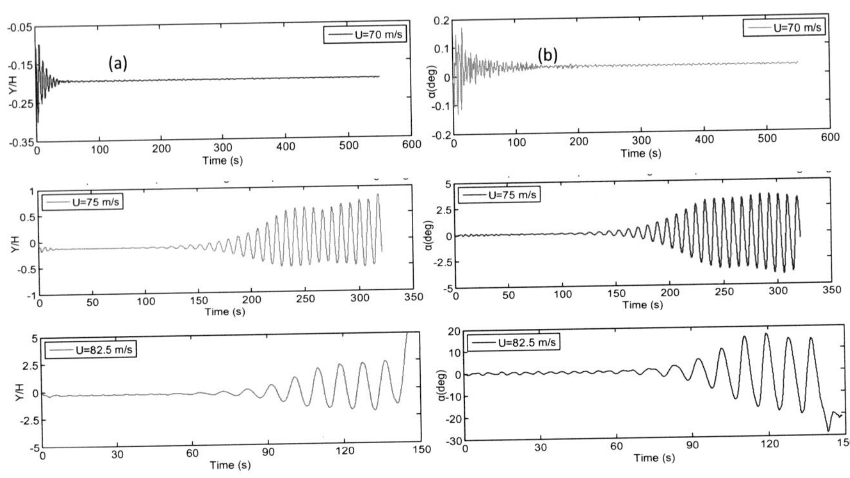

Fig. 4 Time histories of displacement responses: **a** vertical; **b** torsional

3.3 *Displacement Responses and Failure Modes Under Turbulent Flow*

As shown in Fig. 5, the time-dependent torsional displacement response showed a nonlinear growth rate with increasing wind velocity, in which the maximum α was close to 8° under U = 72 m/s and reached 20° under U = 75 m/s. Additionally, the amplitude of torsional oscillation was greatest in three direction vibrations, and the motion configuration of the two main spans was antisymmetrical coupled oscillation. As for the failure mode of the bridge, the failure of the hangers started from those located in the middle ½ L of the left main span of the bridge. Therefore, the whole flutter collapse of the three-tower suspension bridge under turbulent flow directly shifted from the paroxysmal bifurcation to the chaos of bending-torsional coupled divergence.

3.4 *Comparison of Displacement Responses*

Finally, the calculated relationship between the displacement responses and wind velocity using the integrated approach were compared with the experimental results of full-bridge aeroelastic model wind-tunnel tests [22]. It can be seen that all the calculated U_{cr} are generally higher than the checked flutter wind velocity. The minimum experimental U_{cr} was 74.2 m/s at the wind attack angle of +3° under uniform flow, which was lower than the calculated value of $U_{cr} = 82.5$ m/s. However, the experimental measurement (i.e., $U_{cr} = 85.8$ m/s) was higher than the calculated value (i.e., $U_{cr} = 75$ m/s) under turbulent flow. Moreover, Fig. 6 shows the relationship between

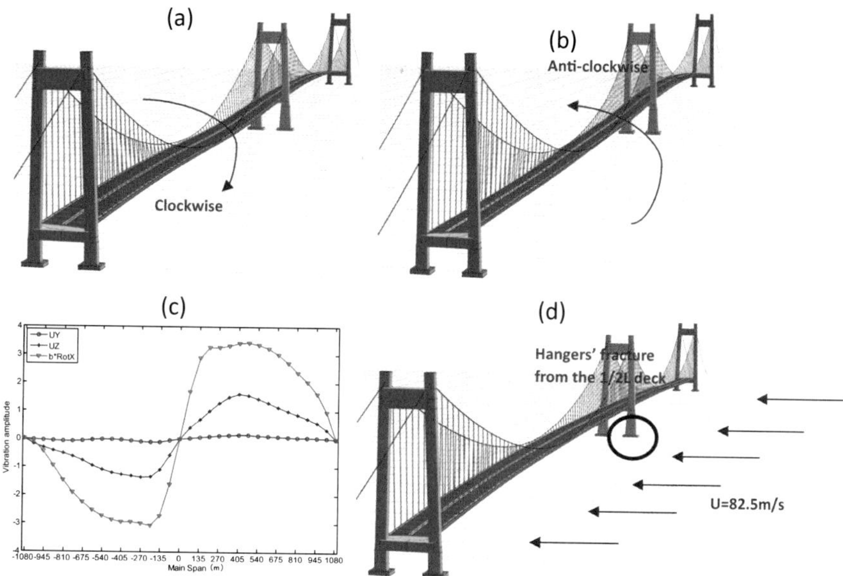

Fig. 5 Flutter collapse of the bridge: **a** clockwise rotation of the left main span; **b** clockwise rotation of the right main span; **c** oscillation configuration; **d** failure mode

the maximum displacement responses and wind velocity at the midspan, the quarter point (1/4 L near the side tower) and the three-quarter point (3/4 L near the middle tower) of the bridge. It demonstrates that all three calculated maximum displacement responses under uniform flow showed a stable upward trend with increasing wind velocity, and the trend became more dramatic when U was over certain threshold (i.e., U = 75 m/s, soft flutter). In addition, the maximum value of torsional displacement (i.e. $U_{cr} = 82.5$ m/s) was the largest among the three dispacements, followed by the vertical displacement. It should also be mentioned that there was a sudden increase in all three experimental maximum displacement responses when the value of U approached 90 m/s under 0° wind attack. Further, all three maximum displacement responses at the midspan of the bridge were the largest compared with other bridge locations, while the values of the displacement responses at ¼ L near the side tower were the smallest. Although the values of all three maximum displacement responses under turbulent flow also increased with increasing the wind velocity, the growth rates under turbulent flow were much higher than those under uniform flow (Fig. 7).

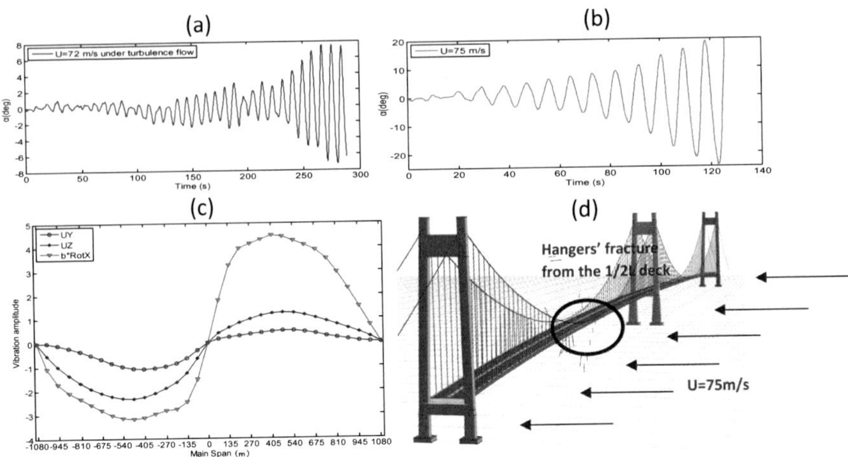

Fig. 6 Flutter collapse of the bridge under turbulent flow: **a**, **b** torsional displacement responses at U = 72 m/s and U = 75 m/s; **c** oscillation configuration; **d** failure mode

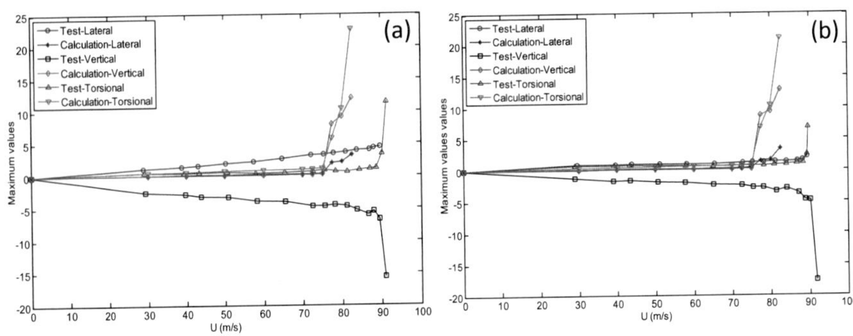

Fig. 7 Comparison of displacement responses: **a**, **b** 1/2 point under uniform flow and turbulent flow

4 Concluding Remarks

In this study, a time-dependent numerical approach was developed to investigate the nonlinear flutter and post-flutter behaviors of a three-tower suspension bridge under different wind excitations. The major findings were:

(1) The oscillation configuration of the bridge was very dependent on the coupled antisymmetrical bending-torsional oscillation of the two main bridge spans. The flutter performance under uniform flow (U_{cr} = 82.5 m/s) was better than that under turbulent flow (U_{cr} = 75 m/s) due to the increase in the vertical degree participation in the coupled motion.

(2) Under uniform flow, the flutter collapse process of the bridge can be described as the change from the stable limit cycle of soft flutter to the unstable limit cycle with disconnection failure of two hangers at the middle of the right main span.
(3) Under turbulent flow, the failure mode of the bridge can be described as the sequential fracture failure of multiple hangers at the middle of the left main span of the bridge, and the flutter collapse process of the bridge can be defined as the direct shift from the paroxysmal bifurcation to chaos.

The present study presents some new sights into nonlinear dynamic behaviors in the flutter collapse process of a three-tower suspension bridge. It should be mentioned that the spatial spanwise effects along the bridge of the aerodynamic forces were not taken into account in this study and should be investigated in future research.

Acknowledgements The authors gratefully acknowledge the support for the research work jointly provided by the Guangdong Province Natural Science Foundation (No. 2023A1515030148, 2019B111106002, 2019A1515012050), National Science Foundations of China (Nos. U2005216, 51908374, and 52178503), and the Shenzhen Science and Technology Program under grant (Nos. JCYJ20220531101609020, KQTD20180412181337494, and ZDSYS20201020162400001).

References

1. Wang H, Tao TY, Zhou R, Hua XG, Kareem A (2014) Parameter sensitivity study on flutter stability of a long-span triple-tower suspension bridge. J Wind Eng Ind Aerodyn 128:12–21
2. Zhou R, Ge YJ, Yang YX et al (2020) Aerodynamic performance evaluation of different cable-stayed bridges with composite decks. Steel Compos Struct 34(5):699–713
3. Zhang WM, Qian KR, Ge YJ (2021) Research on flutter-mode transition of a triple-tower suspension bridge based on structural nonlinearity[J]. Structures 34:787–803
4. Kareem A, Wu T (2013) Wind-induced effects on bluff bodies in turbulent flows: nonstationary, non-Gaussian and nonlinear features. J Wind Eng Ind Aerodyn 122:21–37
5. Chen XZ, Kareem A (2003) Aeroelastic analysis of bridges: effects of turbulence and aerodynamic nonlinearities. J Eng Mech 129(8):885–895
6. Diana G, Resta F, Rocchi D (2008) A new numerical approach to reproduce bridge aerodynamic nonlinearities in time domain. J Wind Eng Ind Aerodyn 96(10–11):1871–1884
7. Wu T, Kareem A (2013) A nonlinear convolution scheme to simulate bridge aerodynamics. Comput Struct 128:259–271
8. Arena A, Lacarbonara W, Marzocca P (2016) Post-critical behavior of suspension bridges under nonlinear aerodynamic loading. J. Comput. Nonlinear Dyn 11(1):011005:1–11
9. Gao GZ, Zhu L, Li JW, et al. (2020) A novel two-degree-of-freedom model of nonlinear self-excited force for coupled flutter instability of bridge decks. J Sound Vib 480:115406
10. Xu FY, Ying J, Zhang MJ et al (2021) Experimental investigations on post-flutter performance of a bridge deck sectional model using a novel testing device. J Wind Eng Ind Aerodyn 217:104752
11. Liu SY (2014) Nonlinear aerodynamic model and non-stationary whole process wind response of long span bridges. PhD dissertation, Tongji University, China
12. Zhou R, Ge YJ, Yang YX et al (2018) Wind-induced nonlinear behaviors of twin-box girder bridges with various aerodynamic shapes. Nonlinear Dyn 94:1095–1115
13. Zhou R, Ge YJ, Yang YX et al (2019) A nonlinear numerical scheme to simulate multiple wind effects on twin-box girder suspension bridges. Eng Struct 183:1072–1090
14. Zhou R, Yang YX, Ge YJ et al (2018) Comprehensive evaluation of aerodynamic performance of twin-box girder bridges with vertical stabilizers. J Wind Eng Ind Aerodyn 175:317–327

15. Zhou R, Ge YJ, Liu SY et al (2020) Nonlinear flutter control of a long-span closed-box girder bridge with vertical stabilizers subjected to various turbulence flows. Thin Wall Struct 149:106245
16. Zhou R, Ge YJ, Liu QK et al (2021) Experimental and numerical studies of wind-resistance performance of twin-box girder bridges with various grid plates. Thin Wall Struct 166:108088
17. Zhou R, Ge YJ, Yang YX et al (2019) Nonlinear behaviors of the flutter occurrences for a twin-box girder bridge with passive countermeasures. J Sound Vib 447:221–235
18. Zhou R, Lu P, Gao XD et al (2023) Role of moveable guide vane with various configurations in controlling the vortex-induced vibration of twin-box girder suspension bridges: an experimental investigation. Eng Struct 281:115762
19. Zhou R, Yang YX, Ge YJ et al (2015) Practical countermeasures for the aerodynamic performance of long span cable-stayed bridge with open deck. Wind Struct An Int J 21(2):223–229
20. Zhou R, Ge YJ, Yang YX,et al. (2023) Effects of vertical central stabilizers on nonlinear wind-induced stabilization of a closed-box girder suspension bridge with various aspect ratios. J Nonlinear dynam 111:9127–9143
21. Yang YX, Zhu JB, Zhou R et al (2023) Aerodynamic performance evaluation of steel-UHPC composite deck cable-stayed bridges with VIV countermeasures. J Constr Steel Res 03:107815
22. Zhang WM, Ge YJ (2014) Flutter mode transition of a double-main-span suspension bridge in full aeroelastic model testing. J Bridge Eng 19(7):06014004

Thermal Transfer Effects of CRTS II Slab Track Under Various Meteorological Conditions

R. Zhou, W. H. Yuan, Y. D. Du, H. L. Liu, and L. H. Zhang

1 Introduction

With the evolution of climate change, the thermal transfer effects of ballastless track in high-speed railways under complicated environmental conditions becomes increasingly important, governed by a number of meteorological factors, including solar radiation, ambient temperature, wind speed and direction, humidity, and many others [1]. Because these meteorological factors are highly site-specific, the huge area traversed by high-speed railway in China is affected by varying meteorological conditions that could have a significant effect on the mechanical behavior of track–bridge systems. The China Railway Track System type II (CRTS II) is a typical ballastless slab track for high-speed railway systems. In order to guarantee good structural performance for long-term operation, it is necessary to study the influence of meteorological conditions on the thermal transfer effects of CRTS II track.

In recent years, more and more researchers have studied the temperature field and the CRTS II track based on monitoring of field data. Dai et al. [2] and Huang et al. [3] investigated the temperature distribution characteristics of the CRTS II track using conventional statistical methods, while Yang et al. [4] and Song et al. [5] revealed the relationship between meteorological factors and the internal temperature of the CRTS II track through finite element models analysis. Using temperature tests of scaled models, Cai et al. [6] and Zhou et al. [7] investigated the influence of cyclic and overall temperature on the displacement, strain, and temperature field of a scaled CRTS II track–bridge structure. Furthermore, Zhu et al. [8] and Zhang et al. [9] explored interfacial damage development of CRTS II track under complex

R. Zhou (✉) · W. H. Yuan · Y. D. Du · H. L. Liu
College of Civil and Transportation Engineering, Shenzhen University, Shenzhen, China
e-mail: zhourui@szu.edu.cn

L. H. Zhang
Department of Infrastructure Engineering, University of Melbourne, Melbourne, VIC, Australia

W. Duan et al. (eds.), *Nanotechnology in Construction for Circular Economy*, Lecture Notes in Civil Engineering 356,
https://doi.org/10.1007/978-981-99-3330-3_2

temperature conditions in a cohesive zone model (CZM) and concrete damaged plasticity model. In addition, Zhou et al. [10–12] studied the mechanical behavior of CRTS II track under the coupling effect of train and environment loads.

2 Meteorological Parameter Collection

Using a typical zone of CRTS II track on a bridge in the Beijing–Shanghai high-speed railway system as a case study, the meteorological data and internal temperature of the track structure were collected for 6 months. The relationships between ambient temperature and the temperature at the midspan and end of the bridge are depicted in Fig. 1, and the relationships among four meteorological parameters (temperature, solar radiation, wind speed, humidity) were also investigated. The change in the temperature within the track structure obviously lagged behind the change in the ambient temperature, and the temperature variation was smaller than that of the ambient temperature. Furthermore, Pearson correlation analysis showed a strong positive correlation between the ambient temperature and solar radiation, with a Pearson correlation coefficient of 0.85, while the correlation coefficient of 0.22 between the ambient temperature and air humidity was the smallest.

3 Heart Transfer Model

Based on the finite element (FE) software of Comsol, the heat transfer numerical models of CRTS II track on a simply-supported box bridge were established (Fig. 2a). The total dimensions of the five slab tracks were 32 m length × 13.4 m width × 3.35 m height, and the track slab, the CAM layer, base plate and box girder were simulated by the solid elements. Two ends of the slab tracks were constrained, and the thermal responses of the track structure under the four meteorological parameters were compared. As shown in Fig. 2b, c, the internal temperature and vertical displacement at the mid-span of the slab tracks become larger with increasing wind speed or solar radiation, especially for wind speeds >6 m/s or solar radiation >750 W/m^2. According to the 3D temperature and stress field shown in Fig. 2d–f, increasing solar radiation or ambient temperature could lead to rapid heat transfer from the slab track to the base plate. The role of wind speed on the heat transfer effect in the track structure was limited.

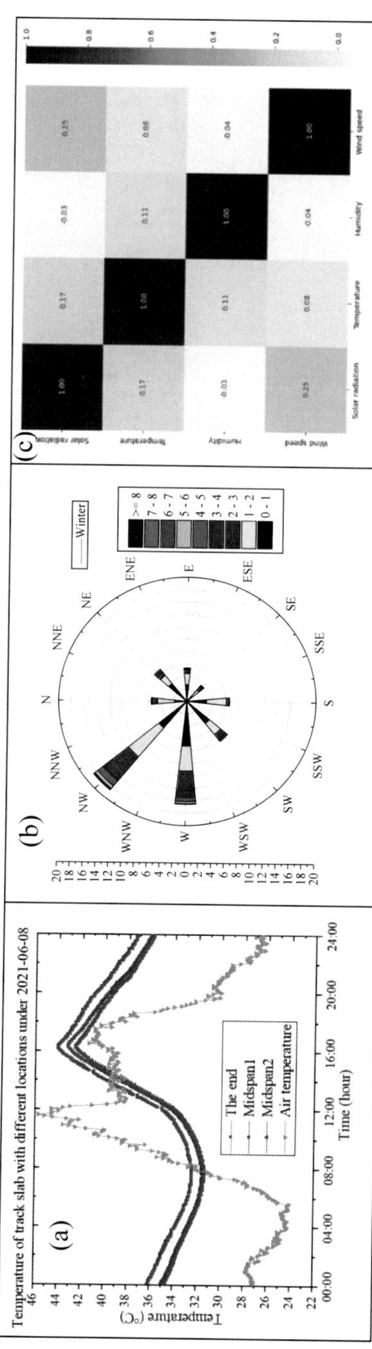

Fig. 1 Four meteorological parameters evaluated

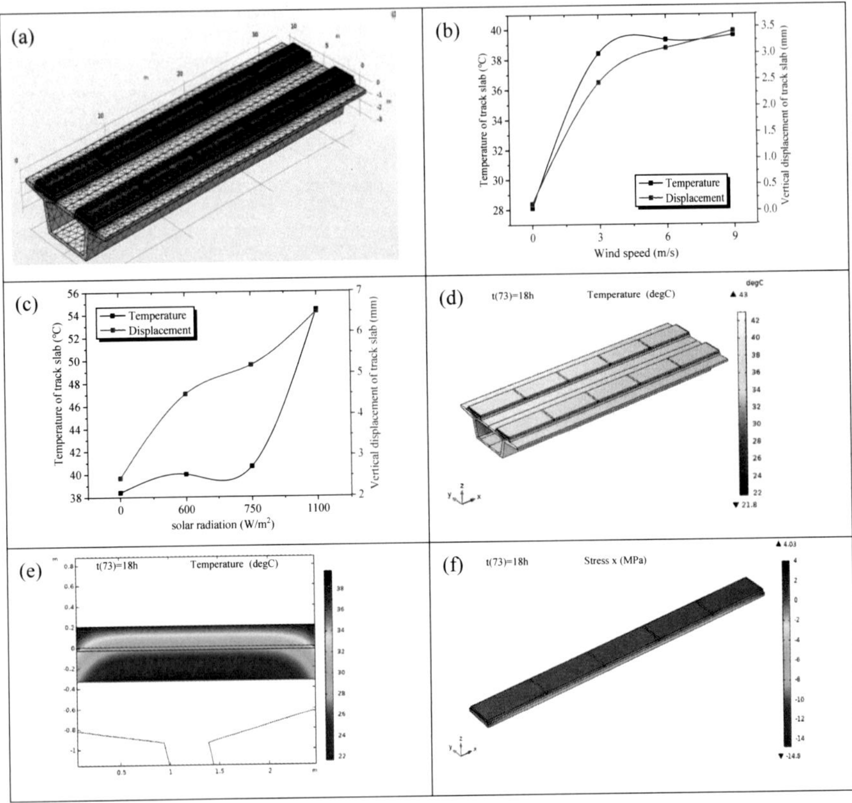

Fig. 2 Thermal responses comparison under meteorological parameters: **a** FE model; **b**, **c** temperature and displacement responses under various wind speeds and solar radiation levels; **d**, **e** temperature field under 1100 W/m^2 solar radiation; **f** stress field under 1100 W/m^2 solar radiation

4 Conclusions

Based on the combination of field measurement data and FE analysis, the thermal transfer effects in a CRTS II slab track–bridge system under various meteorological conditions were studied. The major findings were:

(1) Three meteorological conditions—ambient temperature, solar radiation, and wind speed—had large correlation coefficients, showing they had the greatest influence on thermal transfer in the track structure.
(2) Increasing solar radiation or ambient temperature could lead to increasing deformation and longitudinal stress of the slab track structure, but only wind speeds >6 m/s affected thermal transfer in the track structure.

Acknowledgements The authors gratefully acknowledge support for this research provided by National Natural Science Foundation of China (No.52278311), the National Key Technologies Research and Development Program (No.2022YFB2603300), Guangdong Province Natural Science Foundation (No. 2022A1515010665), the Shenzhen Science and Technology Program under grant (Nos. KQTD20180412181337494, and GJHZ20200731095802007), the Project of Science and technology research and development of China Railway Co., Ltd. (No. K2022G038) and the Open Project of the State Key Laboratory of High-speed Railway Track Technology (No. 2021YJ143) and State Key Laboratory of Mountain Bridge and Tunnel Engineering (No. SKLBT-ZD2101).

References

1. Matias SR, Ferreira PA (2022) The role of railway traffic and extreme weather on slab track long-term performance. Constr Build Mater322:126445
2. Dai G, Tang Y, Liang J et al (2018) Temperature monitoring of high-speed railway bridges in mountainous areas. Struct Eng Int 28(3):288–295
3. Huang YC, Gao L, Zhong YL et al (2022) Study on the damage evolution of the joint and the arching deformation of CRTS-II ballastless slab track under complex temperature loading. Constr Build Mater 309:125083
4. Yang RS, Li JL, Kang WX et al (2017) Temperature characteristics analysis of the ballastless track under continuous hot weather. J Transp Eng A-SYST 143(9):04017048
5. Song L, Liu HB, Cui CX et al (2020) Thermal deformation and interfacial separation of a CRTS II slab ballastless track multilayer structure used in high-speed railways based on meteorological data. Constr Build Mater 237:117528
6. Cai XP, Luo BC, Zhong YL, et al. (2019) Arching mechanism of the slab joints in CRTSII slab track under high temperature conditions. Eng Fail Anal 98:95–108
7. Zhou R, Zhu X, Huang JQ, et al. (2022) Structural damage analysis of CRTS II slab track with various interface models under temperature combinations. Eng Fail Anal 134:106029
8. Zhu SY, Luo J, Wang MZ et al (2020) Mechanical characteristic variation of ballastless track in highspeed railway: effect of train–track interaction and environment loads. Railway Eng Sci 28(4):408–423
9. Zhang Y, Zhou L, Mahunon AD et al (2021) Mechanical performance of a ballastless track system for the railway bridges of high-speed lines: experimental and numerical study under thermal loading[J]. Materials 14(11):2876
10. Zhou R, Zhu X, Ren WX et al (2022) Thermal evolution of CRTS II slab track under various environmental temperatures: experimental study. Constr Build Mater 325:126699
11. Zhou R, Yue H, Du Y, Yao G, Liu W, Ren W (2023) Experimental and numerical study on interfacial thermal behavior of CRTS II slab track under continuous high temperatures[J]. Eng Struct 284:115964
12. Zhou R, Yuan W, Liu W, Zhu X, Yao G, Li F, Zhang L (2023) Thermal performance of CRTS II slab track-bridge structure under extreme temperatures: numerical simulation[J]. Constr Build Mater 377:131147

Investigation on Superhydrophobicity and Piezoresistivity of Self-sensing Cement-Based Sensors Using Silane Surface Treatment

W. K. Dong, W. G. Li, X. Q. Lin, and S. P. Shah

Abstract Cement-based sensors are highly susceptible to the effects of watery environments due to the hydrophilic properties of the cement matrix. In this paper, we applied a surface treatment using a silane/isopropanol solution to graphene/cement-based sensors to achieve superhydrophobicity and mitigate piezoresistive instability in watery environments. After treatment, impressive water contact angles of 163.4° and 142.0° were achieved for the surface and inner cement-based sensors, respectively. Moreover, the piezoresistivity of the coated cement-based sensors exhibited greater stability compared to their untreated counterparts. These results provide valuable insights into the piezoresistivity of hydrophobic cement-based sensors in moist environments, offering promising prospects for future structural health monitoring applications.

Keywords Cement-based sensors · Graphene nanoplates · Piezoresistivity · Silicone hydrophobic powder

1 Introduction

Piezoresistivity-based self-sensing cementitious composites have attracted increasing attention recently, for their ability to achieve structural and pavement health monitoring automatically. Cement-based sensors have been widely

---

W. K. Dong · W. G. Li (✉) · X. Q. Lin
School of Civil and Environmental Engineering, University of Technology Sydney, Sydney, NSW, Australia
e-mail: wengui.li@uts.edu.au

W. K. Dong
Institute of Construction Materials, Technische Universität Dresden, Dresden, Germany

S. P. Shah
Center for Advanced Construction Materials, The University of Texas at Arlington, Arlington, TX, USA

W. Duan et al. (eds.), *Nanotechnology in Construction for Circular Economy*, Lecture Notes in Civil Engineering 356,
https://doi.org/10.1007/978-981-99-3330-3_3

investigated, ranging from their types and content of conductive filler, matrix, additives, curing, and drying methods to field application conditions [1, 2]. However, due to the hydrophilic and porous structure of the cement matrix, the electrical conductivity and piezoresistivity of cement-based sensors can be easily affected by the working environment, especially watery, and humid conditions [3, 4]. Previous studies have attempted to remove the influence of penetrated water on the piezoresistivity of cement-based sensors. The water absorption of cement-based sensors was significantly reduced in the early age, but the efficiency was relatively low, with unstable piezoresistivity in the long term [5].

Basically, using waterproofing materials to treat the surface of cement-based sensors can prevent water penetrating the cement matrix [6, 7], which can reduce the interference of water molecules on electrical resistivity and piezoresistivity. In this study, we propose a special surface treatment of graphene/cement-based sensors by immersing the sensors in a silane/isopropanol solution. The hydrophobic silane is expected to penetrate the cement-based sensors and improve the waterproofing properties.

2 Methods

In this paper, graphene nanoplate (GNP) is used to achieve intrinsic self-sensing ability and piezoresistivity of cement-based sensor. The GNP is commercially available, and its specific properties are documented [8]. The raw materials consisted of general-purpose cement, silica fume, superplasticizer, GNP, silane hydrophobic powder (SHP), and tap water. The addition of SHP can enhance the waterproofing and hydrophobic behaviors of cement-based sensors. The silane used was aqueous trichlorosilane for surface modification, and the solvent isopropanol was chosen to disperse the silane and control the concentration. The physical and chemical properties of silane and isopropanol are listed in Table 1.

To remove surface impurities and smooth the surface, the cement-based sensors were firstly polished and cleaned using sandpaper before coating. The surface modification followed the steps shown in Fig. 1. Isopropanol was prepared in a measuring beaker, and 4% silane by volume of isopropanol was added, followed by 5 min of

Table 1 Physical and chemical properties of silane and isopropanol

Product	Appearance	Color	Relative density	Formula	Molecular weight
Trichloro (1H, 1H, 2H, 2H-perfluorooctyl) silane	Clear liquid	Colorless	1.3 g/cm^3	$C_8H_4Cl_3F_{13}Si$	481.54 g/mol
2-Propanol	Clear liquid	Colorless	0.785 g/mL	C_3H_8O	60.10 g/mol

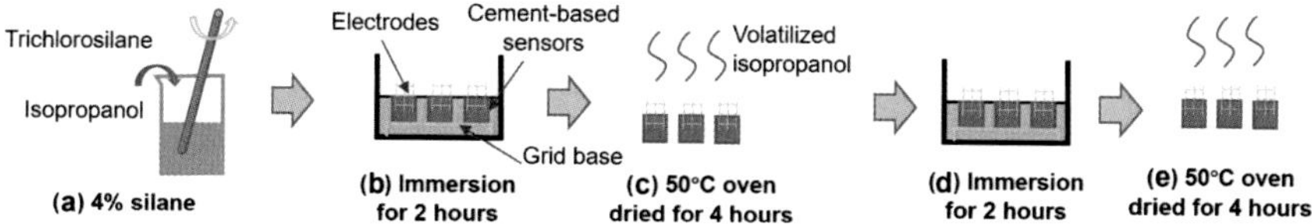

Fig. 1 **a–e** Treatment procedure of cement-based sensors

mechanical mixing to dissolve and disperse the silane. The cement-based sensors were placed above a copper mesh in a plastic container at a distance of 5.0 mm, so all surfaces had continual contact with the silane/isopropanol or silane solution. The mixed solution was gently poured into the container until the top surfaces of cement-based sensors were just covered. It should be noted that the electrodes of the cement-based sensors were not immersed in the silane/isopropanol solution, to ensure excellent conductivity of the electrodes. The container was sealed with plastic film to avoid volatilization of isopropanol. The cement-based sensors were immersed for 2 h to ensure the thorough entrance of the solution into the cracks and pores of the cementitious material. Finally, the cement-based sensors were dried in an oven at 50 °C for 4 h to volatilize the isopropanol.

The water contact angle (CA) measurements of the cement-based sensors before and after surface modification were performed with an optical tensiometer (Attension Theta). The test liquid was deionized water with a volume of 0.2 μL for each water drop. The water CA of the intact surface of the cement-based sensors represents the hydrophobic coating efficiency. To obtain the hydrophobic behavior of the inner sensor, the CA tests were also performed on cross-sections of the cement-based sensors.

3 Results and Discussion

3.1 *Hydrophobic Behavior*

Figure 2 shows the surface water CA of the cement-based sensors before and after surface modification at the time of 0, 1, 5, and 9 s from water dropping to stabilization. The cement-based sensors without a coating shown in Fig. 3a exhibited hydrophilic behavior with an initial CA of 79.2°. Subsequently, the water CA gradually decreased over time until the smallest value of 70.1°, which implied that the water molecules penetrated the cement-based sensors due to the hydrophilic behavior and porous structure of the cement matrix. Consequently, the altered water content would be able to permanently affect the electrical and piezoresistive properties of the cement-based sensors, which indicates the necessity to coat them. For the coated cement-based sensors, it was observed that they exhibited hydrophobic behavior, with a final CA of 163.4°. In addition, the cement-based sensors without a coating showed

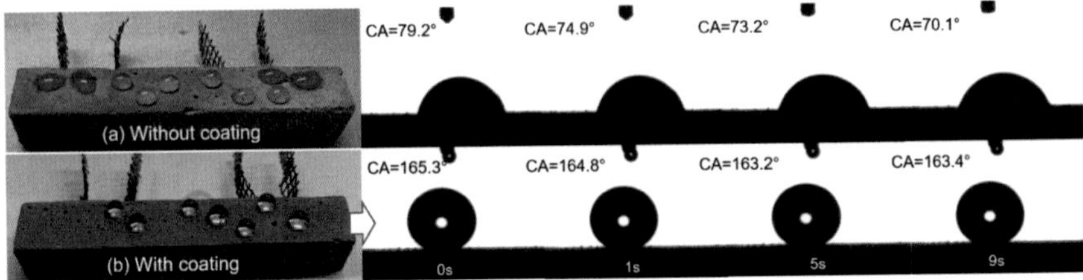

Fig. 2 Water contact angles (CAs) of cement-based sensors **a** without and **b** with surface treatment

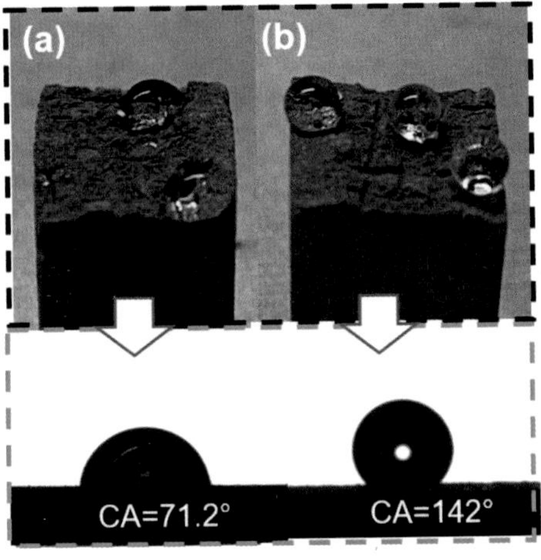

Fig. 3 Water contact angles (CAs) of the interior of cement-based sensors: **a** without coating and **b** with coating

decreasing CA, whereas their coated counterpart only showed a slight fluctuation rather than continual decline.

Figure 3a, b shows the water CAs the cross-sectional surface of the cement-based sensors before and after treatment, to display the hydrophobic or hydrophilic behavior of the interior of the sensors. The cement-based sensor without a coating displayed hydrophilic behavior and a similar CA of 71.2° to that of surface. In contrast, the interior of the sensor became hydrophobic with a CA value of 142.0°, which demonstrated that the silane/isopropanol solution could penetrate into the core of cement-based sensors through micropores and cracks, resulting in hydrophobicity of the cut cross-section.

3.2 *Piezoresistivity*

The stress-sensing performance of the cement-based sensors before and after surface modification is shown in Fig. 4. Fractional changes of resistivity (FCR) of the cement-based sensors exhibited an excellent relationship to compressive stress, with first a

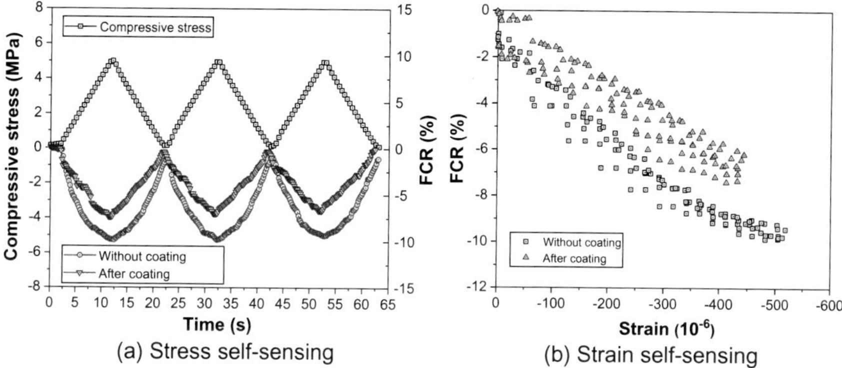

Fig. 4 **a** Stress- and **b** strain-sensing capacities of cement-based sensors before and after surface modification. FCR, fractional change of resistivity

decrease and then returned resistivity in the loading and unloading processes. This finding demonstrated that the silane-based surface modification did not eliminate the piezoresistivity of the cement-based sensors. The graphene-filled cement-based sensor without a coating showed the highest FCR value of 12.6%, followed by an average FCR of 7.2% for the cement-based sensors after surface treatment. These data implied that the stress-sensing efficiency might be weakened by the silane modification. In addition, small fluctuations can be seen for the silane-coated cement-based sensors, mainly due to the brittleness and heterogeneity of the cementitious materials, which led to sudden changes of electrical resistivity. Secondly, the intruded silane aggravated the fluctuation because of its poor electrical conductivity. For the strain-sensing performance, the FCR showed an excellent relationship with compressive strain and showed a similar changing mode to compressive stress.

4 Conclusions

Cement-based sensors can easily absorb water molecules because of their porous structure and hydrophilic behavior. In this study, a silane-based surface modification was applied to improve the waterproofing and superhydrophobic behavior, while maintaining excellent piezoresistivity of the cement-based sensors. The final water CA significantly increased to 163.4°, and the piezoresistivity was relatively well maintained. The piezoresistivity of the cement-based sensors seemed to decrease after the surface modification, with slightly poorer linearity and repeatability, lower gauge factor, and higher hysteresis. Despite this, the cement-based sensors exhibited acceptable linearity and repeatability.

References

1. Dong W, Li W, Tao Z, Wang K (2019) Piezoresistive properties of cement-based sensors: review and perspective. Constr Build Mater 203:146–163
2. Han B, Yu X, Kwon E (2009) A self-sensing carbon nanotube/cement composite for traffic monitoring. Nanotechnology 20(44):445501
3. Dong W, Li W, Lu N, Qu F, Vessalas K, Sheng D (2019) Piezoresistive behaviours of cement-based sensor with carbon black subjected to various temperature and water content. Compos Part B 178:107488
4. del Moral B, Baeza FJ, Navarro R et al (2021) Temperature and humidity influence on the strain sensing performance of hybrid carbon nanotubes and graphite cement composites. Constr Build Mater 284:122786
5. Dong W, Li W, Zhu X, Sheng D, Shah SP (2021) Multifunctional cementitious composites with integrated self-sensing and hydrophobic capacities toward smart structural health monitoring. Cem Concr Compos 118:103962
6. Zhou Z, Li S, Cao J, Chen X et al (2021) The waterproofing effect and mechanism of graphene oxide/silane composite emulsion on cement-based materials under compressive stress. Constr Build Mater 308:124945
7. Li F, Yang Y, Tao M et al (2019) A cement paste–tail sealant interface modified with a silane coupling agent for enhancing waterproofing performance in a concrete lining system. RSC Adv 9(13):7165–7175
8. Dong W, Li W, Wang K, Shah SP (2021) Physicochemical and Piezoresistive properties of smart cementitious composites with graphene nanoplates and graphite plates. Constr Build Mater 286:122943

Use of Brown Coal Ash as a Replacement of Cement in Concrete Masonry Bricks

D. W. Law, C. Gunasekara, and S. Setunge

Abstract Portland cement production is not regarded as environmentally friendly, because of its associated high carbon emissions, which are responsible for 5% of global emissions. An alternative is to substitute fly ash for Portland cement. Australia has an abundance of brown coal fly ash, as it is the main source of primary energy in the State of Victoria. Currently, the majority of this material is stored in landfills and currently there is no commercial use for it in the cement industry because brown coal fly ash cannot be used as a direct replacement material for Portland cement due to the high sulfur and calcium content and low aluminosilicate content. However, the potential exists to use brown coal fly ash as a geopolymeric material, but there remains a significant amount of research needed to be conducted. One possible application is the production of geopolymer concrete bricks. A research project was undertaken to investigate the use of brown coal fly ash from Latrobe Valley power stations in the manufacture of geopolymer masonry bricks. The research developed a detailed understanding of the fundamental chemistry behind the activation of the brown coal fly ash and the reaction mechanisms involved to enable the development of brown coal fly ash geopolymer concrete bricks. The research identified suitable manufacturing techniques to investigate relationships between compressive strength and processing parameters and to understand the reaction kinetics and microstructural developments. The first phase of the research determined the physical, chemical, and mineralogical properties of the Loy Yang and Yallourn fly ash samples to produce a 100% fly ash-based geopolymer mortar. Optimization of the Loy Yang and Yallourn geopolymer mortars was conducted to identify the chemical properties that were influential in the production of satisfactory geopolymer strength. The Loy Yang mortars were able to produce characteristic compressive strengths acceptable in load-bearing bricks (15 MPa), whereas the Yallourn mortars produced characteristic compressive strengths only acceptable as non-load-bearing bricks (5 MPa). The second phase of the research transposed the optimal geopolymer mortar mix designs into optimal geopolymer concrete mix designs while merging the mix design with the optimal Adbri Masonry (commercial partner) concrete brick mix design. The

D. W. Law (✉) · C. Gunasekara · S. Setunge
School of Engineering, RMIT University, Melbourne, VIC, Australia
e-mail: david.law@rmit.edu.au

W. Duan et al. (eds.), *Nanotechnology in Construction for Circular Economy*,
Lecture Notes in Civil Engineering 356,
https://doi.org/10.1007/978-981-99-3330-3_4

reference mix designs allowed for optimization of both the Loy Yang and Yallourn geopolymer concrete mix designs, with the Loy Yang mix requiring increased water content because the original mix design was deemed to be too dry. The key factors that influenced the compressive strength of the geopolymer mortars and concrete were identified. The amorphous content was considered a vital aspect during the initial reaction process of the fly ash geopolymers. The amount of unburnt carbon content contained in the fly ash can hinder the reactive process, and ultimately, the compressive strength because unburnt carbon can absorb the activating solution, thus reducing the particle to liquid interaction ratio in conjunction with lowering workability. Also, fly ash with a higher surface area showed lower flowability than fly ash with a smaller surface area. It was identified that higher quantity of fly ash particles <45 microns increased reactivity whereas primarily angular-shaped fly ash suffered from reduced workability. The optimal range of workability lay between the 110–150 mm slump, which corresponded with higher strength displayed for each respective precursor fly ash. Higher quantities of aluminum incorporated into the silicate matrix during the reaction process led to improved compressive strengths, illustrated by the formation of reactive aluminosilicate bonds in the range of 800–1000 cm^{-1} after geopolymerization, which is evidence of a high degree of reaction. In addition, a more negative fly ash zeta potential of the ash was identified as improving the initial deprotonation and overall reactivity of the geopolymer, whereas a less negative zeta potential of the mortar led to increased agglomeration and improved gel development. Following geopolymerization, increases in the quantity of quartz and decreases in moganite correlated with improved compressive strength of the geopolymers. Overall, Loy Yang geopolymers performed better, primarily due to the higher aluminosilicate content than its Yallourn counterpart. The final step was to transition the optimal geopolymer concrete mix designs to producing commercially acceptable bricks. The results showed that the structural integrity of the specimens was reduced in larger batches, indicating that reactivity was reduced, as was compressive strength. It was identified that there was a relationship between heat transfer, curing regimen and structural integrity in a large-volume geopolymer brick application. Geopolymer bricks were successfully produced from the Loy Yang fly ash, which achieved 15 MPa, suitable for application as a structural brick. Further research is required to understand the relationship between the properties of the fly ash, mixing parameters, curing procedures and the overall process of brown coal geopolymer concrete brick application. In particular, optimizing the production process with regard to reducing the curing temperature to $\leq$80 °C from the current 120 °C and the use of a one-part solid activator to replace the current liquid activator combination of sodium hydroxide and sodium silicate.

Keywords Bricks · Brown coal fly ash · Concrete · Geopolymer

BY

Composition of Alkali–Silica Reaction Products in Laboratory and Field Concrete

M. J. Tapas, K. Vessalas, P. Thomas, N. Gowripalan, and V. Sirivivatnanon

Abstract This study investigated the composition of alkali–silica reaction (ASR) products formed in mortar and concrete that underwent accelerated ASR testing using two test methods: the accelerated mortar bar test (AMBT) and the simulated pore solution immersion test (SPSM). The composition of the ASR products formed in the accelerated tests was compared with those in a 25-year old bridge in New South Wales demolished due to ASR. Results showed that the ASR products inside an aggregate contained calcium (≈20%), silicon (≈60%), and alkalis (≈20%) regardless of the ASR test method used. The ASR products in the AMBT sample only contained sodium, whereas the ASR products in the SPSM test and the demolished bridge both contained significant amounts of sodium and potassium, which indicated that the type of alkali in the ASR product is largely affected by the dominant alkali in the pore solution. However, considering that the total alkali content (Na + K) in the ASR products was similar regardless of the ASR test method used, this suggests that the total alkali content has more influence on the rate of expansion than the type of alkali. The composition of the ASR products also notably varied depending on the location in the concrete. ASR products closer to the cement paste had a higher calcium and lower alkali content than those inside an aggregate, which suggests that the calcium as well as the alkali content of the ASR products plays a significant role in the degree of ASR expansion.

Keywords Accelerated mortar bar test · Alkali–silica reaction (ASR) · ASR products

---

M. J. Tapas (✉) · K. Vessalas · P. Thomas · N. Gowripalan · V. Sirivivatnanon
School of Civil and Environmental Engineering, University of Technology Sydney, Sydney, NSW, Australia
e-mail: mariejoshua.tapas@uts.edu.au

W. Duan et al. (eds.), *Nanotechnology in Construction for Circular Economy*,
Lecture Notes in Civil Engineering 356,
https://doi.org/10.1007/978-981-99-3330-3_5

1 Introduction

The dissolution of the reactive silica components of an aggregate can result in a concrete durability issue known as the alkali–silica reaction (ASR). The reactive silica components of an aggregate, which dissolve when the concrete pore solution's alkali concentration is sufficiently high, can react with calcium and alkali ions in the pore solution to form the ASR products that lead to expansion and deleterious cracking of the concrete [1]. Reactive silica minerals are those that are amorphous, have poor crystallinity, highly strained, or contain high amount of defects making them susceptible to alkali dissolution.

Two test methods are widely used to assess ASR potential: the accelerated mortar bar test (AMBT) and the concrete prism test (CPT). AMBT AS1141.60.1 makes use of 1 M NaOH at 80 °C to accelerate the reaction and an expansion of <0.1% at 21 days indicates that the aggregate is nonreactive [2]. CPT AS1141.60.2, on the other hand, involves boosting the cement alkali content to 1.25% Na_2O_{eq} and storing the concrete prisms in a sealed, high humidity environment at 38 °C. The CPT is a longer test method than the AMBT and has an expansion limit of 0.03% at 1 year for an aggregate to be considered nonreactive [3]. The Australian AMBT and CPT methods are based on ASTM C1260 and ASTM C1293 respectively, with slightly modified test limits.

Both the AMBT and CPT remain widely used although heavily criticized owing to their limitations. The use of high temperature as well as excessive alkali supply in the AMBT can result in false positives (i.e., identifying an aggregate as reactive even if it is not). Moreover, the influence of cement alkalinity, which significantly contributes to the ASR, is hard to determine [4]. CPT on the other hand is prone to alkali leaching, which can lead to underestimation of expansion and is reportedly ≈25–35% in 1 year [5, 6]. Because of the limitations of these test methods, an alternative ASR test, the simulated pore solution method (SPSM), has been developed at the Laboratory of Construction Materials, EPFL Switzerland. This test method involves immersing the concrete prisms in a solution based on the pore solution's alkali concentration for the concrete being assessed for ASR potential and so far has shown promising potential as an alternative ASR test method [7, 8].

Due to the availability of various test methods for assessing the risk of an aggregate for ASR, it is therefore important to assess the effect of these test methods on the type of ASR products. Moreover, the comparison of the composition of the ASR products formed using these methods to those in actual structures affected by ASR is also important to determine whether ASR products formed during accelerated testing resemble those in the field. Therefore, we investigated the composition of the ASR products in both the AMBT, and SPSM and compared them to ASR products in a 25-year-old bridge in Australia, decommissioned because of ASR.

2 Methods

2.1 *ASR Testing as Per AMBT and SPSM*

Expansion tests using AMBT and SPSM were carried out using Australian reactive aggregates and Australian cement. The cement complied with the 0.6% Na_2O_{eq} alkali limit.

For the AMBT, mortar bars composed of 1 part cement to 2.25 parts graded aggregate by mass (440 g cement per 990 g of aggregate) and a water to cementitious materials ratio equal to 0.47 were prepared in accordance with AS1141.60.1. The mortar specimens were prepared in 25 × 25 × 285 mm molds with a gauge length of 250 mm, then cured in a high humidity environment at room temperature (23 ± 2 °C) for 24 h. Next, the specimens were demolded and placed in a water-filled container, before being placed in an oven at 80 °C for another 24 h to allow the specimens to slowly equilibrate to 80 °C. Horizontal comparator was used to obtain zero-hour length measurements before immersing the specimens in 1 M NaOH solution at 80 °C for 28 days. Succeeding expansion measurements were obtained at 1, 3, 7, 10, 14, 21, and 28 days. Three readings were taken per mortar specimen at each age. Total expansion incurred by the aggregate after 10 and 21 days of NaOH immersion was used to classify its ASR potential when used in the field in accordance with AS1141.60.1.

For the SPSM, concrete prisms (70 × 70 × 280 mm) were prepared with a cement content of 410 kg/m^3 and water-to-cement ratio of 0.46. The concretes were cured for 28 days in a high humidity chamber (>90% relative humidity) before storage in simulated pore solution at 60 °C. Expansion measurements were taken before storage and every month thereafter using a vertical comparator. The simulated pore solution was prepared based on the extracted pore solution of an equivalent binder system at 28 days.

2.2 *Analysis of the Composition of the ASR Products*

Polished sections of mortar/concrete that underwent expansion tests using the AMBT, and SPSM, as well as the concrete from the demolished bridge were prepared and subjected to scanning electron microscopy–energy-dispersive spectroscopy (SEM–EDS) analysis. The AMBT samples were sectioned after 28 days in the AMBT bath while the concrete prisms were sectioned after 6 months in the simulated pore solution at 60 °C.

The mortars and concretes were cut to fit a 25-mm diameter mold, then vacuum impregnated with epoxy resin and polished, first with silicon carbide paper until the sample surface had been fully uncovered from the resin, followed by automated polishing using MD Largo Struers discs lubricated with petrol and diamond spray as the polishing agent (9 μm, 3 μm and 1 μm particle sizes). After polishing, the samples were cleaned in an ultrasonic bath for 2 min and then stored in a vacuum desiccator for at least 2 days to dry. The samples were coated with carbon to prevent charging during SEM imaging.

Imaging and elemental analysis of the carbon-coated polished sections were carried out using an FEI Quanta 200 with Bruker XFlash 4030 EDS detector. The microscope was operated in backscattered electron (BSE) mode, 15 kV accelerating voltage and 12.5 mm working distance in a high vacuum.

3 Results and Discussion

The AMBT and SPSM expansion plots are shown in Fig. 1, confirming the high reactivity of the aggregates as indicated by the significant degree of expansion. The AMBT mortars (dacite and greywacke) both exceeded the 0.1% limit at 10 days and 0.3% at 21 days, making them reactive as per AS1141.60.1. There is currently no established limit for the SPSM but the high degree of expansion of the dacite concrete confirmed the reactivity of the aggregate.

Figure 2 presents images of the 25-year-old bridge before it was decommissioned, showing extensive damage due to ASR. The cracks observed have a map crack appearance, which is characteristic of ASR [9].

Figure 3 shows the ASR damage observed in the mortar and concrete that underwent AMBT and SPSM respectively. In both cases, the cracks were concentrated within the aggregate and extend towards the paste, which indicated that deleterious ASR damage originated within the aggregate and explains the characteristic map crack appearance of ASR damage in affected structures. Figure 4 shows the ASR products observed in the AMBT sample, and Fig. 5 shows the ASR products observed in the concrete sample subjected to SPSM with their corresponding EDS maps. For both cases, the presence of calcium, silicon and alkalis are notable confirming the composition of the ASR product (alkali-calcium silicate hydrate). It is however

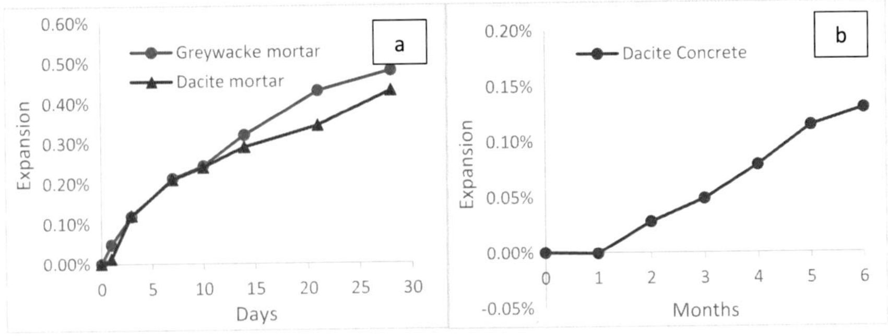

Fig. 1 Expansion plots of the aggregates subjected to **a** accelerated mortar bar test and **b** simulated pore solution immersion test

Fig. 2 A 25-year-old bridge in New South Wales, Australia, suffering from alkali–silica reaction

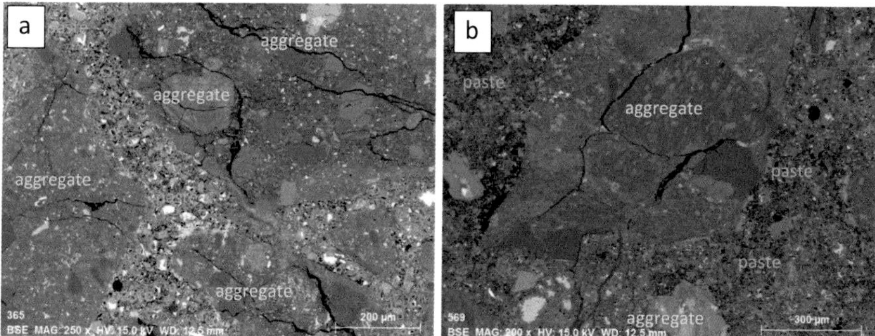

Fig. 3 Scanning electron microscopy images showing extensive damage in the mortar and concrete after **a** accelerated mortar bar test and **b** simulated pore solution immersion test

notable that whereas the alkali present in the AMBT is only sodium, the ASR product in the concrete prism has both sodium and potassium. This indicates that the type of alkali in the ASR product is strongly affected by the dominant alkalis in the pore solution.

Figure 6 shows the ASR products forming around the aggregate and in the paste, which has a notably darker color than the ASR products inside the aggregate, suggesting a difference in composition. Table 1 tabulates the EDS results of the ASR products inside the aggregate (SPSM sample) and the ASR products around the aggregate and near the paste (SPSM sample). As can be seen, the ASR products outside the aggregate have a composition closer to C-S–H and a much higher Ca/Si ratio and lower Na + K/Si than the ASR products inside the aggregate. In general, the silicon content of the ASR product decreased and calcium content increased as the product came in closer contact with the cement paste [10–14]. The role of calcium, however, remains controversial. Although higher calcium content in the ASR product results in higher stiffness [15], as the ASR product becomes more rigid, it also has decreased swelling potential [1, 16]. The substitution of alkalis with calcium suggests there is a competitive reaction between calcium and alkalis and

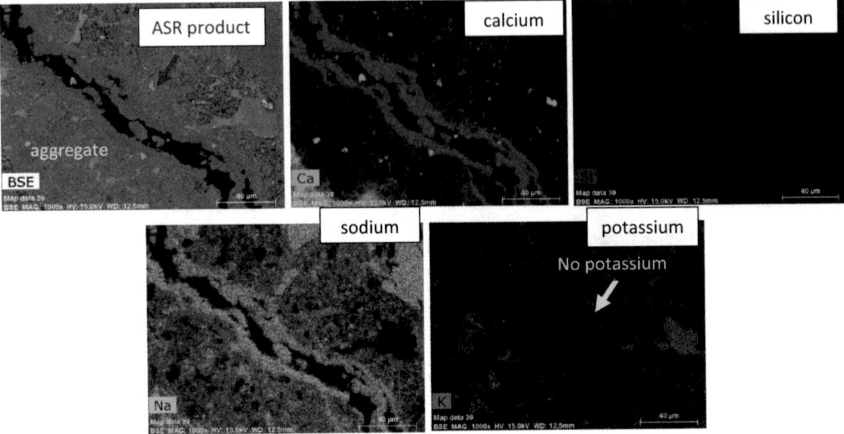

Fig. 4 Alkali–silica reaction (ASR) product observed in the accelerated mortar bar tested mortar with EDS maps showing strong presence of calcium (Ca), silicon (Si) and sodium (Na). BSE, backscattered electron mode; EDS, energy-dispersive spectroscopy

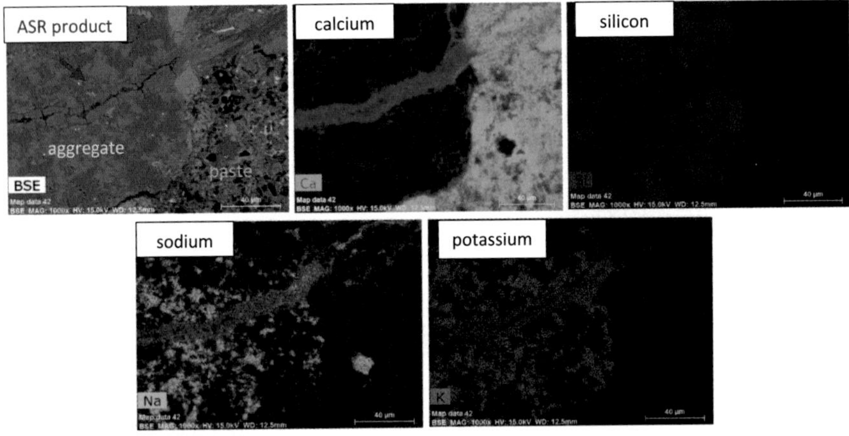

Fig. 5 Alkali–silica reaction (ASR) product observed in the concrete subjected to the simulated pore solution method (SPSM) sample with EDS maps showing strong presence of calcium (Ca), silicon (Si), sodium (Na) and potassium (K). BSE, backscattered electron mode; EDS, energy-dispersive spectroscopy

that calcium is always preferentially absorbed, which supports the alkali recycling theory [1].

Table 2 tabulates the EDS results of the AMBT sample (inside the aggregate). As can be observed, consistent with the EDS mapping results, almost no potassium can be detected in the ASR products. The total alkali content (Na + K) is, however, comparable to the SPSM samples (≈20%), as well as the Ca/Si and (Na + K)/Si. This indicates that the composition of the ASR products inside an aggregate has a

Fig. 6 Alkali–silica reaction (ASR) products observed in the simulated pore solution method sample surrounding the aggregate and located in the cement paste

Table 1 Energy-dispersive spectroscopy results for the Alkali–silica reaction (ASR) product observed in the simulated pore solution method (SPSM)sample (normalized without oxygen)

SPSM sample	Location	Ca	Si	Al	Na	K	Na + K	Ca/Si	(Na + K)/Si
ASR product 1	Inside aggregate	18.03	59.04	1.14	8.50	13.29	21.79	0.31	0.37
ASR product 2	Inside aggregate	20.00	58.31	0.81	6.54	14.34	20.88	0.34	0.36
ASR product 3	Near/in the paste	46.51	40.96	6.02	3.37	3.13	6.51	1.14	0.16
ASR product 4	Near/in the paste	49.88	39.76	6.27	2.89	1.20	4.10	1.25	0.10

Table 2 Energy-dispersive spectroscopy results for alkali–silica reaction (ASR) products observed in the accelerated mortar bar test (AMBT) sample (normalized without oxygen)

AMBT sample	Location	Ca	Si	Al	Na	K	Na + K	Ca/Si	(Na + K)/Si
ASR product 1	Inside aggregate	19.29	61.46	0.70	17.41	1.13	18.55	0.31	0.30
ASR product 2	Inside aggregate	18.88	62.86	0.79	16.19	1.28	17.47	0.30	0.28

similar stoichiometric ratio of calcium, silicon and alkali regardless of the ASR test method. The type of alkali, however, varies depending on the dominant alkali/s in the pore solution.

Figure 7 shows the ASR product in the demolished concrete bridge with corresponding EDS maps. As can be observed, the ASR product was also concentrated inside the aggregate and also contained calcium, silicon and alkali similar to the ASR products in the AMBT and SPSM samples. The ASR product in the bridge, however,

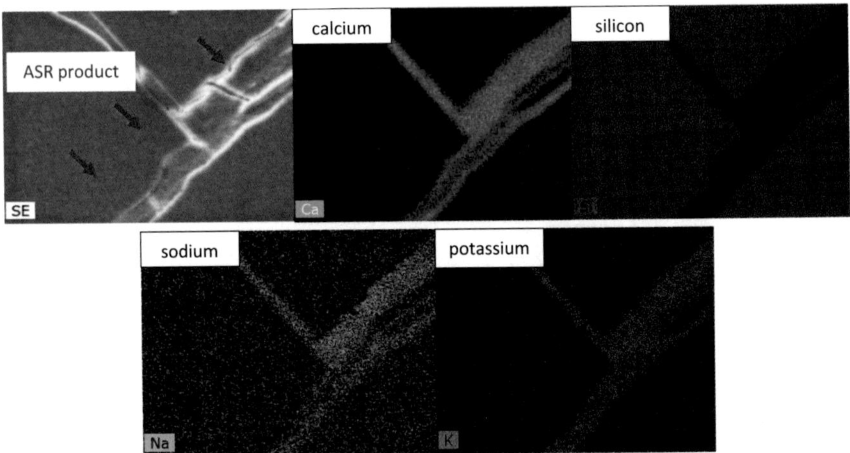

Fig. 7 Alkali–silica reaction (ASR) product in the demolished bridge (NSW, Australia) with energy-dispersive spectroscopy mapping

showed the presence of both sodium and potassium and hence its composition was closer to the SPSM sample than the AMBT sample.

4 Conclusions

This study showed a difference in composition between ASR products found in a mortar subjected to AMBT, concrete subjected to SPSM and a 25-year-old bridge demolished because of ASR. Our results showed the following.

1. The ASR products in the AMBT, SPSM and demolished bridge samples all contained calcium, silicon and alkali. The ASR products in the AMBT sample, however, only contained sodium whereas those in the SPSM and bridge samples contained both sodium and potassium, which indicated that the type of alkali in the ASR product is significantly affected by the dominant alkali in the pore solution.
2. Although the type of alkali in the ASR product was affected by the test method, the total Na + K in the ASR products found inside the aggregate remained the same regardless of the test method, which suggests that the total alkali plays a more significant role in ASR expansion than the type of alkali.
3. The composition of the ASR products varied according to the location in the concrete. ASR products within the aggregate had a lower Si/Ca ratio and higher (Na + K)/Si ratio than ASR products in the cement paste and those surrounding the aggregates. Thus, the closer the ASR product is to the cement paste, the higher the calcium content and the lower the alkali content, which confirms the theory of alkali recycling where Ca^{2+} substitutes for alkali. This finding also suggests

that the composition of the ASR products affects the rate of expansion and that ASR products with lower Si/Ca ratio and higher alkali content (observed inside the aggregate) may be more deleterious.

Acknowledgements This study was part of University of Technology Sydney research funded through the Australian Research Council Research Hub for Nanoscience-Based Construction Materials Manufacturing (NANOCOMM) with the support of Cement Concrete and Aggregates Australia (CCAA) and the Australian Government Research Training Program Scholarship. This work would also not have been possible without laboratory equipment provided by the Laboratory of Construction Materials at EPFL Switzerland, courtesy of Professor Karen Scrivener.

References

1. Rajabipour F, Giannini E, Dunant C, Ideker JH, Thomas MDA (2015) Alkali–silica reaction: current understanding of the reaction mechanisms and the knowledge gaps. Cem Concr Res 76:130–146
2. Standards Australia, AS 1141.60.1 Methods for sampling and testing aggregates method 60.1: potential alkali-silica reactivity-accelerated mortar bar method, standards Australia Limited, Sydney, Australia, 2014
3. Standards Australia, AS 1141.60.2 Methods for sampling and testing aggregates method 60.2: potential alkali-silica reactivity-concrete prism method, standards Australia Limited, Sydney, Australia, 2014
4. Islam MS, Alam MS, Ghafoori N, Sadiq R (2016) Role of Solution Concentration, cement alkali and test duration on expansion of accelerated mortar bar test. Mater Struct 49:1955–1965
5. Rivard P, Berube MA, Ollivier JP, Ballivy G (2007) Decrease of pore solution alkalinity in concrete tested for alkali-silica reaction. Mater Struct 40:909–921
6. Thomas M, Fournier B, Folliard K, Ideker J, Shehata M (2006) Test methods for evaluating preventive measures for controlling expansion due to alkali–silica reaction in concrete. Cem Concr Res 36:1842–1856
7. Chappex T, Sofia L, Dunant C, Scrivener K (2016) A robust testing protocol for the assessment of ASR reactivity of concrete. 15th international conference on alkali aggregate reaction in concrete (ICAAR) São Paulo Brazil
8. Tapas MJ, Sofia L, Vessalas K, Thomas P, Sirivivatnanon V, Scrivener K (2021) Efficacy of SCMs to mitigate ASR in systems with higher alkali contents assessed by pore solution method. Cem Concr Res 142:106353
9. Thomas MDA, Fournier B, Folliard KJ, Resendez YA (2011) Alkali-silica reactivity field identification handbook, federal highway administration Washington, DC
10. Fernandes I (2009) Composition of alkali–silica reaction products at different locations within concrete structures. Mater Charact 60:655–668
11. Shia Z, Geng G, Leemann A, Lothenbach B (2019) Synthesis, characterization, and water uptake property of alkali-silica reaction products. Cem Concr Res 121:58–71
12. Thaulow N, Jakobsen UH, Clark B (1996) Composition of alkali silica gel and ettringite in concrete railroad ties: SEM-EDX and X-Ray diffraction analyses. Cem Concr Res 26:309–318
13. Bleszynski R, Thomas M (1998) Microstructural studies of alkali-silica reaction in fly ash concrete immersed in alkaline solutions. Adv Cem Based Mater 7:66–78

14. Scrivener KL, Monteiro PJM (1994) The alkali-silica reaction in a monolithic opal. J Am Ceram Soc 77:2849–2856
15. Leemann A, Lura P (2013) E-modulus of the alkali–silica-reaction product determined by micro-indentation. Constr Build Mater 44:221–227
16. Juenger MCG, Ostertag CP (2004) Alkali–silica reactivity of large silica fume-derived particles. Cem Concr Res 34:1389–1402

Behavior of Hybrid Engineered Cementitious Composites Containing Nanocellulose

H. Withana and Y. X. Zhang

Abstract Nanocellulose (NC) is a promising reinforcing material for cementitious composites, but its effect on the mechanical properties of hybrid engineered cementitious composites (ECCs) has not been studied. In this paper, we investigated a hybrid polyethylene (PE)-steel fibre ECC reinforced with NC and the effect of NC dosages ranging from 0.1% to 0.4% on the compressive strength of the hybrid ECC. The optimum quantity of NC for the best mechanical property of ECC was determined. Enhancement of compressive strength was observed for all the mixes with NC compared with the reference mix, and the mix containing 0.2% NC showed the maximum improvement.

Keywords Engineered cementitious composites · Nanocellulose · Polyethylene fibers · Steel fibers

1 Introduction

Engineered cementitious composites (ECCs) micromechanically designed by adding microfibers show significant improvements in the ductility, tensile strength, and toughness of cement-based materials. Nevertheless, they have no noticeable influence on compressive and flexural strength [1]. Microfibers such as polymer fibers, due to their relatively small surface areas, can have limited interfacial strength. Thus, they can pose problems in reinforced cementitious composites by entrapping air voids and degrading the strength [2]. In this regard, reinforcement of the ECC at the nanoscale might be an alternative option.

Research of cementitious composites reinforced with nanomaterials is rapidly expanding and recently the growing push for greener construction materials has shifted the attention to plant-based nanomaterials such as Nano cellulose (NC). NC

H. Withana · Y. X. Zhang (✉)
School of Engineering, Design and Built Environment, Western Sydney University, Kingwood, NSW 2747, Australia
e-mail: Sarah.Zhang@westernsydney.edu.au

W. Duan et al. (eds.), *Nanotechnology in Construction for Circular Economy*, Lecture Notes in Civil Engineering 356,
https://doi.org/10.1007/978-981-99-3330-3_6

is a nanofiber derived from cellulose, which is the primary component of plant cell walls. Being an abundant natural ingredient, it is one of the most sustainable raw materials [3]. In addition to its environmental benefits, including renewability, biodegradability, low environmental impact, and low health risks associated with production, it also has a low production cost. NC also has outstanding mechanical and physical benefits such as high modulus and strength, high specific and reactive surface, high aspect ratio and hydrophilic and hygroscopic properties, all of which are favorable for use as a reinforcement for cementitious composites [4].

NC-based composites have been successfully used in areas such as the medical, food, paper, and electrochemical industries [5]. However, their application in the construction industry is limited. Although improvement of the mechanical characteristics of NC-based composites has been established in the literature, research into their use in cementitious composites, notably ECC, has been minimal and none for hybrid ECC.

In this study, a novel NC-reinforced hybrid polyethylene (PE)-steel ECC was developed and the effect of NC on its compressive strength was investigated. A series of uniaxial compression tests were carried out, incorporating different dosages of NC to study the effect on the compressive properties. Compressive strength of the mixes was compared with a reference mix without NC, and the optimum quantity of NC that modifies the performance of ECC was determined.

2 Specimen Preparation

The raw materials used included cement, sand, water, fly ash, silica fume, superplasticizer, steel fibers, PE fibers, and NC. General Purpose Portland cement that conformed to AS3972, fly ash of ASTM class F (SG 2–2.5) and high-grade silica fume were used as the binding materials, while sand with a mean grain size of 225 μm was used as fine aggregate. PE fibers (1.5% by volume) and steel fibers (0.5% by volume) were used as high modulus and low modulus fibers, respectively, and cellulose nanofibrils (CNF) were used as the NC. Additionally, polycarboxylate-based high-range water-reducing admixture Rheobuild 10000N7 was used to attain good workability with consistent rheological properties for uniform fiber dispersion.

The NC was purchased from Cellulose Lab in Canada. It was derived from bleached softwood pulp and prepared by subjecting the pulp to intensive mechanical treatment by a high-pressure homogenizer. NC was provided as a slurry in aqueous gel form with 3.0 wt%.

Six NC-reinforced ECC mixes were prepared by adding NC dosages of 0.1%, 0.2%, 0.25%, 0.3% and 0.4% to the hybrid ECC. ECC with 0% NC was used as the reference mix. The hybrid ECC used in this study was a 0.5% steel and 1.5% PE ECC that was developed by the authors in a previous study [6]. Compressive tests were carried out for all seven mixes including the reference mix.

To make the samples, the dry ingredients including cement, fly ash, silica fume and sand) were mixed for about 2 min. The fibers were then slowly added, and the mixing was continued for a few more minutes until all the fibers were evenly distributed. Water and superplasticizer were combined and gradually added to the dry mix while the mixing continued. CNF was diluted in water using a hand blender and added to the mix and the mixing was continued for 3 min. Once the mix was liquefied and in a consistent and uniform state, it was poured into molds and vibrated for 30 s. Following this, the molds were cling film-wrapped and kept at room temperature for 24 h until demolding. The specimens were then wrapped in plastic sheets and placed in an oven at 23 ± 1 °C and relative humidity of 50% until the age of testing.

3 Results of Uniaxial Compression Test

Uniaxial compression tests were carried out in accordance with AS 1012. The tests were conducted at 28 days after casting the 50-mm cubic specimens. An INSTRON 5500R machine with 1000 KN capacity was used at a loading rate of 20 MPa/min. Three specimens from each mix were tested.

The results of the compression tests are presented in Fig. 1. All the mixes with NC showed an improvement in compressive strength compared with the reference mix without any NC. NC concentrations of 0.1%, 0.2%, 0.25%, 0.3% and 0.4% improved the compressive strength by 29.1% 45.3%, 29.4%. 15.8% and 5.8%, respectively. A threshold was reached at 0.2%, where a maximum compressive strength of 68.4 MPa was achieved. Beyond this threshold concentration, the compressive strength started to decrease.

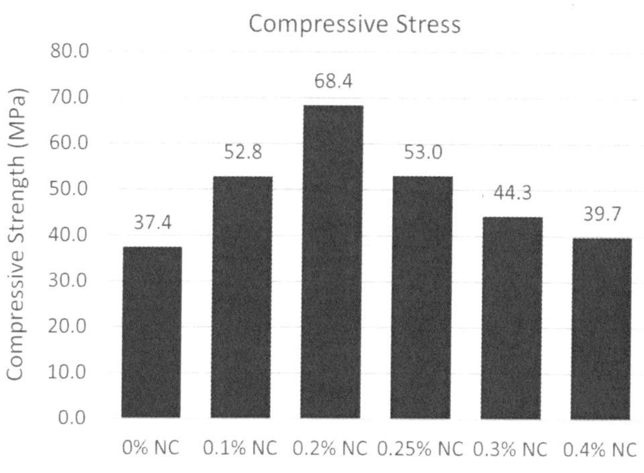

Fig. 1 Compressive strength at 28 days with different dosages of nanocellulose (NC)

Increased interaction between the nanofibers due to a densified matrix can be attributed to the improvement in strength, while the reduction in strength at concentrations above 0.2% could be attributed to fiber agglomeration arising from difficulty of dispersion.

An internal curing effect of NC that stems from the hydrophilic and hygroscopic nature of NC enhances the degree of hydration (DOH) of the matrix [7–9]. When the DOH is enhanced, porosity is reduced, and strength is increased. Moreover, the high aspect ratio and the high specific surface area increase the cellulose hydroxyl groups available for hydrogen bonding in the cementitious matrix [10]. Consequently, strong bonding is promoted, resulting in enhanced fiber–matrix interaction and a densified matrix.

On the other hand, the -OH groups at high density on the surface of cellulose fibers try to bond with adjacent –OH groups by hydrogen bonds causing agglomeration or entanglement of the fibers [11]. Due to the larger surface area, this effect is more significant in nanofibers. Therefore, when NC is applied in excessive quantities, dispersion becomes difficult and the fibers agglomerate, causing high porosity and the fibers acting as stress concentrators, resulting in a reduction in strength.

4 Conclusions

Compressive strength increased as the NC concentration increased until a threshold was reached at 0.2%, where the maximum compressive strength of 68.4 MPa was achieved. Beyond this threshold concentration, the compressive strength started to decrease. All the mixes with NC achieved enhancements of compressive strength compared with the reference mix, with the mix containing 0.2% NC achieving the highest improvement of 45.3%. The decline in strength at high doses of NC was attributed to fiber agglomeration. Microscale studies will be conducted in the future for a more insightful investigation.

References

1. Li VC (2008) Engineered cementitious composites (ECC) material, structural, and durability performance
2. Wang S, Li VC (2004) Tailoring of pre-existing flaws in ECC matrix for saturated strain hardening
3. Eichhorn SJ, Dufresne A, Aranguren M, Marcovich NE, Capadona JR, Rowan SJ, Weder C, Thielemans W, Roman M, Renneckar S, Gindl W, Veigel S, Keckes J, Yano H, Abe K, Nogi M, Nakagaito AN, Mangalam A, Simonsen J, Benight AS, Bismarck A, Berglund LA, Peijs T (2010) Review: current international research into cellulose nanofibres and nanocomposites. J Mater Sci 45(1):1–33. https://doi.org/10.1007/s10853-009-3874-0
4. Claramunt J, Ardanuy M, Fernandez-Carrasco LJ (2015) Wet/dry cycling durability of cement mortar composites reinforced with micro-and nanoscale cellulose pulps. BioResources 10(2):3045–3055

5. Abitbol T, Rivkin A, Cao Y, Nevo Y, Abraham E, Ben-Shalom T, Lapidot S, Shoseyov O (2016) Nanocellulose, a tiny fiber with huge applications. Curr Opin Biotechnol 39:76–88. https://doi.org/10.1016/j.copbio.2016.01.002
6. Withana H, Zhang YX, Zhong T (2022) Material properties of a new hybrid fibre-reinforced engineered cementitious composite with high volume fly ash. Manuscript in preparation
7. Haque MI, Ashraf W, Khan RI, Shah S (2022) A comparative investigation on the effects of nanocellulose from bacteria and plant-based sources for cementitious composites. Cement Concr Compos 125:104316
8. Hisseine OA, Omran AF, Tagnit-Hamou A (2018) Influence of cellulose filaments on cement paste and concrete. J Mater Civil Eng 30(6). https://doi.org/10.1061/(asce)mt.1943-5533.0002287
9. Onuaguluchi O, Panesar DK, Sain M (2014) Properties of nanofibre reinforced cement composites. Constr Build Mater 63:119–124. https://doi.org/10.1016/j.conbuildmat.2014.04.072
10. Cao Y, Zavaterri P, Youngblood J, Moon R, Weiss J (2015) The influence of cellulose nanocrystal additions on the performance of cement paste. Cement Concr Compos 56:73–83
11. Wang B, Sain M (2007) Isolation of nanofibers from soybean source and their reinforcing capability on synthetic polymers. Compos Sci Technol 67(11–12):2521–2527

Investigation of ASR Effects on the Load-Carrying Capacity of Reinforced Concrete Elements by Ultra-Accelerated Laboratory Test

J. Cao, N. Gowripalan, V. Sirivivatnanon, and J. Nairn

Abstract The alkali–silica reaction (ASR) can cause expansion, cracking, and degradation of the mechanical properties of affected concrete. Concerns about the safety of ASR-damaged reinforced concrete structures have driven the demand for studying the effects of ASR on residual load capacity of the deteriorated structure. Conventionally, field load testing methods are used to assess the residual load capacity of ASR-affected structures. In this study, a novel accelerated laboratory test using the LVSA 50/70 autoclave to accelerate ASR was applied to investigate the flexural and shear behavior of small-scale reinforced concrete beams affected by ASR. The specimens were subjected to three cycles of 80 °C steam curing at atmospheric pressure in the autoclave, with 60 h/cycle. Significant expansion and ASR damage were observed. Load carrying capacity tests on the small-scale reinforced concrete beams showed that, at the expansion levels achieved, the flexural capacity of the reinforced concrete beams was not significantly affected. Shear resistance of the reinforced concrete beams, however, was found to increase compared with their 28-day counterparts, which could be attributed to the prestressing effect due to ASR expansion. It appears that the multicycle 80 °C steam-curing autoclave test is suitable for investigating ASR deterioration of actual concrete mixes within a short period of time. ASR effects on the load carrying capacity of reinforced concrete elements at higher expansion levels, however, need further investigation.

Keywords Alkali–silica reaction (ASR) · Accelerated ASR test · Expansion · Load-carrying capacity

---

J. Cao (✉) · N. Gowripalan · V. Sirivivatnanon
School of Civil and Environmental Engineering, University of Technology Sydney, Ultimo, NSW 2007, Australia
e-mail: Jinsong.cao@uts.edu.au

J. Nairn
Cement Concrete and Aggregates Australia, Sydeny, NSW 2020, Australia

W. Duan et al. (eds.), *Nanotechnology in Construction for Circular Economy*, Lecture Notes in Civil Engineering 356,
https://doi.org/10.1007/978-981-99-3330-3_7

1 Introduction

The alkali–silica reaction (ASR) is one of the major durability problems for concrete structures and has been observed and studied for decades worldwide. Concerns about the safety of ASR-damaged reinforced concrete structures have driven the demand for studying the effects of ASR on the performance of the structure and the effect of ASR on residual load capacity of the deteriorated structure [1].

During the past decades of extensive research on ASR, field load testing on real structures was used to assess the residual load capacity of ASR-affected structures [2]. In addition, under controlled laboratory conditions, efforts were made to investigate the flexural and shear behavior of small-scale to full-scale reinforced concrete specimens [3–5]. Large-scale in-situ field exposure testing has also been conducted by different researchers [6]. Some researchers tested specimens in long-term tests with up to 10 years of field exposure to accelerate ASR [7]. These tests provided valuable results and knowledge on evaluating the residual load capacity of ASR-affected members. The long test duration of these field tests is required due to the reality that ASR damage takes a long time to develop in structures, but research needs call for rapidly and reliably producing ASR expansion in the laboratory with appropriate accelerated test conditions [1]. Hence, a reliable and rapid accelerated laboratory test to determine the risk of ASR expansion is needed.

In this study, we applied a novel accelerated test using an autoclave with 80 °C steam curing to study the flexural and shear behavior of small-scale reinforced concrete beams affected by ASR. The beams were longitudinally reinforced with two levels of reinforcement ratios. For simplicity, no shear reinforcement was used for the beams. Load carrying capacity tests on the small-scale reinforced concrete beams were conducted. Moreover, the mechanical properties of ASR-affected concrete under accelerated tests were investigated.

2 Methods

2.1 Materials and Mix Proportions

A general-purpose (Type GP) cement with equivalent alkali content (Na_2O_{eq}) of 0.50%, a nonreactive sand (Sydney sand), and a highly reactive dacite aggregate with a maximum nominal size of 20 mm as coarse aggregate, were used in the concrete mixes. As for the reinforcement, deformed bars with either 5-mm diameter (N5) or 8-mm diameter (N8) were used. In addition, to promote ASR in the accelerated test, technical-grade sodium hydroxide (NaOH) pellets with purity of 98% were used to raise the alkali content to 2.5% Na_2O_{eq} by mass of cement in the concrete. The NaOH pellets were pre-dissolved in a fraction of the mixing water 24 h prior to concrete mixing. The mix proportions for all of the small-scale reinforced concrete beams, cylinders and prisms were: cement: 520 kg/m^3; nonreactive sand: 620 kg/m^3; 20-mm highly reactive dacite aggregate: 1160 kg/m^3; water: 192.5 kg/m^3; and NaOH pellets: 13.69 kg/m^3.

2.2 Specimen Fabrication and Steam-Curing Procedure

We fabricated 12 small-scale reinforced concrete beams with a size of 100 × 100 × 340 mm; 6 beams were reinforced with two N5 deformed bars and the other 6 beams were reinforced with two N8 deformed bars as the main reinforcement. The reinforcement ratios for the two series of small-scale beams were 0.39% and 1.0%, respectively. No transverse reinforcement was provided for the beams. Concrete cover of the beams was maintained at 20 mm. Details of reinforcement of the beams are shown in Fig. 1.

Plain ∅ 100 mm (diameter) × 200 mm cylinders and 75 × 75 × 285 mm prisms were also cast. The cylinders were used for the mechanical properties test and the prisms were for free expansion measurement. Specimens were demolded after 24 h from casting. Thereafter, all specimens were stored in a humidity cabinet with temperature of 23 °C and relative humidity (RH) of 90%.

At the age of 28 days, a testing regime of 80 °C steam curing using an autoclave (Zirbus LVSA 50/70 with a chamber volume of 153 L) was used to accelerate ASR.

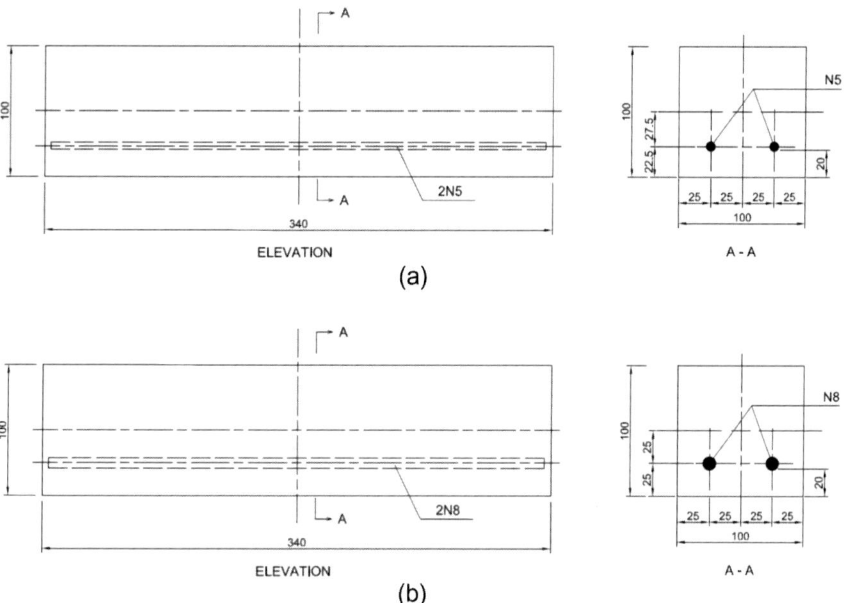

Fig. 1 Reinforcement details of small-scale reinforced concrete beams: **a** with two N5 deformed bars and **b** with two N8 deformed bars (all dimensions are in mm)

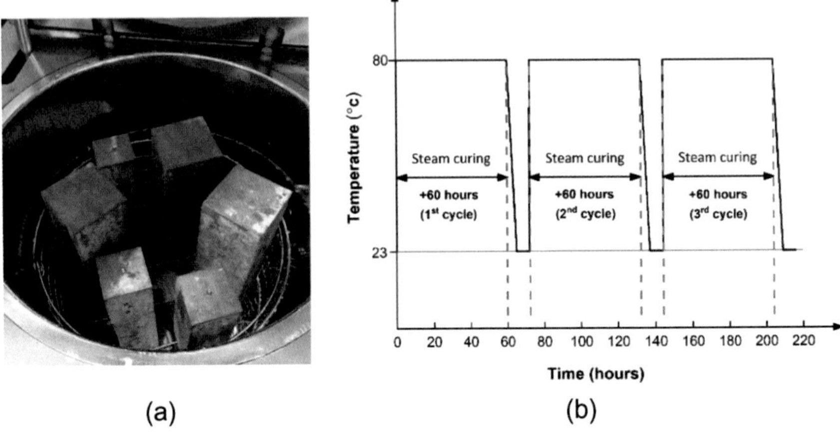

Fig. 2 Acceleration of the alkali–silica reaction (ASR) using the Zirbus LVSA 50/70 autoclave: **a** specimens in the autoclave and **b** time–temperature cycles for accelerating ASR

The inside the autoclave chamber was kept at atmospheric pressure. In total, three cycles of steam curing were applied, with 60 h of steam curing for each cycle. Figure 2a shows test specimens as placed in the autoclave, and Fig. 2b illustrates the temperature–time relationship of steam curing for the three cycles.

At the end of each cycle, the free expansion of the prisms, due to the accelerated ASR, was recorded and the next cycle was applied.

2.3 *Expansion Measurements*

Initial lengths of the prisms were measured and recorded using a digital comparator after demolding. After each cycle of steam curing in the autoclave, the prisms were taken out and stored in sealed plastic bags for 6 h to cool down to room temperature at 23 ± 2 °C, and then the length measurements were taken. Changes in length were used to calculate the expansion of the specimens after 1, 2, and 3 cycles of autoclave steam curing.

2.4 *Mechanical Property Testing*

At the age of 28 days, the modulus of elasticity and compressive strength were measured in accordance with AS1012.17 [8] and AS1012.9 [9] on ∅ 100 × 200 mm cylinders; at the end of each cycle, three cylinders were taken out of the autoclave and mechanical property tests were carried out.

2.5 Load Carrying Capacity Testing Under Four-Point Loading

For each batch, load carrying capacity tests were conducted on two reinforced beams at the age of 28 days under four-point loading; at the end of each cycle, one reinforced beam was taken out, cooled to room temperature, and tested for load carrying capacity. The remaining beam was kept for investigating long-term ASR effects on load carrying capacity. During the load capacity test, a real-time digital image correlation system (Mercury RT®) was used to carry out strain and in-plane displacement measurements.

3 Results and Discussion

3.1 Cracking of Specimens

Figure 3 shows the cracking pattern of the concrete cylinders and prisms after accelerated ASR. Typical external map cracking was observed on the surface of the specimens due to accelerated ASR expansion after three cycles of steam curing in the autoclave.

Fig. 3 External map cracking on cylinders and prisms after three cycles of steam curing in an autoclave

3.2 Expansion of Concrete Prisms

Figure 4 shows the length change of the concrete prisms from the time of demolding and after three cycles of steam curing in the autoclave. A slight shrinkage of approximately 0.019% was recorded during storage in the humidity cabinet up to the age of 28 days. Afterwards, due to accelerated ASR in the autoclave, the length of the prisms increased with each steam-curing cycle. The average expansion of the prisms was recorded as 0.05% after one cycle, 0.13% after two cycles and it reached about 0.18% after three cycles. According to ASTM C1778-20, aggregate having 1-year CPT expansion $\geq$ 0.12% and < 0.24% can be classified as highly reactive. For dacite aggregate, the 1-year CPT expansion result from Cement Concrete & Aggregates Australia is $\approx$0.23%. In the current study, using three cycles of steam curing, the expansion reached $\approx$0.18%. This result shortens the testing period for classifying aggregate reactivity. However, more cycles are needed for slowly reactive aggregate. Further study is suggested to test more aggregates ranging from nonreactive to very highly reactive to establish a standard testing procedure with fine-tuned parameters including temperature, duration, heating and cooling rates and number of cycles, to determine aggregate reactivity, following the ASTM C1778-20 expansion limit criterion.

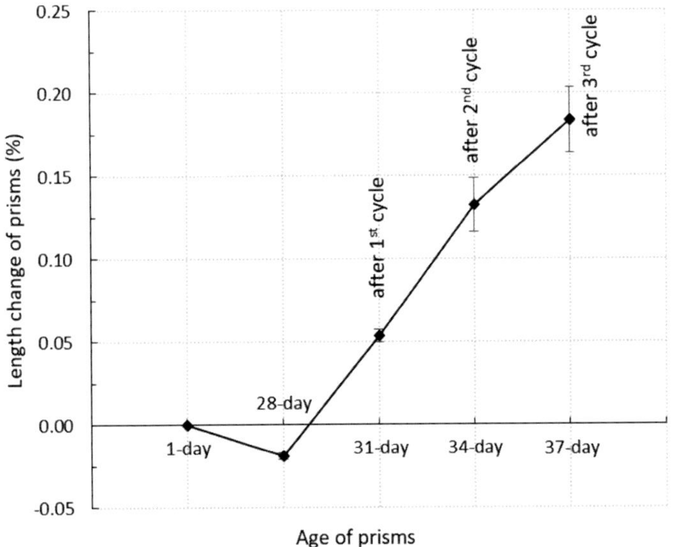

Fig. 4 Expansion of concrete prisms under three cycles of accelerated ASR in the autoclave

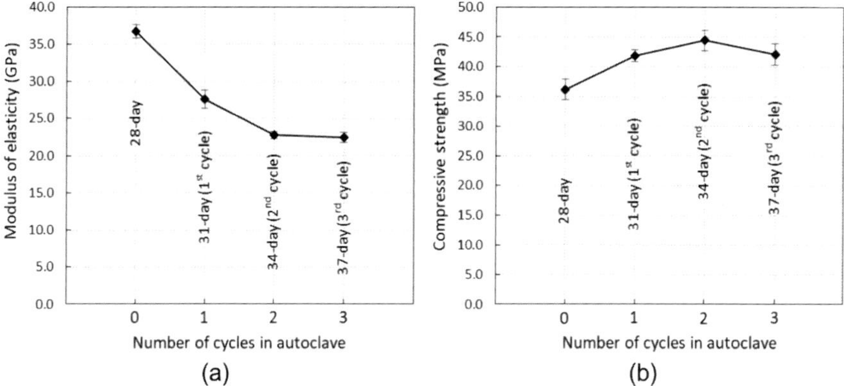

Fig. 5 Modulus of elasticity **a** and compressive strength **b** and before and after three cycles of steam curing in an autoclave

3.3 *Mechanical Properties of Concrete Under Accelerated ASR Test Conditions*

Figure 5a shows the modulus of elasticity test results at 28 days and after 1, 2, and 3 cycles of steam curing in the autoclave. It is generally acknowledged that the modulus of elasticity is the most sensitive mechanical property influenced by ASR. As can be seen, it decreased as expected with the ASR expansion achieved after each cycle of 80 °C steam curing using the autoclave. After the third cycle when average expansion reached 0.18%, a reduction of 39% was recorded in comparison with the initial 28-day value. The reduction was attributed to the microcracking of the concrete caused by accelerated ASR.

Figure 5b shows the compressive strength test results, which demonstrated that the compressive strength initially increased with increasing of number of cycles until the end of the second cycle, at a relatively low expansion level, and thereafter, the compressive strength showed a decreasing trend. With increasing expansion, compressive strength is expected to continue to decrease. This trend had been already observed by Gautam et al. [10]. They boosted the alkali content of concrete specimens to 1.25% Na_2O_{eq}. Samples were stored in hermetically sealed plastic pails and conditioned at 38 °C with RH > 95%. They reported that at age 365 days when ASR expansion reached 0.24–0.35%, the maximum reduction in compressive strength was 4–6%, in comparison with the 28-day compressive strength [10].

3.4 *Load Carrying Capacity of Reinforced Concrete Beams*

To investigate the reduction in load capacity of the reinforced concrete beams after accelerated ASR, load carrying capacity tests were conducted under four-point loading at the age of 28 days and after 1, 2, and 3 cycles of accelerated ASR. All the beams with 2 × N5 reinforcing steel bars failed in flexure and all the beams with 2 × N8 bars failed in shear. Figure 6 shows the initial load capacity of a typical reinforced concrete beam with 2 × N5 tested at the age of 28 days failing in flexure. Figure 7 demonstrates a reinforced concrete beam with 2 × N8 reinforcing bars tested after two cycles of accelerated ASR showing typical shear failure.

The load carrying capacity test results are shown in Fig. 8. It can be seen that, for the reinforced beams with 2 ×N5 bars, the flexural capacity was not significantly

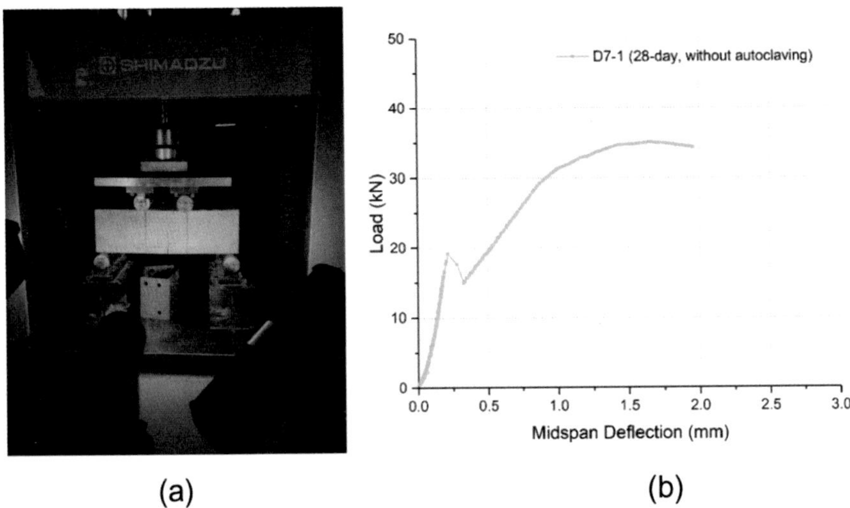

Fig. 6 Reinforced concrete beam with 2 × N5 bars tested at 28 days: **a** test set-up and **b** load–displacement curve

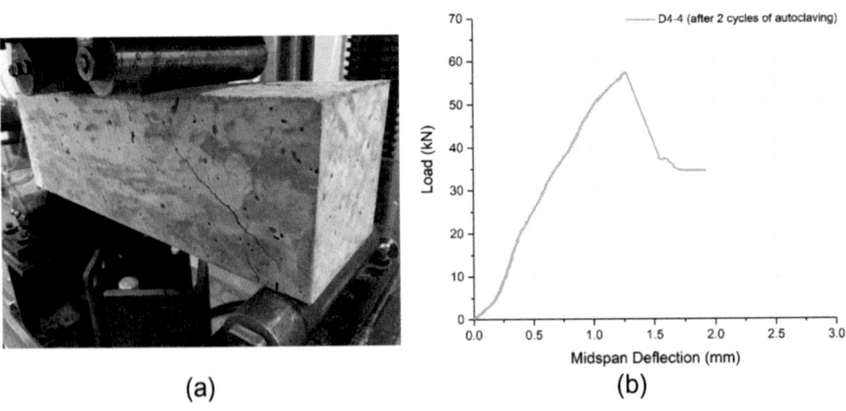

Fig. 7 Reinforced beam with 2 × N8 bars tested after two cycles of accelerated alkali–silica reaction: **a** failure mode and **b** load–displacement curve

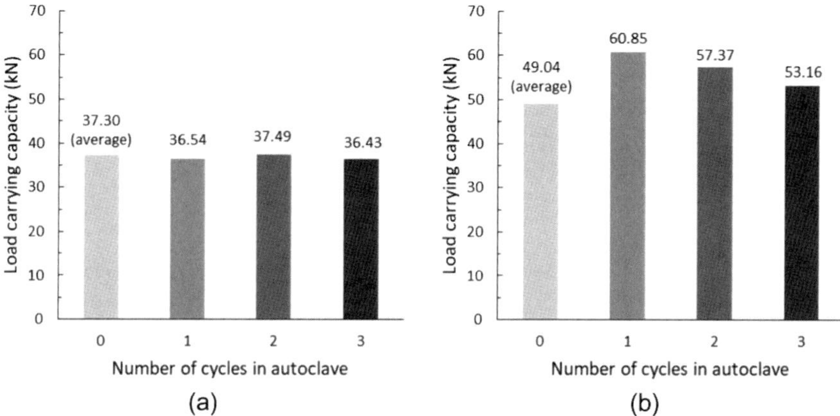

Fig. 8 Reinforced concrete beams with **a** 2 × N5 bars failed in flexure and **b** with 2 × N8 bars failed in shear

influenced by ASR expansion achieved under three cycles of accelerated test. Failure load of the reinforced beam with 2 × N8 bars was found increased after 1, 2, and 3 cycles of accelerated ASR in comparison with the 28-day value. This could be attributed to the prestressing effect of ASR expansion. Meanwhile, some reduction in the failure load after each cycle was recorded due to only one sample being tested for each cycle. Load carrying capacity at higher ASR expansion levels, however, needs further investigation.

4 Conclusions

(1) Using three cycles of 80 °C steam curing in an autoclave, the ultra-accelerated test produced ASR damage and expansion of alkali boosted concrete elements within 9 days.
(2) Modulus of elasticity systematically decreased with increasing number of 80 °C steam-curing cycles. Compressive strength increased until the end of the second cycle, and thereafter showed a decreasing trend.
(3) Shear resistance of reinforced beams with 2 × N8 bars was increased after 1, 2, and 3 cycles of accelerated ASR in comparison with the 28-day value. This could be attributed to the prestressing effect of ASR expansion.
(4) Flexural capacity of the beams reinforced with 2 × N5 bars was not significantly influenced by the extent of ASR expansion achieved in the current accelerated autoclave test.

Acknowledgements This research was funded through an Australian Research Council Research Hub for Nanoscience Based Construction Materials Manufacturing (IH150100006) with the support of Cement Concrete and Aggregates Australia.

References

1. Thomas M (2018) Alkali-silica reaction: eighty years on. In: 5th International fib congress, pp 27–41
2. Blight G, Alexander M, Ralph T, Lewis B (1989) Effect of alkali-aggregate reaction on the performance of a reinforced concrete structure over a six-year period. Mag Concr Res 41(147):69–77
3. Fan S, Hanson JM (1998) Effect of alkali silica reaction expansion and cracking on structural behaviour of reinforced concrete beams. ACI Struct J 95:498–505
4. Swamy RN, Al-Asali M (1989) Effect of alkali-silica reaction on the structural behavior of reinforced concrete beams. ACI Struct J 86(4):451–459
5. Bilodeau S, Allard A, Bastien J, Pissot F, Fourinier B, Mitchell D, Bissonnette B (2016) Performance evaluation of thick concrete slabs affected by alkali-silica reaction (ASR)–part II: structural aspects
6. Deschenes D, Bayrak O, Folliard K (2009) Shear capacity of large-scale bridge bent specimens subject to alkali-silica reaction and delayed ettringite formation, Structures Congress. In: Don't mess with structural engineers: expanding our role, pp 1–9
7. Hamada H, Otsuki N, Fukute T (1989) Properties of concrete specimens damaged by alkali-aggregate reaction, laumontite related reaction and chloride attack under marine environments. In: Proceedings of the 8th international conference on AAR. Kyoto, Japan, pp 603–608
8. AS 1012.17 (2014) Methods of testing concrete, method 17: determination of the static shord modulus of elasticity and poisson's ratio of concrete specimens, Standards Australia Ltd, Sydney, Australia
9. AS 1012.9 (2014) Methods of testing concrete, method 9: compressive strength tests concrete, mortar and grout specimens, Standards Australia Ltd, Sydney, Australia
10. Gautam BP, Panesar DK, Sheikh SA, Vecchio FJ (2017) Effect of coarse aggregate grading on the ASR expansion and damage of concrete. Cem Concr Res 95:75–83

3D printed Ultra-High Performance Concrete: Preparation, Application, and Challenges

G. Bai, G. Chen, R. Li, L. Wang, and G. Ma

Abstract 3D printed ultra-high performance concrete (3DP-UHPC) plays an important role in the realization of ultra-high compressive and tensile strengths. Considering the particular characteristics of UHPC, the conversion of UHPC to 3DP-UHPC is a complex phenomenon and has been the subject of numerous studies. It is very important to be able to design a thixotropic structure in the early hydration stage for bridging the gap between the slow setting of UHPC and the rapid setting of the 3D printing procedure. In the design and application of 3DP-UHPC, requirements such as the ratio of coagulant and flocculant, fiber alignment, reinforced 3D printed no-rebar reinforced concrete, safety, cost etc. need to be taken into account. We present a comprehensive review of 3DP-UHPC in concrete construction from preparation to application, including design method, raw materials, mechanical, reinforced methods, and applications. Finally, recommendations are provided to promote the application of 3DP-UHPC in engineering practice.

Keywords 3D printing · Fiber alignment · Printability · Ultra-high performance concrete

1 Introduction

The rapid development of 3D concrete printing (3DCP) has the potential to greatly reduce labor demand, improve sustainability, reduce construction costs, and effectively overcome the dilemma faced by traditional construction methods [1–4]. In

G. Bai · L. Wang (✉) · G. Ma
School of Civil and Transportation Engineering, Hebei University of Technology, Tianjin, China
e-mail: wangl1@hebut.edu.cn

G. Chen
Zhong Dian Jian Ji Jiao Expressway Investment Development Co., Ltd., Shijiazhuang, China

R. Li
Yaobai Special Cement Technology R&D Co., Ltd., Xian, China

W. Duan et al. (eds.), *Nanotechnology in Construction for Circular Economy*, Lecture Notes in Civil Engineering 356,
https://doi.org/10.1007/978-981-99-3330-3_8

recent years, substantial achievements have been made by 3DCP in the field of architecture and civil engineering. One of the challenges is that conventional steel bar reinforcement cannot be directly integrated into the printed concrete. Researchers and engineers have tried different reinforcement methods by applying continuous fiber, shut fiber, microcable, mesh and U-nails etc., to improve the brittleness of 3D printed concrete [5–8]. Fiber reinforcement is widely applied for printed concrete due to its effectiveness and ease of operation. Small-sized fibers show less interference with the flexible extrusion characteristics of 3D printing. More importantly, the mechanical properties of fiber-reinforced concrete can meet the structural requirements, such as compressive strength and tensile strain exceeding 100 MPa and 4%, respectively [9, 10]. Ultra-high performance concrete (UHPC), as a type of fiber-reinforced concrete, can meet these structural requirements [11, 12] and is currently mainly being used for new structures, reinforcement, and repair of existing infrastructure [13, 14].

The development of 3D printed UHPC (3DP-UHPC) will greatly drive the application of 3D printing technology to structural engineering. UHPC is considered a combination of self-compacting concrete, high-performance concrete, and fiber-reinforced concrete [15]. The good construction performance of UHPC is strong related to the casting procedure, but this advantage is difficult to match with the 3D printing construction of layer by layer stacking. Current research has eliminated the gap between cast and 3D printed UHPC in construction by adding viscosity-modifying admixtures (VMA), such as nano-clay, hydroxypropyl methylcellulose (HPMC), etc. [16–20]. Similar to traditional UHPC, 3DP-UHPC will be widely used in new and existing structures. For example, the construction of a curvilinear bench with free form and light structure [16]. In addition, more attention is being paid to the dynamic performance. Zhou et al. [20] discussed the performance of 3DP-UHPC based on projectile and explosive impacts tests. Yang et al. [21] carried out split-Hopkinson pressure bar (SHPB) tests and analyzed the strain rate effect of 3DP-UHPC

3DP-UHPC has attracted much attention because of its excellent mechanical properties and ongoing research mainly focuses on the preparation of materials, static, and dynamic mechanical properties, and so on. Given the complexity of the preparation method of 3DP-UHPC and the unknowns of the problems that may be faced in its applications, we systematically reviewed the latest research progress on 3DP-UHPC. Finally, some suggestions are put forward to promote the development of 3DP-UHPC based on the current challenges.

2 Preparation of 3DP-UHPC

2.1 *Design of UHPC for 3DCP*

The gap between the printing and casting procedures of UHPC mainly depend on rheological properties. 3D printed concrete is a typical yield stress material; that is, its yield stress first increases and then decreases with increasing shear rate, and finally maintains at a certain yield stress platform, as shown in Fig. 1a. The cast UHPC is self-leveling in the static state because its yield stress makes it difficult to maintain its shape. Moreover, cast UHPC is a shear thickening fluid; that is, its yield stress increases rapidly with increasing shear rate, as shown in Fig. 1a. On the other hand, shear thickening fluid means that the shear rate needs to be continuously increased to overcome the yield stress of UHPC. UHPC strips will be narrowed or even interrupted when the yield stress exceeds the maximum shear stress provided by the 3D printer.

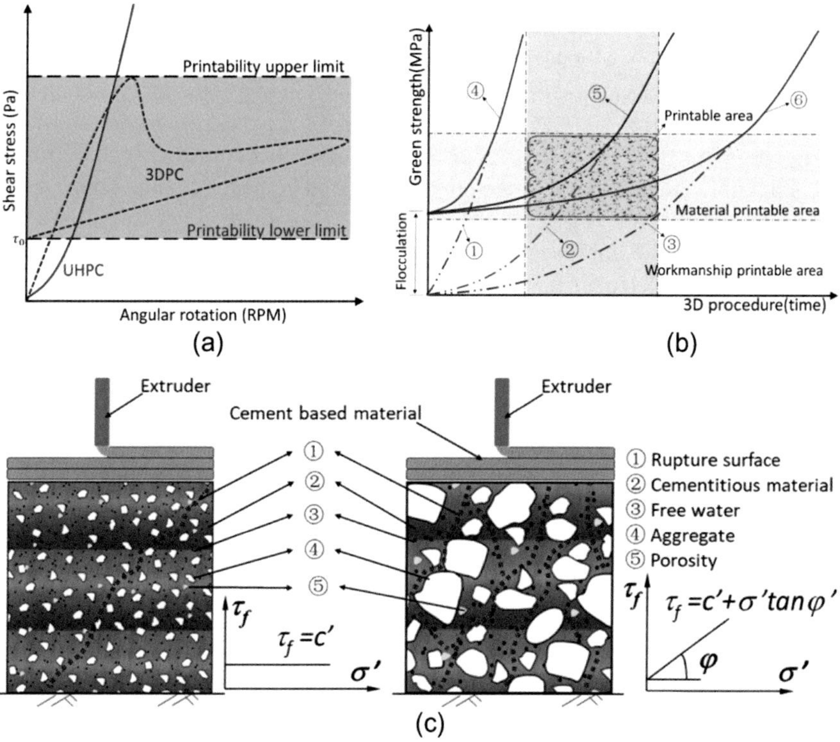

Fig. 1 **a** Fluid types of ultra-high performance concrete (UHPC) and 3D printed concrete (3DPC); **b** printability principle of 3DP-UHPC [22]; **c** schematic of shear resistance of fresh 3DPC without and with coarse aggregate [4]

The methods for modifying UHPC to achieve printability are the chemical hydration accelerated hardening method [20] (CM) and the physical flocculation method (PM) [16, 17]. The CM matches the hydration rate of UHPC with the printing rate by adding materials that change the cement hydration rate. For example, three levels of fast, medium, and slow hydration rates of UHPC can be represented by the curves ①–③ in Fig. 1b. These different rates can be achieved by reducing the accelerator dosage, such as sulfoaluminate cement. The curves ①–③ can match the printing rate of slow, medium, and fast, respectively. However, the matching gap between the hydration rate and printing rate of UHPC usually leads to a short open time of CM. It is easy to cause large deformation and rough surface dry cracking or even collapse and fracture due to insufficient hydration and too rapid printing. PM matches the printing rate by adding VMAs, such as silica fume (SF), nano-clay, and HPMC, to make thixotropic structures in the UHPC before hydration structure formation. For example, based on curves ①–③ in Fig. 1b, the addition of the same VMA gets curves ④–⑥. The addition of VMA makes the material printable earlier. The thixotropic structure will produce obvious deformation when the accumulated weight of the UHPC exceeds its yield stress, affecting the forming accuracy and even causing collapse. Therefore, PM is limited to a certain range of printing rates to allow the hydration rate to follow up.

The potential third printability control method, namely the framing effect provided by raw materials, needs attention because of the current development trend of mixing large-size aggregate into UHPC and the existence of steel fibers. In our previous study on the printability of large-size aggregate, we found that construction deformation and strength were derived from the bonding force of cementitious materials and the aggregate biting force [4], as shown in Fig. 1c. Therefore, the contribution of the framing effect to the improvement of yield stress cannot be ignored.

Designing UHPC for 3DCP will be based on the above principles. Specifically, the purpose of the preparations is to improve the early yield strength of UHPC, reduce its fluidity and improve its shape retention ability by adding a regulator or VMA alone or adding both. Limited by the lack of quantitative preparation theory, the specific dosage can only be obtained by testing. It should be noted that determination of these doses is significantly related to the printing equipment and the selected printing process, which reduces the repeatability of the material mix to a certain extent.

2.2 *Manufacturing of 3DP-UHPC*

Table 1 lists the raw materials and mix proportions of 3DP-UHPC in the existing literature. Zhou et al. [20] also adopted CM to promote the hydration rate to achieve matching with the printing process, specifically by adding slag. Arunothayan et al. [16, 17] used PM to achieve printability of UHPC, by adding HPMC and nano-clay. It cannot be ignored that the proportion of SF is relatively large, accounting for $\approx 30\%$ of the cementitious materials. Excessive SF also contributed to the printability of UHPC. To avoid the defects of the short open time of CM and the low unit construction rate of PM, our group proposed a method of combining CM and PM to achieve printable UHPC[19]. Specifically, sulfoaluminate cement was used to replace the 10% mass fraction of Portland cement to accelerate the hydration rate to preliminarily match the printing rate. Then, the thixotropic material Nano clay and HPMC were added to supplement the insufficient yield stress of UHPC to realize the continuous and stable printing. Here, the mass ratio of the accelerator composed of sulfoaluminate to the flocculant composed of nano-clay and HPMC was ≈11:1. The mix proportions of Yang et al. [18] are the most popularized, because the raw material composition does not different from the cast UHPC base except for the addition of an appropriate amount of nano-calcium carbonate, and its mechanical properties are also the highest, which may be related to the use of a water reducing agent.

The microstructure of UHPC mixed with sulfoaluminate cement, sulfoaluminate cement and VMA was observed by scanning electron microscopy. The preparation time of the two samples was between the initial and final setting. Figure 2 shows the obvious differences in the microstructure of UHPC before and after adding VMA. The microstructure of UHPC without VMA is mainly C-S–H gel with 3D network structures (Fig. 2a); that is, 3DP-UHPC prepared by CM. The microstructure of UHPC doped with sulfoaluminate cement and VMA is mainly flocculent, formed by the adsorption of hydration products by VMA. A similar flocculent microstructure after adsorption was reported by Zhang et al. [23].

Table 1 Mix proportions of 3DP-UHPC (kg/m^3)

Raw materials	References				
	Arunothayan et al.		Yang et al.	Bai et al.	Zhou et al.
	[16]	[17]	[18]	[19]	[20]
Cement	700	273	750	612	500
Fly ash	–	410	165	112	–
Sand	1000	1268	1080	1200	1000
HRWRA	15	18	10	10	15.5
Steel fiber[a]	2% (13 mm)	2% (6 mm)	1% (6 mm)	2.5% (12 mm)	4.5%[b] (6 mm)
Water	160	156	154	180	200
Silica fume	300	293	165	200	200
CSA	–	–	–	68	–
HPMC	3	–	–	1	–
Slag	–	–	–	–	300
NCC	–	–	24	–	–
Nano clay	–	5.5	–	5	–

[a]Fiber volume fraction; [b]fiber composed of 4% steel fiber and 0.5% basalt fiber CSA, sulfoaluminate cement; HPMC, hydroxypropyl methylcellulose; HRWRA, high-range water-reducing admixture; NCC, nano-calcium carbonate

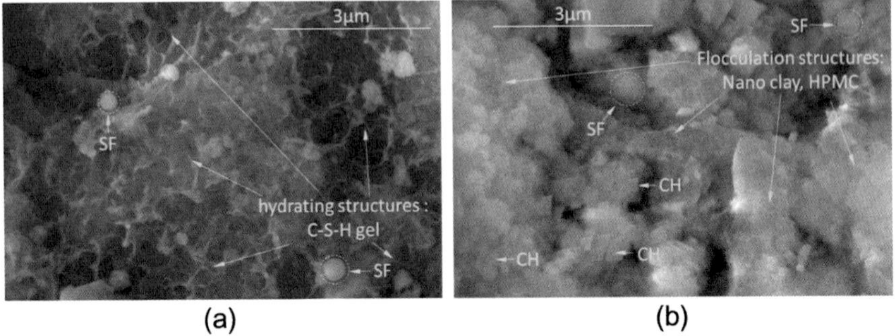

Fig. 2 Scanning electron microscopy images of **a** hydrating structure: C–S–H gel; **b** flocculation structure formed by nano-clay and hydroxypropyl methylcellulose (HPMC)

2.3 Mechanical Anisotropy of 3DP-UHPC

The compression and flexural properties of 3DP-UHPC in the existing literature were statistically analyzed. The average compressive strength of all 3DP-UHPC samples at 28 days was in the range of 120–160 MPa. The flexural strength of specimens was between 30 and 50 MPa. The printing procedure had\s a favorable effect on the orientation of fibers; that is, the orientation of steel fibers was parallel to the printing direction. Figure 3a, b show computed tomography scans of single or interwoven printed 3DP-UHPC samples. The consistency between fiber orientation and printing direction is obvious. The fiber orientation and stress direction should be designed to be consistent to reach optimal application of UHPC materials [24]. For example, the flexural strength of 3DP-UHPC specimens nearly doubled that of cast UHPC specimens [22], as shown in Fig. 3c. The flexural and tensile properties of 3DP-UHPC will be either ductile or brittle in different directions due to fiber orientation [24]. 3DP-UHPC has obvious mechanical anisotropy and generally has weaker mechanical properties along the interstrip or interlayer directions, as shown in Fig. 3d. Therefore, it is necessary to optimize the printing path to fully realize the toughening effect of fiber orientation and achieve the excellent bearing performance of 3DP-UHPC structures [18].

3 Applications of 3DP-UHPC

3.1 Large-Scale Special Components

3DP-UHPC with its ultra-high mechanical properties is expected to alleviate the contradiction that 3D printed concrete structures cannot be directly used in engineering under steel-free reinforcement conditions. Some specially shaped structures with independent bearing capacity have been constructed, although no large-scale

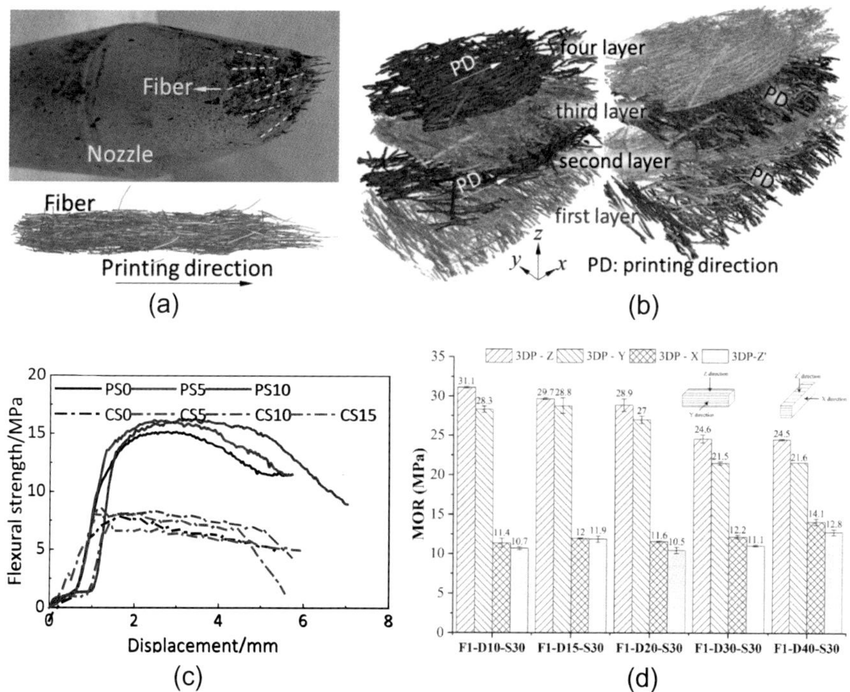

Fig. 3 Computed tomography scans of **a** single printed and **b** interwoven printed 3D printed ultra-high performance concrete (3DP-UHPC) sample; **c** flexural strength [22] and **d** the modulus of rupture (MOR) [24] of 3DP-UHPC. CS, cast sample; PS, printed sample

application of 3DP-UHPC has been found. There are two construction paths for 3DP-UHPC in new structures; one is the construction using 3DP-UHPC alone, such as a curvilinear bench [16] or hollow column [17], as shown in Fig. 4a, b; another is to use 3DP-UHPC to construct permanent templates, such as the specially shaped columns constructed by our group, shown in Fig. 4c. Attention should be paid to the construction of steel-free reinforcement concrete structures with 3DP-UHPC, namely UHPC-reinforced concrete (URC). This has obvious advantages in maintaining the flexibility of 3D printing and reducing the time and labor consumption of rebar implantation.

3.2 As the Reinforcement Materials

The in-process reinforcing method (IRM) was inspired by reinforced concrete materials under the technical constraints of steel bar implantation. As a type of IRM, the dual 3D printing procedure for URC has proved to be a potential and effective reinforcement method [19]. This is because UHPC and concrete are both cement-based

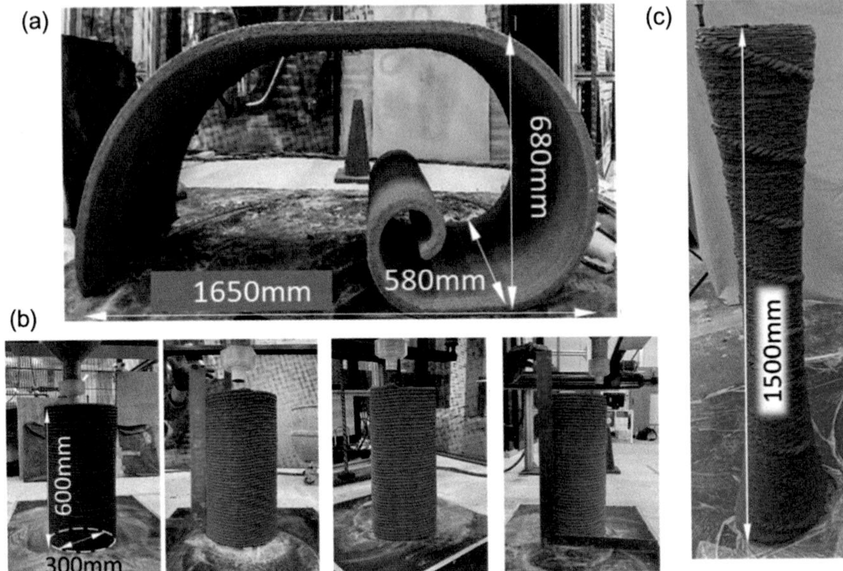

Fig. 4 **a** Curvilinear bench [16]; **b** hollow column [17]; **c** variable diameter hollow column

materials, which means that there are advantages in the printing path following and interface bonding performance [25]. The principle of an extrusion system for dual 3D concrete printing is that UHPC and 3DPC are fused at the nozzle to complete synchronous printing, as shown in Fig. 5a. Figure 5b shows the three-direction profile of the sample printed by this technique. It can be seen that the 3DP-UHPC is similar to the rebar arrangement in concrete. Another advantage of this technique is that the concrete wrapping UHPC blocks air and water to protect steel fibers from corrosion. In addition, it can be predicted that this method will maximize the utilization rate of UHPC to achieve structural strengthening and toughening, such as printing along the stress line of the bridge or in the required position based on the results of topological optimization.

3.3 As Protection Against Explosion and Impact

3DP-UHPC application has been extended to anti-explosion and anti-impact structures due to its good energy absorption capacity. The effectiveness of 3DP-UHPC has been demonstrated by experimental studies on explosion resistance and impact resistance. Yang et al. [21] tested the dynamic compression performance of 3DP-UHPC. Zhou et al. [20] explored the unique behavior of 3DP-UHPC subjected to projectile and explosive impacts and revised an empirical formula for calculating the crater depth of 3DPC. The preparation process is shown in Fig. 6(a). Our recent work

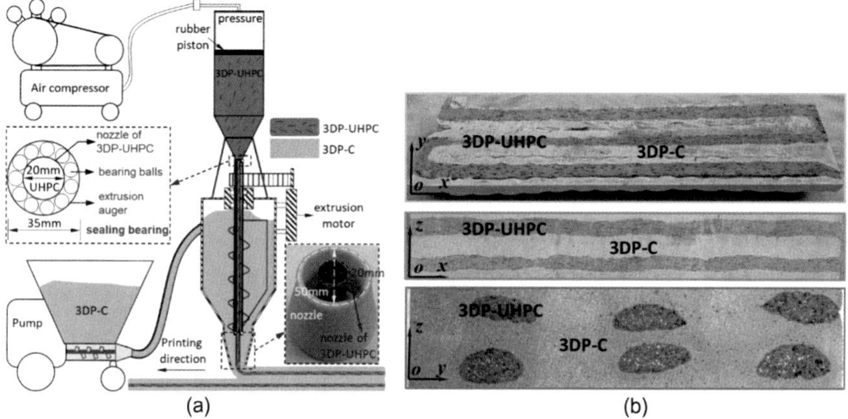

Fig. 5 **a** Schematic of extrusion system for dual 3D concrete printing; **b** three orthogonal cross-sections of 3DP-UHPC reinforced concrete specimen [19]

explored the contact explosion performance of 3DP-UHPC. In that work, 3DP-UHPC with different thicknesses was used to strengthen ordinary concrete. The preparation process is shown in Fig. 6(b). It is worth noting that there is no rebar inside the slabs of the test group. We found that PURN8, the thickness of 3DP-UHPC replacing ordinary concrete was 53.3%, was marginally improved to that of reinforced concrete.

4 Challenges of 3DP-UHPC

Despite the increasing amount of research published thus far on 3DP-UHPC, many challenges and research barriers require further innovative exploration.

4.1 Standardization of Materials

The most fundamental challenge is standardization of 3DP-UHPC. In the preparation process of 3DP-UHPC, our proposed method that combines the CM and PM had qualitative mixing ratio designs based on experience. Following a certain rule of mix ratio design will be more conducive to standardization of 3DP-UHPC and improvement of material properties. The mixing ratio design of 3DP-UHPC has been completed, but regulating the fluidity of UHPC brings a series of challenges to the ultra-high performance of the material, such as mechanical and durability weakening problems due to the density of the material and the bonding performance between the fiber and the matrix. In addition, steel fibers cannot solve the strain-softening

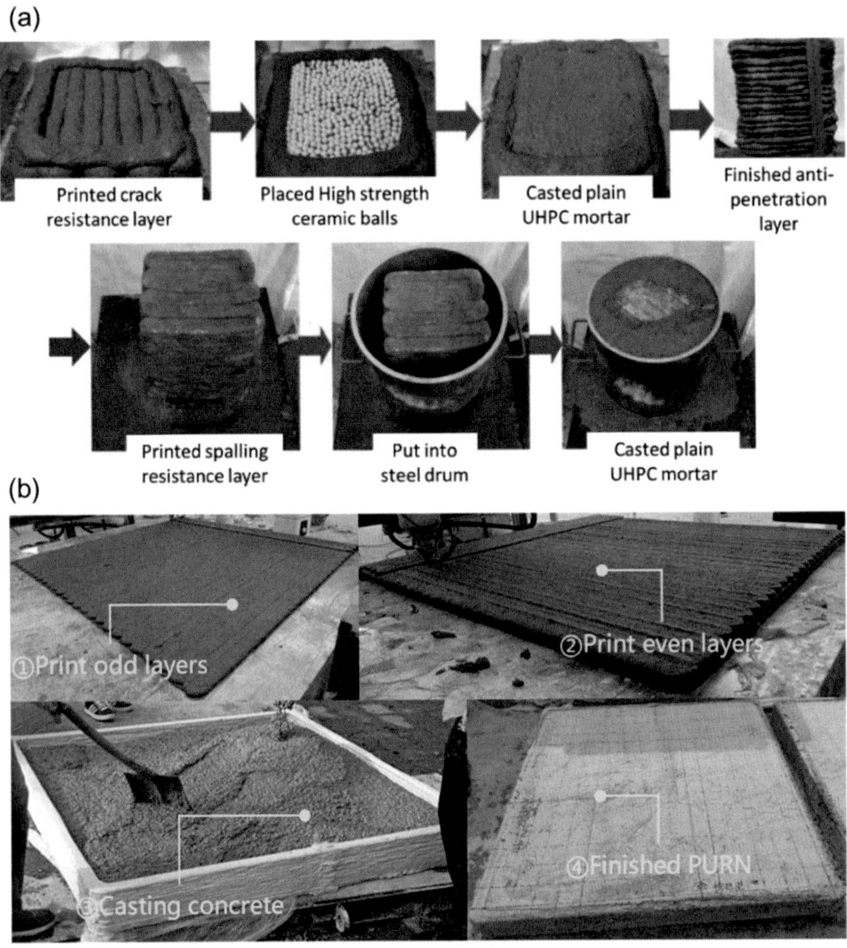

Fig. 6 **a** Technology for preparing 3D printed target [20]; **b** construction process of test slabs. UHPC, ultra-high performance concrete

behavior after peak stress, and it is difficult to achieve large tensile strains of materials. Exploratory studies have found that mixing steel fibers and polyethylene fibers can effectively improve this problem [10]. Therefore, an integrated 3DP-UHPC mix ratio design method combining comprehensive fluid performance regulation, fiber–matrix interface, and synergistic toughening of multiple fibers can contribute to optimization of material properties, and further research on this material is required, such as durability and mechanical properties under different strain rate conditions, etc.

4.2 Flexible Design of Structures

It is very important to manufacture 3DPC structures considering both safety and economy in structural applications. 3DP-UHPC has good mechanical toughness under different strain rate conditions and will be widely used in the construction of new structures and the reinforcement of existing military and civil buildings. Reinforcing plain concrete where it is needed provides a specific way to achieve the construction of 3DPC structures. 3DP-UHPC is regarded as a disruptive method for construction due to the advantages of 3D printing and its economics. Initial exploration has demonstrated the feasibility of 3DP-UHPC to partially replace or even completely replace steel reinforcement for concrete structures, but its application in larger-scale structures requires refined design, such as combining prestressing methods and topological optimization methods, design, etc. The reinforcement of existing buildings with 3DP-UHPC is also a very important structural application scenario. The coordinated deformation of interfaces of 3DP-UHPC and other materials under different working conditions will require special attention, such as the shrinkage deformation performance and load-deformation performance of the two materials. The structural response of 3DP-UHPC under strain rate loading will also be a very interesting research area.

5 Summary and Outlook

Combined with the reviewed research, we have provided a specific understanding of the research progress of 3DP-UHPC, which will be of significance to promote the application of 3DCP.

- Integrated material design combining fluid modification, fiber–matrix interface enhancement, and hybrid fiber toughening can be used to improve the mix design of 3DP-UHPC at the theoretical level, such as quantifying the impact of VMAs on the printability of UHPC, improving fiber–matrix cohesion through nanomaterials, and finding critical fiber lengths, etc.
- Further exploration of the mechanical properties and durability of materials under different strain rates and establish corresponding constitutive models to provide a theoretical basis for structural design.
- Further exploration of the application of 3DP-UHPC in newly constructed 3DPC structures or the reinforcement of existing buildings and explore the structural response laws under different strain rate loads to realize the safety and economy of 3D printed structures.

Acknowledgements We acknowledge the financial support from the National Natural Science Foundation of China (Nos 51878241, 52078181 and 52178198), the Natural Science Foundation of Hebei (Nos. E2021202039 and E2022202041), and the Natural Science Foundation of Tianjin (No. 20JCYBJC00710).

References

1. Ma G, Buswell R, Leal da Silva WR, Wang L, Xu J, Jones SZ (2022) Technology readiness: a global snapshot of 3D concrete printing and the frontiers for development. Cem Concr Res 156:106774
2. Xiao J, Ji G, Zhang Y, Ma G, Mechtcherine V, Pan J, Wang L, Ding T, Duan Z, Du S (2021) Large-scale 3D printing concrete technology: current status and future opportunities. Cement Concr Compos 122:104115
3. Wang L, Yang Y, Yao L, Ma G (2022) Interfacial bonding properties of 3D printed permanent formwork with the post-casted concrete. Cement Concr Compos 128:104457
4. Bai G, Wang L, Ma G, Sanjayan J, Bai M (2021) 3D printing eco-friendly concrete containing under-utilised and waste solids as aggregates. Cement Concr Compos 120:104037
5. Li VC, Bos FP, Yu K, McGee W, Ng TY, Figueiredo SC, Nefs K, Mechtcherine V, Nerella VN, Pan J, van Zijl GPAG, Kruger PJ (2020) On the emergence of 3D printable engineered, strain hardening cementitious composites (ECC/SHCC). Cem Concr Res 132:106038
6. Marchment T, Sanjayan J (2020) Mesh reinforcing method for 3D Concrete Printing. Autom Constr 109:102992
7. Li Z, Ma G, Wang F, Wang L, Sanjayan J (2022) Expansive cementitious materials to improve micro-cable reinforcement bond in 3D concrete printing. Cement Concr Compos 125:104304
8. Wang L, Ma G, Liu T, Buswell R, Li Z (2021) Interlayer reinforcement of 3D printed concrete by the in-process deposition of U-nails. Cem Concr Res 148:106535
9. Korniejenko K, Łach M (2020) Geopolymers reinforced by short and long fibres—innovative materials for additive manufacturing. Curr Opin Chem Eng 28:167–172
10. Wu L-S, Yu Z-H, Zhang C, Bangi T (2022) Design approach, mechanical properties and cost-performance evaluation of ultra-high performance engineered cementitious composite (UHP-ECC): a review. Constr Build Mater 340:127734
11. Bahmani H, Mostofinejad D (2022) Microstructure of ultra-high-performance concrete (UHPC)—a review study. J Build Eng 50:104118
12. Bajaber MA, Hakeem IY (2021) UHPC evolution, development, and utilization in construction: a review. J Market Res 10:1058–1074
13. Zhu Y, Zhang Y, Hussein HH, Chen G (2020) Flexural strengthening of reinforced concrete beams or slabs using ultra-high performance concrete (UHPC): A state of the art review. Eng Struct 205:110035
14. Deng Y, Zhang Z, Shi C, Wu Z, Zhang C (2022) Steel fiber–matrix interfacial bond in ultra-high performance concrete: a review. Engineering
15. Shah HA, Yuan Q, Photwichai N (2022) Use of materials to lower the cost of ultra-high-performance concrete—a review. Constr Build Mater 327:127045
16. Arunothayan AR, Nematollahi B, Ranade R, Bong SH, Sanjayan J (2020) Development of 3D-printable ultra-high performance fiber-reinforced concrete for digital construction. Constr Build Mater 257:119546
17. Arunothayan AR, Nematollahi B, Ranade R, Khayat KH, Sanjayan JG (2022) Digital fabrication of eco-friendly ultra-high performance fiber-reinforced concrete. Cement Concr Compos 125:104281
18. Yang Y, Wu C, Liu Z, Wang H, Ren Q (2022) Mechanical anisotropy of ultra-high performance fibre-reinforced concrete for 3D printing. Cement Concr Compos 125:104310
19. Bai G, Wang L, Wang F, Ma G (2021) In-process reinforcing method: dual 3D printing procedure for ultra-high performance concrete reinforced cementitious composites. Mater Lett 304:130594
20. Zhou J, Lai J, Du L, Wu K, Dong S (2022) Effect of directionally distributed steel fiber on static and dynamic properties of 3D printed cementitious composite. Constr Build Mater 318:125948
21. Yang Y, Wu C, Liu Z, Li J, Yang T, Jiang X (2022) Characteristics of 3D-printing ultra-high performance fibre-reinforced concrete under impact loading. Int J Impact Eng 164:104205
22. Bai G, Wang L, Wang F, Cheng X (2021) Investigation of the Printability and Mechanical Properties of 3D Printing UHPC. Materials Reports 35(12):67–73

23. Zhang T, Wang W, Zhao Y, Bai H, Wen T, Kang S, Song G, Song S, Komarneni S (2021) Removal of heavy metals and dyes by clay-based adsorbents: From natural clays to 1D and 2D nano-composites. Chem Eng J 420:127574
24. Arunothayan AR, Nematollahi B, Ranade R, Bong SH, Sanjayan JG, Khayat KH (2021) Fiber orientation effects on ultra-high performance concrete formed by 3D printing. Cem Concr Res 143:106384
25. Bai G, Wang L, Wang F, Ma G (2023) Assessing printing synergism in a dual 3D printing system for ultra-high performance concrete in-process reinforced cementitious composite. Addit Manuf 61:103338

Nanosilica-Modified Hydrogels Encapsulating Bacterial Spores for Self-healing Concrete

J. Feng and S. Qian

Abstract Microbially induced calcium carbonate precipitation is effective in achieving self-healing of concrete cracks when the bacteria are well protected in concrete with a high pH and dense microstructure. Calcium alginate hydrogels are appropriate for encapsulating bacteria in concrete due to the mild environment with rich moisture in the hydrogels. Nevertheless, the low alkaline tolerance and breakage ratios of the hydrogels after concrete cracking restrict their applications in concrete. To address these problems, nanosilica was doped into calcium alginate hydrogels with encapsulated bacterial spores to react with the $Ca(OH)_2$ surrounding hydrogels in concrete. Due to the modification by nanosilica, the bond of the hydrogels with cement matrix was enhanced as needle-like C–S–H was generated at the interface after hydration for 7 days. Moreover, the urease activity of the encapsulated spores in the modified hydrogels was higher than that in plain hydrogels after submersion in saturated $Ca(OH)_2$ solution or simulated concrete solution for 7 days. Therefore, it can be concluded that nanosilica holds promise for modifying hydrogels to improve the effectiveness of encapsulated bacterial spores for self-healing of concrete.

Keywords Bacterial spores · Hydrogels · Nanosilica · Self-healing concrete

1 Introduction

Concrete is widely used in construction engineering due to its good durability, adjustable strength, and low cost. However, cracking can occur in concrete due to overloading and volume instability [1], which will reduce both the mechanical performance and durability of concrete structures. Hence, to prolong the service life of concrete structures, timely repair of concrete is critical.

J. Feng · S. Qian (✉)
School of Civil and Environmental Engineering, Nanyang Technological University, Nanyang, Singapore
e-mail: szqian@ntu.edu.sg

W. Duan et al. (eds.), *Nanotechnology in Construction for Circular Economy*,
Lecture Notes in Civil Engineering 356,
https://doi.org/10.1007/978-981-99-3330-3_9

Currently, concrete repair is achieved manually using epoxy resin [2] or some cementitious material [3]. Nevertheless, manual repairing cannot be timely because the detection of cracks takes time and is not feasible for inaccessible cracks. Therefore, developing concrete with a self-healing capacity that can effectively heal cracks without human intervention is desired.

Concrete itself has a certain self-healing ability resulting from further hydration of cementitious materials and carbonation [4, 5], but only for cracks of limited sizes. The self-healing capacity can be further improved by incorporating some components specific for self-healing into the concrete. For example, minerals [6, 7], superabsorbent polymers [8, 9], and shape memory alloy [10, 11] have been added to concrete to enhance self-healing. In addition to these materials, bacteria-based self-healing agents for self-healing concrete has been investigated in recent years and promising results have been obtained [12]. For bacterial self-healing agents, carriers to encapsulate or immobilize the bacteria are required as the concrete environment with high pH [13] and dense microstructure is incompatible [14].

Among the carriers to encapsulate bacteria in concrete, hydrogels with moderate pH environment and rich moisture content have high potential for protecting bacteria [14]. Specifically, calcium alginate, which has good biocompatibility [15, 16], can be used to encapsulate bacteria, but its susceptibility to environmental factors [17] and poor bonding with concrete matrix could lead to the ingress of alkalis and low efficiency of releasing bacteria after cracking.

To address these issues, nanosilica was doped into calcium alginate hydrogels to react with the surrounding calcium hydroxide, simultaneously lowering the local pH and generating C–S–H at the interface between the hydrogel and cement matrix, which can enhance the bonding of hydrogels with concrete. A previous report [18] revealed that hydration product could be generated around or within hydrogels that contain silica, indicating the feasibility of the approach in this research. Herein, the microstructure of cement paste with hydrogels was observed and the alkali tolerance of hydrogels encapsulating bacterial spores was evaluated to analyze the effects of nanosilica modification on calcium alginate hydrogels.

2 Methods

2.1 Preparation of Bacterial Spores

One ureolytic bacteria *Lysinibacillus sphaericus LMG 22,257* from Belgian Coordinated Collection of Micro-organisms were used to prepare bacterial spores. The preparation methods were in accordance with the steps in [19].

2.2 Preparation of Hydrogels Encapsulating Bacterial Spores

7.5 g/L nanosilica powder (10–20 nm) was dispersed in sodium alginate solution (15 g/L) by sonication for 15–20 min before 2.0wt% bacterial spores were added to the mixture. Next, the mixture was dropped into calcium nitration solution (0.1 M) using a peristaltic pump with a rotary speed of 3 rpm. The prepared hydrogels were collected by gravity sedimentation then hardened in a fresh 0.1 M calcium nitrate solution for 24 h at 5 °C. Finally, the hydrogels were separated and washed three times with distilled water before being freeze-dried and stored at 4 °C. The plain hydrogels were synthesized using the same procedures except for the incorporation of nanosilica powder.

2.3 Observation of Interface Between Hydrogel and Cement Matrix

Cement paste with hydrogels was prepared by using Portland Cement I 52.5, tap water and hydrogels at a mass ratio of 1:0.5:0.0055. After moist curing (23±1°C, 75±5%°C relative humidity) for 7 and 28 days, the specimens were broken into pieces before being immersed in isopropanol for 24 h to stop hydration. The samples were impregnated in epoxy, then cut with a precision saw to expose the cement paste with hydrogels; the exposed surface was further ground, polished and washed with ethanol. Afterwards, the samples were vacuum dried and coated with gold before being observed under a scanning electron microscope (FESEM, JEOL JSM-7600F) at backscattered electron (BSE) mode at an accelerating voltage of 15 keV.

2.4 Revival of Bacterial Spores Encapsulated in Hydrogels After Exposure to Alkali Environments

The hydrogels with encapsulated endospores were immersed in saturated $Ca(OH)_2$ solution and simulated concrete pore solution containing 0.001 M $Ca(OH)_2$, 0.2 M NaOH and 0.6 M KOH [20] for 7 days respectively. After that, 1 g hydrogels were removed from the solution and washed three times with distilled water before being incubated in 100 mL sterile UYE medium (20 g/L urea and 20 g/L yeast extract). The optical density at a wavelength of 600 nm and the urease activities of the medium were measured by using a UV/Vis spectrometer (UV mini-1240, Shimadzu) and conductivity meter (DDS-11A, Lightning), respectively, after the inoculated mediums were cultivated for 1, 2, and 3 days.

3 Results and Discussion

The morphology of cement paste with modified hydrogels is shown in Fig. 1. After hydration for 7 days, nanosized calcium silicate hydrate (C–S–H) with needle-like morphology was observed on the outer surface of modified hydrogel, as shown in Fig. 1a, b. The generations of C–S–H were due to the reaction of the incorporated nanosilica on the hydrogel surface with calcium hydroxide in the cement paste, suggesting the effect of nanosilica modification on interface enhancement between the hydrogels and cement matrix. After hydration for 28 days, the BSE image shown in Fig. 1c revealed no obvious gaps or cracking at the interface, indicating the modified hydrogels bonded well with the cement matrix. The improved bonding of the hydrogels with cement matrix could facilitate the release of encapsulated bacterial spores after concrete cracking, as the cracks might propagate through the hydrogels rather than the interfaces, leading to breakage of the hydrogels and exposure of bacterial spores to the cracks. After the release of bacterial spores, the endospores can conduct germination and outgrowth with leaching of nutrients from the concrete matrix, contributing to the generation of calcium carbonate within cracks by producing urease to catalyze hydrolysis of urea.

The alkaline tolerance of the hydrogels encapsulating bacterial spores is shown in Fig. 2. After immersion in the alkaline solution for 7 days, the growth of the encapsulated bacteria in the modified hydrogels during incubation was more rapid than in the plain hydrogels, as illustrated in Fig. 2a. The optical densities of the medium with the modified hydrogels reached approximately 2 and 2.5 approximately after cultivation for 2 days respectively, while the optical densities of the medium with bacteria in plain hydrogels were close to 2 after cultivating for around 3 days. With the growth of bacteria, the medium with bacteria encapsulated bacteria in modified hydrogels presented higher urease activities than that with bacteria in plain hydrogels, as shown in Fig. 2b. After incubation for 1 day, the urease activities of medium were 3.7 U/mL and 6.6 U/mL approximately for hydrogels undergoing 7-day submersion in simulated concrete or saturated $Ca(OH)_2$ solution respectively, while those in medium with plain hydrogels was < 1 U/mL. Although the urease activities of medium with plain hydrogels further increased with incubation duration,

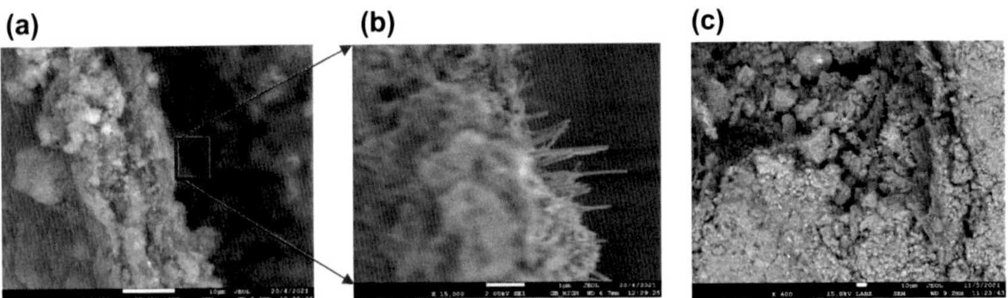

Fig. 1 Backscattered electron images of cement paste with nanosilica modified hydrogels. **a** After hydration for 7 days; **b** close-up view of the red box in (**a**); **c** after hydration for 28 days

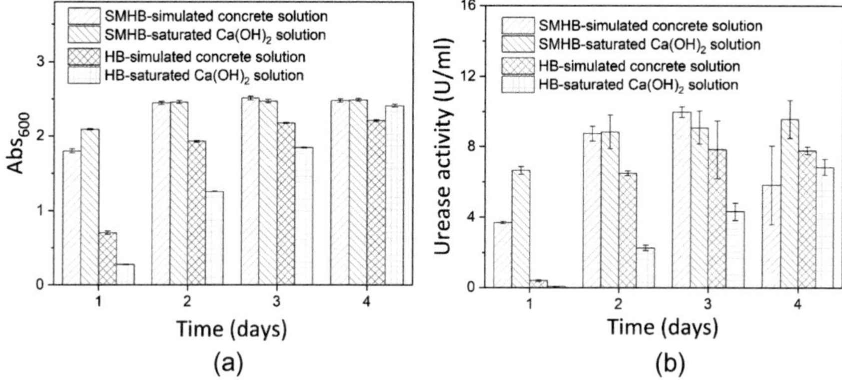

Fig. 2 **a** Optical densities of mediums with modified and plain hydrogels after being immersed in simulated concrete or saturated $Ca(OH)_2$ solution for 7 days; **b** urease activities of mediums with modified and plain hydrogels after being immersed in simulated concrete or saturated $Ca(OH)_2$ solution for 7 days. SMHB denotes the nanosilica modified hydrogels with encapsulated bacterial spores; HB denotes plain hydrogels with encapsulated bacterial spores

they were still lower than those with modified hydrogels after incubation for 2 and 3 days. Collectively, the bacteria in the modified hydrogels grew more rapidly and showed higher urease activities than those in plain hydrogels after exposure to an alkaline environment, which suggested the effectiveness of nanosilica modification on alkaline tolerance improvement of hydrogels for bacterial encapsulation.

4 Conclusions

This research investigated the feasibility of nanosilica modification of calcium alginate hydrogels for bacterial encapsulation in self-healing concrete. The results revealed that the nanosilica modification enhanced the bonding of the hydrogel with the cement matrix due to C–S–H generation at the interface. Moreover, the modified hydrogels with encapsulated endospores showed improved alkaline tolerance, because as the 7-day immersion in simulated concrete and saturated $Ca(OH)_2$ solution caused less inhibition on bacterial growth and urease activity in medium inoculated with modified hydrogels encapsulating endospores. Hence, it is feasible to modify calcium alginate hydrogels with nanosilica to increase the effectiveness of bacterial encapsulation in concrete. In the future, the effects of nanosilica-modified hydrogels encapsulating bacterial spores on self-healing performance (i.e. strength recovery, etc.) of concrete can be investigated.

References

1. Mindess S, Young JF (1981) Concrete. Prentice-Hall
2. Yoo DY, Oh T, Shin W, Kim S, Banthia N (2021) Tensile behavior of crack-repaired ultra-high-performance fiber-reinforced concrete under corrosive environment. J Mater Res Technol
3. Chindaprasirt P, Sriopas B, Phosri P, Yoddumrong P, Anantakarn K, Kroehong W (2021) Hybrid high calcium fly ash alkali-activated repair material for concrete exposed to sulfate environment. J Build Eng:103590
4. Huang HL, Ye G, Qian CX, Schlangen E (2016) Self-healing in cementitious materials: materials, methods and service conditions. Mater Des 92:499–511
5. Reinhardt HW, Jooss M (2003) Permeability and self-healing of cracked concrete as a function of temperature and crack width. Cem Concr Res 33(7):981–985
6. Feng J, Dong H, Wang R, Su Y (2020) A novel capsule by poly (ethylene glycol) granulation for self-healing concrete. Cem Concr Res 133:106053
7. Wu X, Huang H, Liu H, Zeng Z, Li H, Hu J, Wei J, Yu Q (2020) Artificial aggregates for self-healing of cement paste and chemical binding of aggressive ions from sea water. Compos B Eng 182:107605
8. Lee HXD, Wong HS, Buenfeld NR (2016) Self-sealing of cracks in concrete using superabsorbent polymers. Cem Concr Res 79:194–208
9. Snoeck D, Van den Heede P, Van Mullem T, De Belie N (2018) Water penetration through cracks in self-healing cementitious materials with superabsorbent polymers studied by neutron radiography. Cem Concr Res 113:86–98
10. Teall O, Pilegis M, Davies R, Sweeney J, Jefferson T, Lark R, Gardner D (2018) A shape memory polymer concrete crack closure system activated by electrical current. Smart Mater Struct 27(7)
11. Chen W, Lin B, Feng K, Cui S, Zhang D (2022) Effect of shape memory alloy fiber content and preloading level on the self-healing properties of smart cementitious composite (SMA-ECC). Constr Build Mater 341:127797
12. Van Tittelboom K, De Belie N, De Muynck W, Verstraete W (2010) Use of bacteria to repair cracks in concrete. Cem Concr Res 40(1):157–166
13. Wang J, Van Tittelboom K, De Belie N, Verstraete W (2012) Use of silica gel or polyurethane immobilized bacteria for self-healing concrete. Constr Build Mater 26(1):532–540
14. Wang JY, Mignon A, Snoeck D, Wiktor V, Van Vlierberghe S, Boon N, De Belie N (2015) Application of modified-alginate encapsulated carbonate producing bacteria in concrete: a promising strategy for crack self-healing, Front Microbiol:6
15. Dong Y, Zhang Y, Tu B, Miao J (2014) Immobilization of ammonia-oxidizing bacteria by calcium alginate. Ecol Eng 73:809–814
16. Seifan M, Samani AK, Hewitt S, Berenjian A (2017) The effect of cell immobilization by calcium alginate on bacterially induced calcium carbonate precipitation. Fermentation 3(4):57
17. Kim BJ, Park T, Moon HC, Park SY, Hong D, Ko EH, Kim JY, Hong JW, Han SW, Kim YG (2014) Cytoprotective alginate/polydopamine core/shell microcapsules in microbial encapsulation. Angew Chem Int Ed 53(52):14443–14446
18. Bose B, Davis CR, Erk KA (2021) Microstructural refinement of cement paste internally cured by polyacrylamide composite hydrogel particles containing silica fume and nanosilica. Cem Concr Res 143:106400
19. Wang J, Mignon A, Trenson G, Vlierberghe SV, Boon N, Belie ND (2018) A chitosan based pH-responsive hydrogel for encapsulation of bacteria for self-healing concrete. Cem Concr Compos (93):309–322
20. Ghods P, Isgor OB, McRae GA, Cu GP (2010) Electrochemical investigation of chloride-induced depassivation of black steel rebar under simulated service conditions. Corros Sci 52(5):1649–1659

Reusing Alum Sludge as Cement Replacement to Develop Eco-Friendly Concrete Products

Y. Liu, Y. Zhuge, and W. Duan

Abstract Alum sludge is a typical by-product of the water industry. The traditional sludge management method, disposing of sludge in landfill sites, poses a critical environmental and economic concern due to a significant increase in sludge amount and disposal cost. In this paper, the feasibility of reusing sludge as cement replacement is investigated, and the physical performance and microstructure modification of concrete products made with sludge is discussed. The obtained results indicated that a satisfying pozzolanic reactivity of sludge after calcination at high temperatures and grinding to the appropriate size was identified. When 10% cement was replaced with sludge, the reaction degree of sludge was up to 39%, and the obtained concrete blocks exhibited superior mechanical performance. Based on the microstructural analysis, e.g., x-ray diffraction, thermogravimetric analysis, and advanced nanoindentation method, the high aluminum content in sludge was incorporated into C–(A)–S–H gel; the original "Al-minor" C–(A)–S–H gel in pure cement paste was converted to 'Al-rich' C–(A)–S–H gel. Also, sludge promoted the formation of aluminum-bearing hydrates, such as ettringite and calcium aluminate hydrates (C–A–H). Although the Al incorporation had no significant effect on the hardness and modulus of C–(A)–S–H gel, the homogeneous mechanical properties (hardness and modulus measured with nanoindentation) of binder paste degraded with increasing sludge ash content above 10%, attributing to the lower hardness of unreacted sludge than cement clinker and the relatively lower reaction degree. Using sludge in concrete products offers an economical and environmentally friendly way to dispose of sludge and preserve diminishing natural resources. Also, the reduction of cement usage may contribute to achieving carbon neutrality.

Keywords Advanced nanoindentation technique · Cement replacement · Alum sludge · Green concrete · Supplementary cementitious materials · Value-added recycling

---

Y. Liu · Y. Zhuge (✉) · W. Duan
UniSA STEM, University of South Australia, Mawson Lakes, SA 5095, Australia
e-mail: Yan.Zhuge@unisa.edu.au

W. Duan et al. (eds.), *Nanotechnology in Construction for Circular Economy*,
Lecture Notes in Civil Engineering 356,
https://doi.org/10.1007/978-981-99-3330-3_10

1 Introduction

Alum sludge is a typical by-product of the drinking water industry. The heterogeneous sludge waste is formed when the aluminum-based coagulant is combined with suspended solids, dissolved colloids, organic matter, and microorganisms in raw water. It is estimated that global sludge production has exceeded 10,000 tonnes per day, and the rapid population growth and economic development may result in a significant increase in its amount in future decades [1]. In Australia, most sludge is disposed of at landfill sites (see Fig. 1), which may cause severe environmental issues because of land wastage and secondary pollution. In view of the transition toward a circular economy, vast-available sludge should be considered as a resource with the potential to be valorized instead of a waste.

Most alum sludge has 20–63 wt% Al_2O_3 and 17–41 wt% SiO_2 [2]. Its aluminosilicate nature makes sludge can be recycled as cement replacement, proposing a possible solution to reuse sludge in large quantities. Also, reducing cement usage may contribute to achieving the target of carbon neutrality. Some previous studies have already investigated the feasibility of alum sludge as cement replacement in concrete products [3]. In general, raw alum sludge exhibits no pozzolanic reaction, and the high organic matter in sludge may hinder the cement hydration, resulting in deteriorated mechanical and durability performance of concrete products [4]. Treating sludge with high temperatures, ranging from 600 °C to 800 °C, can efficiently improve the pozzolanic reactivity of sludge due to the fact that crystal phases of silicon and aluminum were dehydroxylated to form disordered phases with high reactivity [5]. However, the optimum temperature to activate sludge activity is still controversial. For the performance of sludge-derived concrete products, a moderate cement replacement (e.g., 10%) with calcined sludge is feasible without compromising the mechanical and durability properties [6–8].

Based on the above literature, the ideal temperature (between 600 and 800 °C) to activate the pozzolanic reactivity of alum sludge needs to be clarified. The reaction

Fig. 1 Sludge landfill sites in South Australia

degree of sludge, composition of hydration products, and chemo-mechanical properties of sludge-cement binders need to be in-depth discussed. Therefore, in this study, the pozzolanic reactivity of alum sludge under different temperatures was assessed by the strength index test. The reaction degree of sludge was determined by the selective dissolution method. The chemical composition of sludge-cement binders was assessed by x-ray diffraction (XRD) and thermogravimetric analysis (TGA). The chemo-mechanical properties of blended binders, e.g., indentation hardness and modulus, were investigated by an advanced nanoindentation method. Final, the strength of concrete blocks with cement replaced with sludge at weight percentages of 0%, 10%, 20%, and 30% was studied.

2 Methodology

2.1 Materials

Alum sludge is collected from a local water treatment plant in South Australia. First, raw sludge was crushed and milled to less than 75 μm. Then, milled sludge was calcined at 600°C, 700°C, and 800°C, respectively. General-purpose cement is used according to AS 3972. Fine aggregates and coarse aggregates used to manufacture concrete blocks (CB) were concrete sand and crushed limestone. The particle size distribution of the sludge under different temperatures was determined by Mastersizer 3000. The chemical composition of sludge and cement was investigated by X-ray fluorescence.

2.2 Sample Preparation

CB were manufactured with the dry mix method, and the detailed cast procedures are described in a previous study [9]. The pastes were also cast in plastic molds with dimensions of $10 \times 10 \times 10$ mm^3, which contained the same water to cement ratio and sludge content as CB, assisting in the study of hydration products by eliminating aggregate interference. Table 1 shows the mix design of the CB.

2.3 Experimental Methods

The pozzolanic reactivity of sludge under different temperatures was determined by the strength index (SAI) test according to ASTM C311. The selective dissolution method was used to assess the reaction degree of sludge based on a mixture of ethylenediaminetetraacetic acid–triethanolamine–NaOH [10]. The compressive

Table 1 Mixture design of blocks (kg/m^3)

Samples	Sludge	Cement	Fine aggregate	Coarse aggregate	SP
CB-0	0	335	962	985	2.0
CB-10	34	301	962	985	2.6
CB-20	68	267	962	985	3.0
CB-30	100	235	962	985	3.3

CB, concrete blocks; SP, superplasticizer

strength of the CB was measured according to AS 4456.4. The load was applied at a constant rate of 2 kN/s without any shock until failure.

The hydration products in the cement-sludge binder matrix were characterized by XRD and TGA, and the hydration reaction of samples was stopped by ethanol immersion. An advanced nanoindentation technology (coupling conventional statistic nanoindentation and chemical mapping) was used to characterize the in-situ chemo-mechanical properties. Detailed sample preparation procedures and the data analysis method for the nanoindentation test are described in our previous study [10].

3 Results and Discussion

3.1 *Material Characterization*

Table 2 shows the chemical composition and the particle size distribution of cement and sludge under different calcination temperatures. The main components in raw sludge were Al_2O_3, SiO_2, and organic matter, and minor contents of Fe_2O_3, CaO, and K_2O were also observed. After calcination, most organic matter content was eliminated, resulting in an increase in the proportions of Al_2O_3 and SiO_2. Because the chemical composition of sludge calcined between 600°C and 800°C was similar, only the 800°C-treated one is shown in Table 2. It is worth noting that the sum of the Al_2O_3, SiO_2, and Fe_2O_3 content in calcined sludge was higher than 70%, which satisfied the composition requirement of natural pozzolan material based on ASTM C618. In Table 3, the particle size distribution of calcined sludge and cement is shown. The Blaine fineness of sludge decreased with increasing calcination temperature, which could be explained by the fact that dehydroxylated particles agglomerate under high temperatures to produce new porous grains, especially fine particles [11]. The cement exhibited finer particle size but comparable Blaine fineness to that of calcined sludge.

The results of the SAI test are shown in Table 4. After calcination at 700 °C and 800 °C, sludge exhibited a satisfactory pozzolanic reactivity (SAI $\geq$ 75%). In contrast, the 600 °C-treated sludge could not be used as a pozzolan material. Compared with 700 °C, 800 °C is a better temperature to activate the pozzolanic reactivity of sludge,

Table 2 Chemical composition (LOI, loss on ignition)

Type	Al_2O_3	SiO_2	Fe_2O_3	Na_2O	CaO	K_2O	TiO_2	LOI
Raw sludge	28.3	26.4	7.7	–	5.4	1.2	–	29.5
Calcined sludge	47.7	31.1	5.9	–	4.3	–	–	6.2
Cement	4.6	20.0	3.1	0.2	63.4	0.4	0.3	3.3

Table 3 Particle size distribution

	Cement	600°C-sludge	700°C-sludge	800°C-sludge
D10 (μm)	3.4	5.4	5.8	6.1
D50 (μm)	14.5	29.6	29.2	30.4
D90 (μm)	33.1	61.3	60.3	60.9
Blaine fineness (m^2/kg)	400.0	445.4	429.8	423.4

Table 4 SAI test of sludge under different calcination temperatures (SAI, strength index)

	600°C-sludge	700°C-sludge	800°C-sludge
7-day SAI (%)	15.07	61.69	87.00
28-day SAI (%)	14.59	91.38	113.60

in which the SAI value is up to 113.60%. Therefore, 800 °C-treated sludge was used to cast the paste and CB samples.

3.2 *Reaction Degree and Hydration Mechanism of Sludge*

The selective dissolution method determined the reaction degree of sludge in the blended binders. The paste samples containing 10% sludge by weight exhibited the highest reaction degree, up to 39.0%. Further increasing the sludge content to 20% and 30%, decreased the reaction degree to 25.2% and 24.8%, respectively. The decrease in reaction degree could be attributed to a lack of sufficient portlandite. Figure 2 shows the TGA and XRD. In Fig. 2a, c, a significant endothermic peak occurred ≈ 120°, which was associated with the decomposition of the AFt phase, (e.g., ettringite). The ettringite peak intensity increased with increasing sludge content; thus, sludge might promote the formation of AFt phases. Also, adding sludge resulted in the formation of calcium aluminate hydrate (C–A–H). These results could be related to the high reactive Al content in sludge, leading to the formation of additional Al-bearing phases [12]. In Fig. 2b, d, the XRD patterns of pastes at 28 days and 90 days are shown. The reflection peaks related to C–A–H and stratlingite were only observed in the blended pastes, not the reference ones. The ettringite peak intensity was enhanced with increasing sludge content, which was consistent with the TGA analysis. In both the TGA and XRD analyses, the content of portlandite (CH)

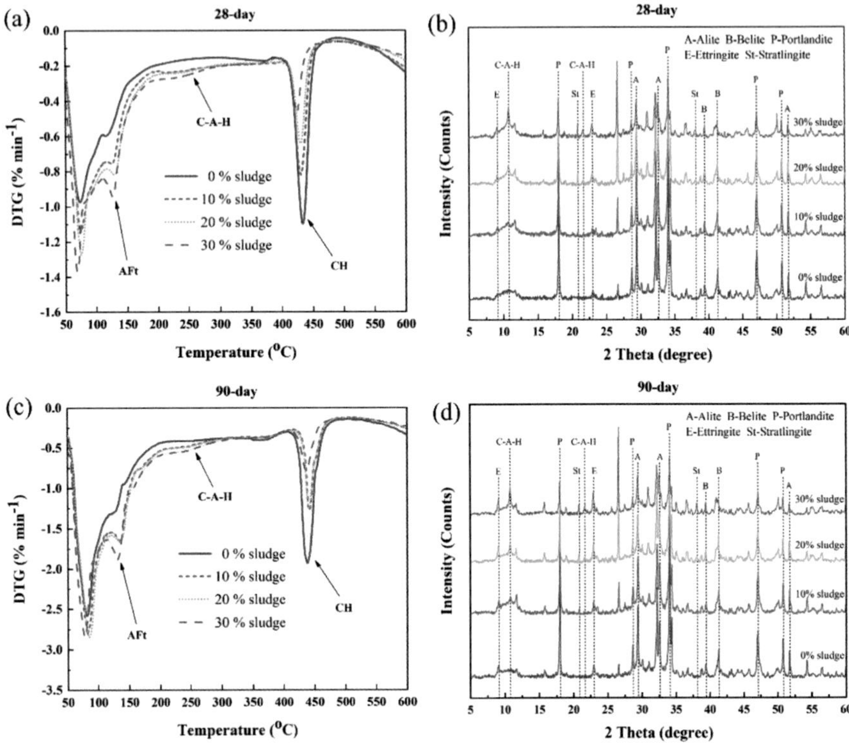

Fig. 2 **a**, **c** TGA and **b**, **d** XRD analysis of blended pastes (modified from Liu et al. [13])

decreased with increasing sludge content, confirming the pozzolanic reactivity of sludge. It is interesting to note that the CH content in the paste with 30% sludge was significantly low compared with the reference or 10% sludge pastes. Such a lower concentration of available CH content might result in a decrease in the reaction degree of sludge.

3.3 Chemo-Mechanical Properties of Pastes and Strength of Blocks

Table 5 shows the chemo-mechanical properties of the pastes and the strength of CB. The homogenized modulus of binders significantly decreased when >10% of cement was replaced with sludge. Such a reduction could be attributed to decreased High-density (HD) C–S–H gel in samples with 30% sludge. However, the total volume of C–S–H gel in the different pastes was almost the same, indicating that the filler effect of sludge might compensate for the cement dilution effect, although the high amount of sludge hindered the transformation of Low-denisty (LD) C–S–H gel to

Table 5 Chemo-mechanical properties of C–A–S–H gel in pastes and the strength of CB

Cementitious pastes	Ref	10% sludge	30% sludge
Homogenized modulus (GPa)	35.3	36.7	30.7
HD C–S–H modulus (GPa)	37.4	36.1	34.7
LD C–S–H modulus (GPa)	21.7	22.7	21.0
Relative Al intensity in HD C-S–H	0.14	0.11	0.27
Relative Al intensity in LD C-S–H	0.14	0.13	0.33
HD C–S–H gel volume (%)	0.52	0.55	0.36
LD C–S–H gel volume (%)	0.28	0.24	0.42
Total C–S–H gel volume (%)	0.80	0.79	0.78
Concrete blocks			
7-day compressive strength (MPa)	18.2	13.4	9.1
28-day compressive strength (MPa)	20.4	20.3	11.0
90-day compressive strength (MPa)	22.3	25.8	12.7

HD C–(A)–S–H gel. Based on the results for Al intensity in the C–S–H gel, with the addition of sludge to 30%, the original "Al-minor" C–(A)–S–H gel in pure cement paste was converted to "Al-rich" C–(A)–S–H gel. There was no significant difference in indentation modulus for C–(A)–S–H gel in the different pastes, indicating that Al incorporation had negligible effect on the mechanical properties of C–(A)–S–H gel.

At a curing age of 7 days, the compressive strength of CB containing sludge was significantly lower than that of the reference samples. However, after curing for 28 days, the samples with 10% sludge exhibited a comparable or even higher compressive strength. The optimum sludge content in blocks was 10%, which was in agreement with the results of the nanoindentation analysis.

4 Conclusions

In general, the feasibility of reusing sludge as cement replacement was confirmed in this study. 800 °C was the best temperature to activate the pozzolanic reactivity of sludge. 10% of cement could be replaced with sludge in concrete blocks without compromising mechanical performance. The reaction degree of sludge in blended pastes could be up to 39%, and the Al content in C–(A)–S–H gel increased with adding sludge. However, the C–(A)–S–H gel modification had no significant effect on the indentation modulus or hardness of the gel. In addition to incorporation into C–(A)–S–H gel, the high content of reactive Al promoted the formation of other Al-bearing phases (e.g., ettringite and stratlingite).

References

1. Liu Y et al (2020) Recycling drinking water treatment sludge into eco-concrete blocks with CO2 curing: Durability and leachability. Sci Total Environ 746
2. Zhuge Y, Liu Y, Pham PN (2022) Sustainable utilization of drinking water sludge. Low carbon stabilization and solidification of hazardous wastes. Elsevier, pp 303–320
3. Gomes SDC et al (2020) Recycling of raw water treatment sludge in cementitious composites: effects on heat evolution, compressive strength and microstructure. Resour Conserv Recycl:161
4. Wang L et al (2018) A novel type of controlled low strength material derived from alum sludge and green materials. Constr Build Mater 165:792–800
5. Tantawy MA (2015) Characterization and pozzolanic properties of calcined alum sludge. Mater Res Bull 61:415–421
6. Frías M et al (2014) Influence of activated drinking-water treatment waste on binary cement-based composite behavior: Characterization and properties. Compos B Eng 60:14–20
7. Bohórquez González K et al (2020) Use of sludge ash from drinking water treatment plant in hydraulic mortars. Mater Today Commun:23
8. Owaid HM, Hamid R, Taha MR (2014) Influence of thermally activated alum sludge ash on the engineering properties of multiple-blended binders concretes. Constr Build Mater 61:216–229
9. Liu Y et al (2020b) Properties and microstructure of concrete blocks incorporating drinking water treatment sludge exposed to early-age carbonation curing. J Clean Prod:261
10. Liu Y et al (2021) Cementitious composites containing alum sludge ash: An investigation of microstructural features by an advanced nanoindentation technology. Constr Build Mater:299
11. Fabbri B, Gualtieri S, Leonardi C (2013) Modifications induced by the thermal treatment of kaolin and determination of reactivity of metakaolin. Appl Clay Sci 73:2–10
12. Tironi A et al (2013) Assessment of pozzolanic activity of different calcined clays. Cement Concr Compos 37:319–327
13. Liu Y et al (2021) The potential use of drinking water sludge ash as supplementary cementitious material in the manufacture of concrete blocks. Resour Conserv Recycl:168

Role of Aggregate Reactivity, Binder Composition, and Curing Temperature on the Delayed Ettringite Formation and Associated Durability Loss in Concrete

L. Martin, P. Thomas, P. De Silva, and V. Sirivivatnanon

Abstract The durability of concrete is critical to its worldwide use as a structural material for buildings and infrastructure, with the lifetime service of concrete greatly affecting its economic, environmental, and social costs. Causes of durability loss in some concrete structures can be attributed to the alkali–silica reaction (ASR) and delayed ettringite formation (DEF). Both are chemical reactions that have the potential to cause expansion and strength loss in affected elements. Significant overlap exists in the factors contributing to ASR and DEF in concrete structures, with widely reported evidence of deleterious DEF frequently occurring in conjunction with mild or moderate ASR. For precast concrete, experiments in mortars have provided limits in the alkali and sulfate content of the binder and maximum curing temperatures used to minimize DEF risk. The role of other constituents in concrete specimens, notably the aggregate, has been overlooked. We investigated the role of reactive aggregates and ASR in the susceptibility of concrete to deleterious DEF.

Keywords Alkali-silica reaction · Delayed ettringite formation · Durability loss · Heat cure · Precast concrete

L. Martin (✉)
School of Mathematical and Physical Sciences, University of Technology Sydney, Sydney, NSW, Australia
e-mail: liam.martin@uts.edu.au

P. Thomas · V. Sirivivatnanon
School of Civil and Environmental Engineering, University of Technology Sydney, Sydney, NSW, Australia

P. De Silva
School of Behavioural and Health Sciences, Australian Catholic University, Sydney, NSW, Australia

W. Duan et al. (eds.), *Nanotechnology in Construction for Circular Economy*, Lecture Notes in Civil Engineering 356,
https://doi.org/10.1007/978-981-99-3330-3_11

1 Introduction

The durability of concrete as a structural material is vitally important to its modern use in buildings, homes, and infrastructure around the world. Concrete and cement-based materials have many desirable properties for use in a variety of structures, allowing in-form construction, low labor cost, low technical threshold, low cost and environmental concerns (excluding high carbon emissions), high strength, good durability, and long-term efficacy [1, 2]. However, the extended service life, multi-faceted application, and complex nature of concrete as a composite material creates many potential causes of durability loss across the structural, mechanical, and chemical fields. These include acid, chloride, or sulfate attack, steel corrosion, leeching, freeze–thaw events, salt and de-icing ingress, humidity cycles, and erosion [3, 4].

Two potential causes of durability loss in concrete structures are the alkali–silica reaction (ASR) and delayed ettringite formation (DEF), both of which are chemical reactions that can lead to deleterious expansion, microcracking, and strength loss in affected elements. DEF is a form of internal sulfate attack in concrete, with the primary mechanism being the dissolution–precipitation reaction of the sulfate mineral ettringite, which can cause expansion and cracking [5]. DEF is of most concern in the precast concrete industry, where the conditions used to accelerate the strength gain of concrete have the potential to increase the risk of deleterious DEF [6]. The ASR is a chemical reaction in concrete involving the alkaline pore solution and aggregate material containing non-crystalline silica, leading to the formation of an expansive ASR–gel phase and the potential for expansion and cracking [7]. In many real-world instances of DEF, ASR was also found in [8–10]. In an Australian context, all reported cases of DEF in concrete have been in conjunction with mild or moderate ASR, and have been linked to cracking in steam-cured precast concrete and massive concrete elements [6]. Thus, manufacturing specifications for precast and/or large concrete have a conservative view to DEF risk, leading to restrictive processing methods and substantial industry costs [6, 11].

The DEF mechanism is based on the solubility of sulfate in the pore solution of concrete. Ettringite normally forms during early cement hydration, with no negative effects on the final concrete material. Elevated temperatures and alkalinity (high pH) result in the decomposition of ettringite, and when the concrete system returns to normal conditions of temperature and alkalinity, ettringite will recrystallize [12]. The slow precipitation of ettringite as an expansive phase in hardened concrete at later ages induces cracking and strength loss. To minimize the risk of deleterious DEF, temperature limits are used in the manufacturing of precast concrete, as based on laboratory mortar tests and chemical composition of the cement [6, 13]. The role of other components of concrete, specifically the aggregate, has been downplayed.

1.1 Project Outline

Available literature and research results concerning DEF have focused on laboratory experiments in mortars with modified cement systems. Guidelines and specifications for cement production and the design of concrete structures are based on these findings by convention. The consideration of additional factors found in true concrete elements, including the potential for ASR, has been overlooked. Evidence-based contributions to this topic will assist in the development of targeted guidelines and standards for the precast concrete industry and mitigation of DEF risk in modern structures. By increasing the confidence in the best use of construction materials, our research will help reduce the environmental, economic, and social costs of steam-cured concrete elements and the risk of catastrophic failure.

The aim of this research work was to investigate the process of DEF and the role of aggregate reactivity, binder composition, and curing temperature on the susceptibility of concrete systems to deleterious DEF and ASR–DEF. The primary objectives of this study were to identify the conditions in concrete containing nonreactive aggregates that can induce deleterious DEF, specifically related to cement composition and curing conditions, and to investigate the role of reactive aggregates on the susceptibility of concrete to deleterious DEF.

2 Methods

The manufacture of concrete specimens for this study was carried out using local sourced materials, complying with relevant Australian standards and industry guidelines. The binder used was an Australian produced, commercial grade general-purpose (GP) cement. The elemental oxide composition of the cement was determined by X-ray fluorescence (XRF) analysis, with results presented in Table 1.

Potable tap water was used for mixing of concrete and preparation of saturated limewater. To retain sufficient workability of freshly mixed concrete, an alkali-free superplasticizer was used (125 g/100 L).

Previous literature has reported a pessimum (worst case) condition for binder compositions of alkali and sulfate content with regards to deleterious DEF [13, 14]. For pessimum concrete specimens, the alkali and sulfate binder content was increased to 1% Na_2O_{eq} (4.5 kg/m^3) and 4% SO_3 (18.0 kg/m^3) by the addition of dissolved sodium hydroxide and powdered calcium sulfate dihydrate respectively. Aggregate materials used were selected according to ASR-reactivity, as classified by Australian

Table 1 Chemical composition of GP cement via XRF

Component	L.O.I	CaO	SiO_2	Al_2O_3	Fe_2O_3	MgO	SO_3	Na_2O	K_2O	Cl	Total	Na_2O_{eq}
Content (w/w %)	4.0	64.1	19.4	4.9	3.0	1.2	2.6	0.17	0.45	0.037	100.1	0.47

Table 2 Concrete mix design

Property	Cement content	Coarse aggregate	Fine aggregate	w/c ratio	Batch size	Slump
Detail	450 kg/m^3	1190 kg/m^3	640 kg/m^3	0.40	100 kg	120 mm

standards AS1141.60.2 for coarse aggregates and AS1141.60.1 for fine aggregates. Coarse material was either a nonreactive basalt (nRe) or reactive dacite (cRe), with a grading of 20 and 10 mm in a 3:1 ratio. The fine material was a washed nonreactive river sand. Concrete prisms (285 × 70 × 70 mm) were manufactured using AS1012.2 as the guide for preparation and mixing, and AS1141.60.2 for the design of prisms, with three prisms for each set of results. Concrete cylinders (100 × 200 mm) were also manufactured, with two cylinders for each set. The concrete mix design, presented in Table 2, was based on a typical large structural element manufactured by the Australian precast industry, utilizing a high cement content and low water content.

Concrete specimens were cured under one of two conditions: curing at ambient temperature (23 ± 2 °C) or heat-cured. The heat-curing process was designed to follow the internal temperature profile of a large precast element [11]; preset of 30 °C for 4 h, heating of 30 °C/h up to 90 °C, soak at 90 °C for 12 h, then cooling to ambient temperature and demolding. During curing, all specimens were stored in sealed plastic bags, with a damp cloth as the moisture source. After demolding, specimens were stored in limewater tanks at ambient temperature. The compressive strength of the concrete specimens was measured at 1 day and 28 days with cylinders, and at 1 year with cubes cut from the prisms (75 mm, 2 cubes), as per AS1012.9. Concrete prisms were measured for linear length and mass, using AS1141.60.2 as the guide, at day 1, day 7 as reference, day 28 and then at monthly intervals up to 1 year.

3 Results and Discussion

Concrete specimens were manufactured as detailed above, with varying binder (pessimum/1N 4$ or control/ctrl), aggregate (reactive/cRe or nonreactive/nRe), and curing (ambient/amb or heat-cured/heat) conditions linked to the occurrence of ASR, DEF, and ASR–DEF. Linear expansion and mass of prisms was monitored over 1 year, and compressive strength was measured at 1 day, 28 days, and 1 year.

"DEF-only" systems had only nRe aggregates present and thus minimal potential for ASR; "ASR–DEF" systems had cRe aggregates present and thus ASR could occur; "pessimum" systems had an altered binder of 1.00% Na_2O_{eq} and 4.0% SO_3, which increased the risk of DEF; "Control" systems used the binder as received, complying with the standards and chemical content limits of Australia.

The threshold for deleterious expansion was arbitrarily set at 0.03% of total linear length, based on the Australian aggregate reactivity test method AS1141.60.2. Note that the specimens were treated differently in this study compared with AS1141.60.2, specifically being immersed in limewater tanks at normal temperature (23 °C) and lower alkali binder content (1% Na_2O_{eq}).

3.1 DEF in Concrete

Concrete specimens were manufactured with selected binder and curing conditions to promote deleterious DEF in the absence of potential ASR. Linear expansion, mass gain, and compressive strength were measured over 1 year, as shown in Fig. 1.

Deleterious expansion and strength loss were observed only in specimens subject to pessimum binder content and sustained heat-curing, and they were attributed to DEF. All other concrete systems, with either curing at ambient temperature or as-received cement, did not show deleterious effects or durability loss. This supports the

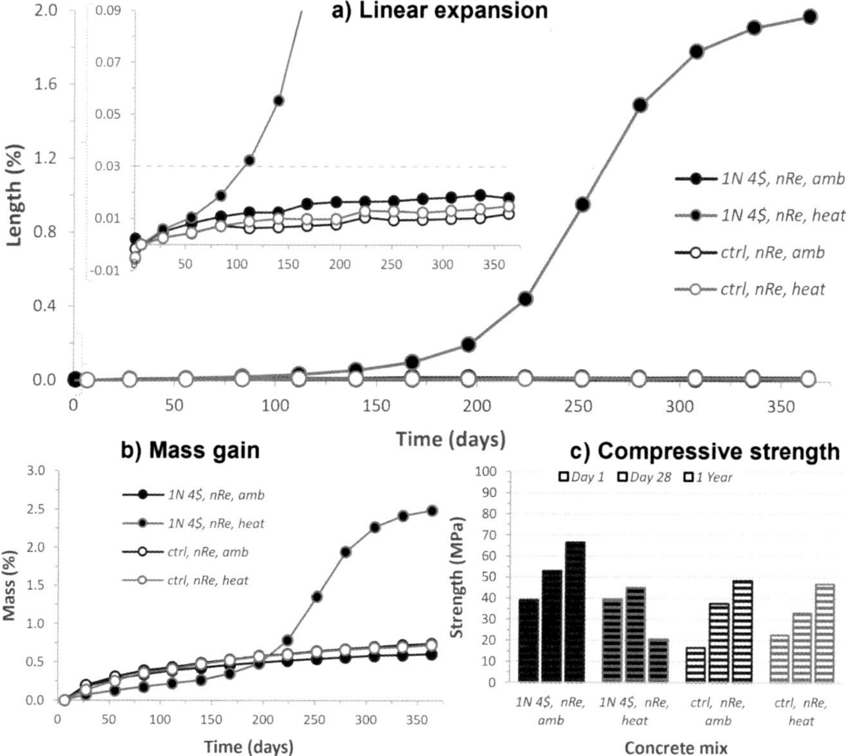

Fig. 1 Physical characteristics of concrete prisms with nonreactive (nRe) aggregates, showing **a** linear expansion; **b** mass gain and **c** compressive strength, over 1 year. Binder used was at pessimum condition (1% alkali, 4% sulfate/1N 4$) or as received (control/ctrl), curing was at ambient temperature (23 °C/amb) or heat-cured (90 °C for 12 h/heat)

current understanding that high temperature, elevated alkali, and elevated sulfate are all essential factors for the occurrence of deleterious DEF in concrete. For structures without any of these factors present, the risk of durability loss due to DEF is predicted to be minimal.

Observed expansion in the heat-cured pessimum prisms was significant, reaching 2% total of original length at 1 year, with expansion slowed but ongoing in the final months of measurement. It is expected that ongoing change would plateau in a further 6 months. This overall expansion was severe but comparable to other reported results of deleterious DEF, ranging from 1.2 to 1.8% [13]. For all other systems, the expansion was below the 0.03% threshold at 1 year, with decreasing expansion of pessimum/ambient, control/heat-cured, and control/ambient as the least. These minor variations were attributed to increased porosity of heat-cured cement paste [6, 14], and the sulfate addition inducing limited early expansion [4].

Changes in the mass of the concrete specimens closely followed that of expansion, with only the pessimum/heat-cured system showing a large expansion of 2.5% at 1 year. All other concrete systems only had minor increases to the total mass after 1 year. This large change was attributed to two causes: the inflection of the expansion curve indicating rapid development of new phases, including ettringite, which retains significant amounts of water in bound hydrates, and the formation of microcracks across the bulk material from internal expansive stress leading to ingress and uptake of water.

Similarly with the strength of the concrete specimens, significant strength loss was observed only in pessimum/heat-cured systems and attributed to DEF, with measured strength at day 28 to 1 year decreasing from 45 MPa to 20.5 MPa, respectively. Again, all other concrete systems showed strength gain as expected. For day-1 strength, heat-cured specimens were equivalent to or better than the corresponding ambient-cured specimens, as expected with precast manufacturing of concrete [15]

3.2 ASR–DEF in Concrete

Concrete specimens containing cRe aggregates were manufactured with selected binder and curing conditions to promote deleterious DEF, with the potential of ASR. Linear expansion, mass gain, and compressive strength were measured over 1 year, shown in Fig. 2.

Measured linear expansion, mass gain, and compressive strength of the ASR–DEF systems followed the same pattern as that of the DEF-only systems, with concrete systems with pessimum conditions and heat-curing exhibiting deleterious results over 1 year, while all other systems did not. For the pessimum/heat system, observed results at 1 year were expansion of 1.4%, mass gain of 2.0%, and compressive strength of 43.0 MPa at 28 days to 32.0 MPa at 1 year. The exception to this was minor expansion in the control/heat-cured system, reaching 0.03% at 1 year, with no corresponding change in mass or strength. This was attributed to increased porosity and alkali release of the cement paste due to the heat-curing process, leading to ASR.

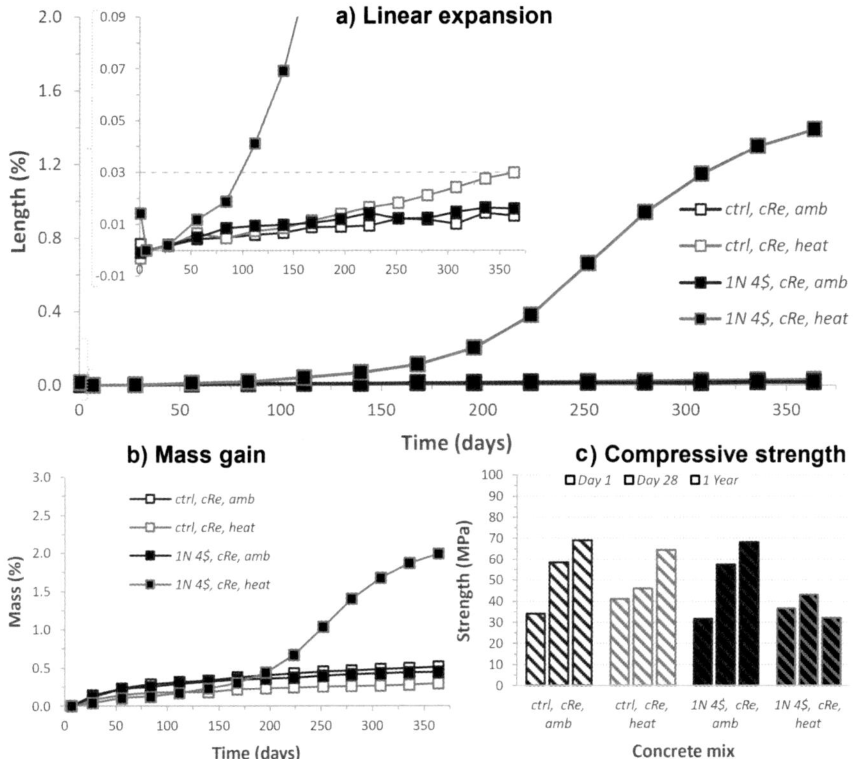

Fig. 2 Physical characteristics of concrete prisms with coarse reactive (cRe) aggregates, showing **a** linear expansion; **b** mass gain; and **c** compressive strength, over 1 year. Binder used was at pessimum condition (1% alkali, 4% sulfate/1N 4$) or as received (control/ctrl), curing was at ambient temperature (23 °C/amb) or heat-cured (12 h at 90 °C/heat)

Comparing the deleterious DEF-only and ASR–DEF systems, the most significant difference was the reduced magnitude of durability loss in the concrete specimens. The pessimum/heat-cured DEF-only system had more expansion, mass gain, and strength loss than the equivalent system with cRe aggregates present at 1 year. One possible cause is the development of ASR and related phases slowing the DEF process and reducing the total amount of microcracking and expansive phases. In this scenario, reactive silica and small amounts of ASR–gel could act as a weak supplementary cementitious material (SCM), blocking transport pores, dispersing ettringite away from the aggregate–paste boundary, and creating micro-voids for ettringite to form in without restraint. Supporting evidence can be found in the efficacy of SCMs such as fly ash to mitigate DEF [11], and the use of finely ground reactive aggregates as replacement SCMs [16]. Otherwise, the trajectory of change in expansion remained the same, with high early expansion (75–150 days), divergence (200 days), and reduced rate in the later months (300–350 days).

4 Conclusions

This research investigated the role of the ASR in the susceptibility of concrete to DEF via an experimental study with concrete specimens. The factors of alkali, sulfate, temperature, and aggregate reactivity were assessed for their contribution to deleterious DEF and ASR–DEF in concrete.

Deleterious DEF was not observed in concrete specimens prepared with locally produced Australian cement, which is linked to the low alkali and low sulfate binder content. Binder systems with elevated alkali and sulfate content increase the risk of deleterious DEF. The combination of binder composition with high alkali (1.00% Na_2O_{eq}) and high sulfate (SO_3%) content, and curing conditions of high temperature (90 °C) and sustained heat (12 h) are necessary for deleterious DEF to occur in concrete. Specimens not subjected to these conditions did not exhibit deleterious expansion during 1 year of measurement.

Reactivity of aggregates influences expansion attributed to DEF, but only in the presence of pessimum conditions and heat-curing. Observed durability loss with expansion, mass gain, and strength loss was greater in DEF-only systems compared with ASR–DEF systems.

Acknowledgements This research was conducted at the University of Technology Sydney and supported by an Australian Government Research Training Program Scholarship. It was funded through an Australian Research Council Research Hub for Nanoscience Based Construction Materials Manufacturing (NANOCOMM) (IH150100006) with the support of Cement Concrete and Aggregates Australia (CCAA).

References

1. Hewlett P (2003) Lea's chemistry of cement and concrete. Elsevier Science
2. Taylor HFW (1997) Cement Chemistry. Thomas Telford
3. Australasian Concrete Repair Association (ACRA), Commonwealth Scientific and Industrial Research Organisation (CSIRO): Division of Manufacturing and Infrastructure Technology, Standards Australia (2018) Guide to concrete repair and protection, *SA HB* 84:2018
4. Scrivener K, Ouzia A, Juilland P, Kunhi Mohamed A (2019) Advances in understanding cement hydration mechanisms. Cem Concr Res 124:105823
5. Taylor HFW, Famy C, Scrivener KL (2001) Delayed ettringite formation. Cem Concr Res 31(5):683–693
6. Shayan A, Quick GW (1992) Microscopic features of cracked and uncracked concrete railway sleepers. ACI Mater J 89(4)
7. Rajabipour F, Giannini E, Dunant C, Ideker JH, Thomas MDA (2015) Alkali–silica reaction: current understanding of the reaction mechanisms and the knowledge gaps. Cem Concr Res 76:130–146
8. Au J, Peek A (2011) A case study review of concrete deterioration by ASR/DEF. 18th Int Corros Congr 2011 2011, 3:1743–54.
9. Bruce S, Freitag S, Shayan A, Slaughter R, Pearson R (2008) Identifying the cause of concrete degradation in prestressed bridge piles. Corros Prev

10. Ma K, Long G, Xie Y (2017) A real case of steam-cured concrete track slab premature deterioration due to ASR and DEF. Case Stud Constr Mater 6:63–71
11. Mak J, Vessalas K, Thomas P, Baweja D (2012) Evaluation of factors pertaining to delayed ettringite formation in steam cured precast concrete members. In: The New Zealand Concrete Industry Conference
12. Glasser FP (1996) The role of sulfate mineralogy and cure temperature in delayed ettringite formation. Cement Concr Compos 18(3):187–193
13. Ramu Y, Thomas P, Sirivivatnanon V, Vessalas K, Baweja D, Sleep P (2019) Non-deleterious delayed ettringite formation in low alkali cement mortars exposed to high-temperature steam curing
14. Ramu YK, Sirivivatnanon V, Thomas P, Dhandapani Y, Vessalas K (2021) Evaluating the impact of curing temperature in delayed ettringite formation using electrochemical impedance spectroscopy. Constr Build Mater 282:122726
15. Concrete Institute of Australia, National Precast Concrete Association Australia (2009) Precast Concrete Handbook
16. Nsiah-Baafi E, Vessalas K, Thomas P, Sirivivatnanon V (2018) Mitigating alkali silica reactions in the absence of SCMs: a review of empirical studies. In: The International Federation for Structural Concrete 5th International fib Congress

Effect of Blending Alum Sludge and Ground Granulated Blast-Furnace Slag as Cement Replacement to Mitigate Alkali-Silica Reaction

W. Duan, Y. Zhuge, and Y. Liu

Abstract The alkali–silica reaction (ASR) is a severe durability issue in cement-based materials. Although using calcium-rich supplementary cementitious materials (SCMs) such as ground granulated blast-furnace slag (GGBS) is beneficial for improving mechanical performance, it can lead to critical ASR-induced damage, primarily when high-reactive aggregates are used. We used alum sludge, a byproduct of drinking water treatment processes, and found it to have high efficiency in mitigating ASR in mortars containing GGBS as cement replacement and waste glass as high-reactive aggregate. The raw alum sludge was calcined for 2 h at 800 °C and ground to pass a 75-μm sieve. Ternary blended binders were made by replacing 10, 20 and 30% of cement with the mixture of alum sludge and GGBS (ratio 1:1). The mortar samples exhibited a considerable compressive strength and significant ASR resistance when 30% of cement was replaced with the mixture of alum sludge and GGBS compared with the reference samples. Microstructural characterization using X-ray diffraction, backscattered electron images and energy-dispersive X-ray spectroscopy indicated that increasing the aluminum content of the alum sludge could prevent the formation of detrimental Ca-rich and low-flowable ASR gels. The hindering effect was attributed to the alkaline binding ability and the extra precipitation of calcium aluminum silicate hydrate phases due to the abundant Al in the binder.

Keywords Alkali–silica reaction · Alum sludge · Ground granulated blast furnace slag · Supplementary cementitious materials · Waste glass aggregates · Value-added recycling

W. Duan · Y. Zhuge (✉) · Y. Liu
UniSA STEM, University of South Australia, Adelaide, SA, Australia
e-mail: Yan.Zhuge@unisa.edu.au

W. Duan et al. (eds.), *Nanotechnology in Construction for Circular Economy*, Lecture Notes in Civil Engineering 356,
https://doi.org/10.1007/978-981-99-3330-3_12

1 Introduction

The alkali–silica reaction (ASR) is one of the most severe durability issues in Portland cement-based materials. Its occurrence in concrete products can cause significant structural damage and shorten service life. ASR initializes when the reactive (also known as amorphous) silica from the aggregates is dissolved by the alkaline from cement [1]. Although using nonreactive aggregates can be the most effective strategy to prevent ASR, the transportation cost in some locations due to less availability of these aggregates can be significantly unprofitable. In addition, re-using waste materials as aggregates, such as crushed glass, which has a high content of amorphous silica, has been of great interest for decades because it provides an environmentally friendly solution to disposal of the wastes [2]. Therefore, an economic ASR control method is necessary to maximize the environmental benefit.

Adding supplementary cementitious materials (SCMs) is a feasible method of mitigating the ASR. Previous studies have reported the ASR mitigation effects attributed to SCMs, including metakaolin, coal fly ash (FA) and ground granulated blast-furnace slag (GGBS) [3, 4]. Alum sludge is a byproduct of drinking water treatment and because alum-based coagulant is usually added to raw water to remove insoluble particles such as sand and microorganisms, the primary chemical composition of alum sludge includes aluminum, silicon and organic compounds. Therefore, alum sludge could be a SCM. Previous studies [5–10] have demonstrated the successful utilization of alum sludge in the manufacture of mortar and concrete blocks. The results indicated that up to 10% cement replacement with calcined and milled alum sludge could improve mechanical performance, but a strength reduction would occur when higher proportions of cement were replaced. Because GGBS-incorporated concrete products exhibit considerable strength even with high volume cement replacement [11], blending GGBS with alum sludge could potentially improve mechanical performance further when more than 10% cement is replaced. In addition, a ternary blended system is expected to have satisfactory ASR resistance due to the high Al content of the sludge.

In the present study, mortars with ternary blended binders containing GGBS and alum sludge were developed to achieve considerable mechanical performance and ASR resistance. Compressive strength, the ASR mitigation performance and microstructural characteristics were investigated for different mixtures of GGBS and alum sludge.

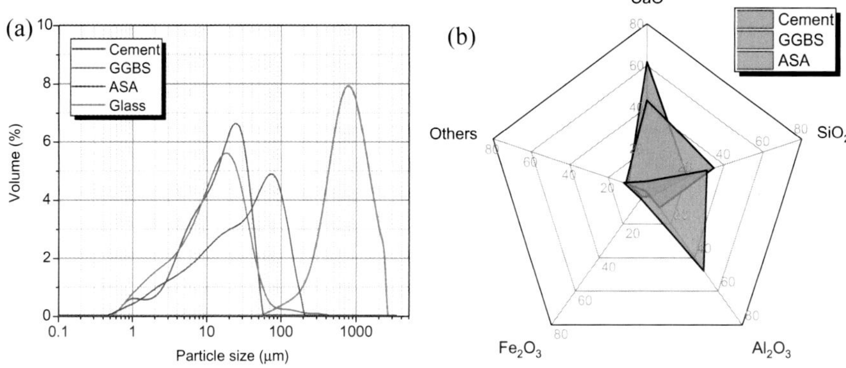

Fig. 1 **a** Particle size of cement, alum sludge ash (ASA), granulated blast-furnace slag (GGBS) and glass aggregate; **b** chemical composition of cement, ASA and GGBS

2 Methods

2.1 Materials

General-purpose cement was used as the binder according to AS3972 [12] and clear crushed glass, which contains high levels of amorphous silica, was used as fine aggregate to accelerate the ASR. The alum sludge was supplied by a drinking water treatment plant located in South Australia. The raw sludge was calcined at 800 °C for 2 h, and then the alum sludge ash (ASA) was milled to pass a 75-μm sieve. The particle size distributions of the crushed glass, ASA and GGBS are shown in Fig. 1a. The chemical composition of the calcined alum sludge was analyzed using X-ray fluorescence (XRF), and the results were compared with those for cement and GGBS (Fig. 1b).

2.2 Experimental Protocol

2.2.1 Sample Preparation

The mix proportions of mortars were designed according to AS1141.60.1 [13], and shown in Table 1. The water to cement and aggregate to cement ratios were 0.47 and 2.25, respectively. All mortar samples except those for the ASR tests were cured in a chamber with temperature and relative humidity controlled at 23 °C and 95%, respectively, for 28 days before testing.

Table 1 Mixture of the mortar samples (kg/m^3)

Sample	Cement	ASA	GGBS	Glass	Sample	Cement	ASA	GGBS	Glass
R0	590	0	0	1328	G20	472	0	118	1328
A10	531	59	0	1328	G30	413	0	177	1328
A20	472	118	0	1328	AG10	531	30	30	1328
A30	413	177	0	1328	AG20	472	59	59	1328
G10	531	0	59	1328	AG30	413	89	89	1328

ASA, alum sludge ash; GGBS, granulated blast-furnace slag

2.2.2 Compressive Strength and Accelerated ASR Tests

The compressive strength was tested according to AS4456.4 [14], and the loading rate was 0.33 MPa/s. The ASR resistance of the mortar samples was evaluated according to AS1141.60.1 [13]. The demolded samples were cured in 80 °C water for 24 h. The initial length of the mortar beams was determined at the end of curing and then the samples were immersed in 1 M NaOH solution in a water bath with the temperature set to 80 °C.

2.2.3 Microstructural Analysis

The microstructural characteristics of the samples after the ASR attack were observed using backscattered electron (BSE) micrographs, and the elemental analysis of the cement matrix was evaluated using energy-dispersive X-ray spectroscopy (EDS) with accelerating voltage at 15 kV. X-ray diffraction (XRD) patterns were obtained using copper Kα radiation at 40 kV and 40 mA.

3 Results and Discussion

3.1 Compressive Strength

The 28-day compressive strength of the control sample was 34 MPa, and the percentage of strength for other samples relative to the control is shown in Fig. 2. For the binder with binary blends (ASA and GGBS group), 10% cement replacement improved the mechanical performance, but greater than this value, a strength reduction was observed. In contrast, for the ternary blended binders (containing both ASA and GGBS, labelled as AG group), the mortar with 30% cement replacement still had a considerable compressive strength compared with the reference, which was attributed to the extra pozzolanic reaction from the synergy of ASA and GGBS. The

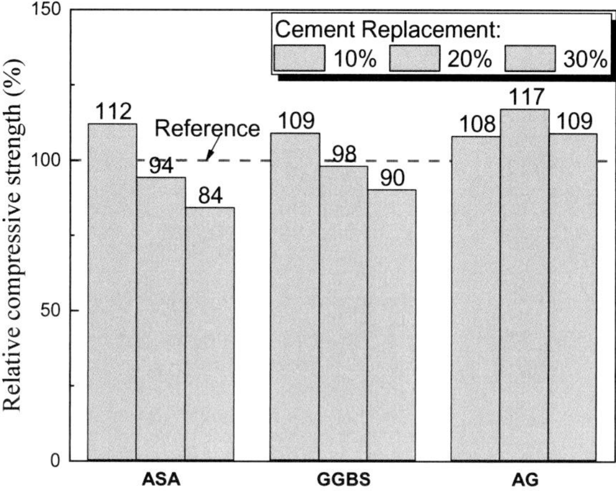

Fig. 2 Compressive strength relative to the control sample. AG, ASA and GGBS; ASA, alum sludge ash; GGBS, granulated blast-furnace slag

excessive portlandite (CH) from the GGBS could react with the silica species in the ASA, contributing to higher mechanical performance at a higher cement replacement level than with the binary blends.

3.2 *ASR-Induced Expansion*

The results of ASR-induced expansion and the surface cracking observed in samples are shown in Fig. 3. Although 30% cement replacement with GGBS kept expansion less than 0.3% for 21 days, the 14-day expansion exceeded the threshold of 0.1%. As a comparison, 20% ASA content effectively prevented the ASR. GGBS exhibited a negligible ASR mitigation effect than ASA for the mortar with binary blended binders due to the higher Ca content in GGBS, which had the potential to promote ASR-induced expansion. In addition, high Al content in ASA was beneficial for binding alkaline. The analysis of the effect of Ca and Al on the ASR is discussed in Sect. 3.4. For the samples in the AG group, 20% cement replaced with a mix of ASA and GGBS suppressed ASR-induced expansion to less than 0.3%, and 30% cement replacement significantly mitigated the ASR. The expansion results were consistent with the surface visual observation of the samples shown in Fig. 3b, where tree-like cracks were found in the reference (R0) sample, and no obvious crack could be identified in AG30.

3.3 *Phase Analysis of the Mortars*

The XRD spectra of the samples before and after the ASR test are shown in Fig. 4. No CH peak can be found for sample A30 before the ASR test, which indicated that the

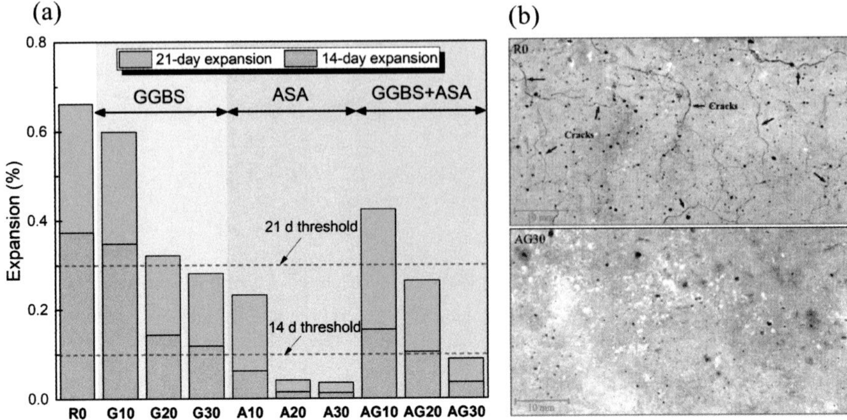

Fig. 3 **a** ASR-induced expansion; **b** surface visual observation of the mortars. ASA, alum sludge ash; ASR, alkali–silica reaction; GGBS, granulated blast-furnace slag

CH from cement hydration participated in the pozzolanic reaction and was consumed by the excess ASA. However, the amount of CH may not be sufficient to generate considerable pozzolanic C-(A)-S–H. Thus, the cement dilution effect dominated the mechanical properties, causing a lower strength than the reference.

Incorporating GGBS into the ASA provided an additional CH source and ensured a high degree of pozzolanic reaction. The cement dilution effect could be overwhelmed by the generation of pozzolanic products, which contributed to the better mechanical performance of the mortars in the AG group than those in the ASA and GGBS groups.

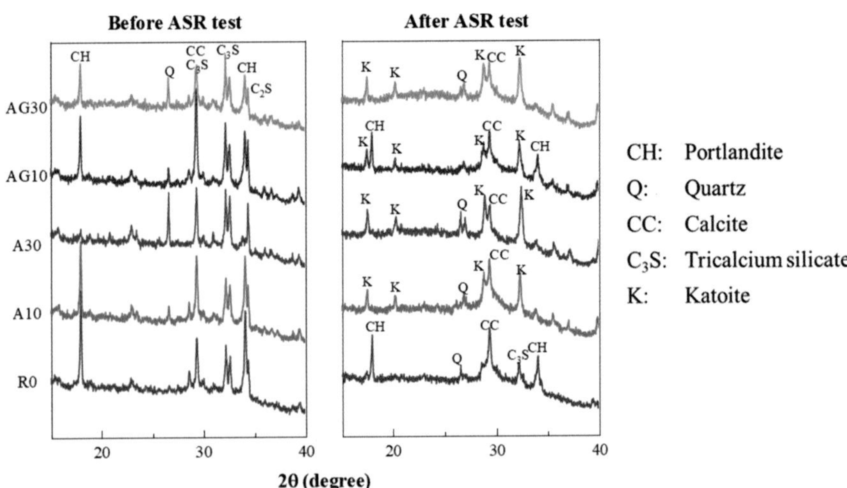

Fig. 4 X-ray diffraction spectra of the samples before and after ASR testing. ASR, alkali–silica reaction

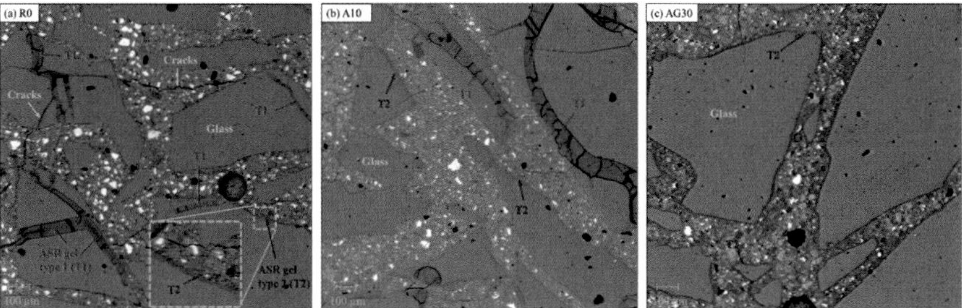

Fig. 5 BSE-SEM images of **a** R0, **b** A10 and **c** AG30 samples. BSE, backscattered electron; SEM, scanning electron microscopy

After the ASR test, kaoite (C_3AH_6) was detected as a new phase in the samples with ASA content. Kaoite is usually precipitated at a high Al/Si ratio, accompanying C-A-S-H formation, especially in high-alkaline solution [15]. The later formed C-A-S–H gels were beneficial for mitigating ASR, attributed to their alkaline binding ability.

3.4 Microstructural Characteristics and Elemental Analysis

Figure 5 shows the BSE images of the R0, A10 and AG30 samples after the ASR attack. ASR gels could be identified in two locations: interior glass aggregate labeled as T1 and surrounding the glass aggregate labeled as T2. Both T1 and T2 were found in R0 and A10, but only T2 was detected in AG30. The major cracks in R0 were collinear to the T1 gels, indicating that the growth of the ASR gels' interior aggregates may be the main factor in cement matrix damage and sample expansion.

The elemental analysis of T1 and T2 was performed using EDS technology to obtain the atomic ratios of Ca, Al, Si and Na in the ASR gels. The results listed in Table 2 show that the Ca/Si ratio of the ASR gels surrounding the aggregates (T2) was significantly higher than that inside the aggregates (T1) for the R0 and A10 samples. Although the initially formed ASR gels had a low Ca content, the gels surrounding the aggregates could be more able to absorb Ca^{2+} than the interior ASR gels and then transform into Ca-rich ASR gels [16]. Incorporating Ca into ASR gels could increase their viscosity, making them difficult to transport in pores. In addition, further taking up Ca from the pore solution could transform ASR gels into C-S-H [17]. The Ca-rich ASR gels and C-S-H layer surrounding the aggregates will limit the extrusion of the gels inside the aggregates but cannot prevent alkaline transportation into T1 and ASR gel precipitation in T1. The growth of the constrained T1 ASR gels finally cracked the aggregates and cement matrix.

Table 2 Atomic ratios in the ASR gel of samples R0, A10 and AG30

Sample	T1			T2		
	Na/Si	Ca/Si	Al/Si	Na/Si	Ca/Si	Al/Si
R0	0.40	0.28	0.02	0.21	0.58	0.03
A10	0.35	0.24	0.02	0.29	0.43	0.01
AG30	–			0.44	0.28	0.07

ASR, alkali–silica reaction

Fig. 6 Energy-dispersive spectrometry mapping of sample AG30

The Ca/Si ratios of the T2 ASR gels in AG20 were much lower than in R0 and A10 and similar to those of T1 in R0 and A10. ASR gels in AG30 were more flowable due to their lower Ca content. The possible reason for this result is that the Al from the ASA prohibited Ca absorption by the ASR gels, thus eliminating the constrain effect. The Al layer surrounding the aggregate observed in the EDS mapping shown in Fig. 1 could be evidence for this explanation (Fig. 6).

4 Conclusions

We developed a mortar with a ternary blended binder containing ASA and GGBS. This newly developed mortar exhibited satisfactory mechanical performance and excellent ASR resistance even with high reactive waste glass aggregates. Based on the results, the following conclusion can be drawn:

- The mechanical performance of mortars can be improved using a mix of ASA and GGBS as a cement replacement. The mortar with 20% cement replaced in the AG group exhibited the highest compressive strength of 40 MPa, and 30% cement replacement resulted in a considerable strength compared with the reference.

- ASA had a more substantial ASR mitigating effect than GGBS when used as the SCM and could effectively mitigate ASR. 10% ASA content in the binder could restrain ASR-induced expansion in mortars within 0.3% in the 21-day test. For the mortars with ternary blended binder, 30% cement replacement could successfully suppress ASR.
- The Al content of ASA played a significant role in mitigating ASR. The XRD result indicated the incorporation of Al in the C-S–H structure, which would limit silica dissolution. The microstructural and elemental analyses also suggested that ASA could prevent ASR gels absorbing Ca, eliminate the restriction effect of ASR gels surrounding the aggregate and impede cracking caused by the growth of ASR gels inside aggregates.

References

1. Rajabipour F, Giannini E, Dunant C, Ideker JH, Thomas MDA (2015) Alkali–silica reaction: current understanding of the reaction mechanisms and the knowledge gaps. Cem Concr Res 76:130–146
2. Fanijo EO, Kassem E, Ibrahim A (2021) ASR mitigation using binary and ternary blends with waste glass powder. Constr Build Mater 280:122425
3. Tapas MJ, Sofia L, Vessalas K, Thomas P, Sirivivatnanon V, Scrivener K (2021) Efficacy of SCMs to mitigate ASR in systems with higher alkali contents assessed by pore solution method. Cem Concr Res 142:106353
4. Meesak T, Sujjavanich S (2019) Effectiveness of 3 different supplementary cementitious materials in mitigating alkali silica reaction. Materials Today: Proceedings 17:1652–1657
5. Liu Y, Zhuge Y, Chow CWK, Keegan A, Li D, Pham PN, Yao Y, Kitipornchai S, Siddique R (2022) Effect of alum sludge ash on the high-temperature resistance of mortar. Resour Conserv Recycl 176:105958
6. Li D, Zhuge Y, Liu Y, Pham PN, Zhang C, Duan W, Ma X (2021) Reuse of drinking water treatment sludge in mortar as substitutions of both fly ash and sand based on two treatment methods. Constr Build Mater 277
7. Liu Y, Zhuge Y, Chow CWK, Keegan A, Ma J, Hall C, Li D, Pham PN, Huang J, Duan W, Wang L (2021) Cementitious composites containing alum sludge ash: an investigation of microstructural features by an advanced nanoindentation technology. Constr Build Mater 299:124286
8. Liu Y, Zhuge Y, Chow CWK, Keegan A, Pham PN, Li D, Oh J-A, Siddique R (2021) The potential use of drinking water sludge ash as supplementary cementitious material in the manufacture of concrete blocks. Resour Conserv Recycl 168:105291
9. Liu Y, Zhuge Y, Chow CWK, Keegan A, Li D, Pham PN, Huang J, Siddique R (2020) Utilization of drinking water treatment sludge in concrete paving blocks: microstructural analysis, durability and leaching properties. J Environ Manage 262:110352
10. Liu Y, Zhuge Y, Chow CWK, Keegan A, Li D, Pham PN, Huang J, Siddique R (2020) Properties and microstructure of concrete blocks incorporating drinking water treatment sludge exposed to early-age carbonation curing. J Clean Prod 261:121257
11. Sun J, Zhang P (2021) Effects of different composite mineral admixtures on the early hydration and long-term properties of cement-based materials: a comparative study. Constr Build Mater 294:123547
12. AS, AS 3972 (2010) General purpose and blended cements
13. AS, AS 1141.60.1 (2014) Method for sampling and testing aggregates, potential alkali silica reactivity–accelerated mortar bar method

14. AS, AS 4456.4 (2003) Masonry units, segmental pavers and flags—methods of test, Method 4: Determining compressive strength of masonry units
15. L'Hôpital E, Lothenbach B, Le Saout G, Kulik D, Scrivener K (2015) Incorporation of aluminium in calcium-silicate-hydrates. Cem Concr Res 75:91–103
16. Leemann A, Münch B (2019) The addition of caesium to concrete with alkali-silica reaction: implications on product identification and recognition of the reaction sequence. Cem Concr Res 120:27–35
17. Shi Z, Lothenbach B (2019) The role of calcium on the formation of alkali-silica reaction products. Cem Concr Res 126:105898

Optimisation of Limestone Calcined Clay Cement Based on Response Surface Method

G. Huang, Y. Zhuge, T. Benn, and Y. Liu

Abstract Limestone calcined clay cement (LC3) is a new type of cement that contains Portland cement, calcined clay, and limestone. Compared with traditional cement clinker, LC3 reduces CO_2 emissions by up to 40%, and is a promising technology for the cement industry to achieve its emission target. We used a numerical approach to predict the optimum composition of LC3 mortar. The experiments were performed using central composite rotational design under the response surface methodology. The method combined the design of mixtures and multi-response statistical optimization, in which the 28-day compressive strength was maximized while the CO_2 emissions and materials cost were simultaneously minimized. The model with a nonsignificant lack of fit and a high coefficient of determination (R^2) revealed a well fit and adequacy of the quadratic regression model to predict the performance of LC3 mixtures. An optimum LC3 mixture can be achieved with 43.4% general purpose cement, 34.16% calcined clay, 20.6% limestone and 1.94% gypsum.

Keywords Limestone calcined clay cement · Optimum LC3 mixture · Response surface method · Value-added recycling

1 Introduction

Concrete is the most widely used construction material, and as an essential component, Portland cement (PC) production accounts for approximately 8% of annual anthropogenic greenhouse gas emissions [1]. Because there is no viable economic alternative to concrete, cement consumption is ever-growing. Currently, using supplementary cementitious materials (SCMs) to achieve a partial clinker replacement is the most effective strategy to reduce cement production. Nevertheless, the shortage of traditional SCMs (e.g., fly ash and slag) requires exploration of other types of cementitious materials [2].

G. Huang · Y. Zhuge (✉) · T. Benn · Y. Liu
UniSA STEM, University of South Australia, Adelaide, SA, Australia
e-mail: Yan.Zhuge@unisa.edu.au

W. Duan et al. (eds.), *Nanotechnology in Construction for Circular Economy*, Lecture Notes in Civil Engineering 356,
https://doi.org/10.1007/978-981-99-3330-3_13

Recently, researchers have worked on limestone calcined clay (LC3) cement, which consists of PC, calcined clay (metakaolin: MK), and limestone (LS). Such a combination of ingredients allows higher clinker replacement and leads to the development of a denser microstructure that gives LC3 excellent mechanical and durability performance [3]. Because LC3 cement has various components, the single-factor method is mainly used in the design of material proportions in LC3 cement, but this can only obtain an optimal mix design for a single performance and cannot achieve multi-objective optimization. In comparison, the response surface method (RSM) is especially advantageous for designing a competitive and more sustainable blended cement binder [4].

In this study, central composite rotation design (CCRD) in RSM was performed. The content of MK, LS and gypsum (GYP) was taken as the variables. The 28-day compressive strength, CO_2 emission, and materials cost were used as response values. The mathematical model was established by analysis of variance (ANOVA) to determine the interaction of various variables and model accuracy. The multi-objective optimization of LC3 mortar was achieved using the numerical-functional optimization model.

2 Methods

2.1 Materials

LS, MK, GYP, and general-purpose cement (GPC) were used to create the LC3 binder mixtures (Table 1). MK was obtained after calcination of kaolin clay at 800 °C for 2 h. The LS, MK and GYP were ground using a horizontal roller mill. Concrete sand with a particle size ranging from 75 to 2.36 mm and a specific density of 2.64 was collected from ResourceCO Australia. A polycarboxylate-based superplasticizer was used to maintain the same consistency for all mixtures (between 185 and 195 mm). Mortar samples with a binder to aggregate ratio of 1:2.75 were cast in accordance with ASTM C305.

2.2 Design of Mixtures

CCRD is the most suitable technique in RSM to obtain a highly effective mathematical experiment design and develop a functional relationship between the variables and responses [5]. For the binder design of LC3 cement, MK content (x_1), LS content (x_2) and GYP content (x_3) were selected as the independent variables. The PC replaced by other ingredients was by mass and defined as a weight percentage of the total binder mass. The variable levels were selected to vary from 0 to 60%

Table 1 Chemical and physical characteristics of the binder materials

Binder	Chemical composition (%)											Physical characteristic
	SiO_2	Al_2O_3	Fe_2O_3	CaO	MgO	Na_2O	K_2O	SO_3	TiO_2	P_2O_5	SrO	d_{50} (μm)
GPC	20	4.6	3.1	63.41	1.6	0.19	0.37	2.6	0.3	0.1	0.1	17
LS	5.5	0.8	0.7	50.8	0.8	0.11	0.19	0.1	0.1	< 0.1	< 0.1	22.2
MK	71.18	25.99	0.42	0.31	0.15	0.11	0.04	0.08	1.38	0.06	0.01	18.3
GYP	3.9	1	0.4	30.8	0.3	0.15	0.16	41.5	0.1	< 0.1	0.4	25.4

GPC, general-purpose cement; GYP, gypsum; MK, metakaolin; LS, limestone

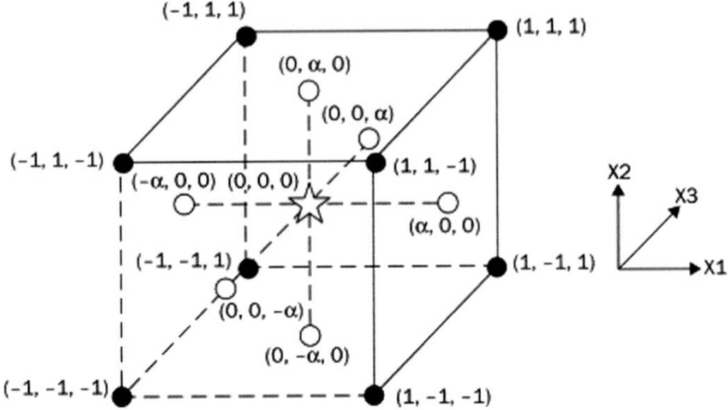

Fig. 1 Schematic illustration of central composite rotational design

for MK, 0% to 30% for LS and 0% to 5% for GYP. The effect of each independent variable was assessed at five levels with a code value of –alpha, –1, 0, 1, + alpha (factorial, axial and central points). The spatial schematic illustration of CCRD is shown in Fig. 1. Three replicates are considered at the central point, resulting in 17 experimental runs. The mix proportions are shown in in Table 2.

The compressive strength of the mortar sample at 28 days (Y_1), CO_2 emissions (Y_2), and materials cost (Y_3) were taken as the response variables. The 28-day compressive strength was recorded experimentally, while the environmental responses, such as CO_2 emissions and materials cost, were evaluated based on per kg of mortar sample. The relevant information on CO_2 emission and the material cost was collected from previous publications or estimations of local materials' market prices. All data used are shown in Table 3. The cost of mortar casting procedures and the emission from mortar service life were not taken into account.

The Design-Expert 12.0.3.0 (Sat-Ease Inc., Minneapolis, MN, USA) software was used to develop the mathematical design and ANOVA of the experiments. The quadratic polynomial model was used to predict the optimal conditions, where Y is the response value, β is the regression coefficient, ε is the random error, X_i and X_j are the independent variables, and k is the number of variables.

$$Y = \beta_0 + \sum_{i=1}^{k} \beta_i x_i + \sum_{i=1}^{k} \beta_{ii} x_i^2 + \sum_{i=1}^{k} \sum_{j>1}^{k} \beta_{ij} X_i X_j + \varepsilon \tag{1}$$

Table 2 Central composite design matrix and LC3 mortar mix proportions

Run no	Coded values			Level of variables (%)			Mixture proportion (kg/m3)						
	x1	x2	x3	MK	LS	GYP	MK	LS	GYP	GPC	Sand	Water	SP
1	−1	−1	−1	12.16	6.08	1.01	7.23	3.61	0.60	48.00	163.44	28.78	0.23
2	1	−1	−1	47.85	6.08	1.01	28.44	3.61	0.60	26.79	163.44	28.78	0.40
3	−1	1	−1	12.16	23.92	1.01	7.23	14.22	0.60	37.39	163.44	28.78	0.26
4	1	1	−1	47.85	23.92	1.01	28.44	14.22	0.60	16.18	163.44	28.78	0.37
5	−1	−1	1	12.16	6.08	3.99	7.23	3.61	2.37	46.23	163.44	28.78	0.27
6	1	−1	1	47.85	6.08	3.99	28.44	3.61	2.37	25.02	163.44	28.78	0.46
7	−1	1	1	12.16	23.92	3.99	7.23	14.22	2.37	35.62	163.44	28.78	0.28
8	1	1	1	47.85	23.92	3.99	28.44	14.22	2.37	14.41	163.44	28.78	0.61
9	−1.68	0	0	0.00	15.00	2.50	0.00	8.92	1.49	49.04	163.44	28.78	0.20
10	1.68	0	0	60.00	15.00	2.50	35.67	8.92	1.49	13.38	163.44	28.78	0.57
11	0	−1.68	0	30.00	0.00	2.50	17.83	0.00	1.49	40.13	163.44	28.78	0.34
12	0	1.68	0	30.00	30.00	2.50	17.83	17.83	1.49	22.29	163.44	28.78	0.28
13	0	0	−1.68	30.00	15.00	0.00	17.83	8.92	0.00	32.69	163.44	28.78	0.33
14	0	0	1.68	30.00	15.00	5.00	17.83	8.92	2.97	29.72	163.44	28.78	0.36
15	0	0	0	30.00	15.00	2.50	17.83	8.92	1.49	31.21	163.44	28.78	0.37
16	0	0	0	30.00	15.00	2.50	17.83	8.92	1.49	31.21	163.44	28.78	0.36
17	0	0	0	30.00	15.00	2.50	17.83	8.92	1.49	31.21	163.44	28.78	0.39

GPC, general-purpose cement; GYP, gypsum; MK, metakaolin; LS, limestone; SP, superplasticizer

Table 3 Relevant data on emission and cost for LC3 mortar manufacture

Response	Binder materials						
	MK	LS	GYP	GPC	Sand	Water	SP
CO_2 emission, kg CO_2/kg mortar	0.25	0.026	0.14	0.79	0.002	0.007	2.96
Materials cost AUD/kg mortar	0.306	0.001	1.320[a]	0.550[a]	0.5	0.003[a]	16.667[a]

[a] Average price from local suppliers

GPC, general-purpose cement; GYP, gypsum; MK, metakaolin; LS, limestone; SP, superplasticizer

3 Mathematical Modelling and Statistical Analysis

3.1 *ANOVA and Adequacy Checking*

The ANOVA for the surface response model revealed the model's accuracy and reliability. Specifically, the p-value and F-value at a 95% confidence level were computed to verify the model and the model's terms significant degree for each response. A p-value < 0.05 and a larger F-value indicated the model terms have a significant influence, while p-values ≥ 0.05 are considered insignificant. In addition, the lack of fit test measured the degree of fitting between the predicted model and input variables. If the p-value for lack of fit is > 0.05 (non-significant), it implies that: (1) the generated model fits the experimental data well, and (2) the independent variables have considerable effects on the response. When the p-value is < 0.05, seriously insufficient fitting is recorded, indicating the model was not well fitted to the input data. The ANOVA results for Y_1 to Y_3 are shown in Table 4. The response for the model showed a p-value < 0.05 indicating a good fit; however, the response for the lack of fit had a p-value > 0.05, indicating an insignificant lack of fit.

The regression coefficients (β_0, β_i, β_{ii}&β_{ij}) were estimated by ANOVA, and the regression models for each response as functions of %MK, %LS and %GYP are summarised in Table 5. The coefficient of determination (R^2) was obtained to ascertain the accuracy of the predicted function. For all responses, the R^2 values were >95%, which indicated a high degree of correlation to the experimental data.

3.2 *Interaction Effects on Response*

The three-dimensional response surface plots (see Fig. 2) were shown to evaluate the influence of interactions of different variables on the corresponding response value. The plot presented the interaction of two variables considering that the third variable was fixed on the central point in the CCRD design. The significance of the variables interaction is denoted by the surface steepness and the variable interaction p-value [6]. Therefore, the variable interaction with the smallest p-value was considered to study the influence of two independent variables on the responses. Based on the

Table 4 Analysis of variables of the test results

Response	28-day compressive strength		CO_2 emission		Materials cost	
Source	F-value	p-value	F-value	p-value	F-value	p-value
Model	19.47	0.0004	3726.21	< 0.0001	26.62	0.0001
Interactions						
X_1	34.24	0.0006	21,471.1	< 0.0001	18.74	0.0034
X_2	77.75	<0.0001	11,880.6	< 0.0001	183	<0.0001
X_3	0.043	0.8414	175.77	< 0.0001	29.81	0.0009
X_1X_2	5.8	0.0468	0.5253	0.4921	0.474	0.5131
X_1X_3	4.41	0.0738	3.43	0.1063	3.34	0.1102
X_2X_3	0.08	0.7861	1.67	0.2367	1.62	0.2431
$X_1{}^2$	51.12	0.0002	0.2767	0.6151	0.227	0.6483
$X_2{}^2$	10.59	0.014	1.73	0.2303	1.71	0.2321
$X_3{}^2$	8.75	0.0212	0.2367	0.6415	0.235	0.6427
Lack of fit	2.14	0.3479	14.73	0.0648	14.6	0.0653

Table 5 Regression models for each response

Response	Predicted regression function	R^2 (%)	R^2– predicted (%)
28-day compressive strength	$Y_1 = 37.664 + 0.272x_1 - 0.0732x_2 + 2.181x_3 + 0.00896x_1x_2 + 0.0468x_1x_3 - 0.0126x_2x_3 - 0.0112x_1{}^2 - 0.0204x_2{}^2 - 0.667x_3{}^2$	96.16	73.50
CO_2 emission	$Y_2 = 0.192 - 0.00127x_1 - 0.00179x_2 - 0.00182x_3 + 8.757\text{E-}07x_1x_2 + 0.000013x_1x_3 - 0.00019x_2x_3 + 2.677\text{E-}07x_1{}^2 - 2.675\text{E-}06x_2{}^2 - 0.000036x_3{}^2$	99.95	99.52
Materials cost	$Y_3 = 0.472 - 0.000546x_1 - 0.00122x_2 + 0.000246x_3 + 4.74\text{E-}06x_1x_2 + 0.000075x_1x_3 + 0.000105x_2x_3 + 1.38\text{E-}06x_1{}^2 - 0.000015x_2{}^2 - 0.000202x_3{}^2$	97.16	77.51

ANOVA results shown in Table 4, the selected variable interaction is X_1X_2, X_1X_3 and X_1X_3 for 28-day compressive strength (Y_1), CO_2 emissions (Y_2) and materials cost (Y_3).

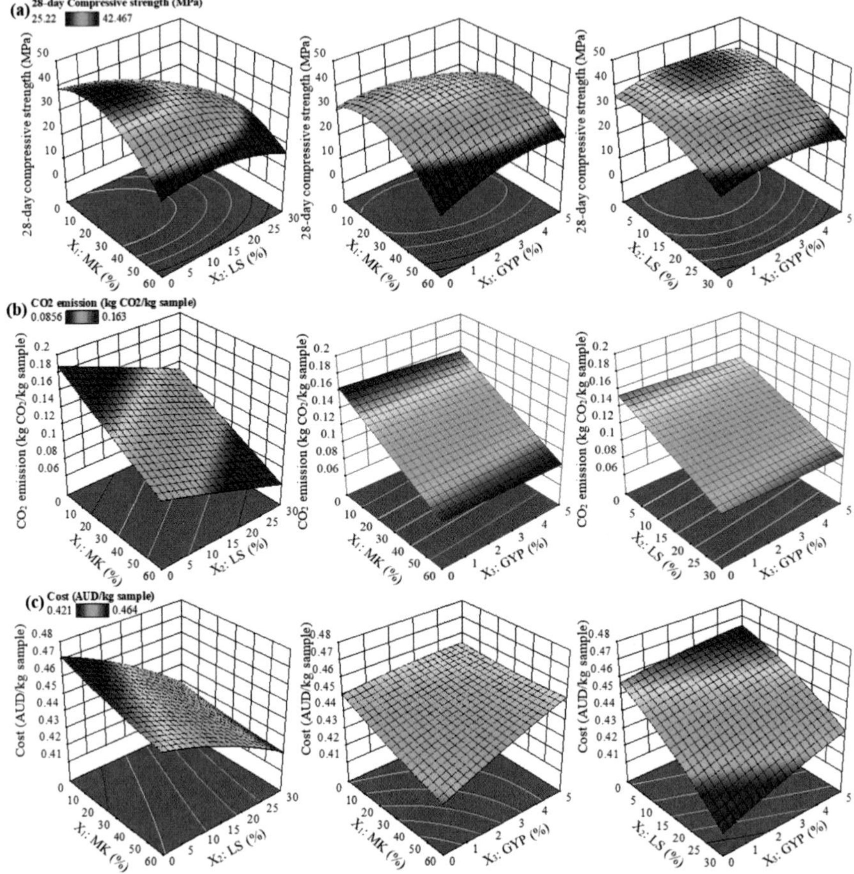

Fig. 2 Three-dimensional response surface plots. **a** 28-day compressive strength, **b** CO_2 emissions, **c** materials cost. GYP, gypsum; MK, metakaolin; LS, limestone

3.3 Response Surface Optimization Analysis

A multi-objective optimization scheme was conducted to determine an optimum formulation of %MK, %LS and %GYP, which satisfied the desired strength specifications while simultaneously minimizing the CO_2 emission and materials cost. The desirability function approach (range from 0 to 1) was used for this purpose. A desirability value of 1 represents the ideal case, while 0 indicates the responses are outside their acceptable limits. The optimal design was verified by the experimental result and shown in Table 6. An overall desirability of 0.72 indicated a reliable result.

Table 6 Optimization results and predicted desirability

Optimum design	Response	Goal	Predicted value	Experimental value	Desirability
%MK = 34.16	28-day compressive strength, MPa	Maximize	34.35	35.63	0.72[a]
%LS = 20.6	CO_2 emission, kg CO_2/kg sample	Minimize	0.109	0.110	
%GYP = 1.94	Materials cost, AUD/kg sample	Minimize	0.436	0.437	

[a]Predicted overall desirability. GYP, gypsum; MK, metakaolin; LS, limestone

4 Conclusion

This study examined the effects of formulation on strength performance, CO_2 emission, and production cost of LC3 cement. The following conclusions were drawn based on the experimental results.

- ANOVA showed satisfactory fit and accuracy of the developed quadratic models.
- The optimization results confirmed that the optimal percentages of MK, LS and GYP were 34.16%, 20.6% and 1.94%, respectively.

References

1. Monteiro PJM, Miller SA, Horvath A (2017) Towards sustainable concrete. Nat Mater 16(7): 698–699
2. Scrivener KL, John VM, Gartner EM (2018) Eco-efficient cements: potential economically viable solutions for a low-CO2 cement-based materials industry. Cem Concr Res 114:2–26
3. Sharma M, Bishnoi S, Martirena F, Scrivener K (2021) Limestone calcined clay cement and concrete: a state-of-the-art review. Cem Concr Res 149.
4. Perez-Cortes P, Escalante-Garcia JI (2020) Design and optimisation of alkaline binders of limestone-metakaolin–a comparison of strength, microstructure and sustainability with portland cement and geopolymers. J Clean Prod 273
5. Montgomery DC (2017) Design and analysis of experiments. John wiley & sons
6. Zhou Y, Xie L, Kong D, Peng D, Zheng T (2022) Research on optimising performance of desulfurisation-gypsum-based composite cementitious materials based on response surface method. Constr Build Mater 341:127874

Designing Waterborne Protective Coatings Through Manipulating the Nanostructure of Acrylic-Based Nanocomposites

S. Ji, H. Gui, G. Guan, M. Zhou, Q. Guo, and M. Y. J. Tan

Abstract Waterborne coatings with intended functionalities have been designed by manipulating acrylic-based nanocomposites with different nanostructures. Taking advantage of the favorable structure of acrylic copolymers, three waterborne coatings with various desired properties were created through molecular engineering either by copolymerizing with other components or through nanocomposite formation. This approach was demonstrated by synthesizing acrylic-based waterborne coatings with three different nanostructures, namely homogeneous, worm-like, and spherical-like nanostructures. The properties of coating samples prepared by this new approach and by traditional physical blending were compared experimentally, which revealed that the incorporation of 3-methacryloxypropyltrimethoxysilane (MPS)-modified nanoparticle TiO_2 in an acrylic base enabled the formation of a nanocomposite with nanoparticles uniformly distributed in the acrylic base. The coating film with this acrylic-TiO_2 nanocomposite showed significantly better UV absorption performance than the coating made by physical blending. The copolymerization of acrylic copolymers with an organic polymer (alkyd) created a worm-like nanostructure of acrylic–alkyd composite that allowed uniform distribution of the acrylic–alkyd nanocomposite in a more closely packed dense coating film, leading to enhanced barrier property and significantly improved corrosion resistance as confirmed by

S. Ji · H. Gui · G. Guan · M. Y. J. Tan (✉)
Institute for Frontier Materials, Deakin University, Geelong Waurn Ponds Campus, Waurn Ponds, VIC, Australia
e-mail: mike.tan@deakin.edu.au

M. Y. J. Tan
School of Engineering, Faculty of Science and Technology, Deakin University, Geelong Waurn Ponds Campus, Waurn Ponds, VIC, Australia

M. Zhou
Jiangsu Shisong New Material Technology Co., Ltd., Changzhou, Jiangsu, China

School of Environmental & Safety Engineering, Changzhou University, Changzhou, Jiangsu, China

Q. Guo
School of Engineering, Huzhou University, Huzhou, Zhejiang, China

W. Duan et al. (eds.), *Nanotechnology in Construction for Circular Economy*, Lecture Notes in Civil Engineering 356,
https://doi.org/10.1007/978-981-99-3330-3_14

electrochemical impedance spectroscopy and salt spray tests. The copolymerization of acrylic monomers with an inorganic polymer (polydimethylsiloxane [PDMS]) led to a spherical-like nanostructure of acrylic–PDMS composite film. The formation of this nanostructure arose from the migration of PDMS segments, and a PDMS-rich phase formed on the film's surface, which resulted in a coating film with PDMS functionalities such as low dirt-picking behavior. Overall, these three cases demonstrated that acrylic copolymer are an excellent base for developing various nanocomposite waterborne coatings with different functionalities through copolymerization and that the nanocomposites with different nanostructures have a significant influence on the coatings' performance.

Keywords Acrylic copolymer · Copolymerization · Multifunctional coating · Nanostructure · Waterborne coatings

1 Introduction

Water-based coatings are eco-friendly because they reduce the emission of environmentally unfriendly volatile organic compounds. Waterborne coatings are usually made by physically blending and emulsifying the binder, pigments, and additives with water. Copolymers of acrylic, vinyl, and styrene compounds are typically used as the binder in this type of coating. Of all the types of polymers, acrylic copolymer is one of most commonly used resins in waterborne coatings. In general, acrylic resins are synthetic copolymers of acrylic and methacrylic acids or their corresponding esters created by free-radical polymerization [1]. Acrylic resin is a low-cost material compared with other resins such as epoxy resins and polyurethane, and acrylic copolymers can be stable in water to form aqueous dispersions for application in eco-friendly coatings [2, 3]. They show excellent performance in film formation, high gloss, good adhesion, fast drying, outdoor durability, high transparency, and so forth [4–11]. Acrylic resins have good compatibility with other components and can be modified by a variety of components, such as polymers, silica nanoparticles, mucin gel, fiber, etc., to form coatings with improved performance [12–15]. In general, acrylic resin is a popular base polymer for developing coatings with desired new properties and for applications in adhesives, construction, automotive and additives, etc. [16–21]. Unfortunately, acrylic-based waterborne coatings have some weaknesses such as low water and corrosion resistance, poor thermal stability and weak chemical resistance, and waterborne acrylic-based coatings are considered inappropriate for application in corrosive environments. If these weaknesses of waterborne acrylic coatings can be overcome, their applicability could significantly broaden.

The monomers used to synthesize acrylic copolymers are all with vinyl groups, which makes the structure of acrylic copolymers easy to manipulate. In this regard, acrylic copolymers can easily react with other components and form covalent bonds between different components. The acrylic-based copolymers can be mixed at a molecular scale and can exhibit new functionalities. For instance, anti-icing performance was gained by the acrylic–silicone copolymer used for wind turbine blades [22]. The introduction of indole derivative groups or tertiary amines into acrylic resins can produce copolymers with anti-fouling performance [23, 24]. A biocompatible copolymer was copolymerized by acrylic copolymer and polyhedraloligosilsesquioxane [25]. These acrylic-based copolymers exhibit biocompatible stability and could be developed as a denture resin. An acrylic–poly(dimethyl siloxane) copolymer with improved gas permselectivity was synthesized by an atom transfer radical polymerization technique [26].

Conventionally, acrylic-based copolymers are prepared with a variety of materials, including organic polymers (polyurethane, polystyrene and epoxy), organic silicone, inorganic nanoparticles (clay, silica), etc. In this study we intended to extend this to the designing of waterborne coatings with intended functionalities through manipulation of acrylic-based copolymers with different nanostructures. The objective was to show that the acrylic copolymer can be an excellent base polymer for developing waterborne coatings by manipulating its nanostructure through the introduction of different components into the acrylic base. Experiments were carried out to demonstrate this approach by synthesizing several acrylic-based waterborne coatings with different nanostructures, including homogeneous, worm-like, and spherical-like nanostructures, by introducing three different components.

2 Methods

2.1 Materials

Acrylic acid (AA), 1,1'-azobis(cyclohexanecarbonitrile) (ACHN), 2-dimethylaminoethanol (DMEA), 1-methoxy-2-propanol, 2-mercaptoethanol, 3-methacryloxypropyltrimethoxysilane (MPS), titanium (IV) oxide (TiO_2), methacryloyl chloride, bis(3-aminopropyl) terminated PDMS and 1-methoxy-2-propanol obtained from Sigma-Aldrich were used directly without further purification. Styrene (St), *n*-butyl acrylate (BA), methyl methacrylate (MMA), 2-hydroxypropyl acrylate (HPA) were supplied by Sigma-Aldrich and were all purified by Al_2O_3 (Sigma-Aldrich) chromatographic column. Alkyd was synthesized as described in our previous study [27]. Curing agent amino resin Resimene 717, a type of melamine resin with alkoxy groups, was supplied by Jiangsu Shisong New Materials Technology Co., Ltd.

2.2 *Preparation of Waterborne Acrylic Copolymer, Acrylic-Based Coatings and Coating Films*

TiO_2 was modified by MPS to introduce reactivity [28]. Bis(3-aminopropyl) terminated PDMS and methacryloyl chloride were used to prepare vinyl-terminated PDMS. The copolymerization process of acrylic and acrylic-based coatings was as follows. The MPS-modified TiO_2 or alkyd or vinyl-terminated PDMS was dispersed in 1-methoxy-2-propanol and added to a 4-necked flask. A mixture of acrylic monomers (BA (50 g), HPA (15 g), AA (15 g), MMA (10 g) and St (10 g)), initiator and solvent was then dropped into the flask and the reaction temperature was 88 °C. The content of MPS-modified TiO_2 or alkyd or vinyl-terminated PDMS was 5 wt% of the acrylic monomers. The reaction was maintained for 3 h. After copolymerization, the solvent was removed by nitrogen gas. DMEA was used to neutralize part of the carboxyl groups at 50–60 °C, and water was added to prepare copolymer aqueous dispersions under mechanical stirring to obtain completely homogeneous dispersions. The copolymers synthesized were named acrylic, acrylic-TiO_2 coating, acrylic–alkyd coating and acrylic–PDMS coating according to the additional component added to the acrylic system. The samples that were a physical blending of acrylic copolymer aqueous dispersion with MPS-modified TiO_2, alkyd and vinyl-terminated PDMS separately were also prepared for comparison, and the loading content of other components was 5%wt of acrylic copolymer. The samples were named PB-acrylic/TiO_2 coating, PB-acrylic/alkyd coating and PB-acrylic/PDMS coating. The coating films were prepared from copolymer aqueous dispersions, which were mixed with the curing agent. After mixing, the mixture was poured onto the substrate and a film adaptor was used to get a wet film of 200 μm. The wet films were dried at room temperature for 1 h, and then cured at 140 °C for 30 min to get dry films of 20 ± 2 μm.

2.3 *Characterization*

The absorption performance of the coating films was measured using an ultraviolet–visible spectrophotometer (Cary Series UV–Vis spectrophotometer, Agilent, USA) in the range of 200–800 nm. The ultraviolet protection factor (UPF) value of the coating films was measured by a UPF and UV penetration/projection measurement system (Model: YG902) from 280 to 400 nm. The mean UPF was calculated automatically by the test system from eight sets of data obtained from different test areas. Transmission electron microscopy (TEM) was used to study the morphologies with a JEOL-2100 microscope under 120 kV. To obtain clear morphologies of micelles, the acrylic and acrylic–alkyd coating samples were negatively stained with a uranyl acetate substitute (UAR-EMS stain) for 30 min at room temperature. Atomic force microscopy (AFM) was used to evaluate the surface morphology and microphases of the coating films. The AFM measurements of the films were performed with a Cypher AFM microscope (Asylum Research). Electrochemical impedance spectroscopy (EIS) tests were conducted with a Bio-Logic electrochemical workstation at ambient temperature (25 ± 2 °C). A three-electrode cell arrangement with 3.5%wt (w/v) NaCl solution was used to conduct the tests. Carbon steel substrates coated with the films were set as the working electrode with a circular tested area of ~ 1 cm^2. An Ag/AgCl (Sat. KCl) electrode was used as the reference electrode and a Pt-coated Ti mesh was used as the counter electrode. The amplitude of the sinusoidal voltage was 10 mV, and the frequency range was 100 kHz to 10 MHz. The EIS data were acquired when the samples were immersed in salt solution after 2 days. Immersion test was performed in 3.5 wt% (w/v) NaCl solution at room temperature. The edges and back sides of the samples used for the immersion test were all covered with epoxy resin. The surface morphologies of the coating films were detected by scanning electron microscopy (SEM) with a Zeiss Supra 55 VP under 5 kV. The samples were coated with 5-nm Au film for good conductivity. The contact angle of the coating films was evaluated by Tensiometer KSV CAM 101 (KSV Instruments Ltd, Finland) at room temperature and water was used as the test liquid. The dirt-picking performance was measured by placing black carbon particles on the coating film and using water to clean the films.

3 Results and Discussion

3.1 Enhanced UV Protection Through the Formation of an Acrylic-Based Nanocomposite

An example of creating a functional waterborne coating with desirable nanostructure is the incorporation of MPS-modified nanoparticle TiO_2 in an acrylic base to enable the formation of a homogeneous nanostructure with nanoparticles uniformly distributed in the acrylic base. UV protection is an important property for waterborne coatings that are exposed to intense UV environmental conditions. It is well known that incorporating UV absorbers into the polymer matrix is a method of improving the UV protection of polymers. Nanoparticle TiO_2 is a commonly used UV absorber for improving the protection performance of acrylics. In the present study, the UV protection performance of the acrylic base coating and the coating with nanoparticle TiO_2 was evaluated by UV absorbance and UPF value tests, as shown in Fig. 1a, b. The absorbance of the PB-acrylic/TiO_2 coating film slightly increased compared with acrylic film, which we believed to be due physical blending being unable to achieve a coating with uniformly distributed nanoparticle TiO_2. The distribution of nanoparticles and compatibility between nanoparticles and polymer matrix need to be improved to obtain well-dispersed mixer for high efficiency of UV protection [29, 30]. This was achieved by the acrylic–TiO_2 coating film that incorporated nanoparticle TiO_2 in the acrylic base by copolymerization, which enabled the formation of a nanocomposite with a homogeneous nanostructure of uniformly distributed nanoparticles in the acrylic base. As shown in Fig. 1a, the coating film with the acrylic–TiO_2 nanocomposite showed significantly better UV absorption performance than the coating film made by physical blending, especially in the wavelength between 315 and 400 nm (UV-A). In addition, the UPF value of the acrylic–TiO_2 coating film was more than threefold higher than that of the PB-acrylic/TiO_2 coating film (Fig. 1b). The remarkable improvements in UV absorbance and UPF value of the acrylic–TiO_2 coating can be explained by the uniformity of the coating film. The appearance of the PB-acrylic/TiO_2 and acrylic–TiO_2 coating films is shown in Fig. 1c. It is obvious that the acrylic–TiO_2 coating film has a homogeneous distribution of nanoparticles.

The difference in UV absorbance can be explained by the acrylic–TiO_2 coating prepared through copolymerization having improved compatibility between the acrylic base and the nanoparticle TiO_2. In this case, the copolymerization of acrylic and MPS-modified nanoparticle TiO_2 achieved the uniform distribution. From the TEM image in Fig. 2b, it is obvious that the TiO_2 nanoparticles are wrapped by the acrylic copolymer. Moreover, from the AFM images in Fig. 3, the acrylic–TiO_2 coating film has a relatively homogeneous nanostructure, therefore the filling of TiO_2 nanoparticles by copolymerization leads to a uniform coating film compared with

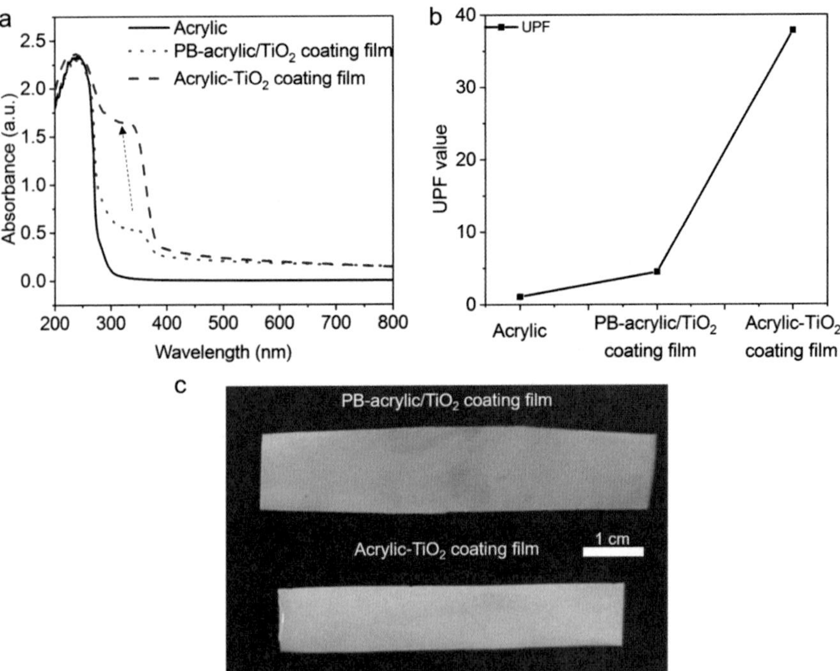

Fig. 1 **a** UV absorption spectra of acrylic base coating film and acrylic-based coating films by UV–vis, **b** UPF value of coating films in the wavelength of 280–400 nm, **c** images of the PB-acrylic/TiO_2 and acrylic-TiO_2 coating films

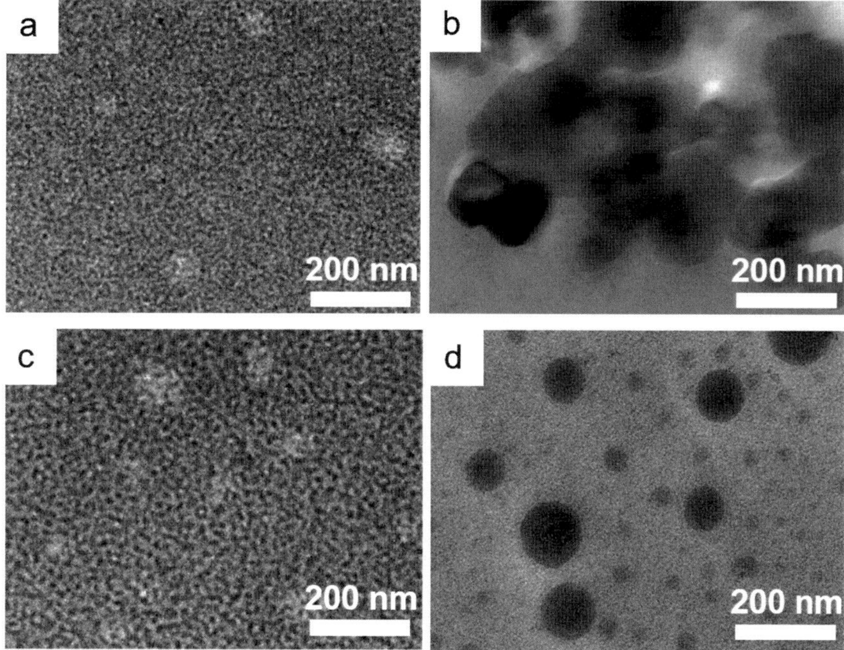

Fig. 2 Transmission electron microscopy images of copolymer water dispersions, **a** acrylic copolymer, **b** acrylic-TiO_2 copolymer **c** acrylic–alkyd copolymer, and **d** acrylic–PDMS copolymer

physical blending. In general, the enhanced UV absorbance and UPF value of the acrylic–TiO_2 coating film suggest that the uniform distribution of nanoparticle TiO_2 in the acrylic base plays an important role in improving UV protection.

3.2 Enhanced Corrosion Resistance of Acrylic–Alkyd Coating with Worm-Like Nanostructure

Acrylic waterborne coatings often have weak corrosion resistance in humid environments [31, 32]. A possibility method of enhancing their corrosion resistance is to incorporate other components with high hydrophobicity such as alkyd in order to produce anti-corrosion waterborne coatings [33]. Unfortunately, waterborne coatings made by physical blending of acrylic and alkyd polymer failed to show increased anti-corrosion performance. As shown in Fig. 4, the corrosion resistance of PB-acrylic/alkyd coating film, as indicated by the impedance values from the Nyquist plots and impedance spectrum of the coating, actually reduced compared with the acrylic coating. This result suggested that physical blending of acrylic with alkyd cannot improve the corrosion resistance of the acrylic base, possibly because physical blending of acrylic and alkyd cannot obtain a uniform coating film with a dense

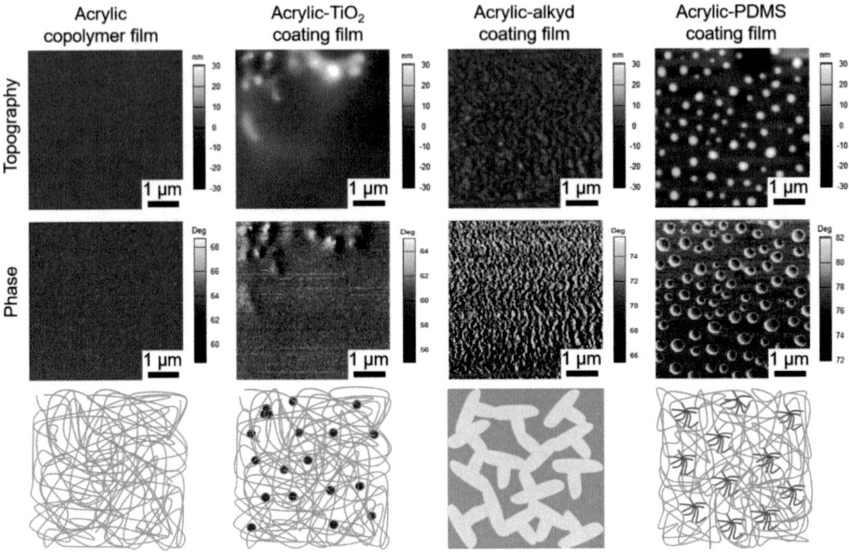

Fig. 3 Atomic force microscopy topographic and phase images of acrylic copolymer film and acrylic-based coating films

film surface due to incompatibility. In contrast, as shown in Fig. 4, the anti-corrosion performance of the acrylic–alkyd coating film was significantly enhanced, which we believe was due to the uniform distribution of alkyd in the acrylic base, better compatibility and the formation of a dense coating film. For the acrylic–alkyd coating film, the aqueous dispersion formed nanoparticles of $\approx$100 nm as shown in Fig. 2c. In addition, the acrylic–alkyd coating film formed a nanocomposite with a uniform and worm-like nanostructure shown in Fig. 3. The worm-like nanostructure was more closely packed, which created a better barrier that reduced the penetration of corrosive salt solution into the coating film, thereby enhancing the anti-corrosion performance of the acrylic base. The dense film surface of the acrylic–alkyd coating film was confirmed in the SEM image shown in Fig. 4f as compared with the loose surface of the acrylic film shown in Fig. 4e. Moreover, the immersion test confirmed the EIS result.

3.3 Multifunctional Acrylic–PDMS Coating with Spherical-Like Nanostructure

Acrylic waterborne coatings are also commonly used to protect surfaces from dirt and graffiti. PDMS has low surface energy and high hydrophobicity, which is often used to creating easy-clean coatings. As another example of creating a desirable waterborne coating through manipulation of acrylic-based nanocomposites with different

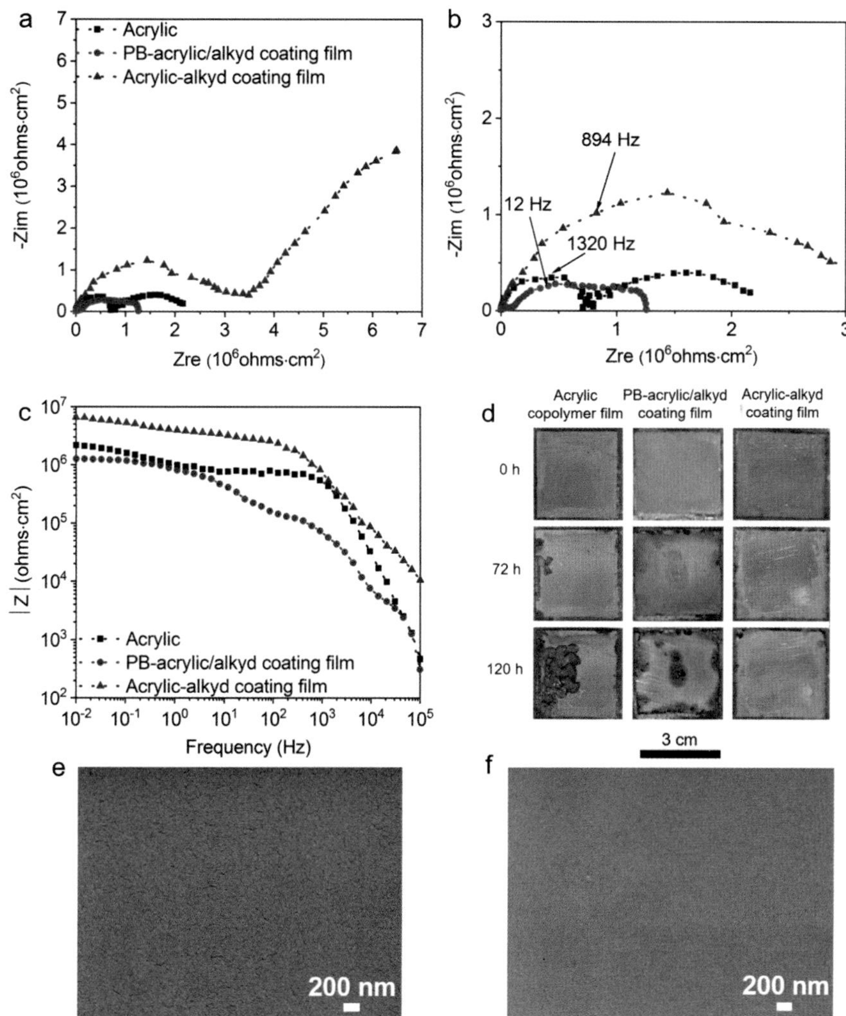

Fig. 4 Impedance spectra of the acrylic copolymer film and acrylic-based coated carbon steel in 3.5%wt NaCl aqueous solution: **a** Nyquist plots, **b** enlarged part of the Nyquist plots, **c** Bode plots, **d** immersion tests of coating films in 3.5 wt% NaCl solution at room temperature; scanning electron microscopy images of **e** acrylic film, and **f** acrylic–alkyd coating film

nanostructures, the PDMS component was incorporated to prepare an acrylic-based copolymer with a high dirt resistance. The surface performance of the coating films was evaluated, as shown in Fig. 5a. The contact angle of the acrylic–PDMS coating film increased compared with acrylic film and the PB-acrylic/PDMS coating film formed by blending the hydrophobic PDMS component. In addition, the dirt-resisting performance of the coating films was tested and shown in Fig. 5b. From the images of the PB-acrylic/PDMS coating film, it is obvious that the black carbon particles

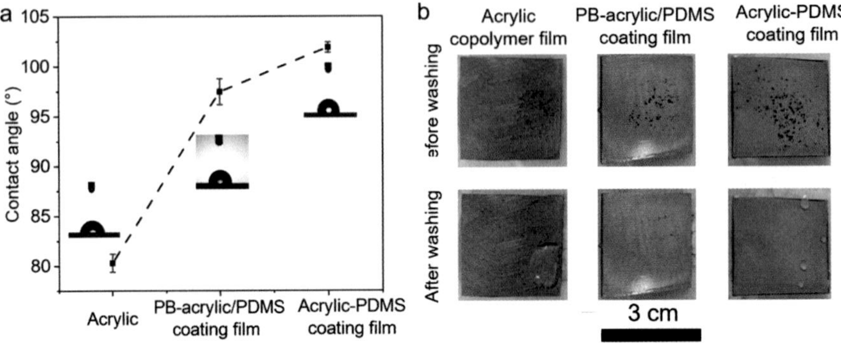

Fig. 5 **a** Water contact angle of acrylic and acrylic-based coating films, and **b** images of dirt-picking property of the coating films

were not washed off completely, whereas the acrylic–PDMS coating film was totally cleaned by water washing.

The dirt-resistance performance of the PB-acrylic/PDMS coating film was unsatisfactory because the PDMS component is hydrophobic and cannot be uniformly distributed in the acrylic base by physical blending due to the poor compatibility between acrylic and PDMS. In contrast, copolymerization improved the compatibility between the acrylic and PDMS components and contributed to the formation of a uniform and spherical-like nanostructure, which can explain the excellent dirt-resistant property of the acrylic–PDMS copolymer. From Fig. 2d, it is clear that the acrylic–PDMS copolymer formed a nanocomposite with particles at the nanoscale. Figure 3 shows spherical-like nanostructure of the acrylic–PDMS coating. In this nanostructure, the PDMS component tended to migrated to the top of the surface due to its low surface energy and hydrophobicity. In this case, the acrylic–PDMS film surface is a PDMS-rich phase that can show similar properties to the PDMS component. These properties contributed to the improvement in the dirt-picking performance of the acrylic–PDMS coating film.

4 Conclusion

We carried out experiments to demonstrate the design of multifunctional waterborne coatings through manipulation of acrylic-based copolymers with different nanostructures. Three components were chosen to study–an inorganic nanoparticle (TiO_2), an organic short polymer (alkyd) and an inorganic polymer (PDMS)–to manipulate the acrylic nanostructure in order to create waterborne coatings with desirable functionalities and improve the compatibility between different components. The copolymerization of MPS-modified TiO_2 and acrylic monomers achieved a uniform acrylic–TiO_2 coating film. The uniform distribution of nanoparticles led to better UV protection property compared with the physical blended sample. Due to thermodynamic incompatibility of acrylic and alkyd, the acrylic–alkyd coating film formed a worm-like nanostructure, which contributed to its dense film surface that had better barrier performance and enhanced the corrosion resistance. The acrylic–PDMS coating film formed a spherical-like nanostructure and the film surface was a PDMS-rich phase, which showed similar performance to the PDMS component compared with PB-acrylic/PDMS. Therefore, the properties of the acrylic–PDMS coating changed significantly: the contact angle increased to 103° and the coating film showed enhanced resistance to dirt-picking. Generally, by copolymerizing acrylic with other components, it is easy to manipulate the nanostructure of acrylic-based copolymers and improve compatibility between different components. Moreover, the nanostructure of acrylic-based copolymers plays an important role in the coating's performance. Acrylic-based copolymers with an organized nanostructure can exhibit desirable functions. This approach to designing acrylic-based copolymers with an organized nanostructure could open opportunities for developing waterborne acrylic-based copolymers with multifunctionality.

5 Data Availability Statement

The raw/processed data required to reproduce these findings cannot be shared at this time as the data also forms part of an ongoing study.

Acknowledgements Financial support from Australian Research Council funding for ARC Research Hub for Nanoscience-Based Construction ARC ITRH-JSNMT-2016-2021 is gratefully acknowledged.

References

1. Tığlı RS, Evren V (2005) Synthesis and characterization of pure poly(acrylate) latexes. Prog Org Coat 52(2):144–150
2. Lewis O, Critchlow G, Wilcox G, DeZeeuw A, Sander J (2012) A study of the corrosion resistance of a waterborne acrylic coating modified with nano-sized titanium dioxide. Prog Org Coat 73(1):88–94
3. Diaconu G, Paulis M, Leiza JR (2008) High solids content waterborne Acrylic/Montmorillonite nanocomposites by niniemulsion polymerization. Macromol React Eng 2(1):80–89
4. Alyamac E, Soucek MD (2011) Acrylate-based fluorinated copolymers for high-solids coatings. Prog Org Coat 71(3):213–224
5. Minari RJ, Goikoetxea M, Beristain I, Paulis M, Barandiaran MJ, Asua JM (2009) Post-polymerization of waterborne alkyd/acrylics. Effect on polymer architecture and particle morphology. Poly 50(25):5892–5900
6. Van HE, Van EJ, German A, Cuperus F, Weissenborn P, Hellgren AC (1999) Oil-acrylic hybrid latexes as binders for waterborne coatings. Prog Org Coat 35(1–4):235–246
7. Yuan Y, Liu R, Wang C, Luo J, Liu X (2014) Synthesis of UV-curable acrylate polymer containing sulfonic groups for anti-fog coatings. Prog Org Coat 77(4):785–789

8. Tang E, Bian F, Klein A, El-Aasser M, Liu S, Yuan M, Zhao D (2014) Fabrication of an epoxy graft poly(St-Acrylate) composite latex and its functional properties as a steel coating. Prog Org Coat 77(11):1854–1860
9. Nollenberger K, Albers J (2013) Poly(meth)acrylate-based coatings. Int J Pharm 457(2):461–469
10. Wang R, Wang J, Wang X, He Y, Zhu Y, Jiang M (2011) Preparation of acrylate-based copolymer emulsion and its humidity controlling mechanism in interior wall coatings. Prog Org Coat 71(4):369–375
11. Jafarzadeh S, Claesson PM, Sundell PE, Tyrode E, Pan J (2016) Active corrosion protection by conductive composites of polyaniline in a UV-cured polyester acrylate coating. Prog Org Coat 90:154–162
12. Patel MM, Smart JD, Nevell TG, Ewen RJ, Eaton PJ, Tsibouklis J (2003) Mucin/Poly (Acrylic acid) interactions: a spectroscopic investigation of mucoadhesion. Biomacromol 4(5):1184–1190
13. Mičušík M, Bonnefond A, Paulis M, Leiza JR (2012) Synthesis of waterborne Acrylic/Clay nanocomposites by controlled surface initiation from macroinitiator modified montmorillonite. Eur Polym J 48(5):896–905
14. Khalina M, Sanei M, Mobarakeh HS, Mahdavian AR (2015) Preparation of Acrylic/Silica nanocomposites latexes with potential application in pressure sensitive adhesive. Int J Adhes Adhes 58:21–27
15. Chen SA, Lee HT (1995) Structure and properties of poly (acrylic acid)-doped polyaniline. Macromolecules 28(8):2858–2866
16. Shiga T, Hirose Y, Okada A, Kurauchi T (1992) Bending of poly (vinyl alcohol)–poly (sodium acrylate) composite hydrogel in electric fields. J Appl Polym Sci 44(2):249–253
17. Zhu K, Li X, Li J, Wang H, Fei G (2017) Properties and anticorrosion application of acrylic ester/epoxy core-shell emulsions: effects of epoxy value and crosslinking monomer. J Coat Technol Res 14(6):1315–1324
18. Zhao F, Fei X, Wei W, Ye W, Luo J, Chen Y, Zhu Y, Liu X (2017) A random acrylate copolymer with epoxy-amphiphilic structure as an efficient toughener for an epoxy/anhydride system. J Appl Polym Sci 134(26)
19. Zhang H, Zhang H, Tang L, Zhang Z, Gu L, Xu Y, Eger C (2010) Wear-resistant and transparent acrylate-based coating with highly filled nanosilica particles. Tribol Int 43(1–2):83–91
20. Zhang SF, Wang RM, He Y-F, Song PF, Wu Z-M (2013) Waterborne polyurethane-acrylic copolymers crosslinked core-shell nanoparticles for humidity-sensitive coatings. Prog Org Coat 76(4):729–735
21. Mai YW, Yu ZZ (2006) Polymer nanocomposites. Woodhead Publishing Limited, Abington
22. Xu K, Hu J, Jiang X, Meng W, Lan B, Shu L (2018) Anti-Icing performance of hydrophobic silicone–acrylate resin coatings on wind blades. Coatings 8(4)
23. Ni C, Feng K, Li X, Zhao H, Yu L (2020) Study on the preparation and properties of new environmentally friendly antifouling acrylic metal salt resins containing indole derivative group. Prog Org Coat 148:105824
24. Hugues C, Bressy C, Bartolomeo P, Margaillan A (2003) Complexation of an acrylic resin by tertiary amines: synthesis and characterisation of new binders for antifouling paints. Eur Polym J 39(2):319–326
25. Kim S, Heo S, Koak J, Lee J, Lee Y, Chung D, Lee J, Hong S (2007) A biocompatibility study of a reinforced acrylic-based hybrid denture composite resin with polyhedraloligosilsesquioxane. JOR 34(5):389–395
26. Semsarzadeh MA, Ghahramani M (2015) Synthesis and morphology of polyacrylate-poly(dimethyl siloxane) block copolymers for membrane application. Macromol Res 23(10):898–908
27. Ji S, Gui H, Guan G, Zhou M, Guo Q, Tan MY (2021) Molecular design and copolymerization to enhance the anti-corrosion performance of waterborne acrylic coatings. Prog Org Coat 153:106140

28. Qi Y, Xiang B, Tan W, Zhang J (2017) Hydrophobic surface modification of tio_2 nanoparticles for production of acrylonitrile-styrene-acrylate Terpolymer/Tio_2 composited cool materials. Appl Surf Sci 419:213–223
29. Liu KQ, Kuang CX, Zhong MQ, Shi YQ, Chen F (2013) Synthesis, characterization and UV-shielding property of polystyrene-embedded CeO_2 nanoparticles. Opt Mater 35(12):2710–2715
30. Alebeid OK, Zhao T (2017) Review on: developing UV protection for cotton fabric. J Text Inst 108(12):2027–2039
31. Wang Y, Qiu F, Xu B, Xu J, Jiang Y, Yang D, Li P (2013) Preparation, mechanical Properties and surface morphologies of waterborne fluorinated polyurethane-acrylate. Prog Org Coat 76(5):876–883
32. Chai CP, Ma YF, Li GP, Ge Z, Ma SY, Luo YJ (2018) The preparation of high solid content waterborne polyurethane by special physical blending. Prog Org Coat 115:79–85
33. Irfan MH (1988) Alkyds in the construction industry. In: Chemistry and technology of thermosetting polymers in construction applications. Springer Science & Business Media, Dordrecht, pp 227–229

Analysis of Categories That Delay Global Construction Projects

M. Abonassrya, M. Alam, and A. Saifullah

Abstract Delay in construction projects is a significant issue and concern for most construction companies. Many studies have addressed this issue by identifying the top-ranked causes, which vary according to project type, location, and the research method used. The combined factors of delay/time overrun need further analysis to understand the top-ranked factors considering the project context. We identified 360 delay/time overrun factors of construction projects from articles published in the past 10 years in top-quality journals ranked as per scientific journal ranking (SJR). The factors were then coded and classified into categories based on their impact and the description in NVIVO software. Finally, the categories were analyzed and ranked by the relative importance index using SPSS software to identify the critical ones in global construction projects. In addition, the developed and developing countries affected by these delay factors were determined. The results revealed that the top five important factors are located under the following categories: orders and requirements; experience and productivity; financial problems; planning; and lastly both external and management categories. These categories were the highest ranking among the five top factors found in the reviewed studies and affect both developed and developing countries.

Keywords Construction · Delays · Economic impact · Overrun

M. Abonassrya (✉) · M. Alam
Department of Civil and Construction Engineering, Swinburne University of Technology, VIC, Australia
e-mail: mabonassrya@swin.edu.au

A. Saifullah
Department of Mechanical Engineering and Product Design Engineering, Swinburne University of Technology, VIC, Australia

W. Duan et al. (eds.), *Nanotechnology in Construction for Circular Economy*,
Lecture Notes in Civil Engineering 356,
https://doi.org/10.1007/978-981-99-3330-3_15

1 Introduction

Delay/time overrun is a common issue in construction projects worldwide, and the success of construction projects can be considered a measure of the political performance of countries because it is related directly to their economy [1]. Faridi and El-Sayegh state that project delays negatively affect the quality and safety of project work and reflect negatively on the services provided to the community [2]. Therefore, in the present study we performed a literature review focused on identifying and ranking the delay factors, although classification of these factors differs among studies. The relative importance index (RII) analysis was commonly used and many types of construction projects were researched and discussed. The delay factors were classified under different categories, but the most prominent were the contractor and client categories, which relate to financial problems, management, materials and equipment, experience, planning, and methods of construction. In addition, the delay factors in developing countries were greater than in developed countries. The delay factors were many and varied according to the project type, location, and the research method used in the study. Our main focus was to identify the critical categories of delay factors found by previous studies over the past 10 years published in journals ranked under the scientific journal rankings (SJR). The factors identified were classified into categories that were analyzed by Cronbach's alpha for reliability verifying and RII using SPSS software for ranking. The top five categories were: orders and requirements; experiences and productivity; financial problems; planning; and external and management.

2 Literature Review

The research methods used in the studies differed, but usually followed two trends: identifying and analyzing the factors [3]. However, some studies also discussed the effects of delay factors [4]. The delay factors identified were generally classified into categories applicable for each study. For example, Bajjou and Chafi identified the delay factors in African countries and classified them into eight categories [5]. Similarly, Wuala and Rarasati analyzed and classified the factors in Southeast Asia into five categories [6]. In addition, studies discussed the delay factors in different types of projects [7]. For example, tunnels [8], residential projects [9–11], roads [12–14], railways [15, 16], sport facilities [17], oil projects [18, 19], as well as transport, power, building, and water irrigation projects [20]. These various delay factors were then also variably categorized. For example, financial problems were identified as the contractor category [21–24], but also as the client category [7, 25–28]. Management factors were placed in the contractor category [29, 30], but also related to shortages in equipment and material factors [6, 31]. Some studies stated that the contractor category related to experience, planning, and construction method factors [32–34], whereas in others the client category was related to decision-making factors [35, 36], or to changes factors [19, 37]. In addition, there were many other categories related to other delay factors, such as the external category related to weather, policy, and security factors [8, 10, 11, 16, 38, 39]. The delay in construction projects affects both developed and developing countries. Rivera et al. studied the delay factors in 25 developing countries and confirmed that 50% of them have similar delay factors [32]. On the other hand, developed countries are also affected by delay factors but less so, according to analysis of three countries: Portugal, the UK, and the USA [40]. In addition, other studies identified global delay factors for both developed and developing countries [17, 23, 41–43]. We reviewed all the studies that addressed and identified delay factors to analyze the top-ranked factors, and then identified the developed and developing countries affected by these factors.

3 Research Method

Figure 1 details the stages in our research. We began with the research scope to select relevant studies, which were then refined. The factors (delay/time overrun) were extracted from the selected studies, classified into categories and ranked based on their importance from the researchers' point of view. Next we analyzed the factor categories by Cronbach's alpha and the RII. Lastly, the resulting categories were also classified to the relevant developed/developing country.

3.1 Research Scope and Select Studies

Using the Scopus database, we collected and sorted studies that addressed the relevant factors (i.e., delay/time overrun) in construction projects. Screening was limited to the title of the study. Next we selected studies published in the 10 years of 2012–2021 with the word "delay" in the title and 410 studies were identified. All steps were repeated for studies with "time overrun" in the title and 41 were identified.

The 451 studies were then screened for publication in a high-quality journal or conference ranked by SJR, resulting in 277 studies, which were further reviewed and refined to choose those that clearly identified and ranked the factors of delay/time overrun in different countries. Finally, 71 studies in total were selected: 62 for delay and 9 for time overrun. Microsoft Excel was used in this stage of the research.

3.2 Factors Identified and Classification into Categories

The studies reviewed in this research depended on delay/time overrun factors identified in previous studies that other researchers had performed through interviews, questionnaires, or case studies. The research area included many developing and

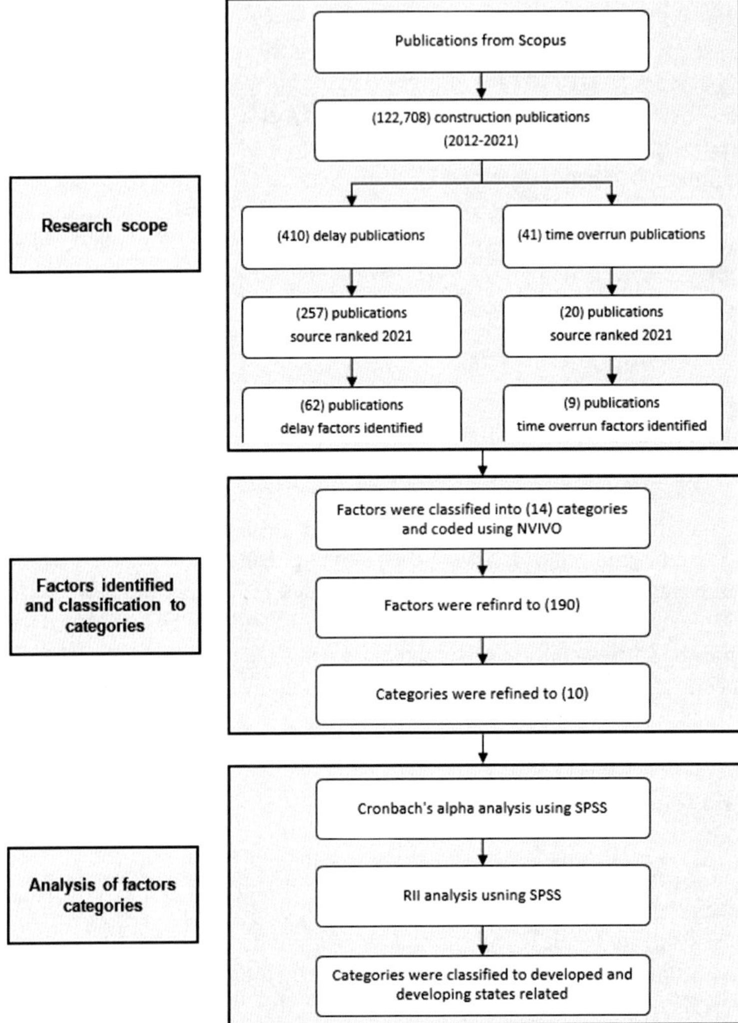

Fig. 1 Research flowchart

developed countries. These studies ranked the top essential factors of delay/time overrun, so the top five factors selected from previous studies were included in this research. The delay factors were not all similar in terms of their importance, as they were based on the project type, location, and the method of identifying the factors used in the study. So, some factors were repeated or had a similar description in some studies, and other factors were utterly different in their description in other studies.

The factors extracted from all the studies reviewed had been classified into categories based on the type and description of each factor and thus to any category it most likely belonged. All categories and their factors were coded using NVIVO software, and factors were ranked inside their categories from 1 to 5 based on the rank that each factor had in the original study. Some categories were combined into one category to reduce the number of categories, such as design and work error; experience and productivity; material and equipment; orders and requirements; and tender and contract category. In addition, some studies contained some factors belonging to the same category.

3.3 Analysis of Factor Categories

We applied two types of analysis: Cronbach's alpha analysis to verify the reliability of data collected regarding delay/time overrun factors and RII analysis to rank the critical categories of the top five factors using SPSS software. The various important factors were identified and ranked based on the project type, location, and research method of the previous studies, which depended on literature reviews to identify the factors, then on questionnaires, surveys or interviews to assess the importance of the factors, and finally, on ranking these factors by analysis. In the questionnaires or interviews, the answers and importance degree of the factors identified in the previous studies were collected from respondents and the answers are ranked to identify the most important using RII analysis. Therefore, we assumed our study was similar to a questionnaire survey of the previous studies that addressed and ranked the top five delay factors, whereby categories classified were questions, studies were respondents, and importance degree of the factors comprised the answers.

We adopted a Likert scale with 5 points to identify the top five essential levels for the top five factors identified in previous studies. The five levels of importance were very high, high, moderate, low, and very low, ranked from 5 to 1 degree of importance. Factors ranked as the first, which was the highest level in the previous studies, took a very high '5' degree of importance, and factors ranked as second, third, fourth, and fifth took a high '4', moderate '3', low '2', and a very low '1' degree of importance respectively. It is good to analyze all factors classified into categories, but the total of factors (i.e., answers) in all categories (i.e., questions) was not equal. Therefore, it was necessary to first reduce the factors for each category to be equal to the category's lowest value of the factors by ignoring the levels of the least important factors. However, the total factors selected after reducing the factors of categories became less than half of the original factors, even with the categories having the lower second, third and fourth values of the total factors. In addition, the total of factors was more than half of the original factors when selecting the following lower fifth and sixth values. However, the sixth lower value of factors had fewer factors and categories than the fifth lower value and there was no point in proceeding with other lower values because the total factors became less, as well as the categories, and it beneficial to analyze the greatest number of factors. So, the best scenario was using the category having the fifth lower value of factors because the total factors selected was more than half the original total. In this process, four categories that had factors with the lowest importance levels were excluded from the analysis. The factors in the remaining categories were analyzed by the two methods stated above using SPSS software.

In addition, the factors extracted from the previous studies included many different countries. Therefore, as the last step in this research, the studies of factor categories resulting from the analysis were classified as developed or developing states. The developing states were identified as per the list of developing countries updated in 2022 on the website of the Australian Government Department of Foreign Affairs and Trade [44]. Finally, the percentage of developed and developing states for each category was calculated and tabulated.

4 Results and Discussion

A total of 360 factors of delay were identified in the 71 studies. The classification of these factors resulted in 14 categories as per the description of each factor and the category to which it belonged. In addition, we found that two or three factors in most studies were mainly classified under the same category. Also, all studies identified 5 factors, except for 5 studies that identified 6 factors due to two factors having the same rank. Each factor's degree of importance in 14 categories is detailed in Table 1. The total factors ranged from 8 in C1, to 49 in C4. There were 72 factors ranked as the first degree of importance, and 71 factors each for the second and third degrees of importance. Lastly, there were 73 factors each in the fourth and fifth importance degrees.

After refining the factors stated in the research method, six scenarios were proposed and selected (Fig. 2). Scenario 5 achieved the highest number of factors, and by getting the factors with the highest degree of importance, the factors for each category was reduced to 19. At the same time, the total of factors was 190, which was more than half of the 360 original factors. Also, the factors in this scenario belonged to 10 categories, because four categories with less than 19 factors were excluded. Table 2 and Fig. 2 detail scenario 5.

Cronbach's alpha analysis was the first analysis applied and the result was 0.974, which indicated an excellent level of reliability (>0.8) and confirmed that the data collected was very acceptable and reliable and can be used for further analysis. Table 3 shows the internal consistency level of Cronbach's alpha analysis.

Table 4 shows the results and ranks of the 10 categories analyzed after excluding the categories with the least importance levels of factors (i.e., C1, C2, C12 & C14). The results of the analysis illustrated the categories of the top critical delay factors ranked in previous studies for the 10 year period. The ranking of categories confirmed that the "orders and requirements" (C9) category had the highest level of importance (RII = 0.905), the "experience and productivity" (C4) category was next (RII = 0.842), the third category was "financial problems" (C6) (RI I = 0.832), followed by the "planning" category (C11; RII = 0.789). Lastly, both the "external" (C5) and "management" (C7) categories came fifth (RII = 0.779). Also, it was noted that the

Table 1 Frequency of degree of importance of factors

No	Category	Code	Degree of importance					Total
			1st	2nd	3rd	4th	5th	
1	Communication	C1	0	4	2	1	1	8
2	Conflict	C2	2	1	5	2	0	10
3	Design and work error	C3	5	7	2	1	4	19
4	Experience and productivity	C4	7	9	12	14	7	49
5	External	C5	8	4	4	5	5	26
6	Financial problems	C6	6	10	6	7	8	37
7	Management	C7	6	5	10	8	3	32
8	Material and equipment	C8	5	2	9	6	12	34
9	Orders and requirements	C9	11	7	1	8	10	37
10	Payment delay	C10	6	4	4	6	3	23
11	Planning	C11	4	10	9	9	7	39
12	Project	C12	1	2	3	0	6	12
13	Response delay	C13	8	3	4	2	4	21
14	Tender and contract	C14	3	3	0	4	3	13
Total			72	71	71	73	73	360

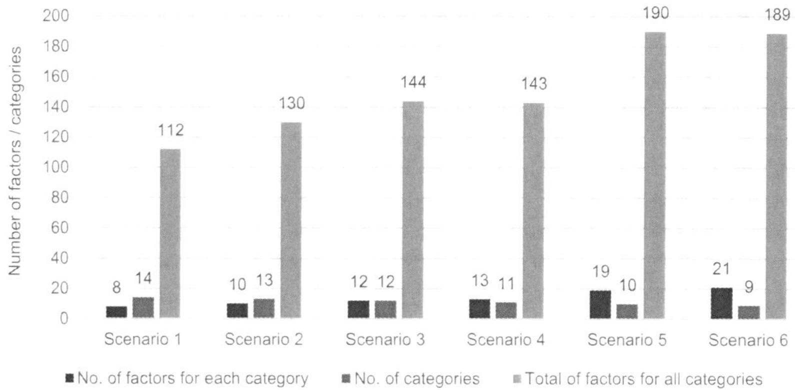

Fig. 2 Proposed scenarios for analyzing the categories of factors

importance levels for all 10 categories were between very high and high. The RII of the first three categories (C9, C4 & C6) was > 0.8, so they all had a very high level of importance. In contrast, the fourth (C11), and fifth categories (C5 & C7) had a high importance level with all other remaining categories. Table 5 illustrates the importance levels in the RII analysis.

Table 2 Frequency of importance degree of 19 factors for each category in scenario 5

No	Category	Code	Degree of importance					Total
			1st	2nd	3rd	4th	5th	
1	Design and work error	C3	5	7	2	1	4	19
2	Experience and productivity	C4	7	9	3	0	0	19
3	External	C5	8	4	4	3	0	19
4	Financial problems	C6	6	10	3	0	0	19
5	Management	C7	6	5	8	0	0	19
6	Material and equipment	C8	5	2	9	3	0	19
7	Orders and requirements	C9	11	7	1	0	0	19
8	Payment delay	C10	6	4	4	5	0	19
9	Planning	C11	4	10	5	0	0	19
10	Response delay	C13	8	3	4	2	2	19
Total			66	61	43	14	6	190

Table 3 Internal consistency level of Cronbach's alpha [7]

Scale	Reliability level	Cronbach's alpha		
1	Poor		α	< 0.5
2	Satisfactory	0.5	α	< 0.7
3	Good	0.7	α	< 0.8
4	Excellent	0.8	α	

Table 4 Ranking of categories of factors by RII analysis

No	Category	Code	N	Mean	RII	Rank	Importance level
1	Design and work error	C3	19	3.4211	0.684	9	High
2	Experience and productivity	C4	19	4.2105	0.842	2	Very high
3	External	C5	19	3.8947	0.779	5	High
4	Financial problems	C6	19	4.1579	0.832	3	Very high
5	Management	C7	19	3.8947	0.779	5	High
6	Material and equipment	C8	19	3.4737	0.695	8	High
7	Orders and requirements	C9	19	4.5263	0.905	1	Very high
8	Payment delay	C10	19	3.5789	0.716	7	High
9	Planning	C11	19	3.9474	0.789	4	High
10	Response delay	C13	19	3.6842	0.737	6	High

Table 5 Importance levels in the RII analysis [7]

Scale	Importance level	RII		
1	Very low	0.0	RII	0.2
2	Low	0.2 <	RII	0.4
3	Moderate	0.4 <	RII	0.6
4	High	0.6 <	RII	0.8
5	Very high	0.8 <	RII	1.0

In addition, the effect of the factors varied according to the different states. Thus, three groups of countries for each category (developed, developing, and global) were classified. In this classification, we considered and calculated all 360 factors, not just those refined and analyzed by RII. Also, some studies discussed global factors and were classified under the global group because they had not specified developed or developing countries. The factors of developed countries were stated in 95 studies, 239 reported factors in developing countries, and 26 studies discussed global factors. Table 6 shows the total factors divided as into those affecting developed and developing countries, as well as global factors.

Table 6 Categories of factors in developed and developing countries

No	Category	Code	Developed	Developing	Global	Total
1	Communication	C1	3	4	1	8
2	Conflict	C2	2	7	1	10
3	Design and work error	C3	10	8	1	19
4	Experience and productivity	C4	14	29	6	49
5	External	C5	5	20	1	26
6	Financial problems	C6	7	28	2	37
7	Management	C7	10	20	2	32
8	Material and equipment	C8	6	23	5	34
9	Orders and requirements	C9	11	25	1	37
10	Payment delay	C10	4	19	0	23
11	Planning	C11	9	26	4	39
12	Project	C12	1	11	0	12
13	Response delay	C13	9	11	1	21
14	Tender and contract	C14	4	8	1	13
Total			95	239	26	360

We also ranked the categories in each group of countries and overall, as shown in Table 7. We excluded global factors in order to identify just those within developed and developing countries groups. The category of "experience and productivity" C4 was the highest rank in terms of the number of factors stated in both developed and developing countries groups and as well as overall rank. Followed by C9 second, C3 & C7 third, C11 & C13 fourth, and C6 as the fifth category in the developed countries group ranking. In the developing countries group, C6, C11, C9, and C8 were ranked one by one. In the overall ranking, the category of "orders and requirements" (C9) was second; "financial problems" (C6) and "planning" (C11) were third, and "management" (C7) and "material and equipment" (C8) were fourth and fifth.

In total, 334 factors were classified within these groups: 95 for developed countries and 239 for developing countries. Therefore, the percentage of developed countries affected by delay factors was 28%, which was less than the percentage of developing countries affected, which was 72%. Table 7 details the ranks of categories affecting developed and developing countries and the overall rank.

Finally, the results clearly showed that the ranks of categories in terms of the top critical factors in construction projects were very close to the overall ranking

Table 7 Ranking of categories in developed and developing countries

No	Category	Code	Developed		Developing		Overall	
			Factor	Rank	Factor	Rank	Factor	Rank
1	Communication	C1	3	9	4	11	7	12
2	Conflict	C2	2	10	7	10	9	11
3	Design and work error	C3	10	3	8	9	18	9
4	Experience and productivity	C4	14	1	29	1	43	1
5	External	C5	5	7	20	6	25	6
6	Financial problems	C6	7	5	28	2	35	3
7	Management	C7	10	3	20	6	30	4
8	Material and equipment	C8	6	6	23	5	29	5
9	Orders and requirements	C9	11	2	25	4	36	2
10	Payment delay	C10	4	8	19	7	23	7
11	Planning	C11	9	4	26	3	35	3
12	Project	C12	1	11	11	8	12	10
13	Response delay	C13	9	4	11	8	20	8
14	Tender and contract	C14	4	8	8	9	12	10
Total			95		239		334	

of developed and developing countries affected by these factors, where the critical categories were the same in both the important factors ranking, and the overall ranking of countries, although these categories were not same as a ranking from 1 to 5.

However, some categories shared the same rank; for example, category C5 was shared as ranking five in categories of essential factors with C7, whereas C6 and C11 shared in the same rank of three in the overall ranking of countries.

5 Conclusion

Delay remains a contentious issue in construction projects, and it is due to many different factors. We identified 14 critical categories classified from the top five delay factors identified in previous studies. The critical categories with the highest ranking were orders and requirements; experience and productivity; financial problems; planning; and both external and management factors. All categories of essential delay factors identified in this research have affected both developed and developing countries, although the effect was 28% for developed countries and 72% for developing countries.

References

1. Idrees S, Shafiq MT (2021) Factors for time and cost overrun in public projects. J Eng, Proj & Prod Manag 11(3)
2. Faridi AS, El-Sayegh SM (2006) Significant factors causing delay in the UAE construction industry. Constr Manag Econ 24(11):1167–1176
3. Doloi H et al (2012) Analysing factors affecting delays in Indian construction projects. Int J Project Manage 30(4):479–489
4. Alfakhri AYY, Ismail A, Khoiry MA (2018) The effects of delays in road construction projects in tripoli, Libya. Int J Technol 9(4)
5. Bajjou MS, Chafi A (2018) Empirical study of schedule delay in Moroccan construction projects. Int J Constr Manag 20(7):783–800
6. Wuala H, Rarasati A (2020) Causes of delays in construction project for developing Southeast Asia countries. In: IOP Conference Series: Materials Science and Engineering. IOP Publishing
7. Fashina AA et al (2021) Exploring the significant factors that influence delays in construction projects in Hargeisa. Heliyon 7(4):e06826
8. Zhang D et al (2020) Causes of delay in the construction projects of subway tunnel. Adv Civ Eng 2020:1–14
9. Durdyev S et al (2017) Causes of delay in residential construction projects in Cambodia. Cogent Eng 4(1)
10. Ramli MZ et al (2019) Causes of construction delay for housing projects in Malaysia. In: Applied Physics of Condensed Matter (Apcom 2019)
11. McCord J et al (2015) Understanding delays in housing construction: evidence from Northern Ireland. J Financ Manag Prop Constr 20(3):286–319
12. Alfakhri A et al (2017) A conceptual model of delay factors affecting road construction projects in Libya. J Eng Sci Technol 12(12):3286–3298
13. Karunakaran S et al (2018) Causes of delay on highway construction project in Klang valley

14. Sharma VK, Gupta PK, Khitoliya RK (2020) Analysis of highway construction project time overruns using survey approach. Arab J Sci Eng 46(5):4353–4367
15. Mohammed Gopang RK, Imran QB, Nagapan S (2020) Assessment of delay factors in Saudi Arabia railway/metro construction projects. Int J Sustain Constr Eng Technol 11(2)
16. Ramli MZ et al (2021) Ranking of railway construction project delay factors in Malaysia by using relative importance index (RII). In: Proceedings of Green Design and Manufacture 2020.
17. Gunduz M, Tehemar SR (2019) Assessment of delay factors in construction of sport facilities through multi criteria decision making. Prod Plan & Control 31(15):1291–1302
18. Kazemi A, Kim E-S, Kazemi M-H (2020) Identifying and prioritizing delay factors in Iran's oil construction projects. Int J Energy Sect Manage 15(3):476–495
19. Bin Seddeeq A et al (2019) Time and cost overrun in the Saudi Arabian oil and gas construction industry. Buildings 9(2):41
20. KV P et al (2019) Analysis of causes of delay in Indian construction projects and mitigation measures. J Financ Manag Prop Constr 24(1): 58–78
21. Akogbe R-KTM, Feng X, Zhou J (2013) Importance and ranking evaluation of delay factors for development construction projects in Benin. KSCE J Civ Eng 17(6):1213–1222
22. Abbasi O et al (2020) Exploring the causes of delays in construction industry using a cause-and-effect diagram: case study for Iran. J Arch Eng 26(3)
23. Sanni-Anibire MO, Mohamad Zin R, Olatunji SO (2020) Causes of delay in the global construction industry: a meta analytical review. Int J Constr Manag 22(8): 1395–1407
24. Hasmori MF et al (2018) Significant factors of construction delays among contractors in Klang valley and its mitigation. Int J Integr Eng 10(2)
25. Amoatey CT et al (2015) Analysing delay causes and effects in Ghanaian state housing construction projects. Int J Manag Proj Bus 8(1):198–214
26. Hoque MI et al (2021) Analysis of construction delay for delivering quality project in Bangladesh. Int J Build Pathol Adapt
27. Ametepey SO, Gyadu-Asiedu W, Assah-Kissiedu M (2018) Causes-effects relationship of construction project delays in Ghana: focusing on local government projects. In: Advances in Human Factors, Sustainable Urban Planning and Infrastructure. pp 84–95
28. Aziz RF (2013) Ranking of delay factors in construction projects after Egyptian revolution. Alex Eng J 52(3):387–406
29. Alhajri A, Alshibani A (2018) Critical factors behind construction delay in petrochemical projects in Saudi Arabia. Energies 11(7)
30. AlGheth A, Ishak Sayuti M (2020) Review of delay causes in construction projects. In: Proceedings of AICCE'19. pp 429–437
31. Chen G-X et al (2017) Investigating the causes of delay in grain bin construction projects: the case of China. Int J Constr Manag 19(1):1–14
32. Rivera L, Baguec H, Yeom C (2020) A study on causes of delay in road construction projects across 25 developing countries. Infrastructures 5(10)
33. Yap JBH et al (2021) Revisiting critical delay factors for construction: analysing projects in Malaysia. Alex Eng J 60(1):1717–1729
34. Ramli M et al (2018) Study of factors influencing construction delays at rural area in Malaysia. J Phys: Conf Ser 2018. IOP Publishing
35. Hwang BG, Leong LP (2013) Comparison of schedule delay and causal factors between traditional and green construction projects. Technol Econ Dev Econ 19(2):310–330
36. Gopang RKM, Alias Imran QB, Nagapan S (2020) Assessment of delay factors in Saudi Arabia railway/metro construction projects. Int J Sustain Constr Eng Technol 11(2): 225–233
37. Vacanas Y, Danezis C (2018) An overview of the risk of delay in Cyprus construction industry. Int J Constr Manag 21(4):369–381
38. Elfi E, Tahir MM, Tukirin SA (2020) Factors affecting the delay in construction at Mentawai Island, Indonesia. In: IOP Conference Series: Materials Science and Engineering. IOP Publishing.
39. Bekr GA (2015) Causes of delay in public construction projects in Iraq. Jordan J Civ Eng 9(2)

40. Choong Kog Y (2018) Major construction delay factors in Portugal, the UK, and the US. Pract Period Struct Des Constr 23(4)
41. Rezaee MJ, Yousefi S, Chakrabortty RK (2019) Analysing causal relationships between delay factors in construction projects: a case study of Iran. Int J Manag Proj Bus
42. Durdyev S, Hosseini MR (2019) Causes of delays on construction projects: a comprehensive list. Int J Manag Proj Bus 13(1):20–46
43. Mbala M, Aigbavboa C, Aliu J (2019) Causes of delay in various construction projects: a literature review. In: Advances in Human Factors, Sustainable Urban Planning and Infrastructure. pp. 489–495
44. Trade D.o.F.A.a (2022) List of developing countries as declared by the Minister for Foreign Affairs. Available from: https://www.dfat.gov.au/about-us/publications/list-of-developing-countries-as-declared-by-the-minister-for-foreign-affairs

Chloride Penetration in Low-Carbon Concrete with High Volume of SCM: A Review Study

C. Xue and V. Sirivivatnanon

Abstract Low-carbon concrete (LCC) uses supplementary cementitious material (SCM) to partially replace cement as a method for reducing its carbon footprint. Previous laboratory and field studies had provided substantial support and experience for using LCC in marine structures, which are the most susceptible to chloride-induced corrosion. Some short-term test methods have provided reliable assessment of the ability of LCC to resist chloride penetration, but the long-term chloride penetration depends on a great many factors and thus could differ from the results obtained from laboratory tests. However, the lack of a correlation between the data from short-term and long-term tests has limited the use of abundant laboratory results for service life design of LCC. This study presents an overview of results obtained when LCCs were exposed to chlorides. The key outcome of this study is a broader synthesis of the available data regarding the relationship between the mix design and the performance of LCCs in various chloride environments, which helps find the possible correlation and fully appreciate the value of the short-term tests.

Keywords Chloride penetration · Low carbon concrete · Marine structures

1 Introduction

Approximately 5–8% of the global CO_2 emissions is attributed to the production of ordinary Portland cement (OPC) [1]. To reduce the carbon footprint of the concrete industry, substitution of cement by supplementary cementitious materials (SCM) without compromising performance is an efficient solution [2]. Another indirect way to contribute to sustainability is prolonging the service life of infrastructure by using durable concrete. From these two viewpoints, ground granulated blast-furnace

C. Xue (✉) · V. Sirivivatnanon
School of Civil and Environmental Engineering, University of Technology Sydney, Ultimo, NSW, Australia
e-mail: Caihong.Xue@uts.edu.au

UTS-Boral Center for Sustainable Building, Sydney, NSW, Australia

W. Duan et al. (eds.), *Nanotechnology in Construction for Circular Economy*, Lecture Notes in Civil Engineering 356,
https://doi.org/10.1007/978-981-99-3330-3_16

slag (GGBFS) and fly ash (FA) are great choices among other SCMs. One of the major benefits of blending OPC with GGBFS or FA is improved resistance to chloride penetration, which has been evidenced by both short-term laboratory tests and long-term field tests. The former includes the widely used rapid chloride migration test method standardized in ASTM C1202 [3] and the chloride diffusion test given in NT Build 492 [4]. However, for application, concrete structures need to be designed for a specific service life, and this requires long-term quantitative field performance assessment, which is not always practical [5]. Therefore, finding the link between the results from laboratory and field tests is important for promoting efficient use of LCCs, but is nevertheless challenging because a high concentration of deleterious species in laboratory tests could have already altered the deterioration processes and the laboratory curing conditions deviate significantly from on-site conditions. The purpose of this paper was to review of the factors affecting chloride penetration in LCC made with GGBFS or FA and the correlations between results from different test methods.

2 Chloride Penetration in LCC

2.1 *Effect of SCM Content on Diffusion Coefficients*

The replacement of cement with SCM is usually no more than 50% for GGBFS and 30% for FA, due to the reduction in strength with increasing SCM content, as shown in Fig. 1, in which the short-term laboratory test results from Dhir et al. [6] were adopted to demonstrate the effect of GGBFS or FA content on *D* as well as the 28 day cube strength. It can be seen that the diffusion coefficients continuously decrease with increasing GGBFS content up to 65%; in the range of 30–50%, *D* is insignificantly affected by FA content but the strength reduction is more pronounced at higher FA dosages. On the other hand, for high-volume FA (HVFA) concrete with >50% FA as cement replacement and a considerably high amount of superplasticizer, the HVFA concrete has proven to yield higher long-term strength and resistance against chloride penetration than OPC concrete, despite the early-age properties of the former being less competitive [7–10]. Thomas et al. [11, 12] found that HVFA concrete with 50% FA had a significantly lower *D* and a slightly higher compressive strength than OPC concrete after being exposed to the field marine environment for up to 10 years. Moreover, Moffatt et al. [13] reported the *D* of a HVFA concrete with 56–58% FA after 24 years of exposure to a harsh field environment where high tides and freeze–thaw cycles occurred, was only 1.5×10^{-13} m^2/s compared with 3.6×10^{-12} m^2/s for the counterpart OPC concrete.

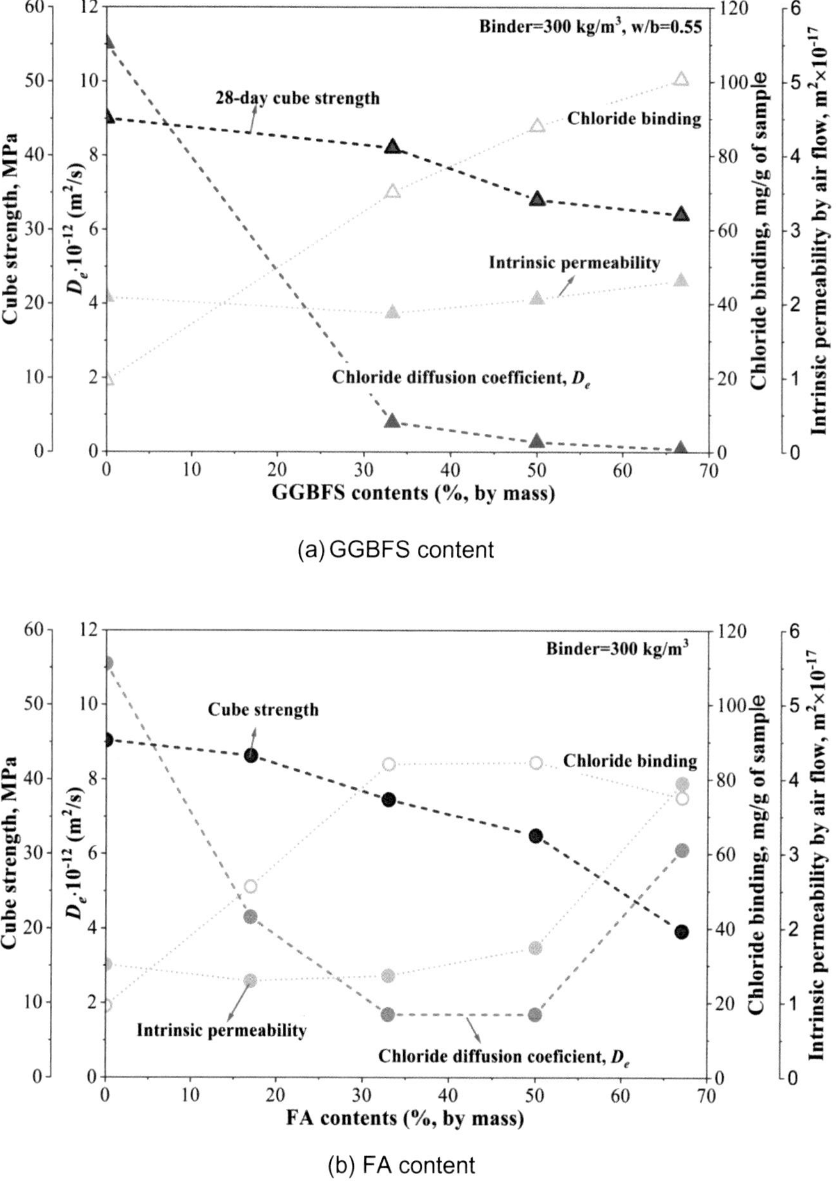

Fig. 1 Effects of **a** ground granulated blast-furnace slag (GGBFS) and **b** fly ash (FA) content as cement replacement on chloride penetration and other properties assessed by short-term laboratory tests [6, 14, 15]

2.2 Effect of Curing Conditions on Chloride Penetration

It has been well established that partial replacement of cement by GGBFS or FA improves the microstructure of concrete and thus the resistance to chloride penetration. In LCC, the GGBFS or FA reacts with calcium hydroxide (formed by the hydration of cement) and water to produce C–S–H and a portion of calcium aluminate phases [16–18]. Additionally, depending on the specific surface area of particles, GGBFS or FA can act as a filler to fill pores and as nucleation sites to enhance hydration [19, 20]. Apart from the products formed by normal cement hydration, the additional hydrates due to either enhanced hydration or hydration of GGBFS/FA reduce capillary porosity (>30 nm) and thus block the chloride diffusion paths; however, these benefits of GGBFS and FA depend on the curing conditions, especially for LCC with GGBFS [21]. Moreover, it is possible that blended cement concrete will perform no better than OPC concrete when structures are exposed to prolonged drying and carbonation [22]. Figure 2 compares the D of concretes cured in wet and dry conditions, from which it can be seen that OPC concrete with a low strength grade is more sensitive to the curing conditions, and the influence of curing conditions diminishes with increasing concrete strength and exposure duration [23, 24]. Irrespective of the curing conditions, at the same grade the blended concretes consistently outperform OPC concrete in resisting chloride penetration. In this regard, Bamforth [25] examined the D of dry-cured (indoor), membrane-cured and water-cured concrete blocks (40 MPa at 28 days) located in the splash/spray zone on the south coast of the UK for 8 years, and found that the effect of curing conditions on the D of different concrete mixes was inconsistent, but that 70% GGBFS-blended concrete yielded the lowest average D compared with 30% FA-blended concrete and OPC concrete. Additionally, a lower grade of blended cement concrete is more durable than a higher grade OPC concrete at the later age, which was also confirmed by Thomas et al. [11].

2.3 Effect of Test Methods on Chloride Penetration

There are only limited published data on relating laboratory test results to long-term field performance of concrete with regard to resistance to chloride penetration. Figure 3 shows the correlation between D from a long-term field test (D_{field} on the x-axis) and counterpart D from laboratory diffusion tests (D_{lab} on the left y-axis) and coulombs (right y-axis) from the rapid chloride permeability test (RCPT), using data from previous studies [13, 26, 27]. Thomas et al. [26] conducted the RCPT as well as diffusion tests (using 16.5% NaCl as per ASTM C1556) on uncontaminated GGBFS-blended concrete exposed in the field to a tidal zone for 25 years, and calculated the chloride diffusion coefficient (D_{field}) from the chloride profiles in the field-exposed concrete. Moffatt et al. [13] obtained D_{field} and coulombs (RCPT) of high-volume FA concrete exposed to the marine environment for 19–24 years. Compared with the results from the RCPT, D_{lab} from laboratory diffusion tests following the procedures

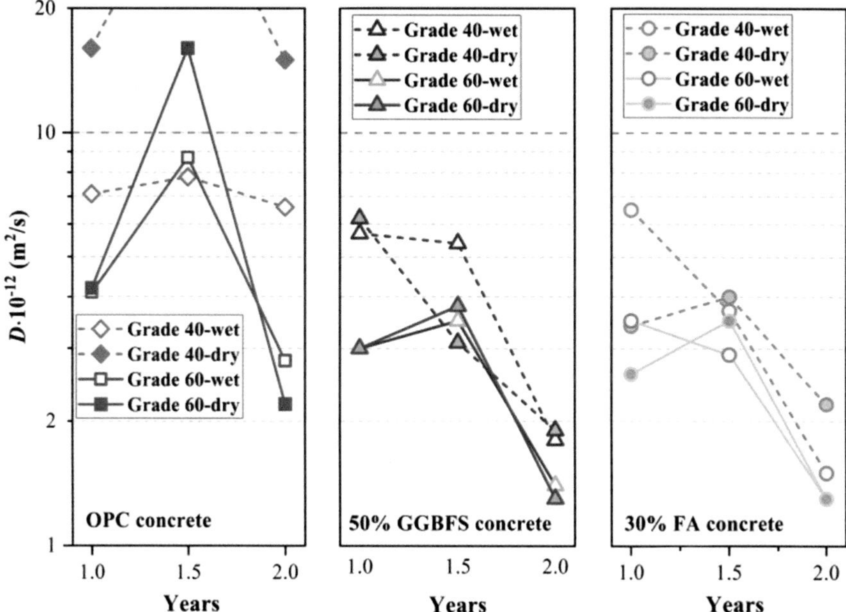

Fig. 2 Effect of curing conditions on the *D* of different grades of concrete exposed to the tidal zone of Cape Peninsula with water temperature between 12 and 15 °C. The wet-cured (wet) concrete was exposed to 6 days' moist curing (23 °C and 90% relative humidity) after demolding, while the dry-cured (dry) concrete was stored in an open area (23 °C and 50% relative humidity). The concrete was exposed to marine environment at age 28 days [23]. GGBFS, ground granulated blast-furnace slag; FA, fly ash; OPC, ordinary Portland cement

given in ASTM C1556, NT Build 443 or other similar procedures, could better indicate the ability of concrete to resist chloride penetration. Although there are synergies between the two diffusion coefficients (D_{lab} and D_{field}), the quantitative relationship between them varies. Figure 4 shows the correlation between D_{lab} (x-axis) and coulombs from the RCPT (left y-axis), and the non-steady-state (D_{nssm}) or steady-state (D_{ssm}) chloride migration coefficient (D_m on the right y-axis) from accelerated migration tests reported in previous studies [28–32]. Note that these previous studies used different NaCl concentrations and exposure durations, which are summarized in Table 1. When D_{lab} was used as the reference, the $D_m > 2 \times 10^{-12}$ (m^2/s) and RCPT coulombs >800 could be more reliable for ranking concretes in terms of the resistance to chloride penetration.

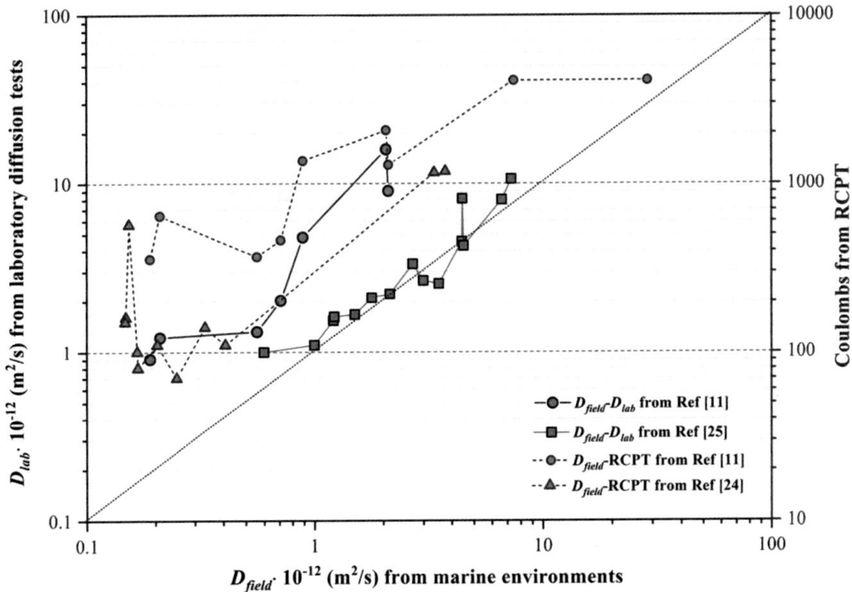

Fig. 3 Correlation between the D from marine exposure (D_{field}) and the D from laboratory diffusion tests (D_{lab}) and coulombs from the rapid chloride permeability test (RCPT) [13, 26, 27]

3 Conclusions

(1) Replacing cement with up to 65% GGBFS or 30% FA improves the resistance of concrete to chloride penetration but decreases early-age strength development. LCC with GGBFS or FA could achieve higher resistance to chloride penetration at equivalent strength or binder content as compared with OPC concrete, indicating that efficient use of LCC requires a performance-based service life design approach.

(2) The influence of curing conditions on chloride diffusion coefficients diminishes with increasing concrete strength grade, but could be significant for low-strength concrete. At strength grade ≥40 MPa, the difference in chloride diffusion coefficients arising from the change in curing conditions is much smaller in LCC than OPC.

(3) Although short-term laboratory diffusion and accelerated migration tests are reliable in distinguishing parameters that affect chloride penetration in concrete, the correlations between results from different methods are difficult to establish.

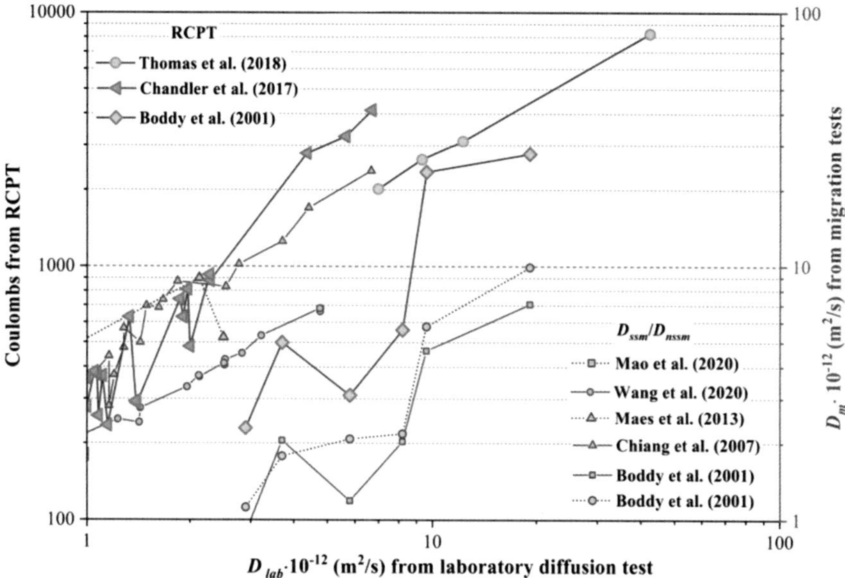

Fig. 4 Correlation between the D from laboratory diffusion tests (D_{lab}) and coulombs from the rapid chloride permeability test (RCPT) and the chloride migration coefficient from migration tests (D_m) [28–32]. The gray lines indicate the RCPT results. The red lines indicate the steady-state chloride migration coefficient (D_{ssm}) and blue lines indicate non-steady-state chloride migration coefficient (D_{nssm})

Table 1 Laboratory test methods for assessing resistance to chloride penetration shown in Fig. 4

Authors	Laboratory diffusion tests			Accelerated migration tests		
	Standard	NaCl	Days	Standard	NaCl	Hours/voltage
Thomas et al. [28]	ASTM C1543	3%	90			
Maes et al. [29]	NT Build 443	16.5%	30	NT Build 492	10%	24/30–60
Boddy et al. [30]	AASHTO T259	3%	90			
Chiang and Yang [31]	AASHTO T259	3%	90	ACMT	0.52 M	24/60
Mao et al. [32]	NT Build 443	16.5%	90	NT Build 492	10%	24/60
Wang and Lui [33]		0.1 M	42–49		0.1 M	24–48/12

Acknowledgements The authors acknowledge the support of the UTS-Boral Centre for Sustainable Building for the opportunity to review the performance-based testing of the resistance of LCC to chloride penetration. This will promote greater and systematic use of LCC in marine environments.

References

1. Habert G, Billard C, Rossi P, Chen C, Roussel N (2010) Cement production technology improvement compared to factor 4 objectives. Cem Concr Res 40(5):820–826
2. Bilodeau A, Malhotra VM (2000) High-volume fly ash system: concrete solution for sustainable development. Mater J 97(1):41–48
3. ASTM-C1202-12 (2012) Standard test method for electrical indication of concrete stability to resist chloride ion penetration. ASTM Annual Book of Standards, V.04.02, ASTM International, 100 Barr Harbour Dr., P.O. box C-700, West Conshohocken, PA USA
4. NT BUILD 492 (1999) Concrete, mortar and cement-based repair materials: chloride migration coefficient from non-steady-state migration experiments, Nordtest method
5. Beushausen H, Torrent R, Alexander MG (2019) Performance-based approaches for concrete durability: state of the art and future research needs. Cem Concr Res 119:11–20
6. Dhir R, El-Mohr M, Dyer T (1996) Chloride binding in GGBS concrete. Cem Concr Res 26(12):1767–1773
7. Giaccio GM, Malhotra V (1988) Concrete incorporating high volumes of ASTM Class F fly ash. Cem, Concr Aggreg 10(2):88–95
8. Dunstan M, Thomas M, Cripwell J, Harrison D (1992) Investigation into the long-term in-situ performance of high fly ash content concrete used for structural applications, vol 132. Special Publication, pp 1–20
9. Mehta PK (2004) High-performance, high-volume fly ash concrete for sustainable development. In: Proceedings of the international workshop on sustainable development and concrete technology. Iowa State University Ames, IA, USA, pp 3–14
10. Zuquan J, Wei S, Yunsheng Z, Jinyang J, Jianzhong L (2007) Interaction between sulfate and chloride solution attack of concretes with and without fly ash. Cem Concr Res 37(8):1223–1232
11. Thomas M (1991) Marine performance of PFA concrete. Mag Concr Res 43(156):171–185
12. Thomas M, Matthews J (2004) Performance of pfa concrete in a marine environment—10-year results. Cem Concr Compos 26(1):5–20
13. Moffatt EG, Thomas MD, Fahim A (2017) Performance of high-volume fly ash concrete in marine environment. Cem Concr Res 102:127–135
14. Dhir R, Hewlett P, Chan Y (1989) Near surface characteristics of concrete: intrinsic permeability. Mag Concr Res 41(147):87–97
15. Dhir R, El-Mohr M, Dyer T (1997) Developing chloride resisting concrete using PFA. Cem Concr Res 27(11):1633–1639
16. Kolani B, Buffo-Lacarrière L, Sellier A, Escadeillas G, Boutillon L, Linger L (2012) Hydration of slag-blended cements. Cem Concr Compos 34(9):1009–1018
17. Sakai E, Miyahara S, Ohsawa S, Lee S-H, Daimon M (2005) Hydration of fly ash cement. Cem Concr Res 35(6):1135–1140
18. Lam L, Wong Y, Poon CS (2000) Degree of hydration and gel/space ratio of high-volume fly ash/cement systems. Cem Concr Res 30(5):747–756
19. Gutteridge WA, Dalziel JA (1990) Filler cement: the effect of the secondary component on the hydration of Portland cement: part 2: fine hydraulic binders. Cem Concr Res 20(6):853–861
20. Berodier E, Scrivener K (2014) Understanding the filler effect on the nucleation and growth of C–S–H. J Am Ceram Soc 97(12):3764–3773
21. Ngala V, Page C (1997) Effects of carbonation on pore structure and diffusional properties of hydrated cement pastes. Cem Concr Res 27(7):995–1007

22. Thomas M, Matthews J (1992) Carbonation of fly ash concrete. Mag Concr Res 44(160):217–228
23. Mackechnie J, Alexander M (1997) Exposure of concrete in different marine environments. J Mater Civ Eng 9(1):41–44
24. Khatib JM, Mangat P (2002) Influence of high-temperature and low-humidity curing on chloride penetration in blended cement concrete. Cem Concr Res 32(11):1743–1753
25. Bamforth P (1999) The derivation of input data for modelling chloride ingress from eight-year UK coastal exposure trials. Mag Concr Res 51(2):87–96
26. Thomas M, Bremner T (2012) Performance of lightweight aggregate concrete containing slag after 25 years in a harsh marine environment. Cem Concr Res 42(2):358–364
27. Sirivivatnanon V, Xue C, Khatri R (2022) Design service life of low carbon concrete in marine tidal conditions (submitted). ACI Mater J
28. Thomas R, Ariyachandra E, Lezama D, Peethamparan S (2018) Comparison of chloride permeability methods for alkali-activated concrete. Constr Build Mater 165:104–111
29. Maes M, Gruyaert E, De Belie N (2013) Resistance of concrete with blast-furnace slag against chlorides, investigated by comparing chloride profiles after migration and diffusion. Mater Struct 46(1):89–103
30. Boddy A, Hooton R, Gruber K (2001) Long-term testing of the chloride-penetration resistance of concrete containing high-reactivity metakaolin. Cem Concr Res 31(5):759–765
31. Chiang C, Yang C-C (2007) Relation between the diffusion characteristic of concrete from salt ponding test and accelerated chloride migration test. Mater Chem Phys 106(2–3):240–246
32. Mao X, Qu W, Zhu P, Xiao J (2020) Influence of recycled powder on chloride penetration resistance of green reactive powder concrete. Constr Build Mater 251:119049
33. Wang J, Liu E (2020) The relationship between steady-state chloride diffusion and migration coefficients in cementitious materials. Mag Concr Res 72(19):1016–1026

A Compact Review on the Waste-Based Lightweight Concrete: Advancement and Possibilities

M. M. U. Islam, J. Li, R. Roychand, and M. Saberian

Abstract Lightweight concrete (LWC) has been used for more than 2000 years, and the technical development of waste-based LWC is still proceeding. Notably, the very first representative concrete mix of infrastructural LWC was introduced for building a family house in Berlin, Germany, a few decades ago. The unique and distinctive combination of waste-based LWC successfully creates an appealing alternative to traditional concrete aggregates in terms of durability, robustness, cost, energy-saving, transportation, environmental advantages, innovative architectural designs and implementations, and ease of construction. Numerous researchers have attempted to utilize waste materials to produce LWC, aiming to bring both ecological and economical solutions to the construction industry over the past few decades. Waste materials, such as crushed glass, waste tire rubber, masonry rubber, chip rubber, plastics, coconut shells, palm oil fuel ash, palm kernel shells, fly ash, and rice husks, possess lower specific gravity than traditional concrete aggregates. Thus waste-based LWC can be a significant replacement for conventional raw materials (cementitious material and aggregates) as it requires less strength than conventional concrete for both structural and non-structural applications. Although waste-based LWC is well recognized and has proven its scientific potential in a broad range of applications, there are still uncertainties and hesitations in practice. Therefore, the primary objective of this study was to demonstrate the current state-of-the-art understanding and advancement of waste-based LWC over the past decades. Furthermore, an equally critical discussion is reported to shed light on the potential benefits of LWC. We highlight how the performance of LWC has been enhanced significantly over the period, and understanding of the properties of waste-based LWC has advanced.

Keywords Lightweight concrete · Lightweight concrete aggregates · Mechanical properties · Structural bond behavior · Thermal conductivity · Waste materials

M. M. U. Islam · J. Li (✉) · R. Roychand · M. Saberian
Discipline of Civil and Infrastructure Engineering, School of Engineering, RMIT University, Melbourne, VIC, Australia
e-mail: jie.li@rmit.edu.au

M. M. U. Islam
e-mail: momeen.structure.um@gmail.com

W. Duan et al. (eds.), *Nanotechnology in Construction for Circular Economy*, Lecture Notes in Civil Engineering 356,
https://doi.org/10.1007/978-981-99-3330-3_17

1 Introduction

Knowledge of lightweight concrete (LWC) dates back almost 3000 years [1]. A number of LWC-based superstructures can be found around the Mediterranean, the most noteworthy of which are the Pantheon Dome and the Port of Cosa, both of which were built in the early age of the Roman Empire. LWC has a considerable number of advantages, such as better fire resistance, thermal insulation, and low density. Notably, implementation of LWC has been extensively explored as a non-structural and structural material.

In the past few decades, LWC has become a significant and resourceful material that has been considerably developed, thanks to scientific endeavors [2, 3]. Indeed, LWC is one of the exciting materials in the contemporary construction sector due to its immense advantages, both ecologically and structurally. In the case of structural advantages, LWC has crucial applications, particularly in heavy structure design, where the dead load governs the total weight and that dead load is substantially greater than the anticipated service load, such as for bridges and multistorey buildings. The lessened self-weight that originates from the utilization of the LWC in structures delivers accountable cost savings and flexibility. LWC also improves fire resistance, and seismic structural response offers longer structural spans, reduces reinforcement ratios, and lowers the cost of foundation materials. Furthermore, precast elements manufactured using LWC reduce placement and transportation costs [4, 5]. For bridges, LWC facilitates longer spans and more lanes. For instance, LWC can be used on one side of a cantilever bridge, while normal-weight concrete (NWC) is used on the other side to facilitate the weight balance for a longer span. From the ecological aspect, LWC possesses lower thermal conductivity than NWC and hence plays a significant role in saving energy when introduced as a thermal insulation material. Moreover, implementation of LWC produced from controlled thermal lightweight materials lessens the energy consumption by air acclimatizing in both warm and cold countries. Nowadays, energy-shortage problems are escalating at an alarming and upsetting rate, and it has become a worldwide concern. Most importantly, the waste materials produced from the agricultural and industrial sectors can be utilized to manufacture the LWC following an eco-friendly and economical approach, which assists in mitigating climate change. In this review, we focused on compiling the standpoints and previous scientific efforts related to LWC to improve knowledge of the entire scenario and identify the research gaps. Notably, this compilation sheds light on the advancement and possibilities of LWC to inspire new researchers for further progress.

2 Lightweight Aggregate Concrete (LWAC)

LWAC can be manufactured using artificial or natural lightweight aggregates (LWAs) to substitute the conventional aggregates in NWC [6]. There are a number of LWAs that possess various physical and mechanical characteristics allowing LWAC to be produced with varying ranges of strengths and densities [7]. Different types of LWAs are also commercially available, which has pushed researchers to investigate and compare them in the production of high-strength LWACs [8]. As an example, several natural materials (perlite or vermiculite), recycled waste materials (crushed glass, waste tire rubber, masonry rubber, chip rubber, and plastics), and argillaceous materials (clay, slate, and shale) have been used by previous researchers in the manufacture of LWAC. Moreover, a significant number of research has been conducted into utilizing agricultural wastes, such as coconut shells, palm oil fuel ash, palm kernel shells, and apricot shells. Including these wastes in the manufacture of LWAC brings sustainable solutions to significant environmental issues [9, 10].

It is well known that the LWAs possess higher porosity than normal-weight aggregates (NWAs) [11] and thus lower strength values together with larger deformations [12]. These characteristics suggest that LWAs will be the weakest constituent, having a significant role in the ultimate performance of the manufactured concrete mix. Furthermore, LWAs usually constitute >50% total volume of the concrete [13]. It is therefore crucial to carefully implement LWAs to improve the overall performance of the LWAC mix in both the fresh and hardened phases. A significant number of researchers have conducted detailed investigations of LWAs properties, such as grading, absorption, and particle size, on the thermal and mechanical behaviors of LWAC. Table 1 shows the requirements for compressive strength and splitting tensile strength of LWAs according to ASTM C331 [14].

Table 1 Requirements for compressive and splitting tensile strength of lightweight aggregates (LWAs)

Concrete type	Dry density, kg/m^3	Minimum splitting tensile strength at 28 days (MPa)	Minimum compressive strength at 28 days (MPa)
LWA	1760	2.2	28
	1680	2.1	21
	1600	2.0	17
Mixture of LWAs and NWAs	1840	2.3	28
	1760	3.1	21
	1680	2.1	17

NWAs, normal-weight aggregates

3 Structural Bond Behavior of LWC

It is necessary to have adequate bonding between the concrete matrix and reinforcing bars for (i) gaining an efficient beam action, (ii) crack control, and (iii) improving ductility [15]. Furthermore, all the empirical equations in the standard codes greatly depend on sufficient bonding between the cement matrix and reinforcing bars [16]. Hence, a reduction in bonding may lead all the design basics towards invalidation. There are two mechanisms for achieving improved bond strength: mechanical (bearing and friction action) and physiochemical (linkage/adhesion) [17]. The adhesion/linkage force originates from the chemical reaction between the surface of the reinforcement and the cementitious matrix. The friction forces come from the bearing force and rough contact, resulting from interlocking between the reinforcing ribs and the concrete matrix [18].

Several researchers have investigated the bond strength behavior of LWC and described the factors that can adversely affect the bond strength. These crucial factors, such as the water–cement ratio (w/c), aggregate types, the diameter of the reinforcement bars, admixtures, surface texture and type of reinforcing bars, types of lateral confinement, and bond strength, significantly control the bond strength [19]. Many equations have been developed to predict the bond strength of LWC, and three are presented below [20–22]:

$$\tau = \left[171.9\left(\frac{h}{d}\right)^2 - 24.2\left(\frac{h}{d}\right) + 1.29\right] f'_c \tag{1}$$

$$\tau = \left[\frac{37.5}{(d + l_d)^{0.25}} - 9.4\right] f_c'^{0.5} \tag{2}$$

$$\tau = K \cdot [44.5 - 60(w/c)] \cdot \frac{\rho_d}{2200} \tag{3}$$

where h is the reinforcing rib height, d is the reinforcing bar diameter, l_d is the length of embedment, f'_c is the compressive strength, w/c is the water–cement ratio, and ρ_d is the dry density of LWC.

4 State of the Art

Extensive research on LWAC (Table 2) has led to many structural applications, such as high-rise buildings, long-span bridges, and buildings where the foundation conditions are vulnerable, and also in highly dedicated applications, such as offshore and floating structures. Moisture content and density are the major factors controlling the thermal conductivity properties of LWAC, whereas the mineralogical properties

Table 2 Engineering properties of waste-based lightweight aggregate concrete

Waste material	Replacement type	Aggregate size	Content (%)	Mechanical properties at 28 days				Thermal conductivity (W/m K)	Remarks	References
				Compressive strength (MPa)	Splitting tensile strength (MPa)	Flexural strength (MPa)	Modulus of elasticity (GPa)			
Rice husks	Cement	Powder	0–35	42–24	2.6–1.8	4.1–2.7	32–29	–	Decreased the mechanical properties by the addition of rice husk	[33]
	Cement	Powder	0–20	68–48	5.1–4.4	8.1–6.9	–	–	20% replacement of cement with rice husk improved the properties	[34]
	Cement	Powder	0–15	36–41	4.5–4.9	–	–	1.21–0.99	Improved thermal conductivity	[35]
	Cement	Ultrafine	10	47	–	–	–	–	Improved compressive strength	[36]
Palm oil fuel ash (POFA)	Cement	300 μm	0–25	41.8–36	3.8–3.1	6.6–4.2	15.8–13.9	–	10% POFA improved the mechanical properties of LWC	[2]

(continued)

Table 2 (continued)

Waste material	Replacement type	Aggregate size	Content (%)	Mechanical properties at 28 days				Thermal conductivity (W/m K)	Remarks	References
				Compressive strength (MPa)	Splitting tensile strength (MPa)	Flexural strength (MPa)	Modulus of elasticity (GPa)			
	FA	0.1–250 μm	0–15	27–35	2.2–3.3	2.1–3.9	60–78	–	Addition of 15% POFA content improved the mechanical properties	[37]
	Cement	300 μm	0–30	32.6–24.9	2.8–2.6	–	–	Varied from 27 to 40 °C for 26 h	Thermal conductivity improved for up to 30% POFA content	[38]
Oil palm shell (OPS)	CA	2.36–9 mm	100	41.8–36	3.8–3.1	6.6–4.2	15.8–13.9	–	Adopting OPS as CA can reduce the consumption of NCA	[2]

(continued)

Table 2 (continued)

Waste material	Replacement type	Aggregate size	Content (%)	Mechanical properties at 28 days				Thermal conductivity (W/m K)	Remarks	References
				Compressive strength (MPa)	Splitting tensile strength (MPa)	Flexural strength (MPa)	Modulus of elasticity (GPa)			
	CA	≤2.36 mm	100	2.1–20	–	–	–	0.4–0.9	Palm shell foamed concrete	[39]
	CA	8 mm	50	34.2–41.3	2.77–3.2	3.8–4.9	–	–	OPS improved the properties	[40]
Waste tire rubber	CA	8–15 mm	100	18	3.34	3.74	15.28	–	Proposed new method to improve the mechanical properties	[4]
	FA	≤4.75 mm	5–25	26–41	3–3.8	3.6–4.4	–	–	Fracture energy increased with the addition of rubber and fibers	[41]

(continued)

Table 2 (continued)

Waste material	Replacement type	Aggregate size	Content (%)	Mechanical properties at 28 days				Thermal conductivity (W/m K)	Remarks	References
				Compressive strength (MPa)	Splitting tensile strength (MPa)	Flexural strength (MPa)	Modulus of elasticity (GPa)			
	FA	1–2 mm	0–20	20.53–33.94	2.25–3.7	3.42–5.5	21.95–33	–	Improved ductile properties	[42]
	FA	1–3 mm	0–15	43–53	3.44–3.8	4.77–6.4	2.52–2.7	–	Addition of rubber decreased properties	[43]
	CA	5–20 mm	40	24.1–28.2	2.4–3.0	4.9–6.1	25.9–28	–	Brittleness index reduced due to the addition of rubber	[44]
	FA	0.075–4 mm	0–20	22–20	–	–	–	1.0–0.9	Improved thermal conductivity	[13]
Waste glass	CA	≤5 mm	0–70	39–45	2.4–3.6	4.4–6.6	–	–	Decreased the properties	[45]
	FA	≤2 mm	100	56–57	–	–	–	1.1–0.9	Improved conductivity	[46]

(continued)

Table 2 (continued)

Waste material	Replacement type	Aggregate size	Content (%)	Mechanical properties at 28 days				Thermal conductivity (W/m K)	Remarks	References
				Compressive strength (MPa)	Splitting tensile strength (MPa)	Flexural strength (MPa)	Modulus of elasticity (GPa)			
Coconut shell	CA	4.7–12.5 mm	100	13.8–24	1.84–2.98	2.95–5.65	44–67	–	Useful to produce structural LWC	[47]

CA, coarse aggregate; FA, fine aggregate

of the aggregates may affect up to 25% of the thermal conductivity for LWC under a similar density value [3, 23]. The concrete matrix penetrates into the LWAs during mixing of the concrete materials [24]. Nevertheless, the penetration rate dramatically depends on the surface layer and microstructure of the aggregates, viscosity of the concrete matrix, and particle size distribution of the cement. Additionally, both the chemical and physical properties of LWAs influence the strength of the LWAC because of the processes occurring at the interfacial transition zone. The compressive strength of LWC increased from 15.5 to 29 MPa for an increase in cement content from 250 to 350 kg/m^3 by maintaining the same density of ≈1500 kg/m^3 [25]. Figure 1a shows the correlation between cube strength and density of LWAC. Another study [26] explored the compressive strength and thermal conductivity of expanded perlite aggregate-based concrete along with the mineral admixtures. They mentioned that using fly ash and silica fume as cementitious materials can reduce the thermal conductivity values by up to 15%, while the compressive strength and density of the concrete were also lowered by 30%. LWC using diatomite as the LWA was manufactured with a density ranging from 950 to 1200 kg/m^3 and compressive strength from 3.5 to 6 MPa, and thermal conductivity was found to increase from 0.22 to 0.30 W/(m K) for a cement content of 250–400 kg/m^3. LWC associated with expanded glass and clay as the LWAs exhibited higher resistance to chloride ion penetration and water with a cement content of 500 kg/m^3 and unit density of 1400 kg/m^3, and the compressive strength of the LWAC reached 24 MPa at 28 days. LWAC produced with dredged silt as the LWA exerted densities from 800 and 1500 kg/m^3 for the varying binder content of 364, 452, and 516 kg/m^3 [27, 28]. The dredged silt-based LWAC exhibited compressive strength from 18 to 42 MPa with thermal conductivity of 0.5–0.7 W/m K at 28 days. It was observed that the density of the LWC crucially affects the strength of the concrete for similar cement and water content. The thermal conductivity of LWC is significantly influenced by various factors such as cement content, water content, and the type and content of LWA [29, 30]. LWC bricks were manufactured using rice husk ash (RHA) and expanded polystyrene as LWAs, and the maximum cement replacement by RHA was 10% by weight [31]. The effect of zeolite inclusion for autoclaved concrete was investigated by using aluminum to introduce a pore-forming agent, where zeolite was used with a total content of 535 kg/m^3. The highest compressive strength of 3.3 MPa was attained with 50% replacement, and the thermal conductivity was 0.18 W/(m K). In another study, autoclaved concrete was produced using bottom-ash as the LWA and as fractional replacement of cement, where the bulk density was ≈1400 kg/m^3 with the increase in compressive strength ranging from 9 to 11.6 MPa, and thermal conductivity varying from 0.5 to 0.61 W/(m K) [32]. Figure 1b shows the correlation between thermal conductivity and dry density of LWAC.

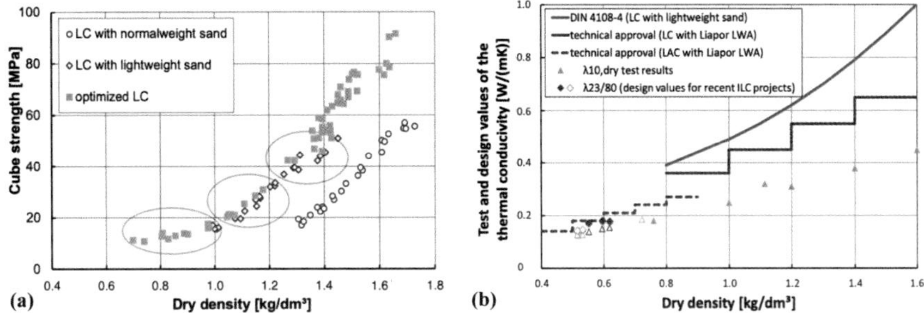

Fig. 1 Correlation between **a** cube strength and dry density for lightweight concretes with different compositions at 28 days [48], and **b** thermal conductivity and dry density for LWC [49]

5 Possibilities

The compilation of previous research gives a clear indication that the thermal and mechanical behaviors of LWAC have been extensively developed and investigated. However, there have been few studies related to the structural applications of LWAC. We suggest widening the scope of LWAC for investigation of its structural application and behaviors, so complete guidelines can be provided to design engineers, assisting them with all required design aids and data. Further studies, especially on the mechanics of lightweight materials, would establish confidence and trust in the potential applications of LWAC as a structural element. Another limitation (i.e., the higher time-dependent deformations of LWAC) requires further investigation. Figure 2 shows some of the superstructures that have been built with LWAs.

Fig. 2 **a** LWC: Schweiz 2003 (Gartmann house), with density 1,100 kg/m^3, strength 12.9 MPa, and thermal conductivity 0.32 W/m K [50], **b** Infra LWC: Berlin 2007 (Schlaich house) with density 760 kg/m^3, strength 7.4 MPa, and thermal conductivity 0.18 W/m K [51], **c** LWC: Stuttgart 2012 (house H36) with density 1,000 kg/m^3, strength 10.9 MPa, and thermal conductivity 0.23 W/m K [50], **d** Infra LWC: TU Eindhoven 2015 (Pavilion) with density 780 kg/m^3, strength 10 MPa, and thermal conductivity 0.13 W/m K [50]

6 Conclusions

LWAC is an exceptionally versatile material that can be implemented in a broad range of applications. Though LWAC has been used for the past two millennia, there are still some limitations that have been reported and discussed in this review. Clear and straightforward definitions have been given for different types of LWAC, and a critical analysis of the mix design, aggregates properties, testing, and classifications of LWC was presented. Several limitations of LWAC have been discussed and presented as crucial information on the state of the art. Infra LWC is a new research area to be explored. We recommend further essential studies to develop a strong design and reliable construction methodology to confirm product quality.

Acknowledgements The authors acknowledge RMIT University, Melbourne, Australia, for financial assistance received through RTP scholarships.

References

1. Chandra S, Berntsson L (2002) Lightweight aggregate concrete. Elsevier
2. Islam MMU et al (2016) Mechanical and fresh properties of sustainable oil palm shell lightweight concrete incorporating palm oil fuel ash. J Clean Prod 115:307–314
3. Jumaat M, Bashar I (2015) Usage of palm oil industrial wastes as construction materials—a review. 5:44–51
4. Islam MMU et al (2022) Design and strength optimization method for the production of structural lightweight concrete: an experimental investigation for the complete replacement of conventional coarse aggregates by waste rubber particles. Resour Conserv Recycl 184:106390
5. Islam MMU, Li J (2022) Pre-treatments for Rubberised Concrete (RuC)—a review
6. Islam MMU et al (2023) Investigation of durability properties for structural lightweight concrete with discarded vehicle tire rubbers: a study for the complete replacement of conventional coarse aggregates. Constr Build Mater 369:130634
7. Islam MMU et al (2016) Durability properties of sustainable concrete containing high volume palm oil waste materials. J Clean Prod 137:167–177
8. Islam MMU Sustainable Approaches Towards Renewing Discarded Vehicle Tire Rubbers.
9. Islam MMU, Alengaram UJ, Jumaat MZ (2021) Fresh and hardened properties of palm oil fuel ash (POFA) based lightweight concrete from palm oil industrial wastes. In: Proceedings of international conference on planning, architecture and civil engineering, 09–11 September 2021. Rajshahi University of Engineering and Technology, Rajshahi, Bangladesh
10. Wang G et al (2022) Use of COVID-19 single-use face masks to improve the rutting resistance of asphalt pavement. Sci Total Environ 826:154118
11. Islam MMU, Alengaram UJ, Jumaat MZ (2022) Compressive strength and efficiency factor of green concrete
12. Saberian M et al (2021) Large-scale direct shear testing of waste crushed rock reinforced with waste rubber as pavement base/subbase materials. Transp Geotech 28:100546
13. Marie I (2017) Thermal conductivity of hybrid recycled aggregate—rubberized concrete. Constr Build Mater 133:516–524
14. ASTM C331M-17 (2017) Standard specification for lightweight aggregates for concrete masonry units. ASTM International, West Conshohocken, PA. www.astm.org

15. Islam, M. M. U. (2019). Investigation of tensile creep and tension stiffening behaviour for Ultra-High-Performance Fiber Reinforced Concrete (UHPFRC) (Doctoral dissertation). doi: https://hdl.handle.net/2440/120660
16. Islam MMU, et al (2022) A comprehensive review on the application of renewable waste tire rubbers and fibers in sustainable concrete. J Clean Prod
17. Islam MMU (2022) Investigation of long-term tension stiffening mechanism for ultra-high-performance fiber reinforced concrete (UHPFRC). Constr Build Mater 321:126310
18. Hosen MA et al (2023) Potential side-NSM strengthening approach to enhance the flexural performance of RC beams: experimental, numerical and analytical investigations. Struct Eng Mech 85(2):179–195
19. Islam MMU (2021) Investigation of tensile creep for ultra-high-performance fiber reinforced concrete (UHPFRC) for the long-term. Constr Build Mater 305:124752
20. Bogas JA, Gomes MG, Real S (2014) Bonding of steel reinforcement in structural expanded clay lightweight aggregate concrete: the influence of failure mechanism and concrete composition. Constr Build Mater 65:350–359
21. Kim D-J et al (2013) Bond strength of steel deformed rebars embedded in artificial lightweight aggregate concrete. J Adhes Sci Technol 27(5–6):490–507
22. Tang C-W (2015) Local bond stress-slip behavior of reinforcing bars embedded in lightweight aggregate concrete. Comput Concr 16(3):449–466
23. Loudon A (1979) The thermal properties of lightweight concretes. Int J Cem Compos Light Concr 1(2):71–85
24. Islam MMU (2023) A study on the integrated implementation of supplementary cementitious material and coarse aggregate for sustainable concrete. J Build Eng 106767
25. Alduaij J et al (1999) Lightweight concrete in hot coastal areas. Cem Concr Compos 21(5–6):453–458
26. Demirboğa R, Gül R (2003) Thermal conductivity and compressive strength of expanded perlite aggregate concrete with mineral admixtures. Energy Build 35(11):1155–1159
27. Wang H, Tsai K (2006) Engineering properties of lightweight aggregate concrete made from dredged silt. Cem Concr Compos 28(5):481–485
28. Siddique M et al (2021) A mini-review on dark-photo fermentation. Int J Environ Monit Anal 9(6):190–192
29. Islam MMU (2015) Feasibility study of ground palm oil fuel ash as partial cement replacement material in oil palm shell lightweight concrete. Jabatan Kejuruteraan Awam, Fakulti Kejuruteraan, Universiti Malaya
30. Mohammad Momeen UI (2015) Feasibility study of ground palm oil fuel ash as partial cement replacement material in oil palm shell lightweight concrete. University of Malaya
31. Ling I, Teo D (2011) Properties of EPS RHA lightweight concrete bricks under different curing conditions. Constr Build Mater 25(8):3648–3655
32. Wongkeo W et al (2012) Compressive strength, flexural strength and thermal conductivity of autoclaved concrete block made using bottom ash as cement replacement materials. Mater Des 35:434–439
33. Padhi RS et al (2018) Influence of incorporation of rice husk ash and coarse recycled concrete aggregates on properties of concrete. Constr Build Mater 173:289–297
34. Panda KC, Behera S, Jena S (2020) Effect of rice husk ash on mechanical properties of concrete containing crushed seashell as fine aggregate. Mater Today: Proc 32:838–843
35. Ferraro RM, Nanni A (2012) Effect of off-white rice husk ash on strength, porosity, conductivity and corrosion resistance of white concrete. Constr Build Mater 31:220–225
36. Das SK et al (2022) Sustainable utilization of ultrafine rice husk ash in alkali activated concrete: characterization and performance evaluation. J Sustain Cem-Based Mater 11(2):142–160
37. Mohamad N, et al (2018) Effects of incorporating banana skin powder (BSP) and palm oil fuel ash (POFA) on mechanical properties of lightweight foamed concrete. Int J Integr Eng 10(9)
38. Mohammadhosseini H, Abdul Awal A, Sam ARM (2016) Mechanical and thermal properties of prepacked aggregate concrete incorporating palm oil fuel ash. Sādhanā 41(10):1235–1244

39. Alengaram UJ et al (2013) A comparison of the thermal conductivity of oil palm shell foamed concrete with conventional materials. Mater Des 51:522–529
40. Shafigh P et al (2014) A comparison study of the mechanical properties and drying shrinkage of oil palm shell and expanded clay lightweight aggregate concretes. Mater Des 60:320–327
41. Noaman AT et al (2017) Fracture characteristics of plain and steel fibre reinforced rubberized concrete. Constr Build Mater 152:414–423
42. Alwesabi EA et al (2020) Experimental investigation on mechanical properties of plain and rubberised concretes with steel–polypropylene hybrid fibre. Constr Build Mater 233:117194
43. Fu C et al (2019) Evolution of mechanical properties of steel fiber-reinforced rubberized concrete (FR-RC). Compos B Eng 160:158–166
44. Chen A et al (2020) Mechanical and stress-strain behavior of basalt fiber reinforced rubberized recycled coarse aggregate concrete. Constr Build Mater 260:119888
45. Park SB, Lee BC, Kim JH (2004) Studies on mechanical properties of concrete containing waste glass aggregate. Cem Concr Res 34(12):2181–2189
46. Sikora P et al (2017) Thermal properties of cement mortars containing waste glass aggregate and nanosilica. Procedia Eng 196:159–166
47. Jerlin Regin J, Vincent P, Ganapathy C (2017) Effect of mineral admixtures on mechanical properties and chemical resistance of lightweight coconut shell concrete. Arab J Sci Eng 42(3):957–971
48. Thienel K, Peck M (2007) Die Renaissance leichter Betone in der Architektur. DETAIL-MUNCHEN- 47(5):522
49. Thienel K-C, Haller T, Beuntner N (2020) Lightweight concrete—from basics to innovations. Materials 13(5):1120
50. Donners M (2015) Warmbeton: De invloed van hydratatiewarmte op de toepassing in de praktijk. Technische Universiteit Eindhoven
51. Walraven JC, Stoelhorst D (2008) Tailor made concrete structures: new solutions for our society. CRC Press (Abstracts Book 314 pages + CD-ROM full papers 1196 pages)

Influence of Reinforcement on the Loading Capacity of Geopolymer Concrete Pipe

S. Dangol, J. Li, V. Sirivivatnanon, and P. Kidd

Abstract Geopolymer concrete is emerging as a sustainable construction material due to utilization of industrial by-products, which greatly reduces its carbon footprint. Past studies of the mechanical properties and resistance to sulfuric acid reaction of cement-less geopolymer concrete indicated its suitability for precast concrete pipes over ordinary Portland cement (OPC) concrete. In the present study, a three-dimensional finite element (FE) model of reinforced concrete pipe was developed using commercial software ANSYS-LSDYNA. The load-carrying capacity of reinforced and non-reinforced geopolymer concrete pipes under the three-edge bearing (TEB) test was investigated and compared with OPC concrete pipes. The results indicated geopolymer concrete with comparable compressive strength to OPC concrete showed higher loading capacity in a pipe structure due to its better tensile performance. The effect of steel reinforcement area on the loading capacity of geopolymer concrete pipes was quantitatively analyzed, and they met the specified strength requirement for OPC concrete in the ASTM standard, with up to 20% reduction in the reinforcement area.

Keywords Geopolymer concrete pipe · Pipe loading capacity · Numerical modelling

1 Introduction

As an integral part of civil infrastructure, concrete pipes are used as conduit for sewage and storm water. Ordinary Portland cement (OPC) concrete pipes have demonstrated reliable long-term performance over years of usage. Their structural

S. Dangol · J. Li (✉) · V. Sirivivatnanon
School of Civil and Environmental Engineering, University of Technology Sydney, Sydney, NSW, Australia
e-mail: jun.li-2@uts.edu.au

P. Kidd
Cement Australia Pty. Ltd., Brisbane, QLD, Australia

W. Duan et al. (eds.), *Nanotechnology in Construction for Circular Economy*, Lecture Notes in Civil Engineering 356,
https://doi.org/10.1007/978-981-99-3330-3_18

performance is evaluated in terms of test load (D_{peak}) corresponding to load causing a 0.3 mm crack and by ultimate load (D_{ult}) corresponding to the load supported before failure of the pipe. The three-edge bearing (TEB) test is a standardized test described in AS/NZS-4058 [1] and ASTM-C76M [2] to examine the mechanical strength of a pipe, wherein a line load is applied to the crown of the pipe while the base of the pipe is supported by two bearers. However, carrying out such destructive tests is uneconomical and often inefficient, considering the need for human judgement of crack formation during the TEB test. Hence, numerous researchers have performed numerical modelling of concrete pipes to investigate the load-carrying capacity and load–deflection behavior of concrete pipes.

de la Fuente et al. [3] and de Figueiredo et al. [4] simulated the TEB test for steel fiber-reinforced concrete (SFRC) pipe using a MAP (mechanical analysis of pipes) model to study mechanical behavior. The results from the numerical simulation were compared with experimental results, concluding the efficiency of the numerical model to design fiber-reinforced concrete pipes because the model gave an average error of 7%, which was within the anticipated contingency. Similarly, Ferrado et al. [5] used the commercial software ABAQUS to simulate SFRC pipes. In their study, the behavior of the pipes was defined by compression and the uniaxial tension curve based on theoretical formulation in the existing literature. The load–deflection curve and stress distribution of the pipes from experimental and numerical analysis were found to be in good agreement. Likewise, a numerical modelling of concrete pipes with different diameter and reinforcement configuration, was conducted by Younis et al. [6] to predict the service load and ultimate load. Following the concrete damage plasticity (CDP) model equation developed by Alfarah et al. [7], the non-linear behavior of concrete in compression and tension was defined. Based on their analysis and experimental results, the average prediction error was $\approx$6% for both service load and ultimate load, suggesting the reliability of numerical modelling for designing concrete pipes.

With the growing interest in sustainable construction materials, numerous studies of geopolymer concrete have been carried out, because it utilizes industrial by-product as its source material. Geopolymer is a cement-less binder formed as a result of reaction between aluminosilicate compounds with alkali [8–10]. Studies exploring the material properties of geopolymer concrete have shown significant development of strength at early age when cured at elevated temperature, high compressive strength, high flexural strength, and resistance to chemical attack [8, 11, 12]. Results of studies conducted to investigate indirect tensile strength and flexural strength of geopolymer have also showed higher indirect tensile strength and flexural strength in comparison with OPC concrete of the same compressive strength [13–16]. It is reported that the geopolymer possessed 1.4- and 1.6-fold higher value for indirect tensile strength and flexural strength, respectively, compared with OPC concrete [17]. Such properties of geopolymer concrete can be used to test the strength of pipes with reduced reinforcement bars. Although much efforts has been made to study the structural behavior of OPC SFRC to enhance the load-carrying capacity, study of the structural performance of geopolymer concrete pipes has not been carried out.

We investigated the load-carrying capacity of geopolymer concrete pipes based on a three-dimensional (3D) finite element (FE) model developed to simulate the TEB test. For this study, fly ash (FA)/slag-based powder form geopolymer Geocem^{TM} developed by Cement Australia with different ratio of FA and slag was utilized. The general purpose geopolymer binder we termed "geocem 1" contained 50% of FA and 32% of ground granulated blast-furnace slag (GGBFS), and the high strength geopolymer binder termed "geocem 2" comprised 30% FA and 50% GGBFS. The developed FE model was updated for both types of geopolymer concrete based on their mechanical properties. Subsequently, the FE model was used to evaluate the load-carrying capacity of the geopolymer pipes and the effect of reduced reinforcement area was evaluated.

2 FE Modelling of TEB Test

Our 3D FE model to simulate the TEB test for the pipe was based on commercial software ANSYS LS-DYNA (Fig. 1). The model comprised three components: concrete part, reinforcement steel bars, and bearing strips. Pipe of diameter 450 mm, length 1000 mm, and wall thickness of 42 mm were modelled. The concrete pipe and bearing strips were modelled using a 3D solid element (SOLID164). Similarly, a beam element (BEAM161) was used to model the reinforcing steel bars. Discrete steel formulation was used and perfect bond condition between the reinforcement bar and concrete was assumed. The bearing strips were modelled to mimic the boundaries in the TEB test: the lower bearing strips were fixed at the bottom to prevent translational and rotational degrees of freedom, and the upper bearing strips were restricted in all directions except for vertical displacement movement to allow for displacement-controlled loading on the pipe. The interaction between the pipe and the bearing strips was defined by an automatic contact surface. For the simulation of the test, displacement-controlled loading was defined as applied downward displacement on the upper bearer. The load–deflection curve was obtained for the analyzed pipe and is presented in terms of design load (N/m/mm) as specified in ASTM-C76M [2] and deflection in millimeters.

3 Material Modelling

Due to the complex material behavior of concrete, which includes elastic, non-linear plastic behavior, and material damage, the available concrete damage models for numerical modelling of concrete structures are often quite complex because these material models often contain parameters for which values are difficult to obtain from simple tests or have only mathematical meaning and no physical meaning [18]. To date, there are a lot of material models available to simulate concrete damage behavior [19, 20]. Among them, one simple concrete damage model implemented in

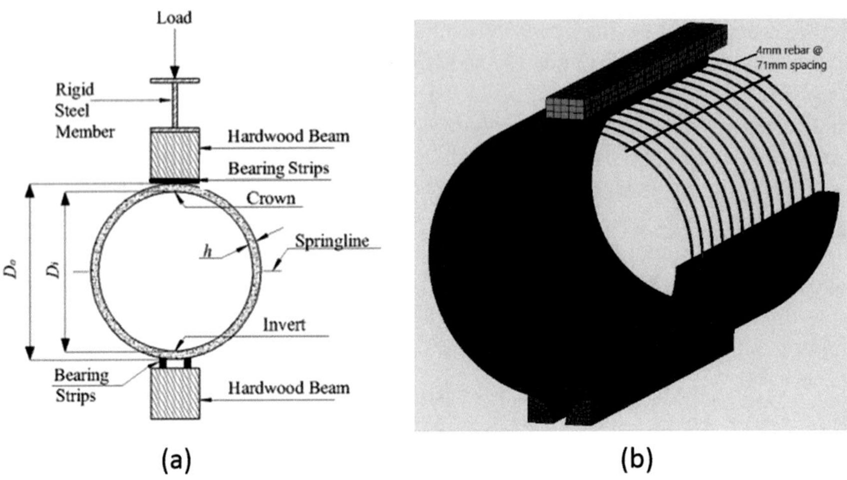

Fig. 1 **a** Three-edge bearing (TEB) test setup; **b** finite element model of concrete pipe for TEB test simulation

LS-DYNA to model concrete behavior is the Karagozian and Case (K&C) concrete model (Fig. 2). A key merit of the K&C concrete model for numerical simulation of concrete behavior is its reliance on just one main input parameter of unconfined compressive strength. Schwer & Malvar [21] stated that the K&C concrete model can be utilized for analysis involving new concrete materials with no detailed information available to characterize the concrete beside its compressive strength, owing to the fact that the unconfined compressive strength of the concrete not only describes the elastic response, but also accounts for the plastic response including shear failure, compression, and tensile failure.

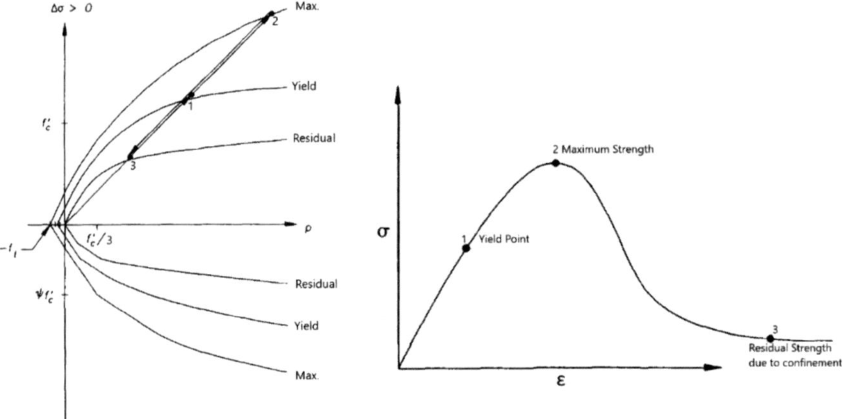

Fig. 2 Three failure surface of K&C concrete model [22]

The material constitutive behavior of the K&C concrete model comprises three parts; for initial loading, the stress is elastic until it reaches the yielding point, after which it increases further till the limit surface, called the maximum yield surface. Following the maximum yield surface, perfectly plastic, or softening behavior up to the residual yield surface is observed. These shear failure surfaces are mutually independent and can be formulated as [22, 23]:

$$F_i(p) = a_{0i} + \frac{p}{a_{1i} + a_{2i}p} \tag{1}$$

where i stands for either yield strength surface (y), maximum strength surface (m) or residual strength surface (r), p is the pressure calculated as $\frac{-I_1}{3}$, and the variables a_{ji} ($j = 0, 1, 2$) are the parameters calibrated from test data.

The resulting failure surface is interpolated between the maximum strength surface and either the yield surface or the residual strength surface as per the following equations:

$$F(I_1, J_2, J_3) = r(J_3)\left[\eta(\lambda)\left(F_m(p) - F_y(p)\right) + F_y(p)\right] \text{ for } \lambda \le \lambda_m \tag{2}$$

$$= r(J_3)[\eta(\lambda)(F_m(p) - F_r(p)) + F_r(p)] \text{ for } \lambda \ge \lambda_\text{m} \tag{3}$$

where I_1, J_2, and J_3 are the first, second, and third invariants of deviatoric stress tensor, λ is the modified effective plastic strain or the internal damage parameter, $\eta(\lambda)$ is the function of the internal damage parameter λ, with $\eta(0) = 0$, $\eta(\lambda_m) = 1$ and $\eta(\lambda \ge \lambda_m) = 0$, and $r(J_3)$ is the scale factor in the form of the William–Warnke equation [24].

The K&C concrete model considers the effect of strain rate, failure, and different mechanical–physical properties in compression and tension and hence is suitable for concrete modelling [18]. Based on the uniaxial compressive strength, material parameters are generated, requiring definition of only a few parameters for the functionality of the material model, and more parameters can be defined if required. The model requires 49 parameters to be defined, as well as equation of state, which is complicated because many parameters have only mathematical meaning. Hence, the developers advocate using parameter generation if the data to define the material are not available. The default parameters in the K&C concrete model were calibrated using uniaxial, biaxial, and tri-axial test data available for well characterized concrete and using the relationship such as tensile strength or modulus of elastic as the function of compressive strength [21]. Hence, the K&C concrete model was used for both OPC concrete and geopolymer concrete modelling for FE analysis (FEA).

For the reinforcement bar, an elastic–plastic constitutive relationship for reinforcement bar, with or without strain hardening, is commonly adopted for numerical analysis. However, the elastic-perfectly plastic assumption shown in Fig. 3a often fails to capture the steel stress at high strain, and accurate assessment of the strength

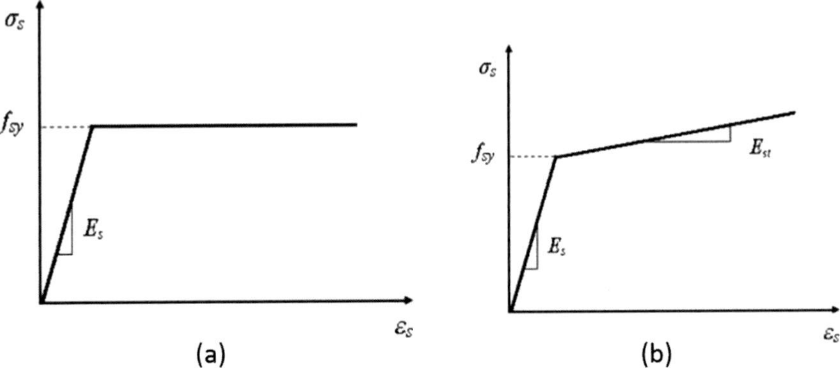

Fig. 3 Typical stress–strain curve of steel reinforcement representing **a** ideal elastic-perfectly plastic model; **b** bilinear elastic–plastic model with linear strain hardening [25]

of structure at large deformation cannot be made [26]. Hence, more accurate idealization of the stress–strain curve as shown in Fig. 3b was used. The Piecewise Linear Plasticity model used to represent the steel reinforcement behavior in LS-DYNA considers the plastic deformation, strain rate effects and failure [19]. In the Piecewise Linear Plasticity model, the stress–strain curve for the reinforcing steel is treated as bilinear by defining the tangent modulus [27]. The steel response is thus defined by parameters such as Young's modulus (E_s), yield strength (f_{sy}) and hardening modulus (E_{st}). The magnitude of E_{st} in the plastic regimen is commonly set at 1% of E_s [28, 29].

4 Load–Deflection Behavior of Concrete Pipes

A concrete pipe model was used for our study of the load–deflection behavior of geopolymer concrete pipes with respect to OPC concrete pipes. For the design of 450 mm reinforced concrete pipe, a reinforcement area of 175 mm^2/m was adopted based on minimum reinforcement area criteria for class II 450 mm concrete pipe defined in ASTM-C76M [2] in order to meet the design load criteria. Table 1 provides the details of the mechanical properties of the materials for the study based on the experimental results.

Comparing the results obtained from numerical analysis with the design requirement specified for OPC concrete of 50 N/m/mm and 75 N/m/mm for peak and ultimate load respectively in ASTM-C76M [2] with the geocem 1 and geocem 2 FEA results, it was evident from the load–deflection curve shown in Fig. 4 that the geopolymer concrete exhibited better load-carrying capacity than OPC concrete. The peak load and ultimate load value of the OPC pipe were 85 N/m/mm and 113 N/m/mm respectively, and for the geocem 1 and geocem 2 pipes, the peak load value was 100 N/m/mm and 105 N/m/mm and the ultimate load value was observed to be 119 N/m/

Table 1 Material properties of the pipe model

Material property	Geocem 1	Geocem 2	OPC
Compressive strength (MPa)	50	50	50
Tensile strength (MPa)	3.6	3.8	2.3
Poisson's ratio	0.2	0.2	0.2
Rebar yielding stress (MPa)	500	500	500

OPC, ordinary Portland cement

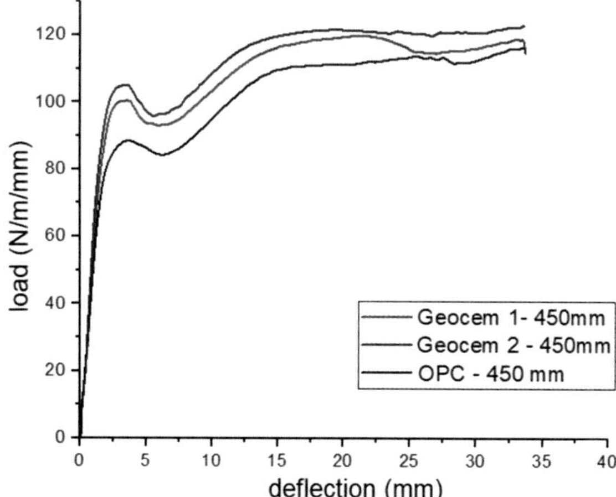

Fig. 4 Load–deflection curves of the 450 mm pipes

mm and 121 N/m/mm respectively. The load-carrying capacity of both geopolymer concretes outperformed the OPC concrete load requirement by ≈15% for peak load capacity and ≈5% for ultimate load capacity. The high tensile strength of geopolymer concrete benefited the load-bearing capacity of the pipe.

In the load–deflection plot of the pipe, a drop in the load was noticed after the pipe reached its peak load capacity. Such a drop in loading capacity for a single-cage model was also observed by Tehrani [30], Peyvandi et al. [31], and Younis et al. [6]. The rise in the load capacity following the drop after peak load signifies the load stress being carried by the steel reinforcement.

5 Effect of Change in Reinforcement Area

To investigate the effect of steel reinforcement area on the load–deflection behavior of the reinforced OPC, geocem 1 and geocem 2 concrete pipes, the steel reinforcement was reduced by 20%, 40% and 50% from the total reinforcement area. The main objective was to the test the load capacity in geopolymer concrete pipes under reduced reinforcement and evaluate against the strength requirement as specified in ASTM-C76M [2].

Figure 5 shows the load–deflection plot of 450 mm pipes under reduced reinforcement conditions. It is obvious for both geopolymer concrete types (Fig. 5a, b) that the load-carrying capacity of the pipes decreased with the reduction in the reinforcement area, especially the ultimate load-carrying capacity. As the development of crack in concrete structures is dependent on the tensile strength of the concrete, changing the steel reinforcement does not change the service load capacity of the pipe, but alters the ultimate load capacity. When sufficient tensile stress develops in the concrete surface causing the concrete to crack, the pipe loses its capacity, which is marked by the drop in the load capacity. As the tensile stress in the concrete pipe is transferred to the steel reinforcement, regaining load capacity is observed until it reaches its ultimate load capacity, after which the pipe fails. With the reduction of reinforcement area by 40 and 50%, a significant reduction in the post-crack loading capacity of the pipe was observed; however, the reduction of reinforcement area by 20% showed decrease ultimate load capacity by 7% and 6% for geocem 1 and geocem 2, respectively, but still satisfied the design requirement specified in the ASTM standard. The strength gained by fully reinforced OPC concrete pipe was approximately equivalent to the strength gained by geopolymer concrete pipes with 20% reduced reinforcement area. This result suggested the geopolymer concrete pipes could resist load without failure with up to 20% reduction in the area of steel reinforcement.

Furthermore, in the comparative study of the load–deflection behavior of unreinforced OPC concrete pipe against unreinforced geocem 1 concrete pipe shown in Fig. 5c, the test load capacity of the unreinforced OPC concrete pipe was 69 N/m/mm and the test load capacity of geocem 1 was 96 N/m/mm. AS/NZS-4058 specifies the test load for 450 mm unreinforced concrete pipe as 30 KN/m (or 67 N/m/mm) [1]. The load capacity of the geocem 1 pipe was 37% higher than that of the OPC concrete pipe. Moreover, considering the AS/NZS-4058 requirement of 67 N/m/mm as the test load and 100 N/m/mm as the ultimate load capacity for class 3 reinforced 450 mm concrete pipe [1], unreinforced geopolymer concrete pipe can be used to meet the design requirement of class 3 reinforced concrete pipe.

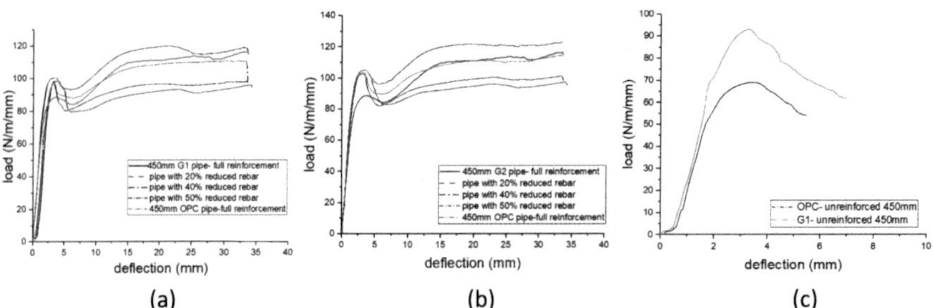

Fig. 5 a–c Effect on load–deflection behavior of 450 mm pipes due to change in reinforcement steel area

6 Conclusion

We developed a 3D FE model of 450 mm concrete pipe to simulate the TEB test in LS-DYNA. The K&C concrete model was used to characterize the OPC and geopolymer concrete behavior, and a bilinear elastic–plastic model with linear strain hardening was used to model steel reinforcement in the pipe FE model. In the comparative analysis of geopolymer and OPC concrete pipes, the peak load and ultimate load capacity exhibited by both types of geopolymer concrete pipes were noticeably higher than by OPC concrete pipe, which can be attributed to the high tensile strength of geopolymer concrete. The FE model was used to study the effect of changing the reinforcement area on the load-carrying capacity of geopolymer and OPC concrete pipes. The change in steel reinforcement area affected the post-crack behavior of the pipes, because reinforcement mostly contributes to strength development in cracked concrete sections. Based on the numerical study, the reinforcement bar area in concrete pipes can be reduced by 20% and still meet the specified design load criteria. Further, comparing the loading capacity of unreinforced geopolymer concrete pipe against OPC concrete pipe revealed that the unreinforced geopolymer concrete pipe can satisfy the load criteria for AS/NZS class 3 450 mm reinforced concrete pipe. Thus, with the use of geopolymer concrete, reinforcement requirement for concrete pipe can be reduced to a certain percentage and still meet the specified design load criteria.

Acknowledgements This research was funded through an Australian Research Council Research Hub for Nanoscience Based Construction Materials Manufacturing (NANOCOMM) with the support of Cement Australia. The authors are grateful for the financial support of the Australian Research Council (IH150100006) in conducting this study.

References

1. AS/NZS-4058 (2007) Precast concrete pipes (pressure and non-pressure). Standards Australia Sydney, NSW
2. ASTM-C76M (2019) Standard specification for reinforced concrete culvert, storm drain, and sewer pipe (Metric) 1
3. de la Fuente A, de Figueiredo AD, Aguado A, Molins C, Neto PJC (2011) Experimentation and numerical simulation of steel fibre reinforced concrete pipes. Mater Constr 61(302):275–288
4. de Figueiredo A, Aguado A, Molins C, Chama Neto PJ (2012) Steel fibre reinforced concrete pipes. Part 2: numerical model to simulate the crushing test. Revista IBRACON de Estruturas e Materiais 5(1):12–25
5. Ferrado FL, Escalante MR, Rougier VC (2018) Simulation of the three edge bearing test: 3D model for the study of the strength capacity of SFRC pipes. Mecánica Computacional 36(6):195–204
6. Younis A-A, Shehata A, Ramadan A, Wong LS, Nehdi ML (2021) Modeling structural behavior of reinforced-concrete pipe with single, double and triple cage reinforcement. Eng Struct 240:112374
7. Alfarah B, López-Almansa F, Oller S (2017) New methodology for calculating damage variables evolution in plastic damage model for RC structures. Eng Struct 132:70–86

8. Davidovits J (1991) Geopolymers: inorganic polymeric new materials. J Therm Anal Calorim 37(8):1633–1656
9. Habert G, De Lacaillerie JDE, Roussel N (2011) An environmental evaluation of geopolymer based concrete production: reviewing current research trends. J Clean Prod 19(11):1229–1238
10. Provis JL, van Deventer JSJ (2014) Alkali activated materials. Springer, Dordrecht
11. Abdullah M, Hussin K, Bnhussain M, Ismail K, Ibrahim W (2011) Mechanism and chemical reaction of fly ash geopolymer cement-A review. Int J Pure Appl Sci Technol 6(1):35–44
12. Wallah S, Rangan BV (2006) Low-calcium fly ash-based geopolymer concrete: long-term properties
13. Hardjito D, Rangan BV (2005) Development and properties of low-calcium fly ash-based geopolymer concrete
14. Neupane K, Chalmers D, Kidd P (2018) High-strength geopolymer concrete—properties, advantages and challenges. Adv Mater 7:15–25
15. Ramujee K, PothaRaju M (2017) Mechanical properties of geopolymer concrete composites. Mater Today: Proc 4(2, Part A):2937–45
16. Sofi M, van Deventer JSJ, Mendis PA, Lukey GC (2007) Engineering properties of inorganic polymer concretes (IPCs). Cem Concr Res 37(2):251–257
17. Raijiwala D, Patil H (2011) Geopolymer concrete: a concrete of the next decade. Concrete solutions, vol 287
18. Kral P, Hradil P, Kala J, Hokes F, Husek M (2017) Identification of the parameters of a concrete damage material model. Procedia Eng 172:578–585
19. Abedini M, Zhang C (2021) Performance assessment of concrete and steel material models in LS-DYNA for enhanced numerical simulation, a state of the art review. Arch Comput Methods Eng 28(4):2921–2942
20. Xu M, Wille K (2015) Calibration of K&C concrete model for UHPC in LS-DYNA. Adv Mater Res (Trans Tech Publ) 1081:254–259
21. Schwer LE, Malvar LJ (2005) Simplified concrete modeling with* MAT_CONCRETE_DAMAGE_REL3. JRI LS-Dyna User Week, pp 49–60
22. Malvar LJ, Crawford JE, Wesevich JW, Simons D (1997) A plasticity concrete material model for DYNA3D. Int J Impact Eng 19(9–10):847–873
23. Wu Y, Crawford JE, Magallanes JM (2012) Performance of LS-DYNA concrete constitutive models. In: 12th international LS-DYNA users conference, vol 1, pp 1–14
24. Chen W-F, Han D-J (1988) Plasticity for structural engineers. Springer-Verlag, New York
25. Du X, Jin L (2021) Methodology: meso-scale simulation approach. In: Size effect in concrete materials and structures, Springer, pp 27–76
26. Supaviriyakit T, Pornpongsaroj P, Pimanmas A (2004) Finite element analysis of FRP-strengthened RC beams. Songklanakarin J Sci Technol 26(4):497–507
27. LSTC (2014) LS-DYNA keyword user's manual: material models. In: LS-DYNA R7.1 edn, vol II. Livermore Software Technology Corporation, Livermore, CA
28. Elnashai AS, Izzuddin BA (1993) Modelling of material non-linearities in steel structures subjected to transient dynamic loading. Earthquake Eng Struct Dynam 22(6):509–532
29. Xiao S (2015) Numerical study of dynamic behaviour of RC beams under cyclic loading with different loading rates. Mag Concr Res 67(7):325–334
30. Tehrani AD (2016) Finite element analysis for ASTM C-76 reinforced concrete pipes with reduced steel cage
31. Peyvandi A, Soroushian P, Jahangirnejad S (2013) Enhancement of the structural efficiency and performance of concrete pipes through fiber reinforcement. Constr Build Mater 45:36–44

Creep of Slag Blended Cement Concrete with and Without Activator

H. T. Thanh, M. J. Tapas, J. Chandler, and V. Sirivivatnanon

Abstract Partly replacing Portland cement (PC) with lower carbon footprint cementitious materials such as ground granulated blast furnace slag (slag) is considered as a practical method for reducing CO_2 emissions in the cement concrete industry. To mitigate the slow reactivity of slag in a cementitious system and enhance early-age strength, the addition of a chemical activator is a solution. However, the effect of the activator on creep behaviour of slag-blended cement concretes remains unclear. This work presents the effect of sodium sulfate (Na_2SO_4) activator on the compressive creep of PC concrete blended with 50 and 70 wt% slag. Four concrete mixes (with and without 2.5% Na_2SO_4 activator) containing 395 kg of cementitious material were prepared. The creep strain measurements were conducted on 150 $\times$ 300 mm cylindrical specimens for 140 days under sustained compressive load. The results showed that the 70% slag concrete had lower creep strain than 50% slag-blended cement concrete. The presence of Na_2SO_4 helped reduce the creep strain of 50% slag concrete but slightly increased that of 70% slag-blended cement concrete. In addition, the applicability of the predictive model in AS3600:2018 for the creep behaviour of high slag content concrete was assessed.

Keywords Activator · Compressive creep · Slag concrete

1 Introduction

Cement-based materials have an essential role in civil infrastructure worldwide. However, the production of clinker, the major constituent of Portland cement (PC), accounts for $\approx$7% of global CO_2 emissions [1], which has a destructive impact on

H. T. Thanh (✉) · M. J. Tapas · V. Sirivivatnanon
School of Civil and Environmental Engineering, University of Technology Sydney (UTS), Ultimo, NSW, Australia
e-mail: Hai.Tran@uts.edu.au

J. Chandler
Boral Limited, North Ryde, NSW, Australia

W. Duan et al. (eds.), *Nanotechnology in Construction for Circular Economy*, Lecture Notes in Civil Engineering 356,
https://doi.org/10.1007/978-981-99-3330-3_19

environmental sustainability. Reducing the clinker to cement ratio by partly replacing PC with lower carbon footprint cementitious materials such as ground granulated blast furnace slag (GGBSF or slag) is considered a practical technique for reducing CO_2 emissions in the cement industry sector. The direct cumulative CO_2 emissions reduction by this method is estimated to reach almost 37% by 2050 [2]. Using slag in cement concrete also enhances its compressive and flexural strengths at later ages and improves the durability properties of concrete [3, 4].

As a pozzolanic-hydraulic activity or latent hydraulic behaviour, slag only provides additional surface area to enhance PC hydration by filler effect in the early days of hydration, without chemical reaction [5]. Hence, at high levels of slag replacement, the slow reactivity and dissolution of slag in the cementitious system causes slower strength gain at early ages. To mitigate this, adding an activator such as sodium sulfate (Na_2SO_4) is adopted [6, 7]. This addition accelerates slag's reactivity and enhances the early strength. However, in the presence of Na_2SO_4, a decrease in the compressive strength growth rate and increased capillary porosity at a later age have been reported [8, 9].

Creep of concrete is a time-dependent property that significantly affects the performance of concrete structures. Generally, creep in concrete includes basic and drying creeps. Under constant stress, gel particles slide and consolidate as water molecules rearrange in capillary pore structures, and microcracks form in the interfacial transition zones between the aggregate and cement paste. These phenomena primarily govern the basic creep of concrete [10]. In terms of creep of cement concrete using slag as a cement replacement, only a few conducted studies are reported. Khatri et al. [11] and Shariq et al. [10] concluded that creep of slag-blended cement concrete increased when the slag replacement increased. In addition, the creep strains were higher in slag cement concrete compared with plain concrete [10]. However, the creep behaviour of high slag-blended cement concretes with the presence of an activator has not yet been reported.

To understand the effect of Na_2SO_4 activator on the creep behaviour of slag-blended cement concrete, we conducted experiments on compressive creep. In this study, four mixes were cast with 50 and 70 wt% slag replacement. A dose of 2.5 wt% by total binder of Na_2SO_4 activator was used for activating two of the four slag-blended cement concrete mixes. The measurement data were analyzed and compared with the obtained values from the predictive model in AS3600:2018 to assess the capability of model prediction for creep behavior of high slag-blended cement concrete.

Table 1 Concrete mix proportions and properties

Mix ID	W/B	Cement (kg/m^3)	Slag (kg/m^3)	Aggregate (kg/m^3)	Sand (kg/m^3)	HRWR mL/m^3	Slump (mm)	f_c 28 days (MPa)	MoE (GPa)
M1_50	0.54	197.5	197.5	1035	850	1685	140	43.0	36.3
M2_50A	0.51	197.5	197.5	1035	850	2130	120	43.0	44.6
M3_70	0.52	118.5	276.5	1035	850	1685	110	38.0	35.8
M4_70A	0.52	118.5	276.5	1035	850	2130	130	35.5	35.7

f_c, compressive strength; HRWR, high-range water reducer; MoE, modulus of elasticity

2 Experiments

2.1 Mixture Proportions and Properties

The four concrete mixes containing 395 kg of cementitious material were prepared using shrinkage limited (SL) cement and slag provided by Boral Cement (Maldon, Australia). The mix proportions and mix IDs are shown in Table 1. In the mix ID, the numbers 50 and 70 signify the replacement percentage of slag, and the letter A denotes the use of 2.5% Na_2SO_4 activator for that mix.

Four cylinders of 150 × 300 mm and seven cylinders of 100 × 200 mm were cast for each concrete mix. Of the four 150 × 300 mm cylinders, two were used for the creep test, and two were used as control samples for measuring free drying shrinkage strain. The compressive strength test and the determination of elastic modulus were conducted on the 100 × 200 mm cylinders. The average results of these two tests are shown in Table 1.

2.2 Creep Test Setup

Before the testing day, demountable mechanical gauge (DEMEC) points were attached with 200 mm gauge length in three-gauge lines that were uniformly drawn around the perimeter of each cylinder. The DEMEC strain gauge is used for measuring creep/shrinkage strains of specimens. The vertical alignment of specimens and creep rigs were carefully checked with water level instruments before applying load. Figure 1a shows the creep rig used to perform compressive creep tests, and Fig. 1b presents the controlled samples for free drying shrinkage measurements.

One creep rig was used to test two mixes of the same replacement percentage of slag with and without activator. The loading started at 28 days, and the applied load was set at 40% of the average 28 day compressive strength of the concretes. Because two mixes were tested within a creep rig, the load was applied at 40% of the lower strength mix. The tests were conducted in a controlled environment (23 ± 2 °C and

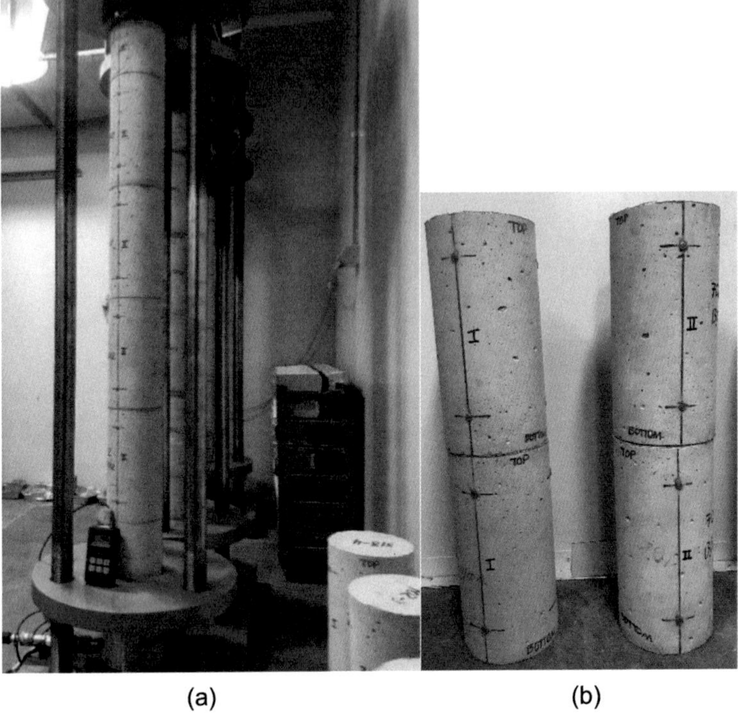

Fig. 1 **a** Creep rigs and loaded samples, **b** control samples

relative humidity 50%). During the testing period, the applied load was monitored by a pressure gauge and maintained at a value not less than 5% of its initial value.

The referenced readings were obtained just before applying load. The creep/shrinkage strains were recorded immediately after loading, 2 and 6 h for the first loading day, daily for the first week, weekly until 2 months, and then monthly. Here we report the findings after sustained load had been applied for 140 days. The creep strains of each mix were averaged from two specimens and three-gauge lines of each specimen.

3 Creep Model in AS3600:2018

As creep of concrete is a time-dependent behavior, using an empirical model to predict the creep strain of concrete is a practical and useful approach for designing concrete structures. From the viewpoint of building construction, it is not practical to perform a long-term creep test for every concrete mix. In AS3600:2018, the predictive creep strain of a sample bearing constant sustained stress σ_o at any time t can be generally expressed as:

$$\varepsilon_{cc}(t) = \varphi_{cc}\left(t, t_{load}, f'_c, t_h, E_{env}, \varphi_{cc.b}\right)\frac{\sigma_o}{E_c} \tag{1}$$

where φ_{cc} is the time-dependent creep coefficient function depending on the loading time t_{load}, characteristic compressive strength f'_c at 28 days, hypothetical thickness of sample t_{h} and the exposure environmental condition E_{env}. The basic creep coefficient $\varphi_{cc.b}$ and elastic modulus E_{c} of a sample loaded at 28 days can be determined based on the characteristic compressive strength f'_c. Assuming that we only know the 28 day compressive strength of the four mixes described in Sect. 2.1, the characteristic compressive strength f'_c was selected equal to 30 MPa and 35 MPa for 50% and 70% slag-blended cement concretes, respectively. Hence, in the predictive model, the values of $E_{\text{c}} = 29.1$ GPa and $\varphi_{cc.b} = 3.6$ were used for 50% slag-blended cement concretes, and $E_{\text{c}} = 31.1$ GPa and $\varphi_{cc.b} = 3.2$ were used for 70% slag-blended cement concretes. In addition, the interior environment ($E_{\text{env}} = 0.65$), which was nearest to the test environment, was also selected.

4 Results and Discussion

4.1 *Total Deformation*

The total deformation due to elastic, creep, and shrinkage strains of the four mixes are plotted in Figs. 2 and 3. It can be observed that the presence of Na_2SO_4 significantly reduced the creep and shrinkage strains of the 50% slag-blended cement concrete. A similar trend can also be observed with the shrinkage of 70% slag-blended cement concrete, but the total deformation of the loaded samples is comparable. A significant difference in deformation at initial loading due to the higher elastic modulus in the activated mix was also notable in the 50% slag-blended cement concrete mixes. The creep strain rate of the two mixes was, however, almost identical.

In addition, it can be seen that the 70% slag-blended cement concrete exhibited lower creep strain than the 50% slag-blended cement concrete. At 140 days, the creep strains of the 50% slag-blended cement concretes was ≈1400 μm/m, while those of 70% slag-blended cement concretes were only slightly greater than 1000 μm/m. This finding contradicts the reports in [10].

4.2 *Creep Strain Measurement and Model Prediction*

The creep strains (including instantaneous elastic strain) were obtained by subtracting the shrinkage strain from the total deformation of the loaded specimens. The development of creep strains of the four tested mixes are shown in Figs. 4 and 5. Although

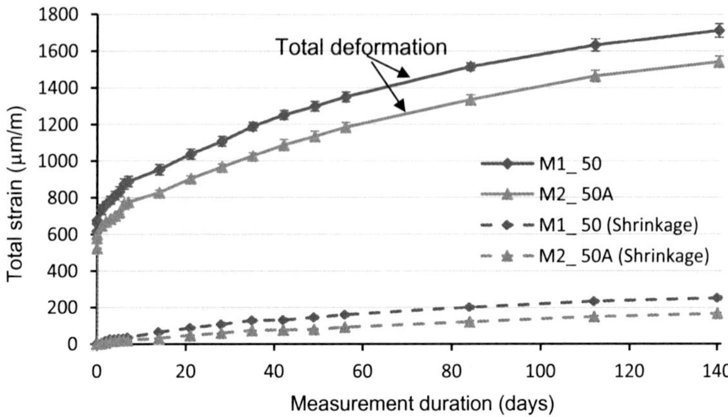

Fig. 2 Total deformation including creep and shrinkage strains of 50% slag-blended cement concrete, with and without Na_2SO_4

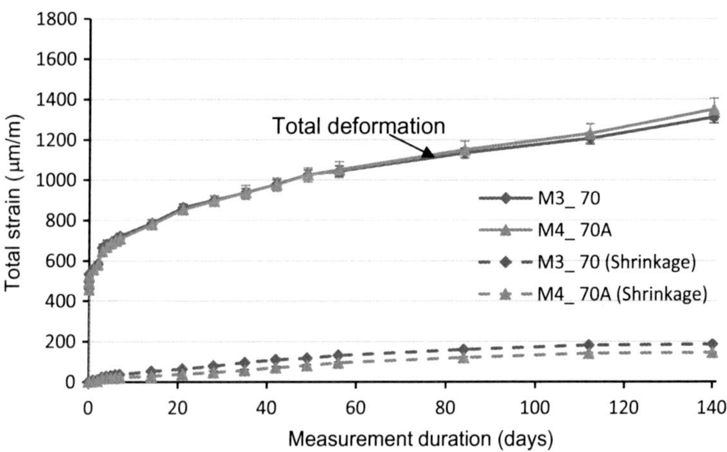

Fig. 3 Total deformation including creep and shrinkage strains of 70% slag-blended cement concrete, with and without Na_2SO_4

the presence of Na_2SO_4 appeared to slightly reduce the creep strains of 50% slag-blended cement concrete, an increase, though within measurement error range, in that of 70% slag-blended cement concrete was observed.

The predictive results from the creep model in AS3600:2018 are also plotted in Figs. 4 and 5. It can be seen that the predictive values from the model are lower than the measured values for all mixes. Except for the initial deformation or instantaneous elastic strain, the model underestimated the development of creep strain for both 50 and 70 wt% slag replacement cement concretes.

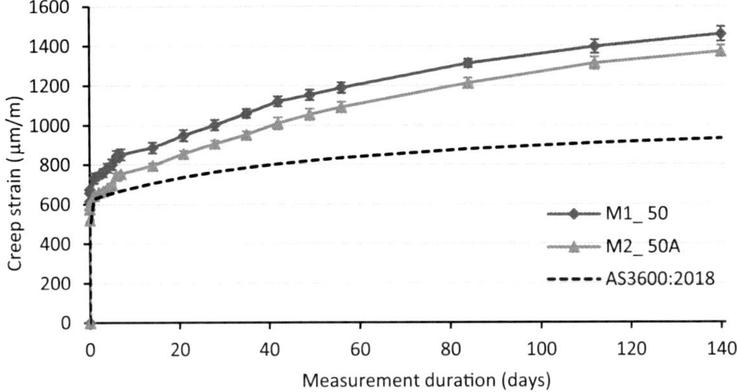

Fig. 4 Creep strain of concretes with 50 wt% slag replacement and predictive results

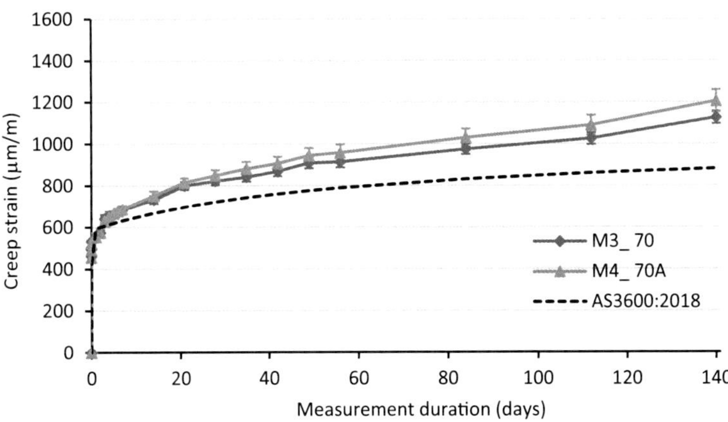

Fig. 5 Creep strain of the concrete with 70 wt% slag replacement and predictive results

5 Conclusions

From the results of 140 day compressive creep testing of PC concrete blended with 50 and 70 wt% slag, the main findings can be summarized as:

(1) 70% slag-blended cement concrete had lower creep strain than 50% slag-blended cement concrete, which suggested that increasing the slag content improved creep performance
(2) adding 2.5% Na_2SO_4 activator had negligible effects on the creep strain of slag-blended cement concretes
(3) the activator slightly decreased the shrinkage strain of slag-blended cement concretes

(4) the predictive model in AS3600:2018 underestimated the creep strain of high slag concretes regardless of the presence of Na_2SO_4.

The monitoring of long-term creep behavior of these slag-cement concretes is ongoing, and the results will be reported in the future.

Acknowledgements This study was supported by the UTS-Boral Centre for Sustainable Building under funding from the Innovative Manufacturing CRC (IMCRC).

References

1. Voldsund M, Gardarsdottir SO, De Lena E, Pérez-Calvo J-F, Jamali A, Berstad D, Fu C, Romano M, Roussanaly S, Anantharaman R (2019) Comparison of technologies for CO_2 capture from cement production—part 1: technical evaluation. Energies 12(3):559
2. Plaza MG, Martínez S, Rubiera F (2020) CO_2 capture, use, and storage in the cement industry: state of the art and expectations. Energies 13(21):5692
3. Özbay E, Erdemir M, Durmuş Hİ (2016) Utilization and efficiency of ground granulated blast furnace slag on concrete properties—a review. Constr Build Mater 105:423–434
4. Giergiczny Z (2019) Fly ash and slag. Cem Concr Res 124:105826
5. Joseph S, Cizer Ö (2022) Hydration of hybrid cements at low temperatures: a study on Portland cement-blast furnace slag—Na_2SO_4. Materials 15(5):1914
6. Zhao Y, Qiu J, Xing J, Sun X (2020) Chemical activation of binary slag cement with low carbon footprint. J Clean Prod 267:121455
7. Lloyd RR, Keyte LM (2011) Low CO_2 cement. Au Patent number 2011245080 B2, Australia Patent Office
8. Fu J, Bligh MW, Shikhov I, Jones AM, Holt C, Keyte LM, Moghaddam F, Arns CH, Foster SJ, Waite TD (2021) A microstructural investigation of a Na_2SO_4 activated cement-slag blend. Cem Concr Res 150:106609
9. Fu J, Jones AM, Bligh MW, Holt C, Keyte LM, Moghaddam F, Foster SJ, Waite TD (2020) Mechanisms of enhancement in early hydration by sodium sulfate in a slag-cement blend—insights from pore solution chemistry. Cem Concr Res 135:106110
10. Shariq M, Prasad J, Abbas H (2016) Creep and drying shrinkage of concrete containing GGBFS. Cement Concr Compos 68:35–45
11. Khatri RP, Sirivivatnanon V, Gross W (1995) Effect of different supplementary cementitious materials on mechanical properties of high performance concrete. Cem Concr Res 25(1):209–220

Partially-Unzipped Carbon Nanotubes as Low-Concentration Amendment for Cement Paste

S. Iffat, F. Matta, J. Gaillard, M. Elvington, M. Sikder, M. Baalousha, S. Tinkey, and J. Meany

Abstract Partially-unzipped multiwalled carbon nanotubes (PUCNTs) combine the chemical structure and basic mechanical properties of multiwalled carbon nanotubes (MWCNTs) and graphene nanoplatelets (GNPs) with exceptionally high graphene-edge content. As a result, PUCNTs can be effectively oxidized for dispersion in aqueous solutions and have a specific surface area that is larger than that of MWCNTs with a comparable aspect ratio. Thus, the incorporation of relatively small concentrations of PUCNTs in cement composites may result in significant physicomechanical enhancements. In the proof-of-concept study presented here, cement paste specimens were manufactured with oxidized PUCNT concentrations of 0, 0.001, and 0.005% in weight of cement (wt%), that is, one order of magnitude smaller than lower-bound concentrations for MWCNTs reported in the literature. Stable dispersion in water was verified through dynamic light scattering analysis. Physicomechanical changes and PUCNT dispersion in the cement matrix were investigated through uniaxial

S. Iffat (✉) · F. Matta
Department of Civil and Environmental Engineering, University of South Carolina, Columbia, SC, USA
e-mail: iffats@farmingdale.edu

S. Iffat
Department of Civil Engineering, Bangladesh University of Engineering and Technology, Dhaka, Bangladesh

Department of Civil Engineering Technology, Farmingdale State College (SUNY), Farmingdale, NY, USA

J. Gaillard · M. Elvington · S. Tinkey
Savannah River National Laboratory, Aiken, SC, USA

M. Sikder
Solenis, Wilmington, DE, USA

M. Baalousha
Center for Environmental Nanoscience and Risk, University of South Carolina, Columbia, SC, USA

J. Meany
Ionic Flask Materials Group, Atlanta, GA, USA

W. Duan et al. (eds.), *Nanotechnology in Construction for Circular Economy*, Lecture Notes in Civil Engineering 356,
https://doi.org/10.1007/978-981-99-3330-3_20

compression tests on 25 × 25 × 76 mm prism specimens, and visual inspection of scanning electron microscopy (SEM) micrographs, respectively. The incorporation of 0.001 wt% and 0.005 wt% of PUCNTs resulted in an average increase in compressive strength of 10% and 29%, respectively, compared with plain cement paste. In both instances, representative SEM micrographs show the preferential formation of cement hydrates in cement paste that accorded with well-dispersed PUCNTs.

Keywords Cement paste · Dispersion · Partially-unzipped multiwalled carbon nanotubes · Nanoparticles · Strength

1 Introduction

Partially-unzipped multiwalled carbon nanotubes (PUCNTs) are synthesized by longitudinal unzipping of the outer walls of multiwalled carbon nanotubes (MWCNTs) [1, 2], resulting in a portion having a morphology similar to that of graphene nanoplatelets (GNPs). MWCNTs are characterized by a relatively high aspect ratio, which may be beneficial when using these nanoparticles as reinforcement in cement composites [3]. Mechanical enhancements have also been reported for nano-amended cement composites with GNP concentrations between 0.03% [4] and 0.5% [5] in weight of cement (wt%). Related mechanisms may be associated with improved cement hydration through the presence of carboxyl and hydroxyl functional groups on the surface of oxidized MWCNTs and GNPs [6, 7], with a reduction in total porosity [8, 9], pore-size refinement [8, 10], and densification of the cement matrix [11].

PUCNTs have appealing properties compared with their fully-unzipped counterparts; elongated strips of graphene with more functional group-edge sites compared with MWCNTs or GNPs, but also a smaller aspect ratio [12]. Partial unzipping of the outer walls of MWCNTs results in a radical increase in the open surface area [12], and oxidation produces a relatively larger amount of functional groups [1], thus enhancing the chemical affinity with the cement matrix. Through this process, a few outer layers of a given MWCNT are partially unzipped without affecting the inner core walls, resulting in a hybrid graphitic structure together with the inner (intact) carbon nanotubes [13]. Therefore, the resulting PUCNTs exhibit: (a) larger surface area and greater functionality than their precursor MWCNTs, which may enhance dispersibility in aqueous solutions and in the cement matrix, and (b) a relatively high aspect ratio, which may enhance the mechanical properties of the cement composite.

We present a feasibility study of cement paste amended with low concentrations of oxidized PUCNTs, namely, 0.001 and 0.005 wt%, which is one order of magnitude smaller than successful lower-bound concentrations reported in the literature for MWCNTs and GNPs.

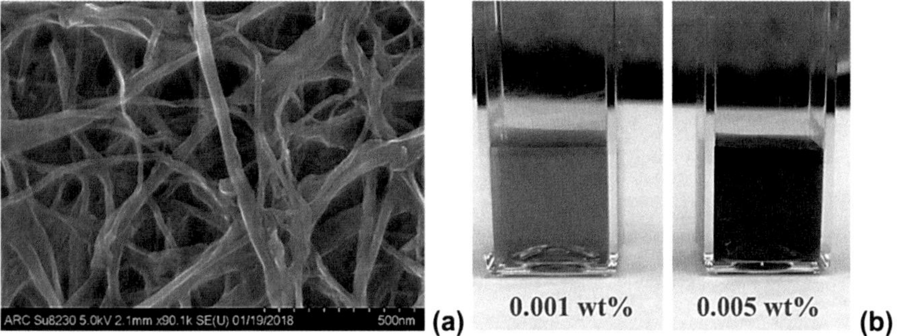

Fig. 1 Partially-unzipped multiwalled carbon nanotubes (PUCNTs) used in cement paste specimens: **a** scanning electron microscopy image showing PUCNTs resulting from oxidative unzipping of multiwalled carbon nanotubes; **b** aqueous suspensions

2 Methods

2.1 PUCNT Aqueous Suspensions

The PUCNTs were prepared at Savannah River National Laboratory (Aiken, SC, USA) using an oxidative unzipping process modified from Kosynkin et al. [14]. Defects in the pristine MWCNTs were created by soaking in concentrated sulfuric acid and o-phosphoric acid, with potassium permanganate serving as the oxidant. By controlling the amount of oxidant, heating time, and soaking time, the PUCNTs shown in Fig. 1a were produced, and then suspended in deionized (DI) water. Stock solution of 3 g/L of these PUCNTs in DI water was diluted with additional DI water to prepare 0.02 and 0.1 g/L suspensions, to be used for the manufacturing of cement paste specimens with PUCNT concentrations of 0.001 wt% and 0.005 wt%, respectively. A small amount of NaOH (<0.01 mol/L) was added to the suspensions to reach an approximate pH of 12, thereby enhancing the stability of acid-treated (i.e., low pH) PUCNTs in water. The suspensions were ultrasonicated for 15 min using an ultrasonic bath sonicator (model CPX 2008, Branson Ultrasonics Corp., CT, USA).

2.2 Cement Paste Specimens

Prismatic specimens with dimensions 25 × 25 × 76 mm were prepared per ASTM C305 [15] using Type I ordinary Portland cement (OPC) and PUCNT aqueous suspensions, with a water-to-cement ratio of 0.5 (in weight). Plain cement paste specimens to be used as the benchmark were also cast using Type I OPC and DI water. All specimens were cast in acrylic molds from which they were removed after 24 h, and then placed in saturated lime water for 28 days.

2.3 Procedure

The dispersibility of the functionalized PUCNTs in DI water solutions was quantitatively assessed via dynamic light scattering (DLS) testing using a particle and molecular size analyzer (model Zetasizer nano ZS, Malvern Panalytical Ltd.).

For each PUCNT concentration, a suspension sample of 1 mL was used to measure the zeta potential. Each measurement was repeated 10 times to assess the variability of the results. Representative aqueous suspensions used for the manufacturing of cement paste specimens with PUCNTs in concentrations of 0.001 and 0.005 wt% are shown in Fig. 1b.

Uniaxial compression tests were performed on the 28 day cured prism specimens using a servo-hydraulic loading frame. The load was applied in displacement-control mode at a rate of 0.3 mm/min and was measured using a 20-kip load cell. Thin polytetrafluoroethylene inserts were placed between the specimen surfaces and the loading platens to minimize friction. A representative test setup is shown in Fig. 2a. Four specimens were tested for each PUCNT concentration (0, 0.001 and 0.005 wt%).

The fractured specimens were used for scanning electron microscopy (SEM) analysis to investigate the incorporation and dispersion of PUCNTs in the cement paste structure. Specifically, SEM micrographs were used to visually assess the presence of undesirable PUCNT agglomerates as a result of ineffective dispersion. To this end, specimens' fragments <5 mm in length in any direction were used. Following the compression tests, these samples were extracted from the prisms, air-dried for 24 h, and placed in a vacuum-suction chamber for 1 h to remove the excess moisture. The samples were gold-spattered before being examined under a SEM (Ultraplus Thermal Field Emission Scanning Electron Microscope, Zeiss).

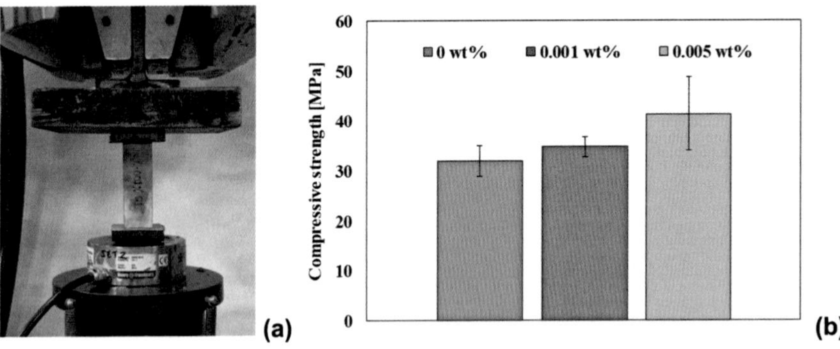

Fig. 2 Uniaxial compression testing of cement paste prism specimens: **a** test setup, **b** summary of results (histogram bars and error bars indicate mean and standard deviation)

3 Results and Discussion

3.1 Dispersibility in Aqueous Suspension

The effective dispersion of PUCNTs in the aqueous suspensions was assessed by zeta potential measurements made as part of the DLS tests. The zeta potential values obtained were –39.5 ± 1.30 mV and –39.6 ± 1.07 mV for the suspensions used to manufacture cement paste specimens with 0.001 wt% and 0.005 wt% of PUCNT concentration, respectively. Because suspensions characterized by a zeta potential smaller than –30 mV are considered stable [16], and also based on visual inspection (Fig. 1b), it was concluded that the oxidized PUCNTs were effectively dispersed in the aqueous solutions used in this research.

3.2 Compressive Strength

The mean and standard deviation of the 28 day compressive strength were 32.0 ± 3.10 MPa, 34.7 ± 1 0.93 MPa, and 41.4 ± 7.4 MPa for cement paste with PUCNT concentrations of 0, 0.001 and 0.005 wt%, as summarized in Fig. 2b.

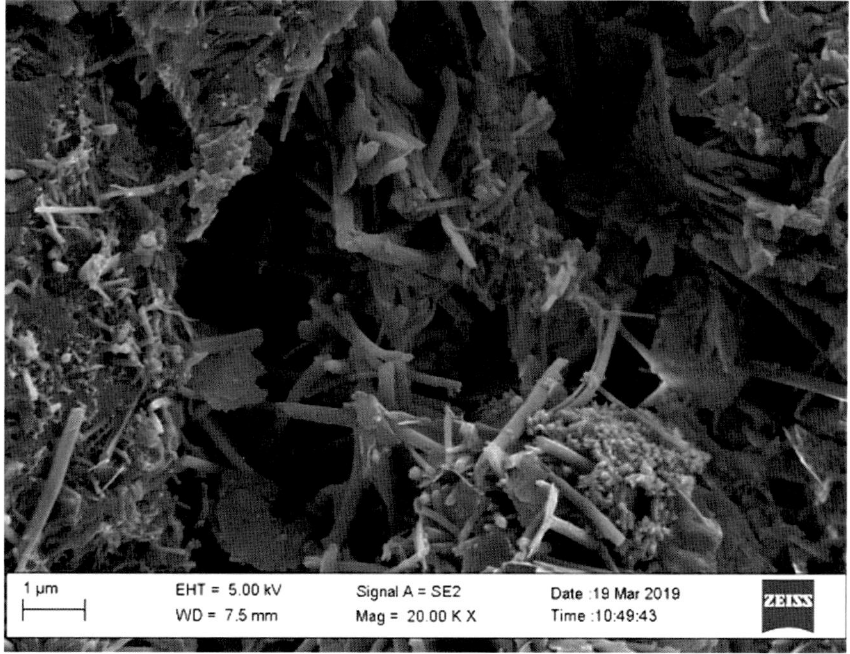

Fig. 3 Scanning electron microscopy image of plain cement paste structure

Our results showed that the incorporation of very small amounts of PUCNTs resulted in significant compressive strength enhancement, namely 29% on average for a concentration of 0.005 wt%, compared with plain cement paste.

Evidence collected through SEM image analysis was used to better understand plausible strengthening mechanisms. In fact, the micrographs shown in Figs. 3, 4 and 5 suggest that the incorporation of PUCNTs resulted in preferential formation of amorphous (C–S–H) and crystalline (calcium hydroxide) cement hydrates (Figs. 4 and 5), with a less porous structure than plain cement paste (Fig. 3).

In addition, the lack of visible clusters of PUCNTs suggested that the methodology used will produce cement composites with well-dispersed and chemically-affine PUCNTs, and thus with a meso-scale homogeneity comparable to or better than that of plain cement paste. However, the significantly larger standard deviation of the compressive strength results for the 0.005 wt% concentration suggested that larger concentrations of PUCNTs may be more difficult to disperse in the cement matrix.

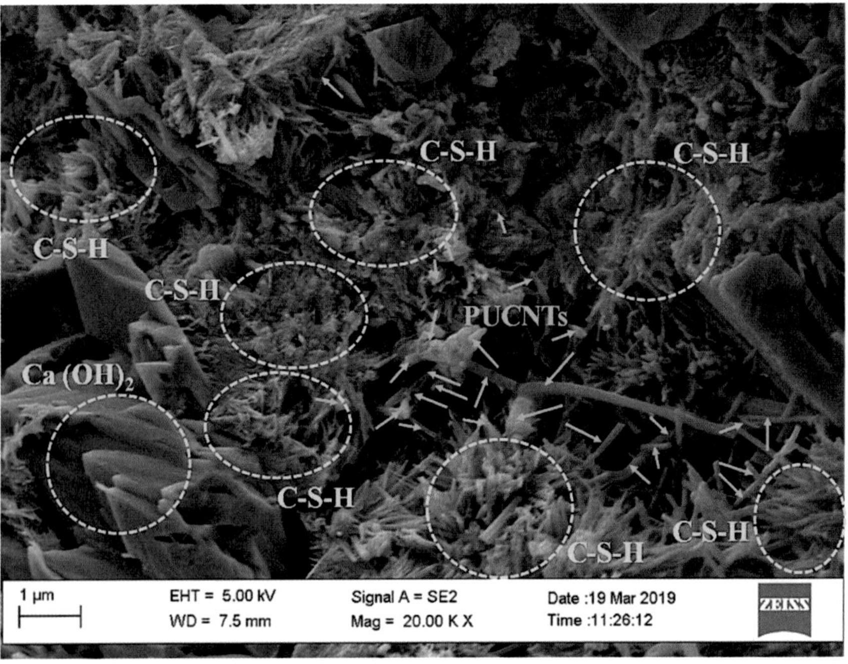

Fig. 4 Scanning electron microscopy image of 0.001 wt% PUCNT-amended cement paste structure. PUCNTs, partially-unzipped multiwalled carbon nanotubes

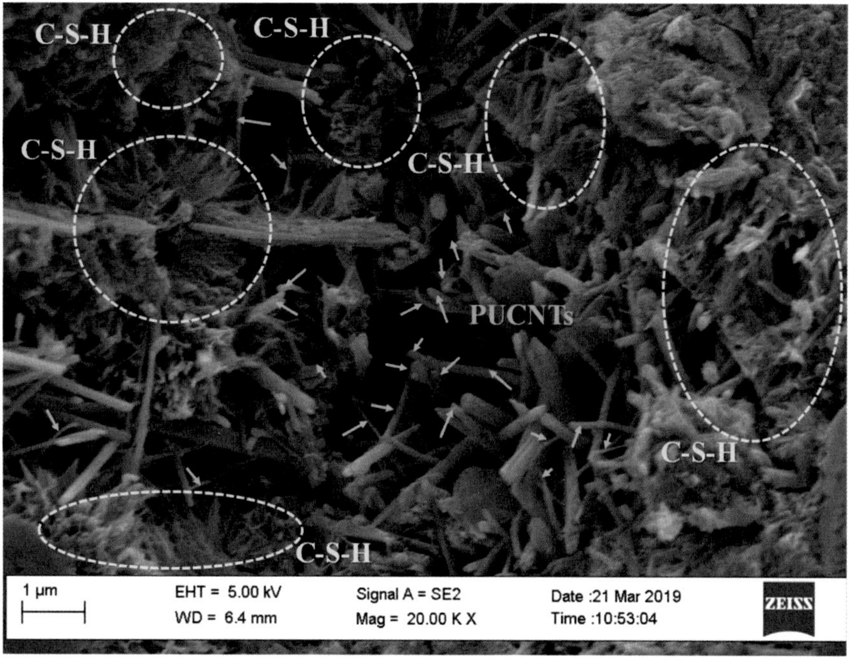

Fig. 5 Scanning electron microscopy image of 0.005 wt% PUCNT-amended cement paste structure. PUCNTs, partially-unzipped multiwalled carbon nanotubes

4 Conclusions

The following conclusions are presented.

- It is feasible to incorporate well-dispersed and chemically-affine PUCNTs to enhance the physicomechanical properties of cement paste. The unzipping process is similar in chemical approach (and thus cost) to the well-known Hummer's method used to generate graphite oxide.
- Due to the morphology and high graphene-edge content of PUCNTs, dispersibility in aqueous solutions was easily achieved.
- A significant increase in the compressive strength of cement paste occurred with a PUCNT concentration of 0.005 wt%, which is one order of magnitude smaller than lower-bound concentrations reported in the archival literature for MWCNT- and GNP-amended cement pastes.
- Microscopic inspection of the cement paste nano- and micro-structures suggests that plausible contributing mechanisms for strength enhancement include the preferential and well-distributed formation of cement hydrates in the vicinity of PUCNTs—also resulting in a less porous structure—compared with plain cement paste.

Acknowledgements This material is based on collaborative work supported by the U.S. Department of Energy Office of Science, Office of Basic Energy Sciences, and Office of Biological and Environmental Research, under award number DE-SC0012530; Savannah River National Laboratory; and, the University of South Carolina (USC) Advanced Support for Innovative Research Excellence (ASPIRE) program. Special thanks are extended to Ms. Erika Rengifo (Ph.D.) and Ms. Sarah Riser (undergraduate research assistant) at the UofSC Department of Civil and Environmental Engineering, and personnel of the Center for Environmental Nanoscience and Risk at the Arnold School of Public Health, and the Electron Microscopy Center, for their technical assistance.

Author Contributions The authors confirm contributions as follows: study conception and design: Iffat S., Matta F., Gaillard J., Baalousha M.; data collection: Iffat S., Levington M., Tinkey S., Meany J.; analysis and interpretation of results: Iffat S., Matta F., Gaillard J., Sikder M., Baalousha M.; draft manuscript preparation: Iffat S., Matta F. All authors reviewed the results and approved the final version of the manuscript.

References

1. Song Y, Feng M, Zhan H (2014) Electrochemistry of partially unzipped carbon nanotubes. Electrochem Commun 45:95–98
2. Cheng Y, Zhang S, Li J, Sun J, Wang J, Qin C, Dai L (2017) Preparation of functionalized partially unzipped carbon nanotube/polyimide composite fibers with increased mechanical and thermal properties. R Soc Chem Adv 7:21953–21961
3. Abu Al-Rub RK, Ashour AI, Tyson BM (2012) On the aspect ratio effect of multi-walled carbon nanotube reinforcements on the mechanical properties of cementitious nanocomposites. Constr Build Mater 35:647–655
4. Lv SH, Deng LJ, Yang WQ, Zhou QF, Cui YY (2016) Fabrication of polycarboxylate/graphene oxide nanosheet composites by copolymerization for reinforcing and toughening cement composites. Cement Concr Compos 66:1–9
5. Zohhadi N (2014) Functionalized graphitic nano-reinforcement for cement composites. PhD thesis, University of South Carolina, Columbia
6. Kaur R, Kothiyal NC (2019) Comparative effects of sterically stabilized functionalized carbon nanotubes and graphene oxide as reinforcing agent on physico-mechanical properties and electrical resistivity of cement nanocomposites. Constr Build Mater 202:121–138
7. Makar JM, Chan GW (2009) Growth of cement hydration products on single-walled carbon nanotubes. J Am Ceram Soc 92(6):1303–1310
8. Li GY, Wang PM, Zhao X (2005) Mechanical behavior and microstructure of cement composites incorporating surface-treated multi-walled carbon nanotubes. Carbon 43:1239–1245
9. Nochaiya T, Chaipanich A (2011) Behavior of multi-walled carbon nanotubes on the porosity and microstructure of cement-based materials. Appl Surf Sci 257:1941–1945
10. Hu Y, Luo D, Li P, Li Q, Sun G, Key S (2014) Fracture toughness enhancement of cement paste with multi-walled carbon nanotubes. Constr Build Mater 70:332–338
11. Liu J, Wang L, Li Q, Xu S (2019) Reinforcing mechanism of graphene and graphene oxide sheets on cement-based materials. J Mater Civ Eng 31(4):04019014
12. Jeong YC, Lee K, Kim T, Kim JH, Park J, Cho YS, Yang SJ, Park CR (2016) Partially unzipped carbon nanotubes for high-rate and stable lithium–sulfur batteries. J Mater Chem A 4:819–826
13. Shende RS, Ramaprabhu S (2016) Thermo-optical properties of partially unzipped multiwalled carbon nanotubes dispersed nanofluids for direct absorption solar thermal energy systems. Sol Energy Mater Sol Cells 157:117–125
14. Kosynkin DV, Higginbotham AL, Sinitskii A, Lomeda JR, Ayrat DA, Price BK, Tour JM (2009) Longitudinal unzipping of carbon nanotubes to form graphene nanoribbons. Nature 458:872–877

15. ASTM International (2014) Standard practice for mechanical mixing of hydraulic cement pastes and mortars of plastic consistency. ASTM C305-14. ASTM International, West Conshohocken, PA
16. Freitas C, Muller RH (1998) Effect of light and temperature on zeta potential and physical stability in solid lipid nanoparticle (SLN™) dispersions. Int J Pharm 168:221–229

Effect of Fine Aggregates and Test Settings on the Self-sensing Response of Cement-Based Composites with Carbon Nanotubes as Conductive Filler

T. C. dos Santos, P. A. Carísio, A. P. S. Martins, M. D. M. Paiva, F. M. P. Gomes, O. A. M. Reales, and R. D. Toledo Filho

Abstract Cement-based self-sensing composites with carbon nanotubes (CNT) have attracted attention due to their multifunctional properties and great potential for their application in the smart monitoring of concrete structures. In this study, the self-sensing properties of one paste and three mortars containing 0.50 and 0.75 wt% of CNT, and 1.5 and 1.0 sand/cement ratio were investigated, aiming to evaluate their impact on the piezoresistive response of the composites. The inclusion of sand in the cement paste with CNT led to a reduced gauge factor and a higher electrical noise response. The inert aggregates modified the compressive loading mechanical response of the composites and possibly acted as barriers to electronic mobility, by increasing the CNT conductive paths' tortuosity or even interrupting them. The mortar containing 0.50% of CNT showed a higher electrical resistivity and, at the same time, greater sensitivity and a more linear self-sensing response than the one with 0.75% CNT, which can be explained by the CNT content being closer to its percolation threshold in the first. In this way, a lower CNT concentration generated a conductive network with a higher capacity to be rearranged under loading, generating significant changes in resistivity, but a higher CNT concentration presented a more stable and conductive network. The results suggested that both the conductive and non-conductive phases affect the detection performance of the composites and, therefore, must be dosed appropriately. Additionally, the test setup modifications positively affected the self-sensing response signal, which is particularly useful to reduce the deleterious effects of the sand additions in the matrix. This overall approach can make the use of self-sensing mortars in structural monitoring a viable option.

T. C. dos Santos (✉) · P. A. Carísio · A. P. S. Martins · M. D. M. Paiva · O. A. M. Reales · R. D. Toledo Filho
Civil Engineering Program, Universidade Federal do Rio de Janeiro, Rio de Janeiro, Brazil
e-mail: thais.santos@numats.coc.ufrj.br

F. M. P. Gomes
Research Center, Eletrobras Furnas, Rio de Janeiro, Brazil

W. Duan et al. (eds.), *Nanotechnology in Construction for Circular Economy*, Lecture Notes in Civil Engineering 356,
https://doi.org/10.1007/978-981-99-3330-3_21

Keywords Carbon nanotubes · Self-sensing mortars · Structural monitoring

1 Introduction

Self-sensing cementitious composites change their electrical properties in response to mechanical loadings and, therefore, are becoming a novel and useful class of material for structural health monitoring when compared with conventional technologies [1–3]. An essential element in these composites is the functional filler that forms an interconnected conductive network across the cement matrix. In this context, carbon nanotubes (CNT) are particularly promising when compared with other conductive fillers, given their excellent electrical, physical, and mechanical properties [4]. Once incorporated into the composites, in an adequate and well-dispersed concentration, it is possible to obtain nanostructured cementitious materials, electrically sensitive to stress–strain that can ultimately be applied as excellent piezoresistive sensors for monitoring structures.

Yoo et al. [5] investigated the performance of cement pastes containing different carbon-based fillers, and CNT were found to be the most effective filler for increasing conductivity when compared with other particles at the same concentration. Han et al. reported the CNT/cement-based materials' ability to detect both stress–strain in the elastic regime, as well as structural damages based on irreversible change in their electrical resistance responses [6]. Nalon et al. verified the ability of mortars to detect fire damage and exhibit a residual piezoresistive response, even after their exposure to high temperatures [7]. Le et al. [8] produced ultrahigh-performance concretes with improved self-sensing properties by combining micro- and nanoscale fibers and functional aggregates. In that case, CNT improved the conductive network at the nano level and, consequently, the conductivity and piezoresistivity of the corresponding composites.

However, the effective application of this material in structural monitoring requires more investigation, which motivated the present study. Several authors report that aggregates, especially the coarse, create obstacles to electrical current flow, extending or interrupting the conductive paths and negatively affecting the self-sensing response [1, 2, 9]. Tian et al. reported that from the paste to the concrete, both the number of publications and the self-sensing performance decreased [3]. On the other hand, the claims regarding the use of aggregates are not unanimous in the literature. Additionally, other researchers suggest that aggregates do not significantly affect the percolation threshold of CNT composites and that their proper incorporation can reduce the matrix voids and promote more contact points between the fillers [4, 10]. Thus, the electromechanical performance of the CNT composites can be improved. The effects of aggregates on the properties of self-sensing composites are complex and therefore require further investigation for better understanding.

Both Han et al. [9] and Wang and Aslani [4] point out that CNT composites with aggregates are much more profitable in real applications than those made of cement paste. According to Dong et al., both the conductive and non-conductive phases are key factors to obtaining self-sensing composites with higher performance [2]. Han et al. reinforced that the proper selection of constituents and the determination of their proportions are crucial for designing self-sensing composites [1]. Han et al. consider that one of the challenges for the development and future implementation of self-sensing composites is their fabrication and, therefore, suggest that upcoming research should include investigations related to the design, optimization, and production of this type of composite, especially those containing aggregates [9].

In this context, aiming to contribute to the smart monitoring of concrete structures, we present the self-sensing properties of paste and mortars containing CNT and evaluate the effects of both conductive filler content and fine aggregates. In addition, the self-sensing test settings and their influence on the quality of the acquired electrical signals were investigated.

2 Methods

2.1 Materials and Sample Preparation

The materials used for this work were high initial strength Portland cement (LafargeHolcim), NC7000 Multi-Walled Carbon Nanotubes (MWCNT; Nanocyl), MasterRheobuild 1000 naphthalene sulfonate-based superplasticizer (BASF Chemicals), crystalline silica flour (S325; Mineração Jundu), and MasterMatrix UW 410 cellulose-based viscosity-modifying agent (VMA; BASF Chemicals). As the fine aggregate, a mixture of fine (#16) and coarse (#100) quartz sands was used in the volumetric fractions of 61.53% and 38.47% respectively.

The MWCNT used in this study consisted of a powder formed by CNT agglomerates, with an average diameter of 9.5 nm, an average length of 1.5 μm, and a surface area between 250 and 300 m^2/g, according to the manufacturer. They were dispersed in deionized water using ultrasonication (Sonics Vibra cell, VCX 500) and superplasticizer as the dispersing agent. In the dispersion procedure, 30 g of suspension with a 2 wt% CNT concentration and 2 wt% of superplasticizer solids was subjected to 20 sonication cycles with an average energy of 110 J/g at an amplitude of 40% and pulses of 20 s. In each cycle, the suspension of CNT was homogenized in a low-temperature bath to cool the mixture. The UV–Vis absorbance spectra of the CNT/superplasticizer aqueous solutions were measured before and after the dispersion procedure. For the measurements, an aliquot of the suspensions was diluted in deionized water in the ratio of 1:2000 (% v/v).

To investigate the influence of sand addition on the self-sensing properties of the composites, samples containing 0.75% of CNT were produced, without and with sand (P1 and M1, respectively) at a 1.5 sand/cement (or s/c) ratio, as shown in Table 1. In addition, to evaluate the effect of the CNT content on the mortars' piezoresistivity, a sample of the same mix design was produced incorporating 0.5% of CNT dispersed by mass of cement (M2). In the second part of the study, to investigate the effects of the test configurations and the sand content, the M3 trait was defined, with an s/c ratio of 1.0.

The P1, M1, and M2 samples were produced as cubic specimens with 50 mm edge and 4 copper electrodes with dimensions of 0.3 × 30 × 50 mm (thickness x width x length) embedded equidistantly (Fig. 1). The M3 sample was cubic with 40 mm edge and copper electrodes with 1050 mm^2 (30 × 35 mm) of embedded area. All mixtures were mixed manually, using a glass rod, into which the previously homogenized fine solids (cement, mineral admixture, and VMA) were incorporated into the aqueous dispersion of CNT and additional water and mixed for 10 min. The sand, when included, was incorporated into the mixture after the previous process and homogenized for another 3 min. Finally, compaction was performed by vibration, before and after the copper electrode insertions. After demolding, the samples were cured in a humid chamber at room temperature.

Table 1 Mix designs of the self-sensing cementitious composites

Sample	w/c	S325 (%[a])	VMA (%[a])	s/c	CNT (%[a])	Mix proportions (kg/m^3)				
						Cement	S325	Sand	Water	CNT
P1	0.5	13.0	0.1	0.0	0.75	1139.42	148.12	0	569.71	5.70
M1				1.5	0.75	669.25	87.01	381.52	334.63	5.02
M2				1.5	0.50	670.00	87.11	381.95	335.00	3.35
M3				1.0	0.50	776.98	101.04	776.98	388.49	3.88

[a] By mass of cement. CNT, carbon nanotubes; VMA, viscosity-modifying agent

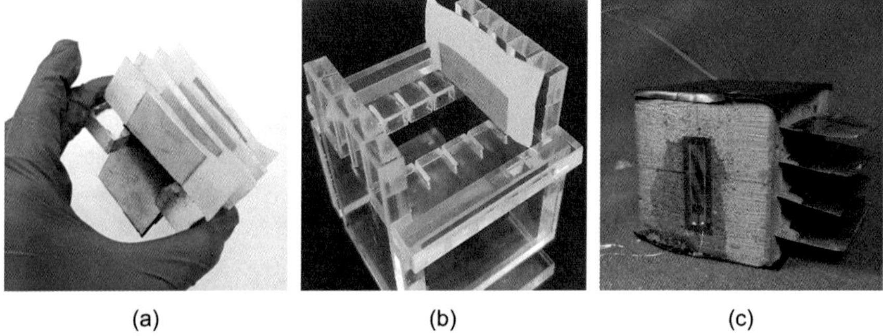

Fig. 1 Copper electrodes (**a**) and views of **b** mold setup and **c** a typical produced sample

2.2 Self-sensing Test

Self-sensing tests were executed by applying cyclic compressive loadings with simultaneous acquisition of electrical signals using the embedded copper plates as electrodes. Before the self-sensing tests, all samples were dried at 60 °C for 3 days to eliminate the effect of humidity. In addition, the specimens were instrumented with strain gauges and electrically isolated from the machine using insulating tape. The electrical resistance of the matrices was evaluated using the four-probe method, in which the DC voltage is applied to the samples through the external electrodes, and the voltage measured in the sample is recorded by a data acquisition system connected to the internal electrodes. The test setup is shown in Fig. 2.

Prior to the mechanical loading, DC voltage was applied to each sample for 20 min, to achieve electrical signal stabilization and to mitigate the sensor capacitive effects. During the test, the electric current of the circuit was acquired by using a shunt resistor with a known electrical resistance connected in series. Using the voltage of the shunt recorded throughout the test and applying Ohm's 1st law (Eq. 1), the electrical current was obtained. With the current and voltage between the internal electrodes, the electrical resistance of the matrix was determined.

For the correlation with the strains, the fractional change in resistance (FCR) in response to the applied loads were obtained according to Eq. 2. The initial electrical resistance of the sample R_0 was measured after a pre-load of 0.5 kN at the end of the electrification time of 20 min. On the other hand, electrical resistance R corresponds to the electrical resistance value over time, under a given compressive loading cycle in the linear elastic regimen of material. Finally, the electrical resistivity ρ of the

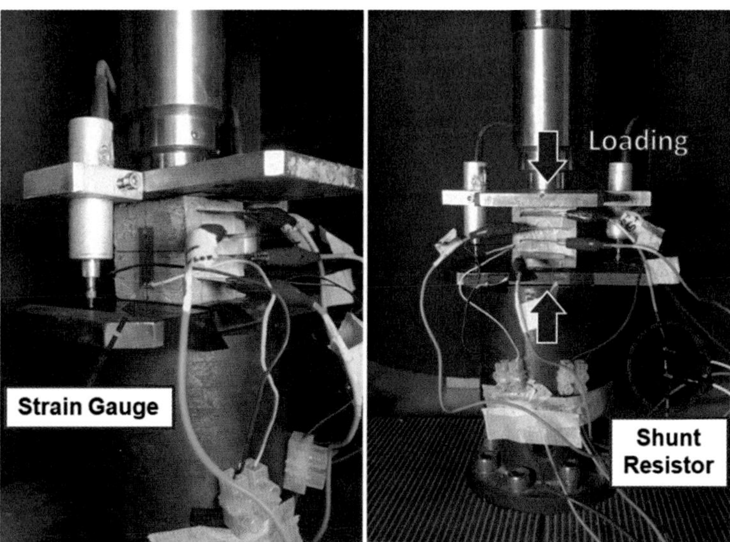

Fig. 2 Experimental setup for the self-sensing test

samples, in Ω.cm, was obtained through Ohm's 2nd law (Eq. 3), considering the distance between the electrodes and their contact area with the cement composite.

$$V = R \times i \tag{1}$$

$$FCR = \frac{R - R_0}{R_0} \tag{2}$$

$$\rho = \frac{R_0 \times A}{L} \tag{3}$$

where: V is the voltage (V); R is the electrical resistance (Ω); i is the intensity of the DC current (A); R_0 is the initial electrical resistance of the sample, in Ω, which corresponds to the value recorded at the end of the electrification period and before the loading cycles; L is the distance between the copper electrodes (cm); and A is the embedded area of the electrodes (cm^2).

In the first sequence of tests, in which samples P1, M1, and M2 were evaluated, the described test setup was used, applying an electrical voltage of 5 V and shunt resistance of 100 Ω. In the second part, using the sample M3 and aiming to investigate the effects of the test configurations, both the applied DC voltage and shunt resistance were varied, respectively, from 4 to 12 V and from 6 to 100 kΩ.

3 Results and Discussion

3.1 Materials Characterization

Figure 3 shows the particle size distribution, obtained by laser diffraction analysis, of the cement, silica flour (S325), and sand fractions. Table 2 lists the chemical composition of the raw materials and their main physical properties. After CNT dispersion by sonication, there was a relative increase in the area under the UV–Vis spectrum and an increase in the absorbance peak, corresponding to the wavelength range around 260 nm (Fig. 4). Before sonication, the absorbance intensity of 0.16 at the peak demonstrated a low degree of dispersion of the nanotubes. After ≈66,000 J of sonication energy, the highest value of 1.31 at the peak suggested effective CNT dispersion in the water/superplasticizer solution.

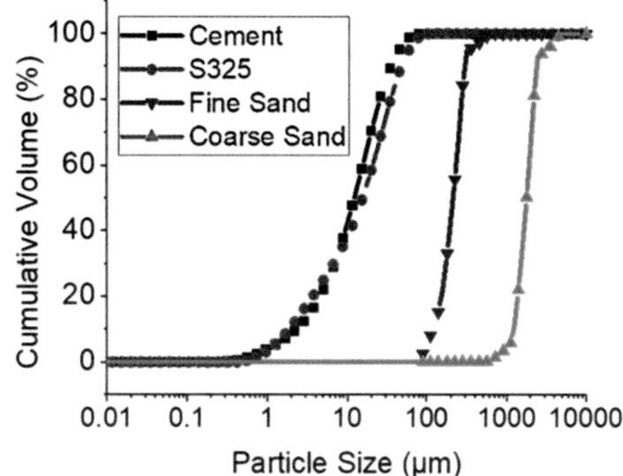

Fig. 3 Particle size distributions of the raw materials

Table 2 Chemical composition and main physical properties of the raw materials

Chemical composition (wt%)	Cement	S325	Fine sand	Coarse sand	CNT
CaO	66.37	–	ND	ND	0.111
SiO_2	14.043	97.140			0.532
Al_2O_3	3.681	1.570			7.013
Fe_2O_3	4.171				0.175
SO_3	3.933	0.813			0.456
K_2O	0.484	–			–
TiO_2	0.294	–			–
SrO	0.275	–			–
MnO	0.082	–			–
ZnO	0.033	0.011			–
CuO	0.011	0.012			–
ZrO_2	–	0.008			–
Co_2O_3	–	–			0.087
P_2O_5	–	–			0.441
Tm_2O_3	–	–			0.003
LOI	6.626	0.445			91.18
Physical properties					
Density (g/cm^3)	3.0676	2.6902	2.50	2.55	1.5
D50 (μm)	12.2	15.5	ND	ND	ND
Maximum size of aggregate (mm)	ND	ND*	0.6	2.4	ND
Fineness module	ND	ND*	0.9	2.9	ND

CNT, carbon nanotubes; ND, not determined

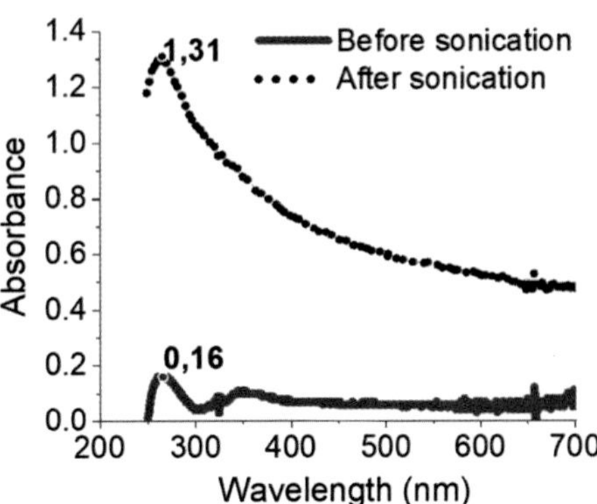

Fig. 4 UV–Vis spectra of carbon nanotubes in the water/superplasticizer solution before and after sonication process

3.2 *Effect of Fine Aggregate and CNT Content on Electrical and Piezoresistive Properties*

The initial resistivity ρ_0 of the composites P1, M1, and M2 was 100, 600, and 700 kΩ.cm, respectively. Figure 5 shows the strain and FCR variations as a function of time for these samples, under cyclic compressive loadings from 2 to 8; and 2 to 12 kN (0.8–3.2 and 4.8 MPa). Figure 6 shows the relationship between strain and FCR values during the self-sensing tests of samples P1, M1, and M2. No smoothing was applied to the obtained data.

A piezoresistive response of all cementitious composites can be seen in Fig. 5, once the FCR values decreased with the increase in both loading and strain, and increased with the reduction of strain, during the unloading step. During cyclic loading, under compression, the CNT came closer and conductive paths formed due to the contact and tunneling conductivities, allowing more electric current flow and reducing the electrical resistance of the sample. Upon unloading, the conductive network returned to its initial state and recovered its electrical resistance, increasing the FCR value [2, 11].

However, it should be noted that there was an increase in the electrical resistivity ρ_0, as well as a change in the amplitude and quality of the piezoresistive response with the addition of fine aggregate and the reduction in CNT content. Comparing paste P1 and mortar M1 (Fig. 5a, b), both with 0.75% of CNT content, it is notable that the aggregate addition affected the FCR variation. In P1, the load cycles generated a higher maximum amplitude of FCR, equal to 0.33, while for M1 a less sensitive response to loading was obtained, with a maximum FCR value of 0.15, besides a greater degree of noise. Naturally, this also affected the gauge factor (GF) values, equal to 1076 for P1 and 375 for M1, which is equivalent to the slope of the FCR–strain curve (Fig. 6) and represents the sensitivity of the composites. In addition, the highest coefficient of determination (R^2) was obtained in paste P1 (0.91), meaning that the FCR–strain curve was less scattered compared with mortar M1—noisier and with a lower R^2 of 0.52.

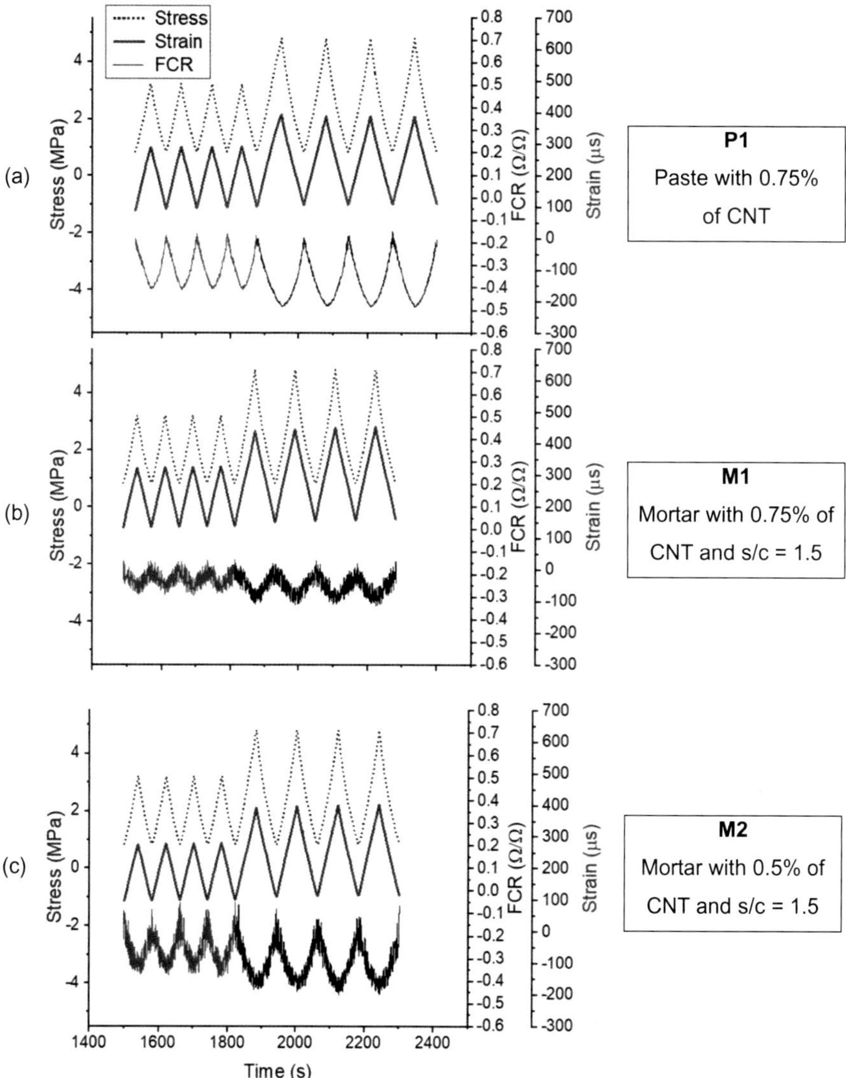

Fig. 5 Fractional change in resistance (FCR), stress, and strain responses as a function of time of **a** P1, **b** M1, and **c** M2 under cyclic compressive loading. CNT, carbon nanotubes; s/c, sand to cement ratio

This behavior is possibly explained by the addition of the fine aggregate, which besides modifying the mechanical response of the composite to loading, disturbed the conductive paths and electron mobility due its insulating nature. Previous evidence from the literature suggests that aggregates can constitute obstacles to current flow, negatively affecting the sensitivity and noise of the self-sensing response [2–4, 9, 12]. Therefore, the results obtained by this study reiterate that both the conductive

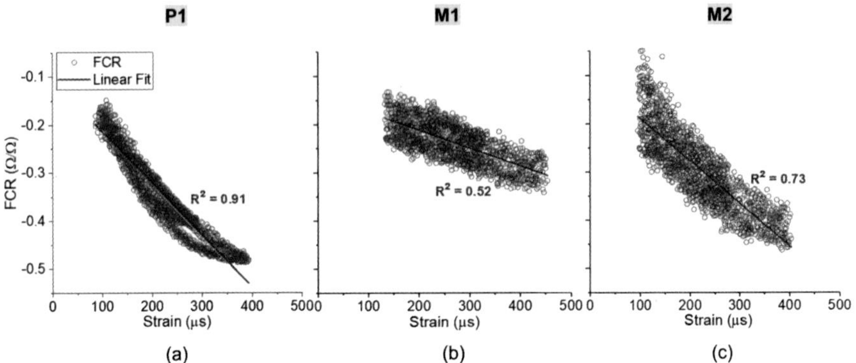

Fig. 6 Fractional change in resistance (FCR) versus strain diagrams of samples **a** P1, **b** M1, and **c** M2 under cyclic compressive loads

(CNT) and non-conductive phases (aggregate) affect the self-sensing performance of the composite.

Chiarello and Zinno observed that the magnitude of the conductivity of the system containing conductive fibers decreases exponentially with the increase of the s/c ratio [13]. García-Macías et al. [14] found variations in the resistivity of paste tenfold greater than those found for mortar and concrete. Dong et al., in a literature review, explained that the presence of the aggregate weakens the sensitivity of the composite, due to the imperfect crack controlling capacity of conductive fibers under the interaction with aggregates [2]. Moreover, due to the tendency of aggregates to separate the connectivity of the CNT conductive network. Both D'Alessandro et al. [12] and Han et al. [15] reported other adverse effects with the use of aggregates, such as reduced sensitivity, repeatability, and signal quality of self-sensing composites containing CNT.

The detection ability of the composites was also affected by the CNT concentration, as can be seen when comparing the mortars M1 and M2, with 0.75% and 0.5% of CNT, respectively. The maximum FCR amplitude obtained for these samples, corresponding to the compressive stress of 4.8 MPa, was equal to 0.15 and 0.29 respectively, suggesting a higher sensitivity of the composite containing 0.5% of CNT. It can be seen that, compared with M1, the self-sensing response of sample M2 showed higher linearity (with $R^2 = 0.73$) and detection sensitivity (GF 883).

This behavior is probably due to the CNT content of M2 being closer to the percolation threshold, which generated a conductive network with a greater capacity to rearrange under loading, resulting in more significant changes in electrical resistivity and greater sensitivity. In the case of samples with 0.75% of CNT, the concentration of conductive filler was closer to the conductivity zone, in which conductive networks are more stable and denser, leading to a lower sensitivity [1–4, 9]. Thus, the self-sensing properties are also dependent on the CNT concentration in the matrix and its distribution, which is negatively affected by the distribution of the aggregates, once they are the non-conductive phase.

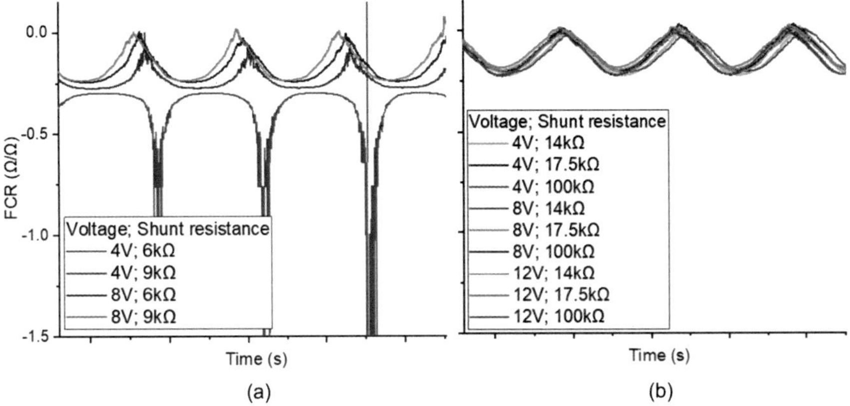

Fig. 7 Fractional change in resistance (FCR) response under 10 kN cyclic compressive load with **a** varying voltage from 4 to 8 V and shunt resistance from 6 to 9 kΩ and **b** varying voltage from 4 to 12 V and shunt resistance from 14 to 100 kΩ

3.3 *Effect of Test Setup on Electrical and Piezoresistive Properties*

Figure 7 shows the variation in FCR for sample M3, under 10 kN of cyclic compressive loading, when both the applied DC voltage and shunt resistance were varied from 4 to 12 V and from 6 to 100 kΩ, respectively. In Fig. 7a the lower values of electrical resistance of the shunt resistor led to noisier, intensified, and even distorted responses. From 14 kΩ of shunt resistance, as seen in Fig. 7b, the amplitude of the FCR variation over time became approximately equal, regardless of the voltage and shunt value adopted, suggesting that the response to loading in those cases was solely dependent on the matrix properties.

Supposedly, lower shunt resistances interfere with their acquired voltages, negatively affecting the current and matrix resistance values derived from these measurements. Higher shunt resistances, compatible with the magnitude of resistance variations observed in the cement matrix, seemed to cause a better resolution of the acquired data. Therefore, the present results suggested that the higher noise level observed was not only a property of the matrix but also dependent on the test configuration. The latter is easily adjustable to allow a more efficient acquisition of electrical signals.

Previous investigations suggest that factors related to the measurement of electrical signals in the self-sensing test can affect the intensity and stability of the piezoresistive response—for example, the configuration of the electrodes or the current type and its magnitude—and therefore should be properly adjusted [2, 10, 16, 17]. Galao et al. applied varying electrical currents to the self-sensing composites (0.1, 1.0, and 10 mA) and found that the piezoresistive response was better with increasing current intensity, achieving better signal stability and correlation with the strain of material [17]. On the other hand, among the applied electrical voltages of 10, 20, and 30 V, Konsta-Gdoutos and Aza obtained an optimal voltage of 20 V, suggesting that high electrical current intensities can also be harmful [16]. Ding et al. evaluated self-sensing cementitious composites using reference resistors with 1000 Ω of electrical resistance [18], and other researchers used the same circuit model but without mentioning the value of shunt resistance [19].

Based on these results, our setup was defined with a shunt resistor of 27 kΩ and a DC voltage of 4 V to evaluate mortar M3. Figure 8a shows the results under cyclic loading of 3.2 MPa for these conditions. Comparing an excerpt of the M2 test using the first setup (Fig. 8b), a clear change in the self-sensing response can be seen, with considerable noise reduction of the signal. The better self-sensing response suggested that not only the concentrations of conductive filler and fine aggregate affect the electrical signals acquired in self-sensing tests. The improved response can be explained in part by the decrease in the s/c ratio but seems to be mainly due to the change in the test settings.

The R^2 coefficient observed in Fig. 9b, equal to 0.96, was even higher than that obtained for the paste P1, equal to 0.91, which had more CNT and no aggregate. The mortar M3 also showed good detection sensitivity, represented by GF of 693.

Yoo et al. [5] investigated the self-sensing performance of cement pastes and verified that the sample containing 1.0% by volume of CNT presented a maximum amplitude of the FCR equal to 0.26 under a compressive load of 40 kN. The GF of the composites ranged from 77.2 to 95.5 with a minimum R^2 of 0.9382. Yin et al. [11] produced cement pastes with 1.7% CNT, by volume, with a high coefficient of sensitivity to deformation (1500), high linearity ($R^2 = 0.97$), and maximum FCR equal to 0.19 under cyclic compressive loading of 10 MPa. The hybrid combination of

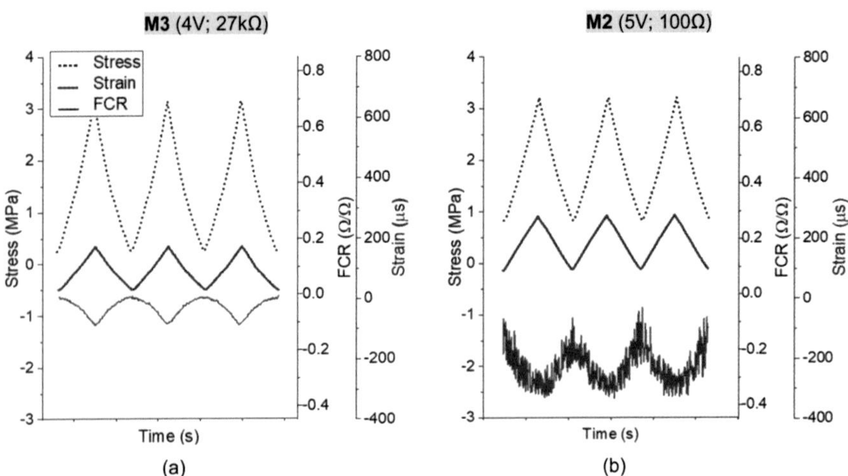

Fig. 8 Fractional change in resistance (FCR), stress and strain responses as a function of time of samples **a** M3 and **b** M2 under cyclic compressive loading

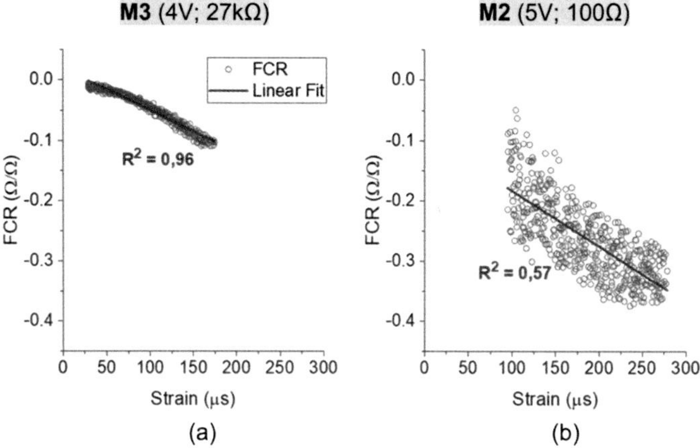

Fig. 9 Fractional change in resistance (FCR) versus strain diagrams of **a** M3 and **b** M2 mortars under cyclic compressive loading

CNT and nickel nanofibers resulted in the best piezoresistive sensitivity and response linearity under the same load, with a maximum FCR equal to 0.24, GF of 1880, and R^2 of 0.99.

4 Conclusions

In this study, the self-sensing properties of cementitious composites containing fine aggregate and varied content of CNT were evaluated under compressive cyclic loading. In the first step, the paste P1 with 0.75% CNT led to higher linearity and sensitivity of the self-sensing response and a low degree of noise in the electrical signal, when compared with mortars M1 and M2. The electrically inert aggregate possibly modified the mechanical response of the composite to compressive load and constituted an obstacle to electron flow, increasing tortuosity or even interrupting the conductive paths of the CNT. Comparing the mortars M1 and M2, with 0.75% and 0.5% of CNT respectively, higher detection sensitivity and linearity were verified for the composite with lower CNT content—closer to the percolation threshold. At this CNT concentration, a conductive network formed with a greater capacity to rearrange under loading, resulting in more significant changes in electrical resistivity. On the other hand, higher concentrations of CNT, especially closer to the conductivity zone, formed denser and more stable conductive networks, leading to lower detection sensitivities. These results confirmed that both the conductive and non-conductive phases affect the electrical and self-sensing behavior of cementitious composites containing CNT. The conductive network is, therefore, formed by an appropriate filler concentration, and the choice of aggregate, as well as the appropriate proportions, is extremely important for the fabrication of self-sensing cementitious composites with good detection sensitivity and signal stability.

Additionally, the test setup configuration should be also taken into account when investigating the self-sensing performance of cementitious composites. We verified that the test settings can allow more efficient acquisition of electrical signals and even reduce the noise caused by the addition of aggregates. In a circuit with a reference resistor, lower values of shunt resistance led to noisy, scattered, and distorted electrical responses, whereas shunt resistances from 14 to 100 kΩ led to a better self-sensing response signal, with reduced noise and higher linearity. Under these conditions, the FCR under loading was always less scattered and with the same amplitude, reinforcing its capability of reflecting the piezoresistive properties of the matrix. This suggests that strategic adjustments in the self-sensing test settings may be particularly useful to reduce the deleterious effects of added sand in the matrix, enabling its use as aggregate in self-sensing composites while maintaining satisfactory detection performance. This approach may benefit the implementation of smart mortars in structural health monitoring. In addition, it can contribute to the massive use of aggregates in this type of composite, consequently lowering the consumption of raw materials with higher cost and environmental impacts such as CNT and Portland cement.

References

1. Han B, Yu X, Ou J (2014). Self-sensing concrete in smart structures. https://doi.org/10.1016/C2013-0-14456-X
2. Dong W, Li W, Tao Z, Wang K (2019) Piezoresistive properties of cement-based sensors: Review and perspective. Constr Build Mater 203:146–163
3. Tian Z, Li Y, Zheng J, Wang S (2019) A state-of-the-art on self-sensing concrete: Materials, fabrication and properties. Compos Part B Eng 177:107437
4. Wang L, Aslani F (2019) A review on material design, performance, and practical application of electrically conductive cementitious composites. Constr Build Mater 229:116892
5. Yoo D-Y, You I, Zi G, Lee S-J (2019) Effects of carbon nanomaterial type and amount on self-sensing capacity of cement paste. Meas J Int Meas Confed 134:750–761
6. Han B, Yu X, Zhang K, Kwon E, Ou J (2011) Sensing properties of CNT-filled cement-based stress sensors. J Civ Struct Heal Monit 1:17–24
7. Nalon GH et al (2021) Residual piezoresistive properties of mortars containing carbon nanomaterials exposed to high temperatures. Cem Concr Compos 121:104104
8. Le HV, Kim MK, Kim SU, Chung SY, Kim DJ (2021) Enhancing self-stress sensing ability of smart ultra-high performance concretes under compression by using nano functional fillers. J Build Eng 44:102717
9. Han B, Ding S, Yu X (2015) Intrinsic self-sensing concrete and structures: a review. Meas J Int Meas Confed 59:110–128
10. Abedi M, Fangueiro R, Gomes Correia A (2021) A review of intrinsic self-sensing cementitious composites and prospects for their application in transport infrastructures. Constr Build Mater 310:125139
11. Yin T, Xu J, Wang Y, Liu L (2020) Increasing self-sensing capability of carbon nanotubes cement-based materials by simultaneous addition of Ni nanofibers with low content. Constr Build Mater 254:119306

12. D'Alessandro A, Rallini M, Ubertini F, Materazzi AL, Kenny JM (2016) Investigations on scalable fabrication procedures for self-sensing carbon nanotube cement-matrix composites for SHM applications. Cem Concr Compos 65:200–213
13. Chiarello M, Zinno R (2005) Electrical conductivity of self-monitoring CFRC. Cem Concr Compos 27:463–469
14. García-Macías E, D'Alessandro A, Castro-Triguero R, Pérez-Mira D, Ubertini F (2017) Micromechanics modeling of the uniaxial strain-sensing property of carbon nanotube cement-matrix composites for SHM applications. Compos Struct 163:195–215
15. Han B, Yu X, Kwon E, Ou J (2010) Piezoresistive multi-walled carbon nanotubes filled cement-based composites. Sens Lett 8:344–348
16. Konsta-Gdoutos MS, Aza CA (2014) Self sensing carbon nanotube (CNT) and nanofiber (CNF) cementitious composites for real time damage assessment in smart structures. Cem Concr Compos 53:162–169
17. Galao O, Baeza FJ, Zornoza E, Garcés P (2014) Strain and damage sensing properties on multifunctional cement composites with CNF admixture. Cem Concr Compos 46:90–98
18. Ding S et al (2022) In-situ synthesizing carbon nanotubes on cement to develop self-sensing cementitious composites for smart high-speed rail infrastructures. Nano Today 43:101438
19. Xiao H, Liu M, Wang G (2018) Anisotropic electrical and abrasion-sensing properties of cement-based composites containing aligned nickel powder. Cem Concr Compos 87:130–136

Effect of Carbonation on the Microstructure and Phase Development of High-Slag Binders

M. J. Tapas, A. Yan, P. Thomas, C. Holt, and V. Sirivivatnanon

Abstract The drive for sustainable concrete production favors the use of high replacement levels of supplementary cementitious materials (SCMs) in the concrete mix. The use of SCMs such as fly ash and slag, however, although they improve the sustainability of concrete production as well as most concrete durability properties, increases the carbonation rate. Carbonation decreases the pH of the concrete pore solution, making the steel reinforcement susceptible to corrosion. The effect of carbonation is, however, not confined to the change in pH of the pore solution. We investigated changes in the microstructure and phases of high-slag binders due to carbonation. The carbonation resistance of mortars with 50 and 70% slag replacement were investigated at exposure conditions of 2%CO_2, 50%RH, 23 °C. The carbonated and non-carbonated parts of the mortars were subjected to various characterization techniques to investigate the effect of carbonation on microstructure and phase development. Results confirmed the absence of portlandite in all the carbonated regions ("colorless" by phenolphthalein test, which indicated that the change in color of the phenolphthalein solution was due to the absence of portlandite to buffer the pH). Significant reduction in the amount of C-S-H, as well as increase in the amount of calcium carbonate, were been observed in the carbonated regions. Aragonite, a polymorph of $CaCO_3$, was very prominent in all the carbonated mortars.

Keywords Carbonation · Microstructure · Slag

M. J. Tapas (✉) · A. Yan · P. Thomas · V. Sirivivatnanon
School of Civil and Environmental Engineering, University of Technology Sydney, Ultimo, NSW, Australia
e-mail: mariejoshua.tapas@uts.edu.au

C. Holt
Boral Innovation, Maldon, NSW, Australia

W. Duan et al. (eds.), *Nanotechnology in Construction for Circular Economy*, Lecture Notes in Civil Engineering 356,
https://doi.org/10.1007/978-981-99-3330-3_22

1 Introduction

Concrete, the most utilized construction material and the second most used substance on earth after water, creates significant amount of CO_2 emissions during production [1]. The CO_2 emissions primarily originate from the calcination of limestone ($CaCO_3$) to produce cement (main binder material in concrete), releasing CO_2 in the process. As CO_2 contributes to global warming, there has been a strong focus on the decarbonation of concrete, and reducing the amount of cement per cubic meter of concrete is a proven approach to deliver on the target CO_2 reduction. Reducing the cement content can be achieved through partial cement substitution with supplementary cementitious materials (SCMs) such as slag or fly ash. Slag, being hydraulic in nature like cement, can be used at higher substitution rates than other SCMs (usually $\geq 50\%$), translating to better CO_2 reduction. However, although slag notably improves the later-age strength as well as most concrete durability properties, including chloride ingress [2], the alkali–silica reaction [3–5], sulfate resistance [6, 7], and delayed ettringite formation [8], its use results in increased susceptibility of the concrete to carbonation [9–11].

Carbonation refers to the ingress of CO_2 into the binder system, which results in the formation of carbonic acid (H_2CO_3) that further dissociates into H^+ and CO_3^{2-} and reacts with calcium ions in the pore solution, resulting in the precipitation of calcium carbonate ($CaCO_3$) and a decrease in the pH of the pore solution. Low concrete pH (≤ 9.5) resulting from carbonation is detrimental to steel-reinforced concrete because steel begins to lose its passivation layer at low pH, making it susceptible to corrosion [10]. Thus, carbonation is a serious durability concern, particularly for steel-reinforced concrete [12].

Phenolphthalein is an indicator used to assess the depth of carbonation. It is colorless at lower pH values (≤ 9), whereas at pH > 10.5, it presents a characteristic purple or magenta. Because mortar or concrete that has been carbonated has pH ≤ 9, phenolphthalein is used to visually confirm the drop in pH due to carbonation [13]. The effect of carbonation is, however, not confined to the change in pH of the pore solution. Although it has been reported that carbonation in general is beneficial and results in an increase in compressive strength due to the conversion of $Ca(OH)_2$ to $CaCO_3$, which increases the volume of the binder and reduces porosity [10], it appears that this may not be true for all binder systems. It has been reported that although moderate carbonation can improve the mechanical properties of the concrete, excessive carbonation impairs mechanical strength due to the decalcification of the C-S-H [14]. High-slag concrete is also particularly susceptible to carbonation shrinkage [9]. Therefore, due to variability in the reported effect of carbonation and considering the increasing levels of slag being used in concrete production, a better understanding of the effect of carbonation on microstructure and phase development is required.

We investigated the effect of carbonation on the microstructure and phase development of high-slag mortars (50% and 70% slag replacement). The mortars were characterized after being subjected to accelerated carbonation conditions (2%CO_2, 23 °C, and 50% relative humidity (RH) for 112 days.

Table 1 Mortar mixes

Mix details	Normen sand (g)	Cement (g)	Slag (g)	Total binder (g)	Water (g)
OPC	1350	450.0	0.0	450.0	171.0
OPC + 50% slag	1350	225.0	225.0	450.0	171.0
OPC + 70% slag	1350	135.0	315.0	450.0	171.0

OPC, ordinary Portland cement

2 Methods

2.1 *Raw Materials*

We used General Purpose cement and slag that complied with AS3972 and AS3582.2 respectively. Normen sand (CEN Standard Normsand according to EN196-1) was used as fine aggregate. The maximum moisture content of sand was 0.2%.

2.2 *Carbonation Test*

The 40 × 40 × 160 mm mortars were prepared using Normen sand in combination with ordinary Portland cement (OPC) and OPC + slag binders (slag at 50% and 70% replacement) at 0.45 water to cement ratio. Table 1 shows the mortar mixes investigated in this study.

The mortars were demolded after 1 day, cured for 28 days inside a sealed moisture bag and then transferred into the shrinkage room (50%RH, 23 °C) to air cure for 7 days in preparation for the accelerated carbonation test. At age 35 days (28 + 7 days), the mortars were transferred into the carbonation chamber running at 2%CO_2, 50%RH and 23 °C for the accelerated carbonation test. Carbonation depth measurements using phenolphthalein (1% solution) were carried out after 1 week (7 days), 4 weeks (28 days), 9 weeks (63 days), and 16 weeks (112 days) exposure of the mortars in the carbonation chamber.

2.3 *Characterization of the Mortars*

After 112 days carbonation, the "colorless" and "pink regions" of the mortars were subjected to thermogravimetric analysis (TG; SDT-Q600 Simultaneous TGA/DSC equipment, TA Instruments) and scanning electron microscopy (SEM). The mortar specimens were ground and 50-mg samples of the ground material were transferred to a platinum crucible, which was placed inside the TG instrument. The thermal analysis was performed in a nitrogen gas atmosphere, within a temperature range of 23–900 °C and at a heating rate of 10 °C/min.

All imaging and elemental analyses of the fractured mortars were performed using a Zeiss Supra 55VP SEM fitted with a Bruker SDD EDS Quantax 400 system and FEI Quanta 200 with Bruker XFlash 4030 EDS detector. The microscopes were operated at 15 kV accelerating voltage and 12.5 mm working distance.

3 Results and Discussion

Figure 1 shows the carbonation depths at 16 weeks (112 days). As expected, the plain OPC mortar showed the best resistance to carbonation (largest pink region) while mortars with slag exhibited poorer resistance. The higher the slag replacement, the poorer the carbonation performance.

Figure 2 shows the increase in carbonation depth over time, with the mortar with 70% slag consistently having the highest carbonation depth and fully carbonated at 63 days.

Figure 3 shows the TG curves of the non-carbonated (pink) and carbonated parts (colorless) of the plain OPC mortar, mortar with 50% slag and the mortar with 70% slag. Mass loss in the range of calcium hydroxide (CH) decomposition at ≈400–500 °C corresponds to the dehydroxylation of CH, ($Ca(OH)_2 \rightarrow CaO + H_2O$) [15]. Therefore, the area under the curve at ≈400–500 °C corresponds to the amount of CH, and thus a larger area means more CH. Comparing the CH content of plain OPC mortar and the 50% slag mortar (pink regions), OPC notably has more CH, as may be expected, which explains its better carbonation resistance. During carbonation, portlandite, which is the most soluble source of calcium in the binder, serves as a buffer and maintains the pH of the pore solution by dissolving and releasing OH^- and Ca^{2+} ions. The OH^- neutralizes the H^+ while Ca^{2+} binds $CO_3{}^{2-}$, precipitating $CaCO_3$ [11]. Therefore, due to the lower amount of portlandite, high-slag binders carbonate much faster (i.e. pH drops faster) than pure cement. Absence of portlandite in the

Fig. 1 Photos of the mortars (OPC, OPC + 50% slag and OPC + 70% slag) at 112 days exposure in the carbonation chamber showing the carbonation depth determined using phenolphthalein

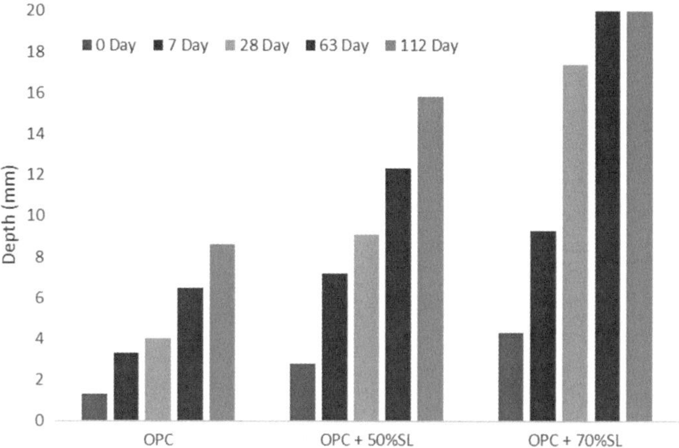

Fig. 2 Carbonation depth measurements before carbonation (Day 0) and after 7, 28, 63 and 112 days exposure to 2%CO_2 at 23 °C and 50% relative humidity. OPC, ordinary Portland cement; SL, slag

"colorless regions" of the plain OPC and 50% slag mortars are notable indicating the full consumption of portlandite in the carbonated regions. There is also no portlandite remaining in the 70% slag mortar, consistent with it being fully carbonated (top and middle areas of the mortar were tested). Moreover, consequent to the full consumption of portlandite in the "colorless regions" of all mortars (i.e., fully carbonated regions), there was a drastic increase in the amount of $CaCO_3$. A notable decrease in the amount of C-S-H, carboaluminates, and ettringite (TG region 0–300 °C) can also be seen, because once portlandite has been fully consumed, the other calcium-bearing phases start to react with CO_2 and carbonate as well [10]. Decalcification of the C-S–H can occur, resulting in carbonation shrinkage [12].

Figure 4 shows the amount of $CaCO_3$ in the different binder systems (carbonated regions) calculated from the decarbonation region ($CaCO_3 \rightarrow CaO + CO_2$). The higher the amount of $CaCO_3$ formed, the higher the CO_2 binding capacity. CO_2 binding capacity is related to the amount of CaO in the binder and because the higher the slag replacement, the lower the CaO available, the CO_2 binding capacity also decreases.

SEM images of the fractured "carbonated" 50% and 70% slag mortars are shown in Figs. 5 and 6 respectively. The presence of aragonite (a $CaCO_3$ polymorph) is very prominent in both systems. The microstructure also appears to be porous, although the change in porosity due to carbonation should be quantified.

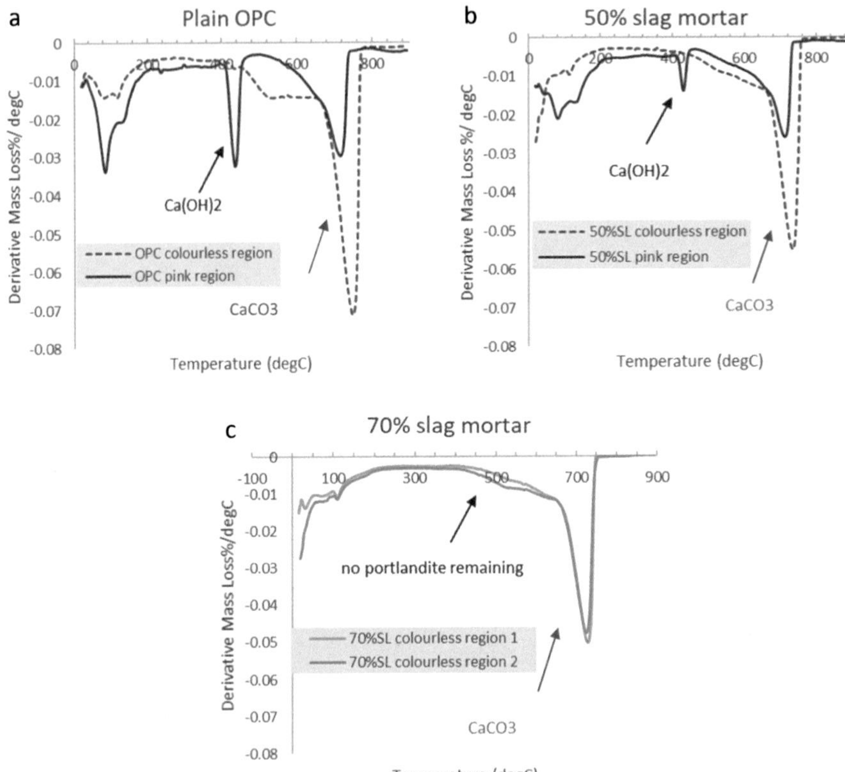

Fig. 3 Derivative thermogravimetric curves of the carbonated and non-carbonated part of plain OPC, 50%SL mortar and 70%SL mortar. OPC, ordinary Portland cement; SL, slag

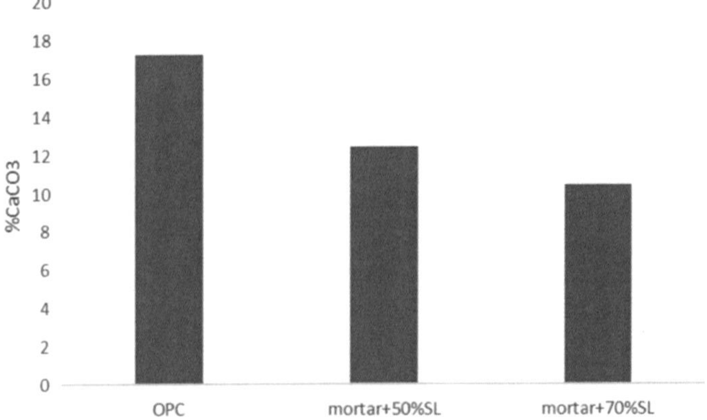

Fig. 4 Percentage $CaCO_3$ in the plain OPC, 50%SL mortar and 70%SL mortar determined from the thermogravimetric mass loss measurements. OPC, ordinary Portland cement; SL, slag

Fig. 5 Scanning electron microscopy images of the carbonated 50%slag mortar

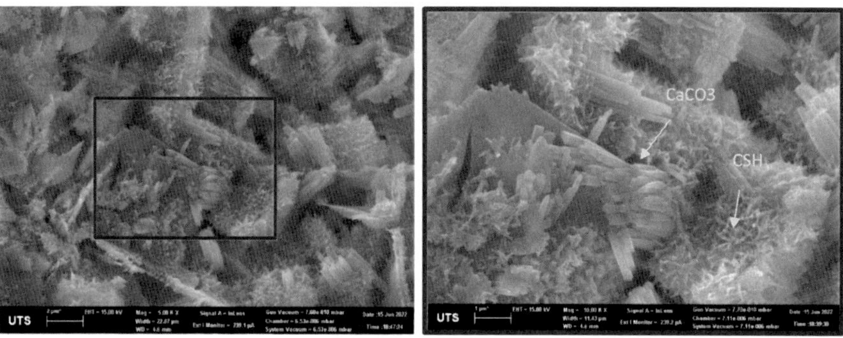

Fig. 6 Scanning electron microscopy images of the carbonated 70%slag mortar

4 Conclusions

In this study we investigated the effect of carbonation on the microstructure and phase development of high-slag binders. Relevant results are as follows.

1. High-slag binder carbonates faster than plain OPC binder. The higher the slag content, the higher the carbonation rate.
2. TG analysis confirmed the absence of portlandite and increased $CaCO_3$ in the carbonated regions ("colorless" regions), which suggests that the trigger for the drop in the pH (i.e., the change in color of phenolphthalein from "pink" to "colorless") is the absence of portlandite to buffer the pH.
3. Significant reduction in the amount of C-S-H was also observed, suggesting decalcification of C-S-H, and that all calcium-bearing phases are prone to carbonation.
4. The higher the slag replacement, the lower the CO_2 binding capacity, which is consistent with the reduced CaO content of high-slag binders as well as their higher carbonation rate.

5. SEM images confirmed the prominent presence of aragonite (a $CaCO_3$ polymorph) in the carbonated mortars.

Acknowledgements This study was carried with the support of the UTS-Boral Centre for Sustainable Building under funding from the Innovative Manufacturing CRC (IMCRC).

References

1. Scrivener KL, John VM, Gartner EM (2018) Eco-efficient cements: potential economically viable solutions for a low-CO_2 cement-based materials industry. Cem Concr Res 114:2–26
2. Thomas MDA (2013) Supplementary cementing materials in concrete. Taylor and Francis Group, LLC, Boca Raton, Florida
3. Thomas M (2011) The effect of supplementary cementing materials on alkali-silica reaction: a review. Cem Concr Res 41:1224–1231
4. Tapas MJ, Sofia L, Vessalas K, Thomas P, Sirivivatnanon V, Scrivener K (2021) Efficacy of SCMs to mitigate ASR in systems with higher alkali contents assessed by pore solution method. Cem Concr Res 142:106353
5. Duchesne J, Berube M-A (2001) Long-term effectiveness of supplementary cementing materials against alkali–silica reaction. Cem Concr Res 31:1057–1063
6. Menéndez E, Matschei T, Glasser FP (2013) Sulfate Attack of Concrete. In: Alexander M, Bertron A, De Belie N (eds) Performance of cement-based materials in aggressive aqueous environments: state-of-the-art report, RILEM TC 211—PAE. Springer, Netherlands, Dordrecht, pp 7–74
7. Van Tittelboom K, De Belie N, Hooton RD (2013) Test methods for resistance of concrete to sulfate attack—a critical review. In: Alexander M, Bertron A, De Belie N (eds) Performance of cement-based materials in aggressive aqueous environments: state-of-the-art report, RILEM TC 211 - PAE. Springer, Netherlands, Dordrecht, pp 251–288
8. Snellings R, Scrivener KL (2016) Rapid screening tests for supplementary cementitious materials: past and future. Mater Struct 49:3265–3279
9. Gruyaert E, Van den Heede P, De Belie N (2013) Carbonation of slag concrete: effect of the cement replacement level and curing on the carbonation coefficient—effect of carbonation on the pore structure. Cement Concr Compos 35:39–48
10. von Greve-Dierfeld S, Lothenbach B, Vollpracht A, Wu B, Huet B, Andrade C, Medina C, Thiel C, Gruyaert E, Vanoutrive H, Saéz del Bosque IF, Ignjatovic I, Elsen J, Provis JL, Scrivener K, Thienel K-C, Sideris K, Zajac M, Alderete N, Cizer Ö, Van den Heede P, Hooton RD, Kamali-Bernard S, Bernal SA, Zhao Z, Shi Z, De Belie N (2020) Understanding the carbonation of concrete with supplementary cementitious materials: a critical review by RILEM TC 281-CCC. Mater Struct 53:136
11. Saillio M, Baroghel-Bouny V, Pradelle S, Bertin M, Vincent J, d'Espinose de Lacaillerie J-B (2021) Effect of supplementary cementitious materials on carbonation of cement pastes. Cem Concr Res 142:106358
12. Borges PHR, Costa JO, Milestone NB, Lynsdale CJ, Streatfield RE (2010) Carbonation of CH and C-S–H in composite cement pastes containing high amounts of BFS. Cem Concr Res 40:284–292
13. Chinchón-Payá S, Andrade C, Chinchón S (2016) Indicator of carbonation front in concrete as substitute to phenolphthalein. Cem Concr Res 82:87–91
14. Chen T, Gao X (2019) Effect of carbonation curing regime on strength and microstructure of Portland cement paste, Journal of CO2 Utilization, 34:74–86

15. Lothenbach B, Durdziński P, Weerdt KD (2016) Thermogravimetric analysis. In: Scrivener K, Snellings R, Lothenbach B (eds) A practical guide to microstructural analysis of cementitious materials. Taylor and Francis

A New Dispersion Strategy to Achieve High Performance Graphene-Based Cement Material

Z. Zhang, Y. Yao, H. Liu, Y. Zhuge, and D. Zhang

Abstract The addition of graphene and its derivatives can enhance the mechanical and functional properties of cement-based composites, but most of the current technologies have limited dispersion and are costly. The creation of a cost-effective graphene-reinforced cement material with uniform graphene dispersion remains difficult. We used glucose as an economical carbon source to induce the in-situ formation of graphene on cement particles. Our proposed method is approximately 80% less expensive than commercial techniques. Evaluation of the microscopic morphology demonstrated uniform distribution of graphene in the cement matrix, which improved the mechanical properties of the cement paste. The compressive strengths of the test groups with 3% carbon source improved by almostly 38% and 48.9%, respectively, compared with pure cement paste. This newly established technique is essential for the future design of excellent graphene-based cement materials and the achievement of multifunctional cementitious applications.

Keywords Cement · Dispersion effect · Graphene · In-situ growth strategy · Mechanical properties

Z. Zhang · Y. Yao (✉)
School of Civil Engineering, Xi'an University of Architecture and Technology, Xi'an, China
e-mail: yaoy@xauat.edu.cn

H. Liu
School of Chemistry and Chemical Engineering, Xi'an University of Architecture and Technology, Xi'an, China

Y. Zhuge
UniSA STEM, University of South Australia, Adelaide, SA, Australia

D. Zhang
College of Civil Engineering, Fuzhou University, Fuzhou, China

W. Duan et al. (eds.), *Nanotechnology in Construction for Circular Economy*, Lecture Notes in Civil Engineering 356,
https://doi.org/10.1007/978-981-99-3330-3_23

1 Introduction

Over the past decades, a significant amount of research has been devoted to manipulating the structure of cement hyadration products and the mechanical properties of cement at the nanoscale by using a wide range of nanomaterials such as nanoscale silicon dioxide [1], carbon nanotubes (CNTs) [2], and graphene-based materials [3–7]). Because graphene-based materials are two-dimensional, they possesses good physical and chemical characteristics, making them a suitable option for the next generation of improved cement-based material [8–12]. Many studies [13–17] have demonstrated that graphene and its derivatives, such as graphene oxide (GO), can effectively improve the mechanical characteristics of cementitious materials by increasing the hydration process of the cement and altering the pore distribution in the matrix. Increased durability of GO-reinforced cement mortar can be achieved with only a small amount of additional GO [18]. However, the high cost and poor dispersion of graphene-based compounds prevent their future practical application. The dispersion of graphene materials in the cement matrix [19, 20] is the primary factor that determines how well graphene can reinforce cementitious materials.

It is possible to attribute the aggregation of graphene and its derivative GO to the powerful van der Waals force that exists between nanomaterials as well as the linking effect that Ca^{2+} and Mg^{2+} ions have on GO in the cement environment. This is because van der Waals forces are known to exist between nanomaterials [21–23]. Graphene and GO in aqueous solution have been prestabilized using a variety of chemical and physical techniques in order to address the issue of their poor dispersion. Graphene-modified cement can be manufactured by combining cement with the prestabilized aqueous solution [14, 23–26]. However, treatment with ultrasonication for extended periods of time and functionalization with strong acids have adverse impacts on graphene materials, which might cause flaws in the graphene structure.

Here, we describe a novel and uncomplicated technique for the synthesis of graphene–cement (GC) composite. This strategy involves the in-situ development of graphene in the cement matrix (Fig. 1) by carbonization and calcination [27, 28]. In the course of the synthesis procedure, the glucose used as the carbon source is thoroughly combined with the cement [27, 29]. In order to obtain advanced GC material, the mixture is heated further at 800 °C for 2 h, during which the glucose is converted into graphene on the cement particles, which inhibits aggregation and ensures well-dispersed graphene in the cementitious matrix.

2 Methods

2.1 Materials

Glucose ($C_6H_{12}O_6$, 98%; Sigma Aldrich) and GO (Suzhou TANFENG graphene Tech) acted as the carbon source and reinforcement material, respectively. Ordinary Portland cement (P.O. 42.5; Jiuqi Building Components) and ethanol (C_2H_5OH, AR.; Sinopharm Chemical Reagent) were also used in this work.

2.2 Sample Preparation

It is a well known that incorporating a negligible quantity of GO (0.05% by weight) into cement materials can dramatically improve the mechanical characteristics of these materials [5, 13, 30–32]. As a result, we carried out a control experiment of adding 0.05 wt% of GO to 100 g of cement. After completely mixing the GO with water, the mixture was sonicated for 30 min by ultrasonic equipment at the highest possible power setting (500 W). Next, the GO solution was included in the cement mix. The production of the new cement paste infused with GO (GOP) followed the same procedure as that of the GC material. In terms of the overall weight of binder, the quantity of GO that was utilized was equal to 0.05 wt%. These techniques provide assurance that the cementitious materials will be capable of meeting the casting standards at the location of the construction project. The GO that is sold commercially has a thickness of roughly 1 nm and a diameter that may reach a maximum of 10 μm. The substance known as GC was created by heating a combination of cement and glucose (carbon source) as shown in Fig. 1. The ratio of cement to glucose powder varied as follows: 100:1, 100:3, and 100:6. The cement and glucose were homogeneously mixed manner at room temperature for 1 h at a rotational speed of 1000 rpm. The mixture was then placed in a furnace to undergo the reaction under a

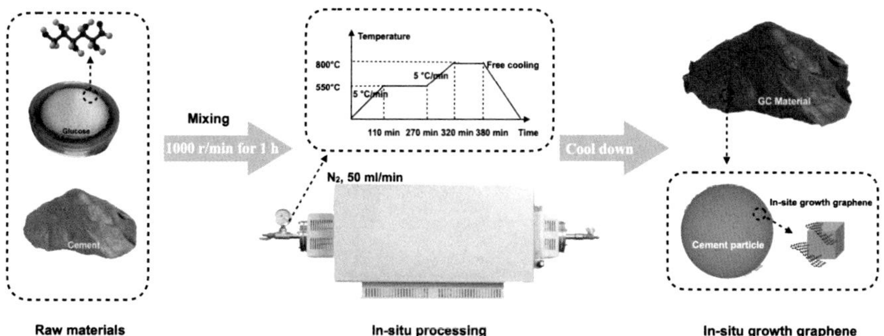

Fig. 1 Diagram of in-situ graphene grown on cement particles

Table 1 Mix proportions of different groups

Group	Cement	Water	GO/glucose
GP-0	100	35	/
GP-1	100	35	1
GP-3	100	35	3
GP-6	100	35	6
GOP	100	35	0.05

* The names of the samples were based on the graphene oxide (GO) or graphene concentration. For example, GP-0 denotes pure cement paste. GOP refers to cement paste containing 0.05% GO. GP-1 refers to the cement paste containing 1% carbon source prior to graphitization

specified high-temperature program while being shielded by N_2 as shown in Fig. 1. In this particular investigation, the oven was initially purged with nitrogen at a rate of 50 mL/min before being heated from 25 to 550 °C at a rate of 5 °C/min over 60 min. The temperature inside the furnace then increased by 5 °C every minute until it reached 800 °C, where it remained for 2 h before gradually decreasing to the temperature of the surrounding air. The amount of graphene could be changed by adjusting the bulk amount of glucose, and the yield of pure graphene was determined by using the same techniques. The percentage of graphene that can be yielded was 7%. In order to investigate the effect that in-situ grown graphene development had on the mechanical characteristics of cement paste, we divided the samples into 4 groups: GP-0, GP-1, GP-3 and GP-6 according to the addition amount of glucose before heat treatment as presented in Table 1. Fresh cement paste containing the in-situ growing graphene was pressed into plastic molds with dimensions of 50 × 50 × 50 mm (for the compressive test) and 40 × 40 × 160 mm (for the flexure test). The molds were removed after 24 h and the samples placed in an environment with a relative humidity of 95% for 3, 7, and 28 days, respectively.

2.3 Experimental Methods

In the current investigation, a porosimeter capable of detecting pores ranging in size from 5 nm to 100 μm was utilized (equivalent to pressures of 206 MPa and 345 kPa, respectively, which are the highest and minimum that were applied). Raman spectra were obtained with a Renishaw spectrometer that included an excitation laser wavelength of 532 nm. The instrument was also fitted with ×50 lens and was focused to a spot size that was 1 m in diameter. The scanning electron microscopy (SEM) examination was performed in order to investigate the microstructure. The compression test of cubic specimens with dimensions of 50 × 50 × 50 mm and the three-point flexure test of beam members with dimensions of 40 × 40 × 160 mm are both common methods for determining the mechanical properties of cementitious materials. Both of these tests were performed on specimens with dimensions of 50 × 50 × 50 mm. When calculating the amount of compressive stress, the applied force is divided by the area that is being loaded. The loading rate was set at 1.2 mm/min for the compressive strength test, and at 0.05 mm/min for the flexure test.

3 Results and Discussion

3.1 Characterization of GC Material

The conversion of glucose (carbon source) into graphene on cement particles, which took place during the manufacturing process of GC material, was an essential step in our process of dispersing graphene evenly throughout the cement composites. Figure 2 illustrates the morphology of graphene as well as GC. The results of tests using energy dispersive X-ray spectroscopy (EDS) showed a distinct distribution of carbon (Fig. 2b, c). EDS mapping of the GC material showed that the elements carbon, oxygen, calcium, and silicon were all equally distributed throughout the material (Fig. 2e). SEM and atomic force microscopy revealed that the wrinkled nanosheets of graphene generated by glocuse had a thickness of 1.1 nm (Fig. 2e, f). Very thin sheets were positioned on the surface of the cement particles and had a wrinkly appearance that was analogous to the morphology shown in Fig. 2e. Additional methods of characterization indicated beyond a shadow of a doubt that glucose was effectively transformed to graphene. The X-ray diffraction (XRD) patterns of the GC composite (Fig. 3a) revealed a new peak around 27° representing as-formed graphene sheets. According to the Raman spectra of the GC material, two additional peaks were discovered at 1578 and 1360 cm^{-1}, corresponding to the G-peak and D-peak of graphene, respectively. These peaks are found at these specific frequencies. The development of graphitic carbon was further supported by the G-band of the samples' 532 nm Raman spectra (Fig. 3b), which was located at 1578 cm^{-1}. The sample displayed a wide D-band with its center at 1360 cm^{-1}, which was indicative of nanoscale graphite particles and chemically modified graphene flakes. The center of the band was at 1360 cm^{-1}. This property, representing the existence of disorder as well as the boundaries of graphene domains, was detected with high-resolution SEM. The results of the studies suggested that the GC composite was composed of cement and graphene. Additionally, the results indicated that the graphene was equally distributed throughout the cement matrix and was confirmed by the fact that the GC material passed the GC test.

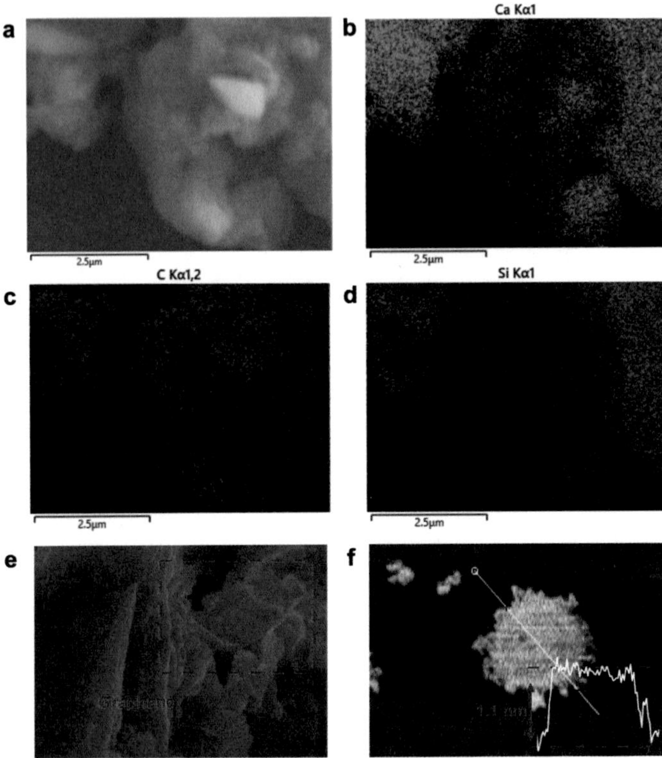

Fig. 2 Microscopic shape and structure of graphene, cement, and graphene–cement (GC) composite. **a** SEM image of fabricated GC composite in this study. **b–d** EDS of GC material. **e** High-resolusion SEM of GC material. **f** AFM of fabricated graphene sheet. AFM, atomic force microscopy; EDS, energy dispersive X-ray spectroscopy; SEM, scanning electron microscopy

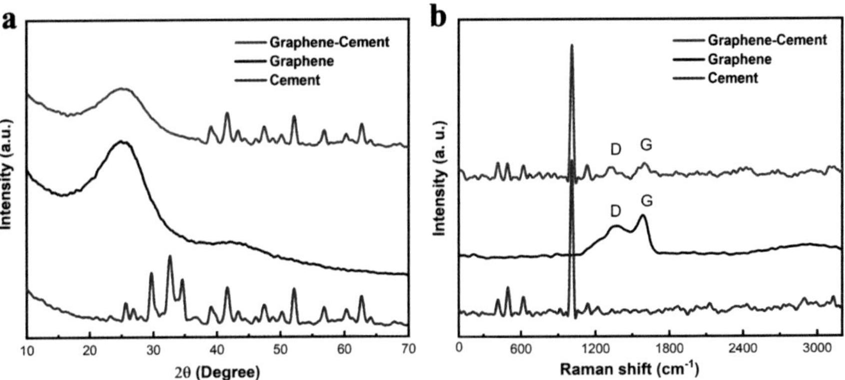

Fig. 3 **a** X-ray diffraction results, **b** Raman results of graphene, cement and graphene–cement composites

3.2 Mechanical Properties of GC Material

After being cured for 28 days, the compressive strengths of GOP paste, GP-0, GP-1, GP-3, and GP-6 were evaluated, and the results are depicted in Fig. 4. When calculating each reported compressive strength, the average of three duplicate specimens was used as the basis for the calculation. The compressive strengths were affected by the different content of the reinforcing materials. The compressive and flexural strengths of the GC paste were superior to those of GP-0 and GOP. The compressive strength of GC paste increased in direct proportion to the graphene content. The GP-3 group demonstrated the strongest compressive strength of all of the groups. In comparison with the GC paste, GOP had a somewhat lower compressive strength. After curing for 28 days, the compressive of GP-3 rose by 38.18% in comparison with GP-0. In contrast, after curing for 28 days, the compressive strength of GOP fell by almost 0.75%, as shown in Fig. 4.

3.3 Dispersion Effect

The large-scale SEM study of GP-3 and GOP, as well as the related element scanning tests, were carried out as shown in Fig. 5 for the purpose of confirming consistent distribution of the graphene. The carbon element distribution map of GCP-3 is shown in Fig. 5b, d, and the element distribution map of GOP is shown in Fig. 5f, both at the same scale as Fig. 5b. As can be seen in Fig. 5f, the aggregation of GO resulted in the formation of carbon element facula. On the other hand, graphene with uniform dispersion does not create an aggregation zone at the same scale (Fig. 5b), which demonstrated that the graphene was uniformly disseminated throughout the cement matrix.

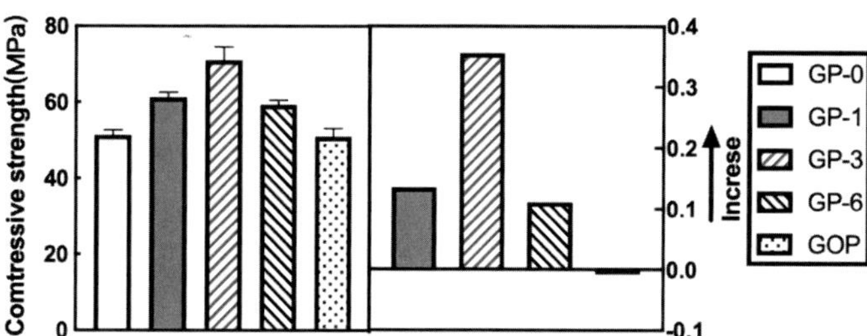

Fig. 4 The 28-day compressive strength tests of GP-0, GP-1, GP-3, GP-6, and GOP and corresponding rising rates of the samples compared with pure cement paste after 28 days

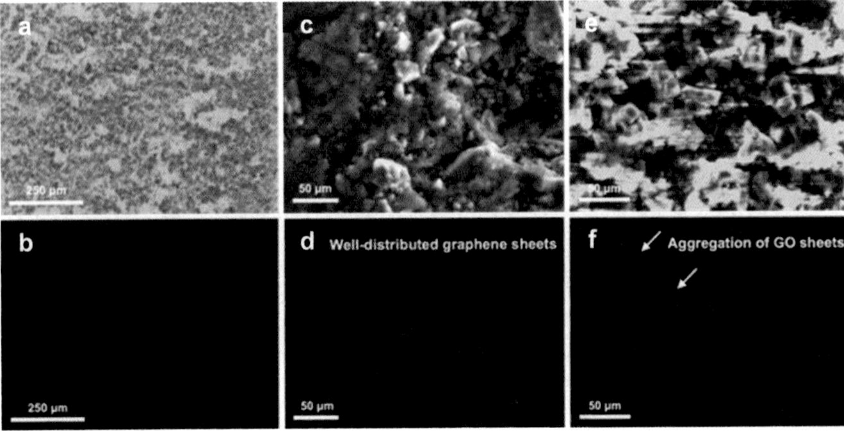

Fig. 5 Microscopic morphology and structure of GP-3 and GOP samples. **a**, **c** SEM image of GP-3 with different magnification. **b** EDS map of (**a**). **d** EDS map of (**c**). **e** SEM image of GOP. **f** EDS map of (**e**). GOP

4 Conclusions

By heating a mixture of glucose powder and cement powder, a new in-situ growth approach has been devised with the goal of successfully dispersing graphene throughout the cement matrix in a homogeneous manner. In order to manufacture high-quality graphene in situ, glucose was used as the carbon source because it reduces the overall cost of the process. This recently developed synthetic technique is extensible to the rational design of additional cement-based materials, and it has already been done. The in-situ growing process that was developed may produce a low-cost product and improve the dispersion effect of graphene sheets in the cement matrix. This in turn improves the mechanical properties of cement paste and makes it more amenable for graphene-based reinforced cement composites to be used in civil engineering.

References

1. Rong Z, Sun W, Xiao H, Jiang G (2015) Effects of nano-SiO2 particles on the mechanical and microstructural properties of ultra-high performance cementitious composites. Cem Concr Compos 56:25–31. https://doi.org/10.1016/j.cemconcomp.2014.11.001
2. Isfahani FT, Li W, Redaelli E (2016) Dispersion of multi-walled carbon nanotubes and its effects on the properties of cement composites. Cem Concr Compos 74:154–163. https://doi.org/10.1016/j.cemconcomp.2016.09.007
3. Rafiee MA, Narayanan TN, Hashim DP, Sakhavand N, Shahsavari R, Vajtai R, Ajayan PM (2013) Hexagonal boron nitride and graphite oxide reinforced multifunctional porous cement composites. Adv Funct Mater 23:5624–5630. https://doi.org/10.1002/adfm.201203866

4. Dimov D, Amit I, Gorrie O, Barnes MD, Townsend NJ, Neves AIS, Withers F, Russo S, Craciun MF (2018) Ultrahigh performance nanoengineered graphene-concrete composites for multifunctional applications. Adv Funct Mater 28:1705183. https://doi.org/10.1002/adfm.201705183
5. Pan Z, He L, Qiu L, Korayem AH, Li G, Zhu JW, Collins F, Li D, Duan WH, Wang MC (2015) Mechanical properties and microstructure of a graphene oxide–cement composite. Cem Concr Compos 58:140–147. https://doi.org/10.1016/j.cemconcomp.2015.02.001
6. Jing G, Ye Z, Wu J, Wang S, Cheng X, Strokova V, Nelyubova V (2020) Introducing reduced graphene oxide to enhance the thermal properties of cement composites. Cem Concr Compos 109:103559. https://doi.org/10.1016/j.cemconcomp.2020.103559
7. Yao Y, Zhang Z, Liu H, Zhuge Y, Zhang D (2022) A new in-situ growth strategy to achieve high performance graphene-based cement material. Constr Build Mater 335:127451. https://doi.org/10.1016/j.conbuildmat.2022.127451
8. Sun C, Huang Y, Shen Q, Wang W, W. Pan, P. Zong, L. Yang, Y. Xing, C. Wan, Embedding two-dimensional graphene array in ceramic matrix, Sci Adv. 6 (2020) eabb1338. https://doi.org/10.1126/sciadv.abb1338.
9. Gómez-Navarro C, Burghard M, Kern K (2008) Elastic properties of chemically derived single graphene sheets. Nano Lett 8:2045–2049. https://doi.org/10.1021/nl801384y
10. Zhu Y, Murali S, Cai W, Li X, Suk JW, Potts JR, Ruoff RS (2010) Graphene and graphene oxide: synthesis, properties, and applications. Adv Mater 22:3906–3924
11. Hou D, Lu Z, Li X, Ma H, Li Z (2017) Reactive molecular dynamics and experimental study of graphene-cement composites: structure, dynamics and reinforcement mechanisms. Carbon 115:188–208
12. Liu C, Huang X, Wu Y-Y, Deng X, Zheng Z, Xu Z, Hui D (2021) Advance on the dispersion treatment of graphene oxide and the graphene oxide modified cement-based materials. Nanotechnol Rev 10:34–49. https://doi.org/10.1515/ntrev-2021-0003
13. Liu J, Li Q, Xu S (2019) Reinforcing mechanism of graphene and graphene oxide sheets on cement-based materials. J Mater Civil Eng. 31:04019014. https://doi.org/10.1061/(asce)mt.1943-5533.0002649
14. Lu Z, Hou D, Hanif A, Hao W, Li Z, Sun G (2018) Comparative evaluation on the dispersion and stability of graphene oxide in water and cement pore solution by incorporating silica fume. Cem Concr Compos 94:33–42. https://doi.org/10.1016/j.cemconcomp.2018.08.011
15. Roy R, Mitra A, Ganesh AT, Sairam V (2018) Effect of graphene oxide nanosheets dispersion in cement mortar composites incorporating metakaolin and silica fume. Constr Build Mater 186:514–524. https://doi.org/10.1016/j.conbuildmat.2018.07.135
16. Sun H, Ling L, Ren Z, Memon SA, Xing F (2020) Effect of graphene oxide/graphene hybrid on mechanical properties of cement mortar and mechanism investigation. Nanomater-Basel 10:113. https://doi.org/10.3390/nano10010113
17. Ho VD, Ng C-T, Coghlan CJ, Goodwin A, Guckin CM, Ozbakkaloglu T, Losic D (2020) Electrochemically produced graphene with ultra large particles enhances mechanical properties of Portland cement mortar. Constr Build Mater 234:117403. https://doi.org/10.1016/j.conbuildmat.2019.117403
18. Liu C, Huang X, Wu Y-Y, Deng X, Zheng Z (2021) The effect of graphene oxide on the mechanical properties, impermeability and corrosion resistance of cement mortar containing mineral admixtures. Constr Build Mater 288:123059. https://doi.org/10.1016/j.conbuildmat.2021.123059
19. Li X, Lu Z, Chuah S, Li W, Liu Y, Duan WH, Li Z (2017) Effects of graphene oxide aggregates on hydration degree, sorptivity, and tensile splitting strength of cement paste. Compos Part Appl Sci Manuf 100:1–8. https://doi.org/10.1016/j.compositesa.2017.05.002
20. Li X, Liu YM, Li WG, Li CY, Sanjayan JG, Duan WH, Li Z (2017) Effects of graphene oxide agglomerates on workability, hydration, microstructure and compressive strength of cement paste. Constr Build Mater 145:402–410. https://doi.org/10.1016/j.conbuildmat.2017.04.058
21. Wang M, Niu Y, Zhou J, Wen H, Zhang Z, Luo D, Gao D, Yang J, Liang D, Li Y (2016) The dispersion and aggregation of graphene oxide in aqueous media. Nanoscale 8:14587–14592. https://doi.org/10.1039/c6nr03503e

22. Lu Z, Chen B, Leung CKY, Li Z, Sun G (2019) Aggregation size effect of graphene oxide on its reinforcing efficiency to cement-based materials. Cem Concr Compos 100:85–91. https://doi.org/10.1016/j.cemconcomp.2019.04.005
23. Lin J, Shamsaei E, de Souza FB, Sagoe-Crentsil K, Duan WH (2019) Dispersion of graphene oxide–silica nanohybrids in alkaline environment for improving ordinary Portland cement composites. Cem Concr Compos 106:103488. https://doi.org/10.1016/j.cemconcomp.2019.103488
24. Sabziparvar AM, Hosseini E, Chiniforush V, Korayem AH (2019) Barriers to achieving highly dispersed graphene oxide in cementitious composites: an experimental and computational study. Constr Build Mater 199:269–278. https://doi.org/10.1016/j.conbuildmat.2018.12.030
25. Zhao L, Zhu S, Wu H, Zhang X, Tao Q, Song L, Song Y, Guo X (2020) Deep research about the mechanisms of graphene oxide (GO) aggregation in alkaline cement pore solution. Constr Build Mater 247:118446. https://doi.org/10.1016/j.conbuildmat.2020.118446
26. Wang J, Tao J, Li L, Zhou C, Zeng Q (2020) Thinner fillers, coarser pores? A comparative study of the pore structure alterations of cement composites by graphene oxides and graphene nanoplatelets. Compos Part Appl Sci Manuf 130:105750. https://doi.org/10.1016/j.compositesa.2019.105750
27. Li X, Kurasch S, Kaiser U, Antonietti M (2012) Synthesis of monolayer-patched graphene from glucose. Angewandte Chemie Int Ed. 51:9689–9692. https://doi.org/10.1002/anie.201203207
28. Zhang B, Song J, Yang G, Han B (2014) Large-scale production of high-quality graphene using glucose and ferric chloride. Chem Sci 5:4656–4660. https://doi.org/10.1039/c4sc01950d
29. Zhang Y, Zhang L, Zhou C (2013) Review of chemical vapor deposition of graphene and related applications. Accounts Chem Res 46:2329–2339. https://doi.org/10.1021/ar300203n
30. Krystek M, Pakulski D, Patroniak V, Górski M, Szojda L, Ciesielski A, Samorì P (2019) High-performance graphene-based cementitious composites. Adv Sci 6:1801195. https://doi.org/10.1002/advs.201801195
31. Wang B, Jiang R, Wu Z (2016) Investigation of the mechanical properties and microstructure of graphene nanoplatelet-cement composite. Nanomater-Basel 6:200. https://doi.org/10.3390/nano6110200
32. Han B, Ding S, Wang J, Ou J (2019) Nano-engineered cementitious composites. Princ Pract 459–518. https://doi.org/10.1007/978-981-13-7078-6_4

Accelerated Mortar Bar Test to Assess the Effect of Alkali Concentration on the Alkali–Silica Reaction

B. Boyd-Weetman, P. Thomas, P. DeSilva, and V. Sirivivatnanon

Abstract We report the outcomes of a study into the influence of alkali concentration on expansion induced by the alkali–silica reaction (ASR), a deleterious reaction that causes cracking and durability loss in concrete structures. We assessed the effect of alkali concentration on mortar bar expansion using a modified form of AS1141.60.1, the accelerated mortar bar test (AMBT). Mortar prisms were prepared with a reactive aggregate and immersed in alkali solutions of varying concentrations (from 0.4 to 1.0 M NaOH) and saturated limewater at 80 °C. Expansion was monitored for 28 days. The degree of expansion was observed to increase with increasing alkali concentration and an induction period prior to expansion was observed for the 0.4 M NaOH. No expansion was observed for mortar bars immersed in the control saturated lime water bath. Additionally, no expansion was observed for mortars using blended cements containing fly ash (FA) and ground granulated blast furnace slag, suggesting the AMBT is a viable technique for demonstrating the efficacy of mitigation strategies.

Keywords Accelerated mortar bar test · Alkali-silica reaction · Durability · Mortar

B. Boyd-Weetman (✉)
School of Mathematical and Physical Sciences, University of Technology Sydney (UTS), Sydney, NSW, Australia
e-mail: brendan.boyd-weetman@uts.edu.au

P. Thomas · V. Sirivivatnanon
School of Civil and Environmental Engineering, University of Technology Sydney (UTS), Sydney, NSW, Australia

P. DeSilva
Faculty of Health Sciences, Australian Catholic University (ACU), Sydney, NSW, Australia

W. Duan et al. (eds.), *Nanotechnology in Construction for Circular Economy*, Lecture Notes in Civil Engineering 356,
https://doi.org/10.1007/978-981-99-3330-3_24

1 Introduction

The alkali–silica reaction (ASR) is a deleterious reaction affecting the long-term durability of concrete structures. ASR proceeds in the pore solution of concrete when alkali hydroxides and reactive forms of silica are present. It can be produced in any cementitious system provided that the three reaction components are available: alkali, reactive silica, and water. Reactive silica, such as strained quartz or amorphous opal, is introduced by aggregates. These silica forms dissolve in the alkaline pore solution and subsequently precipitate to form an alkali–silica gel, the ASR gel. The ASR gel, once precipitated in the concrete, generally within an aggregate, exerts mechanical stress from within the concrete, causing the development of cracking. To reduce the potential for deleterious ASR, a number of mitigation strategies are deployed in Australia, including the use of nonreactive aggregate materials, a limit on the maximum available alkali, and the incorporation of supplementary cementitious materials (SCMs). In Australia, the concrete alkali limit is 2.8 kg/m^3 Na_2O_{eq} [1] and by specifying a cement alkali content limit of 0.6% Na_2O_{eq} [2].

Two standardized test methods have been adopted for use in Australia to determine an aggregate's propensity to cause ASR: AS1141.60.1, the AMBT, and AS1141.60.2, the concrete prism test (CPT). These tests are carried out under accelerated conditions designed to provide aggressive reaction environments to promote ASR gel development [2, 3]. An alkali threshold is the lowest alkali level in concrete where deleterious expansion is found [3]. Threshold testing has largely been the topic of investigation in CPT-based studies, for which RILEM has a recommended testing method published as protocol AAR-3.2 [3]. Although threshold focus has been on CPT-based studies, the AMBT could prove valuable as a tool for identifying threshold behavior. The benefit of assessing aggregate reactivity using the AMBT procedure is that it enables a rapid evaluation of an aggregate's reactivity within a 21-day timeframe (Table 1) and these benefits could also apply to threshold investigation should the method show viability. ASR gels in AMBT specimens have been observed to have compositions that vary compared with gel produced in concretes, with AMBT-induced gels having a higher sodium content compared with similar binder concretes, likely due to the external 1 M sodium hydroxide bath [4]. Regardless of gel compositional discrepancies, the AMBT remains valuable to industry for aggregate reaction screening due to its rapid timeframe and this warrants additional investigation of the test and the properties of the ASR gel it produces. The AMBT is also used to assess aggregate reactivity in the presence of SCM-incorporated binders to identify the level of SCM required to mitigate ASR for a particular aggregate. SCMs mitigate ASR through a variety of mechanisms that reduce the available alkali of the pore solution [5, 6].

The focus of this study was the alkali concentration of the bath solution in modified AMBT experiments and we report expansion tests for reactive aggregates as a function of the bath solution concentration. In addition, expansion of mortars incorporating SCM-blended cements is reported for cements containing fly ash (FA) and slag (S).

Table 1 Reactivity classification for the AMBT as defined by AS1141.60.1

Mean mortar bar expansion %		AS1141.60.1 aggregate reactivity classification
Duration of specimens in 1 M NaOH 80 °C		
10 days	21 days	
–	E < 0.10[a]	Nonreactive
E < 10[a]	0.10 ≤ E < 0.30	Slowly reactive
E ≥ 0.10[a]	–	Reactive
–	0.30 ≤	Reactive

[a] Value for natural fine aggregates is 0.15%

2 Methods

The primary aggregate of focus was reactive river sand classified as reactive by AS1141.60.1. Reactive river sand contains 10.7% moderately strained quartz, 2% heavily strained quartz and 1.3% fine microcrystalline quartz within fragments of indurated meta-greywacke/siltstone and acid volcanic rock. Three binder material combinations were used to prepare the prisms: GP cement, FA and ground granulated blast furnace slag (S). The cement used was a Portland-type GP cement that met the specified requirements of AS3972, with an alkali content of 0.47% Na_2O_e determined by X-ray fluorescence analysis. For the FA and S incorporated mixes, cement replacement percentages of 25% and 65%, respectively, were chosen because they are representative of the recommended SCM substitution rates for mitigating ASR by AS HB79 [1]. To mix the mortar, the procedure outlined in AS1141.60.1 was followed. Gauge studs were placed within the mold prior to mixing and calibrated to have a gauge length of 250 ± 1 mm. Fine aggregate was prepared in its natural unaltered grading by oven drying at 110 °C before cooling for mixing. Potable tap water was used for mixing. The mortar prisms were cured in three gang molds for 24 h before demolding and were then immersed in tap water at room temperature prior to heating to 80 °C for 24 h for equilibration prior to zero day length measurement and subsequent immersion in respective alkali baths equilibrated at 80 °C.

To assess the effect of different external alkali environments and to observe potential threshold behavior, four alkali concentrations were used as immersion baths for the mortar bars: 0.4, 0.7 and 1 NaOH and a bath of saturated $Ca(OH)_2$ solution, which was used as the control bath with no external source of alkali. Distilled water was used to prepare the immersion solutions. The baths were kept at 80 °C throughout the duration of the test. Mortar bars were vertically oriented within the bath, supported by a stainless steel grid so that no contact with the gauge pins occurred.

To determine the comparative length change of the specimens, all comparative expansion measurements were conducted on a steel frame comparator equipped with a Mitutoyo digital micrometer. All expansion measurements are in reference to a 295-mm invar reference bar that was placed with identical positioning for each measurement and checked on a regular basis between mortar bar measurements. The mortar bars' comparative length measurements were recorded at day 0 (immediately after removal from 80 °C water bath), then at 1, 3, 7, 10, 14, 21 and 28 days following immersion in the alkali baths. To measure relative expansion, mortar bars were removed from the alkali baths, placed in the comparator for recording to a precision of 1 micron. These comparative length measurements were carried out within 10 s of removal from the bath and were measured in the same orientation within the comparator at each age.

3 Results and Discussion

An overview of the GP cement mortar bars (no SCM) is shown in Figs. 1 and 2 displays the expansion curve plots for each binder composition over time while immersed in the respective alkali immersion baths listed. The concentrations of the solutions in each bath were measured by titration and the pH calculated is listed in Table 2, which also lists the expansion for each GP cement mortar bar (no SCM). Expansion is a strong function of the alkali content of the solution concentration. Little or no expansion was observed over the timeframe for mortar bars exposed to the saturated $Ca(OH)_2$ solution. Expansion for the alkali solutions increased with increasing bath concentration over the time frame of the experiment. An induction period was apparent for 0.4 M NaOH where expansion was negligible up to 14 days followed by a notable increase in expansion. Expansion appeared to be increasing at 28 days. Further measurements will yield a limit for the expansion and discriminate the concentration effects on ASR. Threshold definitions have yet to be applied to AMBT-based studies. For standard reactivity assessment of the AMBT, if expansion is equal to, or exceeds 0.1% at 10 days (0.15% for natural sands such as the river sand used in the present study) or 0.3% at 21 days in bars exposed to 1 M NaOH, then the aggregate is classified as reactive (Table 1). For the GP mix immersed in 1 M and 0.7 M NaOH we observed the mortar bars exceed this expansion limit, indicating a classification of reactive. For the mortar bars exposed to 0.4 M NaOH, the expansion was within the limit, with an expansion of 0.230% observed at 21 days. Although the expansion was below the expansion limit designated for classification as reactive, the test was nonstandard and some expansion due to the reactivity of the aggregate was observed. A delay in significant expansion was observed after 14 days, suggesting an induction period prior to expansion (a delay in the onset of expansion) for the 0.4 M NaOH bath, rather than an alkali threshold, because expansion tends to be >0.3% at 28 days. For the mortar bars exposed to 0.7 M and 1 M NaOH, it appeared that, within the resolution of the measurements taken, the induction period was relatively similar, with the onset to significant expansion occurring between 3 and 7 days. When the expansion rates were compared (Fig. 1), it could be seen that increasing the NaOH concentration increased the expansion rate. It remains to be seen with this aggregate whether threshold behavior is seen in maximum expansion with respect to the change in alkali bath concentration.

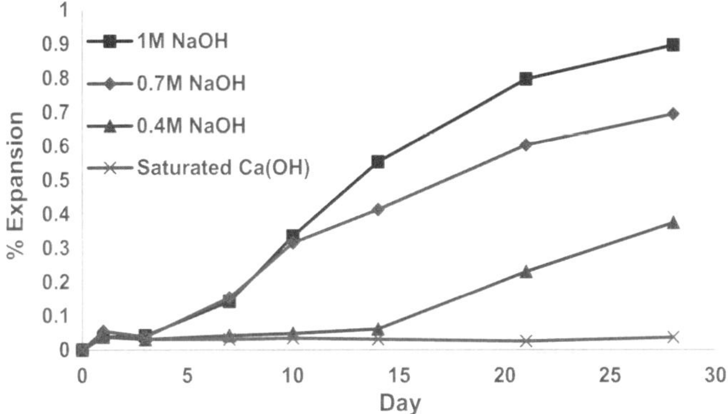

Fig. 1 Percentage expansion of natural sand mortar bars immersed in 0.4 M, 0.7 M and 1.0 M NaOH solutions and saturated $Ca(OH)_2$ solution over 28 days

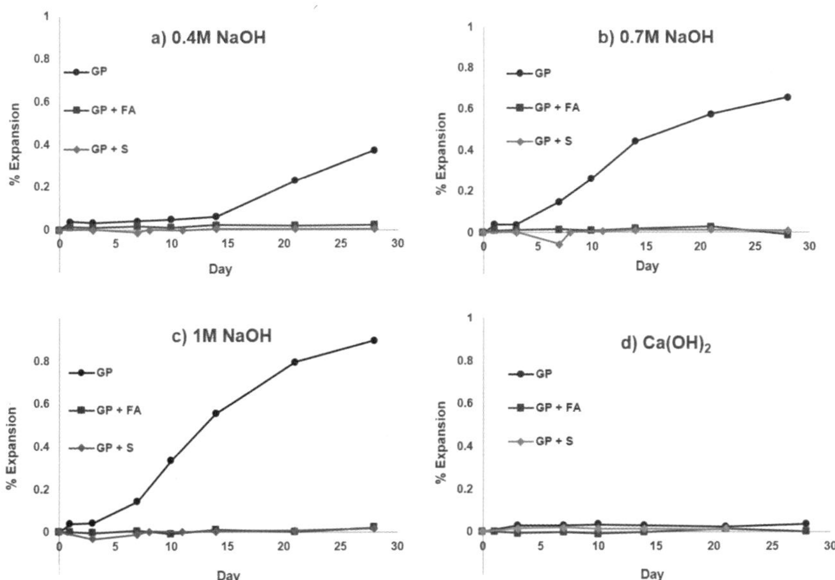

Fig. 2 Expansion of mortars submerged in **a** 0.4 M NaOH, **b** 0.7 M NaOH, **c** 1.0 M NaOH and **d** saturated $Ca(OH)_2$ solutions at 80 °C for 28 days

No significant expansion was observed for the reactive aggregate mortar bars prepared with blended cements containing FA or S (Fig. 2). These SCMs appeared to sufficiently mitigate ASR within the timeframe of the experiment. Measurements to extended ages will identify whether this is the result of complete mitigation or if the expansion-free region is an induction period where the SCM acts as an inhibitor, delaying the onset of expansion once consumed in the pozzolanic reaction.

Table 2 Percentage expansion of mortar bars at age 28 days for 100% GP cement binder and calculated pH for each alkali immersion bath

		Alkali bath		
	Saturated $Ca(OH)_2$	0.4 M NaOH	0.7 M NaOH	1 M NaOH
pH (calculated)	12.45	13.6	13.85	14
% Expansion of GP mix at 10 days	0.034	0.048	0.048	0.335
% Expansion of GP mix at 21 days	0.0248	0.230	0.576	0.796
% Expansion of GP mix at 28 days	0.0356	0.374	0.658	0.895

In summary, expansion increased with increasing concentration of NaOH solution, demonstrating that reactivity increases with increasing availability of alkali. Additionally, the rate of expansion (slope of the expansion curves) was observed to increase with increasing alkali concentration in the bath solution. An induction period was observed for bars exposed to 0.4 M NaOH, and although this is not traditional threshold behavior, it does indicate that there is a mechanism affecting the time for expansion to occur at lower alkali testing environments. As data collection was limited to 28 days, the maximum possible expansion has not been determined. Further measurements at greater ages will be carried out to identify any long-term influence of alkali solution concentration on expansion to attempt to identify the origins of the expansion behavior observed, with the aim of differentiating between threshold behavior and the degree of reaction controlled as the reactivity or concentration of the immersion solution. The incorporation of SCMs in the mortar mix significantly reduced expansion, which indicated mitigation, but further measurement is required to ensure that SCMs are not mitigating the reaction only in the short term and delaying the onset of expansion.

4 Conclusions

As the alkali solution concentration increases, mortar bar expansion increases, which indicates a relationship between alkali hydroxide concentration and ASR reaction severity. Threshold behavior, as defined by RILEM, was not observed for this aggregate at lower alkali levels; however, an induction period before the onset of significant expansion was observed at the lower alkali bath concentration of 0.4 M. The expansion data presented here is limited to the timeframe of the AMBT testing criteria, so further extending AMBT experiment timeframes may clarify the relationship between alkali concentration and extent of the ASR reaction with time, including the maximum observed expansion. The investigation demonstrated that both FA and S mitigate ASR-induced expansion in aggressive accelerated reaction environments during testing at incremental concentrations up to 1 M NaOH.

Acknowledgements This research was funded through an Australian Research Council Research Hub for Nanoscience Based Construction Materials Manufacturing (NANOCOMM) with the support of Cement Concrete and Aggregates Australia. The authors are grateful for the financial support of the Australian Research Council (IH150100006) in conducting this study. This research was supported by an Australian Government Research Training Program Scholarship.

References

1. Standards Australia (2015) Alkali Aggregate reaction—guidelines on minimising the risk of damage to concrete structures in Australia. In: HB79
2. Australian Technical Infrastructure Committee (2019) ATIC-SP43. In: Section SP43(2019) Cementitious Materials for Use with Concrete. ATIC, ATIC-SPEC
3. Nixon PJ, Sims I (2016) RILEM recommendations for the prevention of damage by Alkali-aggregate reactions in new concrete structures, vol 17. Springer Netherlands, pp XVI, 168
4. Gavrilenko E et al (2007) Comparison of ASR-gels in concretes against accelerated mortar bar test samples. Mag Concr Res 59(7):483–494
5. Vollpracht A et al (2016) The pore solution of blended cements: a review. Mater Struct 49(8):3341–3367
6. Shafaatian SMH et al (2013) How does fly ash mitigate alkali-silica reaction (ASR) in accelerated mortar bar test (ASTM C1567)? Cement Concr Compos 37:143–153

Development of High-Strength Light-Weight Cementitious Composites with Hollow Glass Microspheres

X. Li, Y. Yao, D. Zhang, Z. Zhang, and Y. Zhuge

Abstract Achieving both light-weight and high-strength cementitious composites (HSLWCCs) is challenging. In this study, hollow glass microspheres (HGMs) were used to develop a HSLWCC. Different amounts of HGMs were incorporated in the cement mixture and the associated effects on the engineering properties and microstructure were investigated. The results showed that the density and strength decreased with increasing HGM content. Compressive strength of the HSLWCC decreased significantly when the HGM content increased from 30 to 40% and decreased slightly with further increasing HGM content, while the density generally reduced linearly with increasing HGM content. Structural efficiency of the HSLWCC increased when the HGM content was 30% and then decreased significantly at HGM content of 40%. In particular, a floatable cementitious composite with a density of ~970 kg/m^3 and compressive strength of ~31 MPa was developed by incorporating 60% of HGMs. Additionally, two failure modes (i.e., (i) debonding of interface and (ii) crush of HGM) were found in the high-strength light-weight cementitious composite (HSLWCC), with the former dominating in HSLWCC with high HGM content and the later dominating in HSLWCC with low HGM content.

Keywords High-strength light-weight cementitious composites · Hollow glass microspheres · Structural efficiency

---

X. Li · Y. Yao (✉) · Z. Zhang
School of Civil Engineering, Xi'an University of Architecture and Technology, Xi'an, China
e-mail: yaoy@xauat.edu.cn

X. Li · Z. Zhang
XAUAT UniSA An De College, Xi'an University of Architecture and Technology, Xi'an, China

D. Zhang (✉)
College of Civil Engineering, Fuzhou University, Fuzhou, China
e-mail: zhangdong_ce@fzu.edu.cn

Y. Zhuge
UniSA STEM, University of South Australia, Adelaide, SA, Australia

W. Duan et al. (eds.), *Nanotechnology in Construction for Circular Economy*,
Lecture Notes in Civil Engineering 356,
https://doi.org/10.1007/978-981-99-3330-3_25

1 Introduction

To date, cementitious composites are still one of the most widely used construction materials in the world [1, 2]. With increasing demand for long-span structures, offshore platform, and precast prefabricated structures, the development of high-strength, light-weight cementitious composites (HSLWCCs) is attracting attention. The common methods used to produce them are the incorporation of air/bubbles or the use of light-weight aggregates. However, instability of the air bubbles leads to poor mechanical properties of the HSLWCC. Therefore, the use of light-weight aggregates is considered to be a more promising approach [3, 4]. Although some light-weight aggregates have good bonding to the cement matrix because of their porous structure and pozzolanic reactivity [5–7], the developed light-weight aggregate cementitious composites are still compromised by the lesser mechanical properties of the light-weight aggregates [6]. This has prompted researchers to look for more suitable alternatives.

Hollow glass microspheres (HGMs) are a high-performance, ultra-light-weight material consisting of a thin outer shell and an inert gas inside [8–10]. Compared with conventional light-weight aggregates, HGMs have the advantages of superior particle size distribution, lower density, higher strength, and sustainability [11], which creates a new impetus to producing HSLWCCs. HGMs may have pozzolanic reaction in the cement matrix due to their high content of SiO_2 and Al_2O_3 [12, 13]. It has been shown that HGM can react with the surrounding cement matrix at low addition amounts to form new gel particles and increase the bonding properties of HGM with the matrix [14]. Nevertheless, the contribution of low additions of HGM to density is not obvious, and the pozzolanic reactivity of HGMs at different additions has not been well investigated.

Building on previous studies, the density and property changes of cementitious composites produced with larger additions of HGM need to be clarified. The pozzolanic reactivity of HGMs at different additions also needs to be discussed in depth. Therefore, in this study, different additions of HGMs at 30, 40, 50 and 60% by weight of cement were used to develop a HSLWCC, with a special aim of producing a high-strength floatable cementitious composite. The variation in the density and compressive strength of HSLWCC were investigated and the engineering properties of HSLWCC were evaluated by structural efficiency. Finally, the dispersion of HGM in the matrix and pozzolanic reactivity were evaluated by scanning electron microscopy (SEM).

2 Methods

2.1 Raw Materials

In this study, ordinary Portland cement (OPC, CEM I 52.5R) was used as the main binder for the mixture. The HGMs were incorporated into the cement paste to reduce the weight. The HGMs used in the mixture were ≈1–100 μm in size and had an average particle density of ≈460 kg/m^3. They had a high compressive strength of ≈55 MPa. Figure 1 shows the SEM images and X-ray diffraction results of the HGMs. As can be seen, the HGMs had a good non-crystalline structure and potential pozzolanic reactivity. Highly efficient polycarboxylate superplasticizer (SP) was used to obtain suitable workability. The particle size distributions of both the OPC and HGMs were obtained by a laser diffraction particle size analyzer, as shown in Fig. 2. The chemical composition of the raw materials was determined using X-ray fluorescence spectroscopy (XRF) and listed in Table 1.

2.2 Mix Proportions

The mix proportions are listed in Table 2. The amounts of HGMs added were 30, 40, 50, and 60% by weight of cement. Samples without HGMs were marked as REF, while samples with HGMs were denoted by HSL and the replacement ratio of HGMs. For example, the content of HGMs in HSL-30 was 30% of cement by weight. The water-to-cement ratio was kept at 0.5. As extra HGMs were added to the mixture, the SP content was adjusted to ensure similar workability of the different samples.

To prepare the samples with HGMs, SP was first solved in the water and the solution was then added to the cement. The mixture was stirred at high speed for 3 min. The HGMs were added to the fresh mixture and the mixture were stirred for

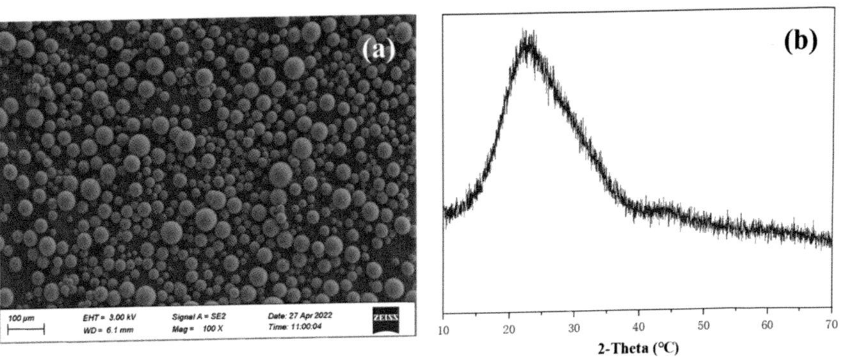

Fig. 1 Scanning electron microscopy image **a** and X-ray diffraction **b** of hollow glass microspheres

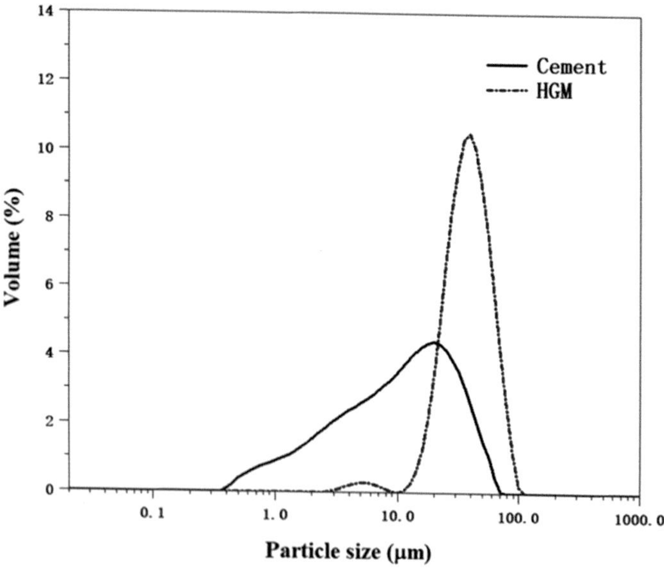

Fig. 2 Particle size distribution of ordinary Portland cement and hollow glass microspheres (HGMs)

Table 1 Chemical composition of materials, wt%

Type	SiO_2	Al_2O_3	CaO	MgO	K_2O	P_2O_5	Fe_2O_3	Na_2O	SO_3
Cement	19.39	4.0	63.61	2.4	0.72	0.13	3.82	0.59	4.66
HGM	80.3	0.85	6.31	0.25	0.05	0.05	0.04	11.5	0.25

HGM, hollow glass microsphere

Table 2 Mix proportions, wt%

Specimen	Cement	SF	HGM	SP	Water
REF	1	0.2	0	–	0.5
HSL-30	1	0.2	0.3	–	0.5
HSL-40	1	0.2	0.4	0.01	0.5
HSL-50	1	0.2	0.5	0.02	0.5
HSL-60	1	0.2	0.6	0.03	0.5

another 6 min to achieve good workability. Subsequently, the mixture was poured in to the mold on a vibrating table. After casting, samples were covered with plastic films and stored under laboratory condition. After 24 h, the samples were demolded and placed in a standard curing chamber until age 28 days under standard curing conditions (temperature: 20 °C, humidity: >90%).

2.3 Experimental Methods

Cubes of 50 × 50 × 50 mm^3 were prepared for the measurement of compressive strength and density. The volume of the sample was tested by the water displacement method, from which the density of the sample could be calculated. For the compression test, the samples were loaded at a rate of 1.5 kN/s. Three samples were used for each mixture and the average value is reported. The microstructure of the different samples was observed by SEM. The samples were soaked in the ethanol to stop the hydration process and then dried in a vacuum chamber.

3 Results and Discussion

3.1 Density, Strength and Structural Efficiency

Figure 3a shows the density and compressive strength of the different samples. As expected, the density of the HSLWCC decreased linearly with increasing HGM content. With the inclusion of 30% of HGM, the density of HSL-30 decreased by 30% compared with the REF. When the HGM content increased to 60% of cement, the density of HSL-60 reduced to 970 kg/m^3 and it became floatable.

Figure 3a also shows that the inclusion of HGMs reduced the compressive strength of the HSLWCC. However, it was interesting to discover that the decreasing rate of compressive strength was not linear with increasing HGM content. With 30% HGMs, the compressive strength of HSL-30 only reduced slightly by ~ 10.1% compared with the REF samples. This might be caused by good bonding between the HGMs and the cement matrix (see SEM results and the associated discussion in Sect. 3.2) and the high compressive strength of HGM as mentioned in Sect. 2.1. With further increasing of HGM content from 30 to 40%, the compressive strength of the HSLWCC reduced

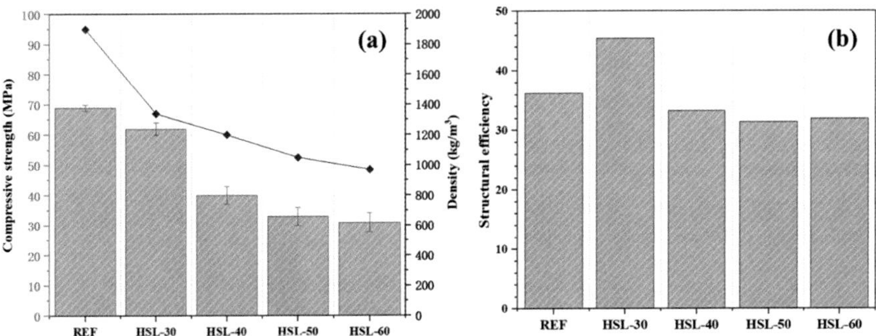

Fig. 3 **a** Density, strength and **b** structural efficiency of high-strength light-weight cementitious composite

significantly by 42% compared with the Ref samples. This may be due to the large amount of HGMs disrupting the continuity of the cement paste [14]. However, the decreasing rate of compressive strength slowed when the content of HGMs exceeded 40%. When the HGM content increased from 50 and 60%, the compressive strength only reduced slightly. Although HSL-60 showed an ultra-low density <1000 kg/m^3, its compressive strength was still maintained at >30 MPa.

Structural efficiency, which is the ratio between compressive strength and density (unit: kN·m/kg), is the main factor evaluating the lightness and strength of concrete. A higher value of structural efficiency represents a higher specific strength (i.e., high strength and light weight). Figure 3b shows the structural efficiency of the samples with different additions of HGMs. As evident, the HSL-30 specimens showed the highest structural efficiency. However, the structural efficiency decreased significantly in HSL-40, caused by the significant reduction in compressive strength as the density reduced linearly. When the HGM content was >40%, the changes in structural efficiency were only slight. However, it should be noted that with high HGM content, the structural efficiency of HSL-40, HSL-50 and HSL-60 did not reduce significantly compared with the REF. HSL-50 showed the lowest structural efficiency among the samples in this study, at 31.4, compared with the structural efficiency of the REF at 36.3.

3.2 SEM Analysis

Figure 4 shows the SEM images of the samples. The fracture surface of the HSL-30 sample shown in Fig. 4a indicates that the HGMs were uniformly dispersed in the matrix. Besides, Fig. 4b also shows that the HGMs were well bonded to the matrix and the hydration product was still partially attached to the surfaces of the HGMs after fracture. As shown in Fig. 1b, HGMs had potential pozzolanic reactivity. The reaction between HGMs and the cement matrix could further enhance the bonding [15, 16]. However, it should be noted that the bonding between the HGMs and cement matrix was still weak, as the reaction was limited, resulting in debonding between the HGMs and the cement matrix under loading, as shown in Fig. 4c. Besides, crushing of HGMs was also found in the samples after the compression test, as shown in Fig. 4d. Generally, two failure modes occurred in the HSLWCC: (i) debonding of the interface and (ii) crushing of HGMs. In the samples with lower HGM additions, failure mode (i) was the dominant damage mode, and in the samples with high HGM additions, damage mode (ii) dominated.

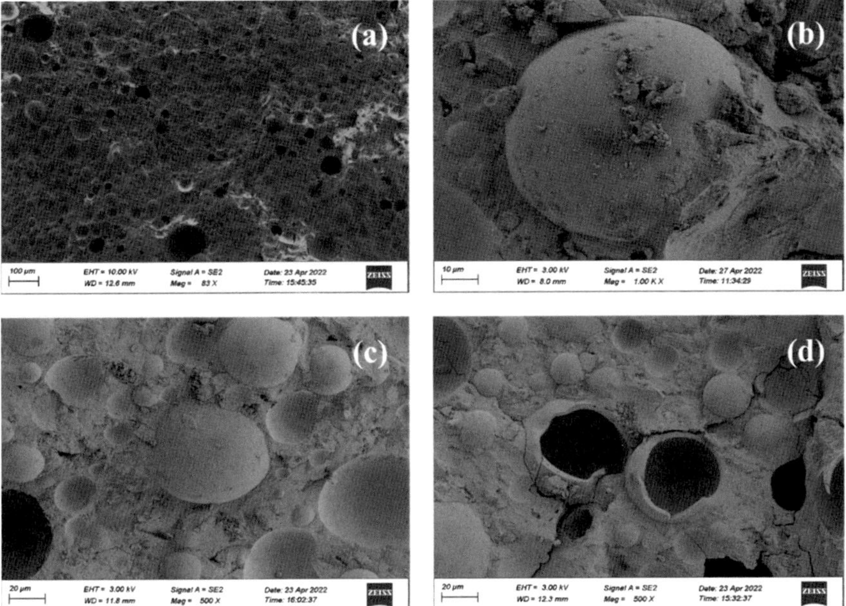

Fig. 4 **a** Fracture image of HSL-30; **b** HGM nucleation reaction; **c** morphology of HSL-60; **d** failure mode of the samples

4 Conclusions

In this study, a HSLWCC was developed using HGMs as the light-weight filler. In particular, a novel floatable cementitious composite with an apparent density of ~970 kg/m^3 and compressive strength of ~31 MPa was developed.

By incorporating HGMs with a content of 30–60% by weight of cement, the density of the developed HSLWCC ranged from 970 to 1340 kg/m^3 and the compressive strength ranged from 31 to 62 MPa. The compressive strength of the HSLWCC decreased significantly when the HGM content increased from 30 to 40%. With further increasing of HGM content, the compressive strength only reduced slightly. The density of the HSLWCC almost decreased linearly with increasing HGM content. The structural efficiency of the HSLWCC showed a sudden significant increase at a HGM content of 30%, while the structural efficiency of other HSLWCC samples was slightly lower than that of the reference sample.

Debonding of the interface and crushing of the HGM were both found at the fracture surface. The debonding of interface dominated in mixtures with high HGM content, whereas crushing of HGMs was usually found in the mixtures with low HGM content.

References

1. Yao Y, Zhang Z, Liu H, Zhuge Y, Zhang D (2022) A new in-situ growth strategy to achieve high performance graphene-based cement material. Constr Build Mater 335:127451
2. Zhang D, Shahin MA, Yang Y, Liu H, Cheng L (2022) Effect of microbially induced calcite precipitation treatment on the bonding properties of steel fiber in ultra-high performance concrete. J Build Eng 50:104132
3. Liu X, Chia KS, Zhang M-H (2011) Water absorption, permeability, and resistance to chloride-ion penetration of lightweight aggregate concrete. Constr Build Mater 25(1):335–343
4. Tandiroglu A (2010) Temperature-dependent thermal conductivity of high strength lightweight raw perlite aggregate concrete. Int J Thermophys 31(6):1195–1211
5. Chen R, Mo KH, Ling T-C (2022) Offsetting strength loss in concrete via ITZ enhancement: From the perspective of utilizing new alternative aggregate. Cement Concr Compos 127:104385
6. Lu J-X, Shen P, Ali HA, Poon CS (2021) Development of high performance lightweight concrete using ultra high performance cementitious composite and different lightweight aggregates. Cement Concr Compos 124:104277
7. Duan W, Zhuge Y, Pham PN, Liu Y, Kitipornchai S (2022) A ternary blended binder incorporating alum sludge to efficiently resist alkali-silica reaction of recycled glass aggregates. J Clean Prod 349:131415
8. Liu B, Wang H, Qin Q-H (2018) Modelling and characterization of effective thermal conductivity of single hollow glass microsphere and its powder. Materials 11(1):133
9. Zhang X, Wang P, Zhou Y, Li X, Yang E-H, Yu TX, Yang J (2016) The effect of strain rate and filler volume fraction on the mechanical properties of hollow glass microsphere modified polymer. Compos Part B Eng 101:53–63
10. Shahidan S, Aminuddin E, Mohd Noor K, Ramzi Hannan NIR, Saiful Bahari NA (2017) Potential of hollow glass microsphere as cement replacement for lightweight foam concrete on thermal insulation performance. In: MATEC web of conferences, vol 103. EDP Sciences, Les Ulis, p 1014
11. Aslani F, Dehghani A, Wang L (2021) The effect of hollow glass microspheres, carbon nanofibers and activated carbon powder on mechanical and dry shrinkage performance of ultra-lightweight engineered cementitious composites. Constr Build Mater 280:122415
12. Krakowiak KJ, Nannapaneni RG, Moshiri A, Phatak T, Stefaniuk D, Sadowski L, Abdolhosseini Qomi MJ (2020) Engineering of high specific strength and low thermal conductivity cementitious composites with hollow glass microspheres for high-temperature high-pressure applications. Cement Concr Compos 108:103514
13. Torres ML, García-Ruiz PA (2009) Lightweight pozzolanic materials used in mortars: evaluation of their influence on density, mechanical strength and water absorption. Cement Concr Compos 31(2):114–119
14. Aslani F, Wang L (2020) Development of strain-hardening lightweight engineered cementitious composites using hollow glass microspheres. Struct Concr J FIB 21(2):673–688
15. Lu J-X, Shen P, Zheng H, Ali HA, Poon CS (2021) Development and characteristics of ultra high-performance lightweight cementitious composites (UHP-LCCs). Cem Concr Res 145:106462
16. Martín CM, Scarponi NB, Villagrán YA, Manzanal DG, Piqué TM (2021) Pozzolanic activity quantification of hollow glass microspheres. Cement Concr Compos 118:103981

Co-effects of Graphene Oxide and Silica Fume on the Rheological Properties of Cement Paste

D. Lu, Z. Sheng, B. Yan, and Z. Jiang

Abstract Polycarboxylate superplasticizer is typically used to prepare a high-quality graphene oxide (GO) solution before mixing with cement grains. However, even if GO is well dispersed in water, they tend to re-agglomerate in the alkaline cement hydration environment, thus seriously decreasing the workability of the fresh mixture. In this study, we propose a more targeted method by synthesizing GO-coated silica fume (SF) to promote the utilization of GO in cement-based materials. Specifically, the surface of pristine SF was modified to convert their zeta potential (modified SF: MSF), then GO-coated SF (i.e., MSF@GO) was prepared via electrostatic adsorption of GO onto the MSF surface. The experimental results revealed that adding 5MSF@GO hybrid (0.04% GO and 5% MSF, by weight of binder) significantly reduced yield stress and plastic viscosity by 51.5% and 26.2%, respectively, relative to the 0.04% GO-modified sample. These findings indicated that application of GO-coated SF is an effective and environmentally friendly way to develop sustainable cementitious composites.

D. Lu
Department of Civil and Environmental Engineering, The Hong Kong Polytechnic University, Hong Kong, China

School of Civil Engineering, Harbin Institute of Technology, Harbin, China

Z. Sheng
Key Laboratory of Concrete and Prestressed Concrete Structures of Ministry of Education, School of Civil Engineering, Southeast University, Nanjing, China

B. Yan
Wuhan Harbor Engineering Design and Research Institute Co., Ltd, Wuhan, China

Hubei Key Laboratory of New Materials and Maintenance and Reinforcement Technology for Offshore Structures, Wuhan, China

Z. Jiang (✉)
Department of Civil and Environmental Engineering, The Hong Kong University of Science and Technology, Clear Water Bay, Hong Kong, China
e-mail: zhenliang.jiang@connect.ust.hk

W. Duan et al. (eds.), *Nanotechnology in Construction for Circular Economy*, Lecture Notes in Civil Engineering 356,
https://doi.org/10.1007/978-981-99-3330-3_26

Keywords Cement composites · Graphene oxide · Rheological property · Silica fume · Surface modification

1 Introduction

Cement-based materials are the most extensively used construction materials because of their low cost, high compressive strength, and durability [1, 2], with an estimated yearly consumption of >30 billion tonnes [3]. However, their brittle nature limits broad application in some structures [4]. Additionally, the cement/concrete industry is energy intensive with a substantial environmental footprint [5]. According to previous studies [5, 6], enhancing the microstructure of cement-based materials and improving their mechanical strengths/durability is considered to be a candidate way to alleviate carbon emissions. Calcium–silicate–hydrate (C–S–H) gel, the principal hydrate of cement grains, is composed of nanocrystalline with an atomic structure similar to that of tobermorite and/or jennyite (i.e., it is a nanoscale material) [7]. Advancements in nanomaterials and nanotechnology have been providing great opportunities to enhance the structure of cement composites at the nanoscale, eventually improving the macroscale properties [7, 8].

As a typical two-dimensional nanomaterial, graphene oxide (GO) has been considered a favorable additive for improving the mechanical strengths and durability of cement-based materials [5, 9]. However, GO tends to agglomerate in an alkaline hydration environment when cement grains dissolve in water [10], which dramatically deteriorates the workability of fresh mixtures. Note that, the fresh properties of mixtures greatly affect the mechanical and durability properties of the hardened composites.

Polycarboxylate superplasticizer (SP) is generally used both for preparing high-quality GO solution and for improving the fresh properties of the cement pastes [10]. However, obtaining a high-quality GO solution before mixing with cement does not directly result in well-dispersed GO in the alkaline cement matrix. It has been demonstrated [9, 11] that using silica fume (SF) can predisperse GO in the cement matrix. These studies claim that the electrostatic repulsion between negatively charged SF and GO is primarily responsible for the improved GO dispersion. Nevertheless, the electrostatic repulsion theory may not apply to such a system because of the remarkably larger lateral size of GO than that of SF [11]. Inspired by GO-coated sand, which can reduce the migration resistance in water and improve the adsorption ability in water treatment applications [12], we developed GO-coated SF via electrostatic adsorption of GO onto the surface of the modified SF (MSF), aiming to better utilize GO in cementitious composites and exert the co-effects of the two materials. Especially, this work focused on investigating the properties of fresh cement pastes incorporating GO-coated SF, aiming to promote the application of GO in environmentally friendly high-performance cement composites.

2 Methods

2.1 Materials

Cement (P·O 42.5) and SF were used to prepare cement paste. GO was synthesized by modified Hummers' method, leading to a specific surface area (SSA) of ~2600 m^2/g. The chemical bonds in the GO used in this study mainly contained C–O, C = O, C = C, and O–H. The XPS data revealed that the C/O was ~1.97.

2.2 Preparation of GO-Coated SF and Cement Paste

To improve the compatibility of GO and the cement matrix, the concept of GO-coated SF was proposed, enabling the co-effects of GO and SF in the cement composite. Specifically, MSF particles were obtained by treating the surface of SF with $Ca(OH)_2$ solution. To eliminate potential chemical reactions on the SF surface, the modification process was optimized according to previous experience [5], where the SF was added to the $Ca(OH)_2$ aqueous solution at a weight ratio of 1:10. After that, the SF/MSF was mixed with the GO solution and stirred for 10 min to synthesize SF-GO or GO-coated SF. All pastes were fabricated by mechanically stirring for 4 min. Finally, the fresh pastes were used for rheological properties tests.

2.3 Testing Methods

The micromorphology was observed using a ZEISS electron microprobe. The surface functionality of particles was assessed via a Nano ZS zeta potential analyzer. SmartLab XRD with an incident beam of Cu-Ka radiation ($\lambda = 1.54$ Å) for a 2θ scanning range of 15–65° was used to examine the crystalline phase analyses of the powder samples. A high-resolution FEI-TEM was usd to compare the morphology of SF-GO and GO-coated SF. A Brookfield RST-SST rheometer equipped with a rotating vane (VT20-10) was used to perform the rheological tests. During testing, the shear rate increased from 5 to 150 s^{-1} in 60 s and a corresponding decrease in shear rate from 150 to 5 s^{-1} in the following 60 s. The yield stress and plastic viscosity can be obtained as follows:

$$\tau = \tau_0 + \mu\gamma + c\gamma^2 \tag{1}$$

where τ is the shear stress (Pa), γ is the shear rate (1/s), μ is the plastic viscosity (Pa·s), τ_0 is the yield stress (Pa), and c is the second-order coefficient (Pa·s^2).

3 Results and Discussion

3.1 Evaluation of the Surface Properties of MSF Particles and the MSF@GO

Considering SF particles have the potential to react with $Ca(OH)_2$ solution, as such the modification process was optimized, aiming to ensure that any chemical reaction on the SF surface could be discounted.

As presented in Fig. 1a, b, SF and MSF both exhibited a spherical shape with a size ranging from 50 to 300 nm. The XRD patterns of the SF and MSF were almost indistinguishable, without the broad peaks of C–S–H gel found in MSF samples (Fig. 1c). As suggested in Fig. 2d, the zeta potential of the SF (–23 mV) converted to ≈+ 3 mV (MSF), thanks to some calcium ions grafted onto the SF surface. These findings all support that MSF maintained the surface morphology and crystalline phase after treatment, its surface only achieving ion exchange.

As indicated in Fig. 2, a thin layer of GO was found uniformly and tightly adsorbed onto the surface of the MSF (Fig. 3b). In contrast, several SF particles were merely interspersed between GO layers (Fig. 3a). Such a strong MSF@GO interaction showed great potential to exert their co-effects in cement-based materials.

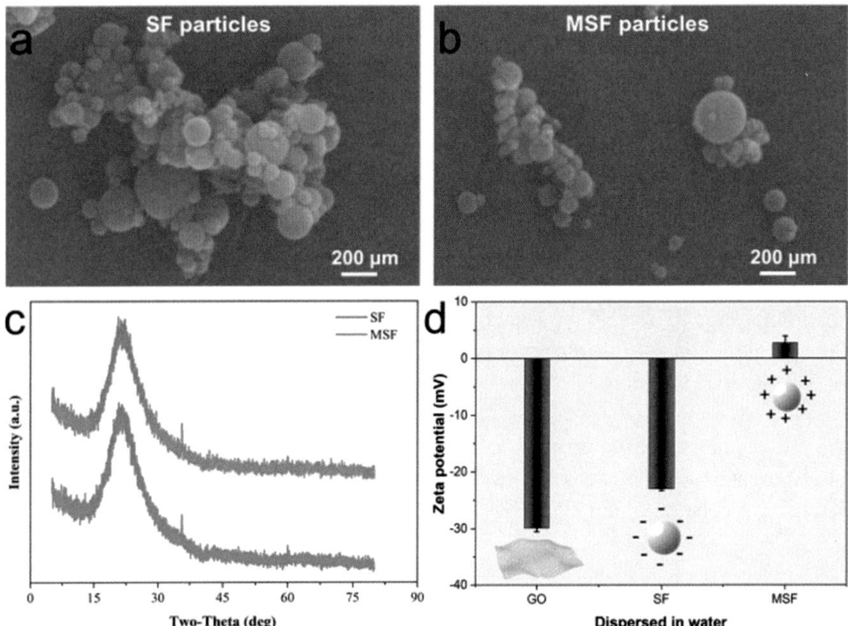

Fig. 1 Properties of the silica fume (SF) particles before and after modification: scanning electron microscopy images of **a** SF, **b** modified SF (MSF); **c** X-ray diffraction patterns; and **d** zeta potential

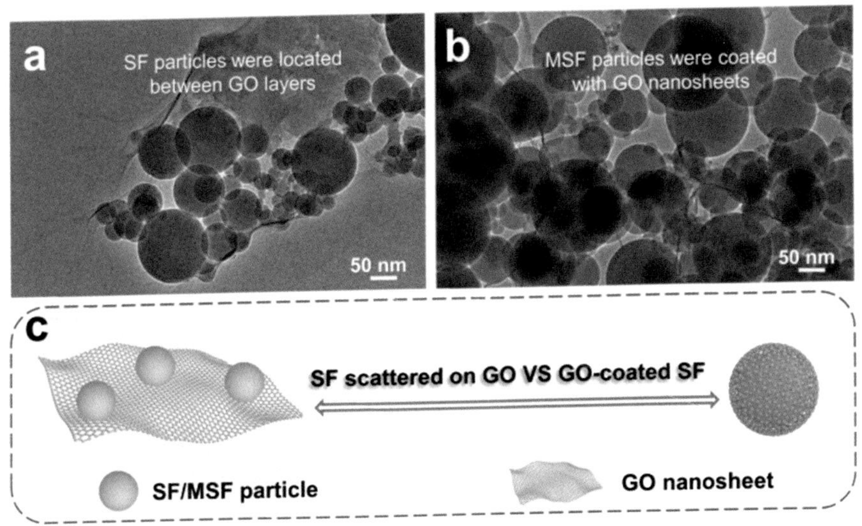

Fig. 2 Morphology of **a** SF-GO and **b** MSF@GO; **c** schematic diagram of SF particles located between GO layers (SF-GO) and GO-coated SF (MSF@GO). GO, graphene oxide; MSF, modified SF; SF, silica fume

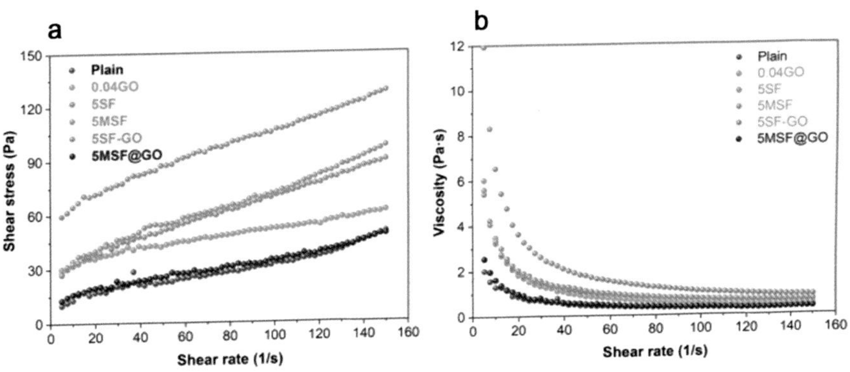

Fig. 3 Rheological parameters of the paste: **a** shear stress–shear rate curves and **b** viscosity–shear rate curves. GO, graphene oxide; MSF, modified SF; SF, silica fume

3.2 Rheological Properties

The shear stress–shear rate curves shifted upwards after admixing 0.04 wt% GO, as compared with plain paste (Fig. 3a). Simultaneously, the viscosity of the 0.04 wt% GO-modified paste also increased at both high and low shear rates (Fig. 3b). Specifically, the yield stress and plastic viscosity of the pastes were calculated based on an improved Bingham's model: adding 0.04 wt% GO to the paste increased the yield stress and plastic viscosity by 92.5% (Fig. 3b) and 88.1% (Fig. 3c), respectively.

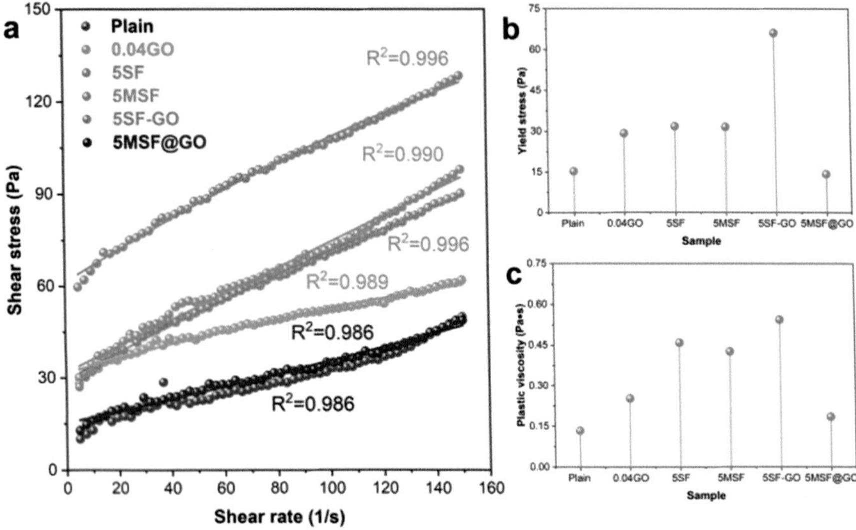

Fig. 4 **a** Linear regression of shear stress–shearrate curves; calculated **b** yield stress and **c** plastic viscosity. GO, graphene oxide; MSF, modified SF; SF, silica fume

Adding SF or MSF decreased the yield stress and plastic viscosity of the pastes (Fig. 4b, c). The mixture of 5SF-GO exhibited the highest shear stress–rate curve (Fig. 3), with the highest yield stress (66.1 Pa) and plastic viscosity (0.54 Pa·s), which implied that this formulation of cement composite has poor workability and is not suitable for practical applications. Additionally, the yield stress and plastic viscosity of the 5MSF@GO mixture decreased by 51.5% and 26.2%, respectively, relative to the 0.04 wt% GO-modified sample (Fig. 4). The MSF@GO hybrid adsorbed GO onto the MSF, thereby making the spherical particles easily migrate. In addition, the negatively charged GO coated onto the surface of the MSF, which provided electrostatic repulsion among particles, enabled better dispersion of GO (and MSF) and released entrapped water to turn into free water.

4 Conclusions

In this study we developed a high-performance cement composite incorporating GO-coated SF. The surface properties of GO-coated SF, the rheological properties, and compressive strength of composites were studied.

The surface of pristine SF particles was modified to convert their zeta potential; while the surface morphology and chemical composition remained unchanged, the SF surface merely achieved ion exchange. The positively charged MSF allowed negatively charged GO to adsorb on its surface via electrostatic adsorption. The strong bonding showed great potential for exerting a cooperative improvement effect in cement composites.

Different from the traditional direct introduction of GO or SF into cement paste, which seriously increased the yield stress and plastic viscosity of the mixtures. Adding 0.04 wt% GO together with 5 wt% MSF decreased the yield stress and plastic viscosity by 51.5% and 26.2%, respectively. The improved fresh properties provide guarantees for the transportation and construction of the cement composites.

References

1. Lu D, Ma LP, Zhong J, Tong J, Liu Z, Ren W, Cheng HM (2023) Growing nanocrystalline graphene on aggregates for conductive and strong smart cement composites. ACS Nano
2. Lu D, Zhong J (2022) Carbon-based nanomaterials engineered cement composites: a review. J Infrastruct Preserv Resil 3(1)
3. Monteiro PJM, Miller SA, Horvath A (2017) Towards sustainable concrete. Nat Mater 16(7):698–699
4. Lu D, Shi X, Zhong J (2022) Understanding the role of unzipped carbon nanotubes in cement pastes. Cem Concr Compos 126
5. Lu D, Shi X, Zhong J (2022) Interfacial nano-engineering by graphene oxide to enable better utilization of silica fume in cementitious composite. J Clean Prod
6. Pan Z, He L, Qiu L, Korayem AH, Li G, Zhu JW, Collins F, Li D, Duan WH, Wang MC (2015) Mechanical properties and microstructure of a graphene oxide–cement composite. Cem Concr Compos 58:140–147
7. Singh NB, Kalra M, Saxena SK (2017) Nanoscience of cement and concrete. Mater Today: Proc 4(4):5478–5487
8. Lu D, Shi X, Zhong J (2022) Interfacial bonding between graphene oxide coated carbon nanotube fiber and cement paste matrix. Cem Concr Compos 134
9. Lin J, Shamsaei E, Basquiroto de Souza F, Sagoe-Crentsil K, Duan WH (2020) Dispersion of graphene oxide–silica nanohybrids in alkaline environment for improving ordinary Portland cement composites. Cem Concr Compos 106
10. Zhao L, Guo X, Liu Y, Ge C, Chen Z, Guo L, Shu X, Liu J (2018) Investigation of dispersion behavior of GO modified by different water reducing agents in cement pore solution. Carbon 127:255–269
11. Lu Z, Hou D, Hanif A, Hao W, Li Z, Sun G (2018) Comparative evaluation on the dispersion and stability of graphene oxide in water and cement pore solution by incorporating silica fume. Cem Concr Compos 94:33–42
12. Hou W, Zhang Y, Liu T, Lu H, He L (2015) Graphene oxide coated quartz sand as a high performance adsorption material in the application of water treatment. RSC Adv 5(11):8037–8043

Automated 3D-Printer Maintenance and Part Removal by Robotic Arms

K. Andrews, K. Granland, Z. Chen, Y. Tang, and C. Chen

Abstract 3D printing by means of fused filament fabrication involves extruding and depositing melted material in layers to produce a 3D part. Current 3D printing requires manual intervention from a human operator between prints, leading to inefficiency. The focus of this study was facilitating the automation of the additive manufacturing process. Based on suggestions for future works in this field, this study extended on automated 3D-part removal systems by implementing additional operations to automate the production process. The proposed system uses robotic arms and grippers to operate and maintain 3D printers; specifically, the removal of 3D-printed parts, the cleaning of printer beds, the application of glue to the printer beds to assist with print adhesion, and the monitoring of bed levelness. The importance of this contribution is the improved efficiency of 3D-printing production, allowing for continuous 3D-printer operation and decreasing the requirement for human interaction and monitoring in the production process. The system is demonstrated using a 7 degrees of freedom KUKA robotic arm and ROBOTIQ gripper to autonomously operate and maintain an Ender 3 V2 printer. Sensor data and information from the 3D printers was used to determine the required operation or function to be performed by the robotic system. Tasks were performed by automated movement sequences of the robotic arm and gripper using supplied data. System status was recorded for monitoring and alerting human operators when intervention was required. The implementation of these functions using an automated robotic system allows 3D-printing production to operate continuously for longer periods, increasing production efficiency as downtime and human involvement for maintenance between prints is minimized.

Keywords 3D printing · Automated robotic system · Smart manufacturing

K. Andrews · K. Granland · Z. Chen · Y. Tang · C. Chen (✉)
Digital Twin Laboratory, Monash Smart Manufacturing Hub, Monash University, Clayton, VIC, Australia
e-mail: chao.chen@monash.edu

W. Duan et al. (eds.), *Nanotechnology in Construction for Circular Economy*, Lecture Notes in Civil Engineering 356,
https://doi.org/10.1007/978-981-99-3330-3_27

1 Introduction

Monash Smart Manufacturing Hub/Digital Twin Lab (DTL) incorporates automated processes and data collection to efficiently manufacture, store and monitor 3D-printed parts. Data and information from the physical systems are used to create a digital twin that allows for all relevant information to be monitored, displayed and controlled via a computer. 3D printing is performed by melting and extruding material in layers to form a 3D part. 3D printers require external interaction for the removal of parts, and other maintenance tasks need to be regularly performed to allow for continuous printer operation. In this study, we propose the use of a KUKA robotic arm and ROBOTIQ 2F-85 gripper to automate the 3D-printing process of Ender 3 V2 printers. The system uses information regarding printer and 3D-part status from the printers and outputs data on task completion status for use by the DTL. This integration of physical processes and information sharing will allow the system to be incorporated into the Monash Smart Manufacturing Hub.

Related works have explored the application of robotic arms to remove prints by exchanging flexible printer beds after each print [1, 2] or by using specialized end-effector mechanisms to directly pick up complex 3D-printed parts [3]. Consumer products [4, 5] allow for continuous printing with conveyor-belt beds or attached pushing mechanisms [6, 7] to slide parts from the end of the print bed. Here, we propose to build on functionalities from these implementations to create an automated system that can address the operation and maintenance requirements that arise from 3D-printing production. Although related works have focused on developing standalone automation features such as part removal, we propose an integrated system that can perform a range of functions so that the 3D-printing workflow can be fully automated.

Our objective was to demonstrate the implementation of an automated system using a robotic arm and gripper to enable the workflow shown in Fig. 1 to remove 3D-printed parts from the printers, apply glue to print beds for improved layer adhesion, clean the print beds and check bed levelness. These functionalities were implemented by programming the robotic arm and gripper to perform movement sequences and interact with the 3D printer to achieve continuous 3D-printer operation.

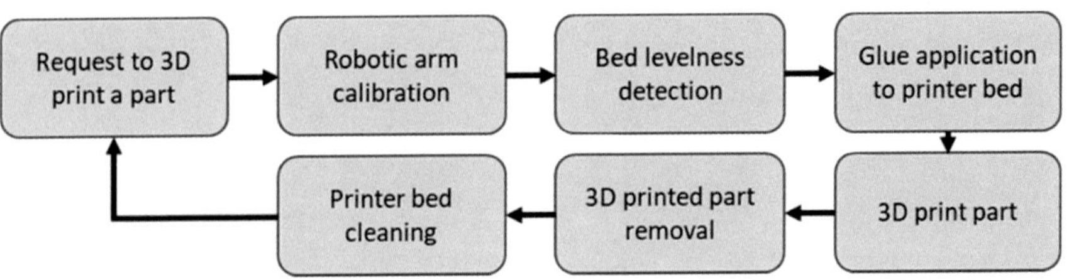

Fig. 1 Automated printer maintenance and operation workflow

2 Methods

2.1 Literature Review

3D printing has the advantage of requiring minimal setup compared with other manufacturing techniques, enabling parts to be printed on demand quickly and cheaply. The ability to continuously print parts is an important aspect of efficient 3D printing, and requires maintenance and operation tasks to be performed regularly on the 3D printers.

Removing finished 3D prints by use of robotic arms has been addressed [1, 2] with a method consisting of a detection algorithm for failed prints and a part removal function consisting of flexible magnetic beds that are removed and replaced. The advantage of using flexible magnetic beds is that complex parts can be removed easily because the robotic arm does not interact directly with the printed parts.

Becker [3] implemented a system capable of analyzing complex 3D-printed parts and determining a satisfactory part removal process using a robotic arm and customized end-effector/gripper for part removal. A CAD-based implementation [8] used grasping and motion planning simulation to determine valid end-effector paths for removing 3D-printed objects, which allowed model geometry to be used to determine optimal gripping points. Vision-based implementation [9], uses a depth camera and reinforcement learning methods to pick and place objects in a simulation environment. Such approaches provide solutions to automated 3D-part removal that can adapt to various part geometries. Becker [3] found that this saved time and decreased the requirement for human involvement in the part removal process, while noting that future works should include transitioning to a robotic arm system on a mobile base to enable interaction with other machines and increasing the flexibility of automated tasks that can be performed.

Aroca et al. [10] implemented a 2-degree of freedom manipulator and monitoring system to remove 3D-printed parts to enable continuous 3D printing. Future works are proposed for the use of a robotic arm to also apply glue to the printer bed for improved print adhesion.

Consumer products [4, 5] allow for continuous printing with a conveyor-belt setup, which is advantageous for prints that require a long z-axis because the part can be moved along the belt to be extended, and parts are pushed off the end of the bed, so additional mechanisms are required for the handling of parts after the printing stage in an automated process. Numerous third-party products and mechanisms [6, 7, 11] are also available that sweep across the print bed to push parts forward and off the print bed. However, these do not check if parts are unstuck properly because it is a 'blind' process and does not consider if the bed is sufficiently clean. Also, additional mechanisms are required for handling of parts after the printing stage as they are pushed to fall off the edge of the printer into a pile. Another aspect of enabling continuous 3D printing is ensuring workflow efficiency. Jim and Lees [12] demonstrate how task sequencing efficiency improvements of a robot can be implemented in the automation of 3D printing and post processing. By optimizing the efficiency of handling parts and moving through a sequence of functions a workflow can be created for producing parts that requires minimal human interaction. However, they noted that further work is required to consider when 3D printers or other post-processing machines are broken or require maintenance.

These works formed the basis for the functionalities to be implemented in this study. As highlighted, the additional automation of 3D-printer maintenance tasks and other supporting operations can further improve continuous 3D printed part production and efficiency [10, 12]. We addressed the gap in this field by demonstrating numerous automated tasks that are required for continuous 3D printing. The automated application of glue to the printer bed and cleaning of the bed after prints are removed allows for continual 3D printing and decreasing the need for human input and monitoring. The physical checking of bed levelness can be used to alert a human operator when maintenance tasks are required before printing begins. Automated part removal can minimize printing downtime so that continuous 3D-printing production can be better achieved.

2.2 Automated System

The automated system set up is shown in Fig. 2 with the KUKA robotic arm and gripper on a mobile base positioned in front of the workbench. An Ender printer is mounted with 3D-printed brackets to the workbench. The tool holder mounted next to the printer contains a glue stick and sponge.

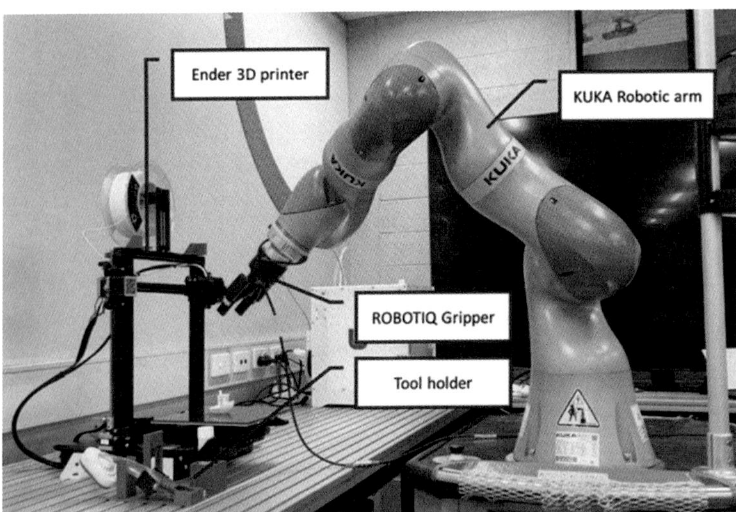

Fig. 2 Automated system setup

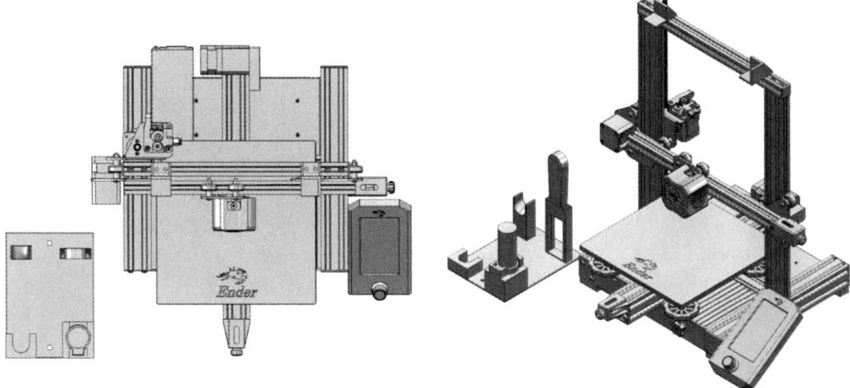

Fig. 3 The 3D printer and tool holder CAD environment

In order to automate the maintenance and operation tasks when 3D printing, print completion status, print size and print location information is required. This data is provided as variables in the KUKA robot's code and used as parameters that dictate robotic arm and gripper movements. It is proposed that the printer data is sent from the DTL to the robotic system in future works.

The movement sequences for the KUKA robotic arm were programmed in Java using Sunrise Workbench. The gripper end-effector was moved with translation and rotation commands and combined with gripper position commands to create complex movement sequences. Figure 3 shows the layout of the 3D printer and tool holder modeled in CAD. This is used to determine the path for the robot.

Coordinate frames consisting of gripper position and rotation values were saved for frequently used positions. The movement of the gripper and arm was also controlled by torque sensing on the robotic arm joints and force sensing in the gripper. This feedback was used to dictate the start and end of certain movement sequences.

3 Results and Discussion

3.1 Robot–Printer Calibration

In order to achieve precise movements and interactions between the robotic arm and 3D printer, calibration was required so that all arm movements were performed relative to the printer. The gripper was calibrated using two calibration points mounted to the top of the printer frame. As shown in Fig. 4, calibration was performed by initially positioning the gripper approximately in front of the first calibration point

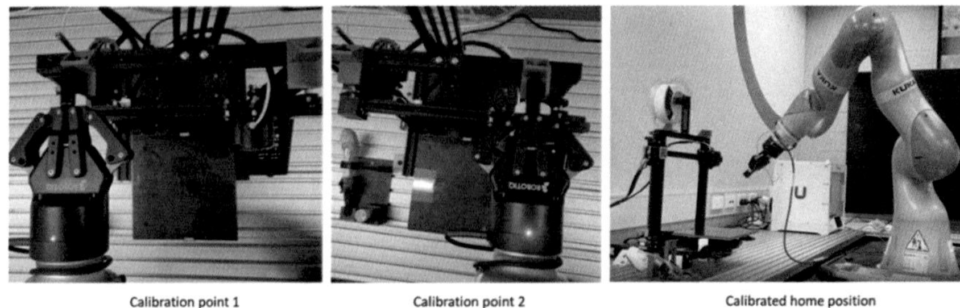

Fig. 4 Calibration between the robotic arm and 3D printer

then translating the gripper in x and z directions until a force was detected; the positions where the force was detected were saved, providing a known point for the robot to reference. The second calibration point was used to measure the printer's rotation relative to the base of the robot. By comparing the difference in positions between both calibration points the required gripper rotation of the gripper was calculated. The gripper was rotated to be square with the printer so that all arm movements were linear relative to the printer. This calibration process allowed positions on the printer to be programmed relative to the known reference frame at the calibration points for accurate arm movements when performing tasks. This process could be substituted for a computer vision system that detects the position of the gripper relative to the printer in future. A demonstration video is available at https://github.com/Kai-andrews/-Automated-3D-printer-maintenance-and-part-removal-by-robotic-arms.

3.2 Bed Levelness Monitoring

The levelness of the printer bed was monitored by an automated sequence moving the robotic gripper across each corner of the printer bed and lowering the gripper onto the bed while recording the relative heights where a force threshold was detected, as shown in Fig. 5. The height differences are displayed on the KUKA console to alert human operators as to which side of the printer bed requires adjusting to achieve a level printer bed. In future the bed levelness data will be sent to the DTL to alert operators of maintenance requirements if the bed is not sufficiently level. The printer bed was manually set to different states of levelness and detected using the gripper. As shown in Table 1, the robotic arm and gripper could detect all bed levelness configurations accurately. Please see the demonstration video.

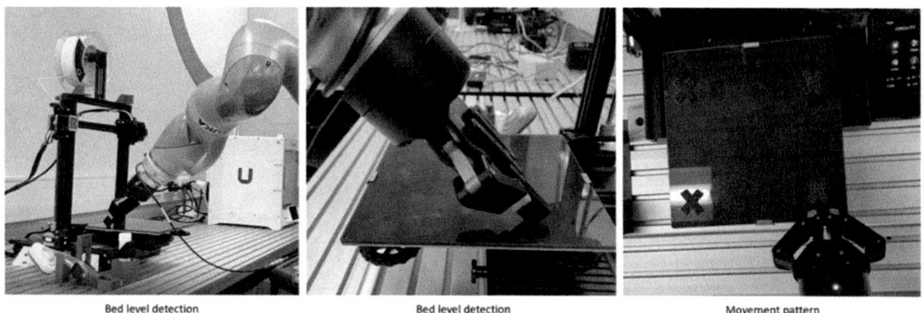

Fig. 5 Automated bed level detection implementation

Table 1 Bed level detection position values

Bed setup	Front left (mm)	Front right (mm)	Back left (mm)	Back right (mm)
Level	60.49	60.64	60.77	60.76
Front higher than back	60.51	60.57	63.42	63.51
Left higher than right	63.33	60.55	63.83	60.81

3.3 *Application of Glue for Improved Part Adhesion*

Automated application of glue to the printer bed was achieved using a glue stick held in a 3D-printed mount consisting of a round hole and slot cut-out for the glue stick to rest on. As shown in Fig. 6, the glue stick had an attached flat section that slotted into the holder to prevent the glue stick from rotating as the robotic arm rotated the glue stick knob to extend the glue. A 3D-printed attachment was mounted on the glue stick, providing a suitable surface for the gripper to hold the glue stick when in use. The glue stick was raised from the holder and lowered onto the print bed. As depicted in Fig. 7, glue was applied to the print bed by moving the glue stick over the area where a part was to be printed. Force was monitored to ensure glue was applied evenly across the bed by moving the arm down to maintain 8 N of vertical force to ensure contact as glue was used. The area that glue was applied to was determined using position and size information of the part that was to be printed. After glue application, the glue stick was returned to the holder. As shown in Fig. 7 the arm was able to apply glue evenly, fully covering the print area and in the correct position where the part was to be printed. Please see the demonstration video.

3.4 *Part Removal*

The automated removal of 3D-printed parts used software inputs that described the center point position of the part and its base length and width. The printer bed was

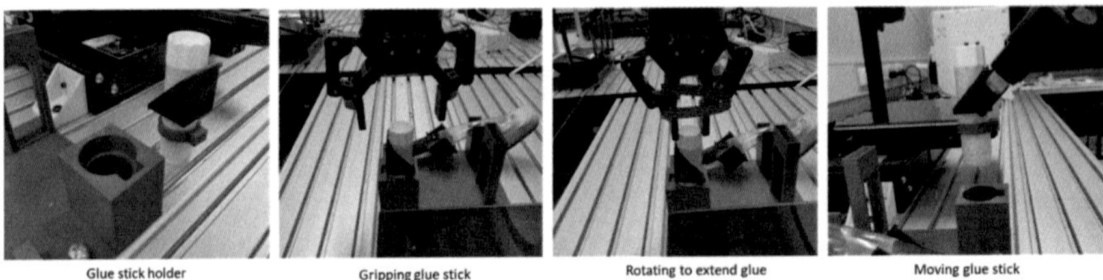

Fig. 6 Automated glue extension process

Fig. 7 Automated glue application and glue coverage

allowed to sufficiently cool before the robotic gripper was positioned above the part and opened. As shown in Fig. 8, the gripper was then lowered around the part until contact with the printer bed was detected and the gripper closed using force sensing so that the base of the part was securely held. The robotic arm then performed a sequence of rotations sideways, forward and backwards as shown in Fig. 9 while torque and position monitoring were used to determine if the part was stuck (indicated if torque was in the joints above a threshold) to prevent damage to the part or printer bed. The part was lifted from the printer bed and placed in a designated location next to the printer. This method allowed for parts with simple geometry and <85 mm (the grippers open width) to be removed successfully from the printer bed. Future improvements for the automated removal of more complex printed parts include using a more specialized end-effector and gripping techniques customized for each part. Please see the demonstration video.

3.5 Bed Cleaning

The cleaning of the printer bed was performed using a sponge with a handle containing a cleaning solution, which rested on a 3D-printed holder. As shown in Fig. 10, an attachment with flat sides was mounted to the handle to provide a suitable gripping point for the robotic gripper. The sponge was picked up by the robotic arm

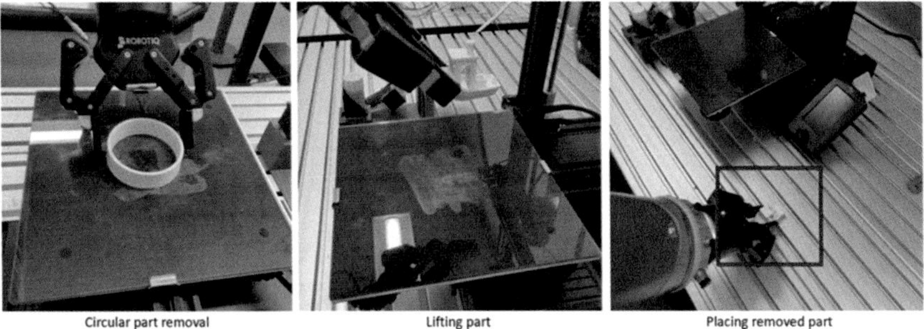

Fig. 8 Automated part removal demonstration

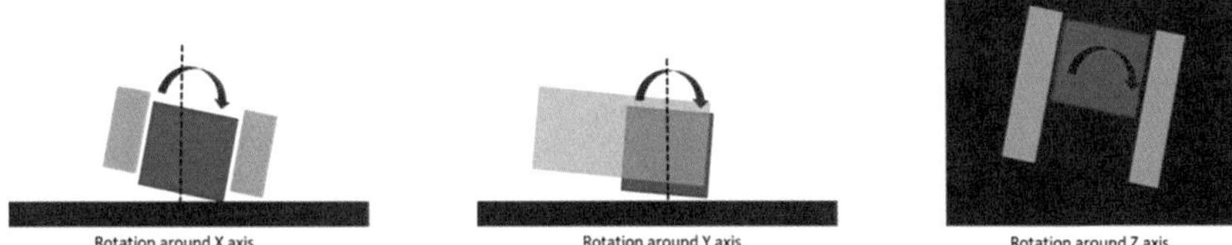

Fig. 9 3D-printed part removal

and moved over the print bed, then lowered until a vertical force of 7 N was detected, indicating that the sponge was fully in contact with the printer bed. The sponge was then moved forwards and backwards repetitively to remove dried glue and plastic build-up. Cleaning solution was applied to the sponge by moving the sponge up and down against the printer bed to squeeze the cleaning solution from the handle and sponge. The sponge was returned to the holder after printer bed cleaning. As shown in Fig. 11, the automated cleaning of the printer bed was able to successfully remove glue and residue build-up. This process could be further improved by implementing an automated scraping process using a scraping tool to remove larger amounts of physical contamination stuck to the printer bed. Please see the demonstration video.

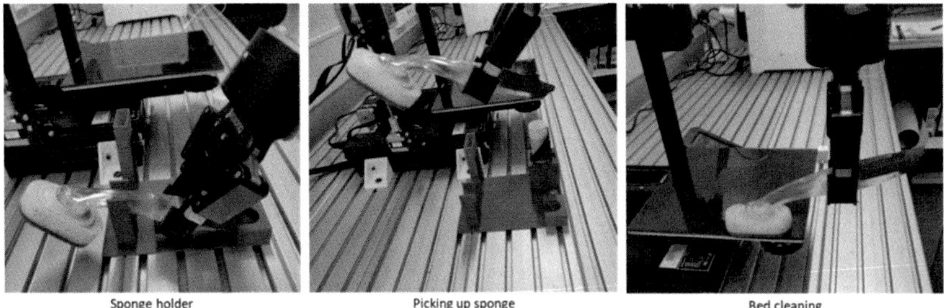

Fig. 10 Automated printer bed cleaning process

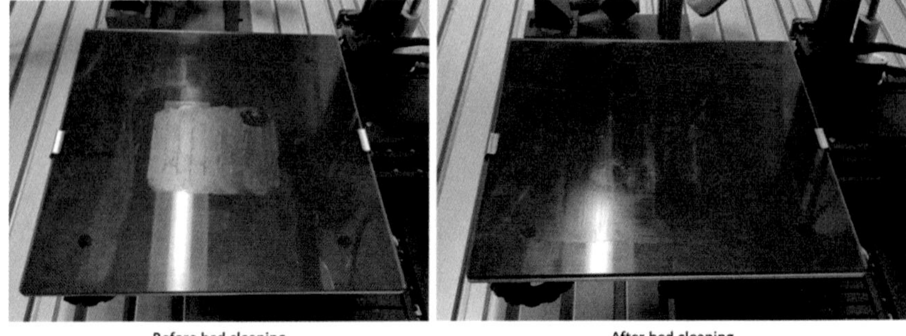

Fig. 11 Comparison between before and after automated bed cleaning

Table 2 Execution times for automated functionalities

Function	Completion time (s)	Success rate
Arm calibration	26	10/10
Bed level detection	20	10/10
Glue application	43	10/10
Part removal	26	10/10
Bed cleaning	64	10/10
Total	179	100%

3.6 Overall System Discussion

As shown in Table 2, the robotic arm and gripper repeatedly performed all automated functionalities consistently, completing 100% of test runs. Successful execution was defined as the robotic arm and gripper being able to correctly determine bed levelness, apply sufficient glue over the entire required area for printing, successfully remove the 3D-printed part and place it in a designated location and sufficiently clean the print bed so that another print can take place. The execution time for each automated process demonstrates the ability of the system to efficiently perform the required operation and maintenance tasks for 3D printing with a total execution time of 2 min 59 s, which can be improved in future with further velocity increases to the robotic arm movements.

4 Conclusions

In this paper, automation of the additive manufacturing process using a robotic arm and gripper is proposed and described. An automation sequence was developed to facilitate bed levelness detection with the ability to record data for monitoring by human operators, glue application to assist in print adhesion covering 100% of the print area, 3D-printed part removal and printer bed cleaning sufficiently to allow for another print to occur. The contribution of this research is the demonstration of an automated robotic system that performs required 3D-printer operation and maintenance tasks for continuous unmanned 3D-printing production. Recommendations for improvements and future work include creating custom tools for glue application and bed cleaning that allow for improved workflow using the automated robotic arm and gripper. Calibration between the robotic gripper and 3D printer can be improved using a computer vision system. The automated removal of parts can be optimized using adaptive or specialized gripper mechanisms for smart gripping of more complex 3D-printed parts. The data outputs on system status such as bed levelness or tasks completion can also be integrated into the DTL. We have demonstrated the ability of a robotic arm and gripper to enable continuous unmanned 3D printing through automation of key processes.

References

1. Li H-Y, Ma Y, Bin Miswadi MNA, Luu Nguyen Nguyen L, Yang L, Foong S, Soh GS, Sivertsen E, Tan U-X (2021) Detect-remove-replace: a robotic solution that enables unmanned continuous 3D printing. IEEE robotics & automation magazine 2–14
2. roboticsbusinessreview (2022) Case study: voodoo manufacturing triples 3D printing production with cobots. [Online]. https://www.roboticsbusinessreview.com/manufacturing/case-study-voodoo-manufacturing-triples-3d-printing-production-cobots/
3. Henger E, Roennau A, Dillmann R, Becker P (2019) Flexible object handling in additive manufacturing with service robotics. In: IEEE 6th international conference on industrial engineering and applications (ICIEA), pp 121–128
4. BlackBelt 3D (2022) Blackbelt 3D. [Online]. https://blackbelt-3d.com/. Accessed 2022
5. 3DPrintMill 3D Printer (2022) Creality, [Online]. https://www.creality.com/goods-detail/creality-3dprintmill-3d-printer. Accessed Jan 03 2022
6. Quinly for Ender 3/Pro/V2 Automation for Printers & Farms. 3DQue, 2022. [Online]. https://shop.3dque.com/products/quinly-ender-3?variant=41243357642909. Accessed 2022
7. d-weber, "prusa-chain-production," (2020) [Online]. https://github.com/d-weber/prusa-chain-production
8. Heider M, Nordsieck R, Angerer A, Dietrich S, Hahner J, Wiedholz A (2021) CAD-based grasp and motion planning for process automation in fused deposition modelling. ICINCO
9. Martins FN, Lima J, Wörtche H, Gomes NM (2021) Deep Reinforcement learning applied to a robotic pick-and-place application. In Optimization, learning, algorithms and applications. Springer International Publishing, pp 251–265
10. Aroca RV, Ventura CE, De Mello I, Pazelli TF (2017) Sequential additive manufacturing: automatic manipulation of 3D printed parts. Rapid Prototyp J 23:653–659
11. Automation to Scale 3D Printing (2022) 3DQue [Online]. https://www.3dque.com/. Accessed 2022
12. Kim H-J, Lee J-H (2021) Cyclic robot scheduling for 3D printer-based flexible assembly systems. Ann Oper Res 339–359

Machine Vision-Based Scanning Strategy for Defect Detection in Post-Additive Manufacturing

S. Zhang, Z. Chen, K. Granland, Y. Tang, and C. Chen

Abstract The surge in 3D printer availability, and its applications over the past decade as an alternative to industry-standard subtractive manufacturing, has revealed a lack of post-manufacturing quality control. Developers have looked towards automated machine learning (ML) and machine-vision algorithms, which can be effective in developing such additive manufacturing (AM) technologies for industry-wide adoption. Currently, most research has explored in-situ monitoring methods, which aim to detect printing errors during manufacturing. A significant limitation is the single, fixed monitoring angle and low resolution, which fail to identify small or hidden defects due to part geometry. Therefore, we investigated a novel ex-situ scanning strategy that combines the advantages of robotics and machine vision to address the limitations; specifically, the viability of image-recognition algorithms in the context of post-fabrication defect detection, and how such algorithms can be integrated into current infrastructure by automatically classifying surface faults in printed parts. A state-of-the-art and widely accepted ML-based vision model, YOLO, was adapted and trained by scanning for prescribed defect categories in a sample of simple parts to identify the strengths of this method over in-situ monitoring. An automated scanning algorithm that uses a KUKA robotic arm and high-definition camera is proposed and its performance was assessed according to the percentage of accurate defect predictions, in comparison with a typical in-situ model.

Keywords 3D printing · Additive manufacturing · Defect detection · Machine vision · Neural networks

S. Zhang · Z. Chen · K. Granland · Y. Tang · C. Chen (✉)
Digital Twin Laboratory, Monash Smart Manufacturing Hub, Monash University, Clayton, VIC 3800, Australia
e-mail: chao.chen@monash.edu

W. Duan et al. (eds.), *Nanotechnology in Construction for Circular Economy*, Lecture Notes in Civil Engineering 356,
https://doi.org/10.1007/978-981-99-3330-3_28

1 Introduction

1.1 Machine Learning in Additive Manufacturing

Machine learning (ML), a sub-branch of artificial intelligence (AI), has seen steady adoption and popularity in applications that require adaptive decision-making for their algorithms. Such applications include but are not limited to, data modeling in mathematics, decision making and logic structures in systems automation, and the focus of this study, defect detection in additive manufacturing (AM) [1]. The strength of ML in real-world applications is that it allows the software to intelligently interpret data from connected hardware and telemetry such as sensors, cameras, and radars. AM, known also as 3D printing or rapid prototyping (RP) [2], is based on the principle of sequentially layering material filament according to a computer-aided design (CAD) model. This is the opposite to subtractive manufacturing, currently the prevalent fabrication method that involves reducing a stock of material down to shape by cutting, milling, lathing etc. Research at the intersection of these two technologies has become popular in the past 5 years, with the focus being directed towards improving part quality, minimizing operational and material costs, and optimizing the fabrication process [3]. A key step in achieving these outcomes is ensuring components can be reliably produced, known as printability [4]. This is what ML integration aims to do, using both printer and camera data to automatically predict and diagnose sources of error.

1.2 Applications of AM Defect Detection

With the rapid adoption and prevalence of AM, a method of ensuring consistency and defect detection in 3D-printed parts is, however, yet to be universally adopted [5]. This is a major challenge because it is not only highly dependent on the printing parameters of the model, but also the machine operator's expertise and the reliability of the printer itself. Primary among the failure modes in 3D printing is filament entanglement, also known as spaghettification, warping, and surface defects after the part has been fabricated. Ideally such errors are mitigated during manufacturing, and indeed endeavors have been made with the use of live print camera-feed (in-situ) integration [6], but there is still no doubt that the value lies in informing smarter part designs or operational methods from failure diagnosis data.

ML in 3D-printing applications to date has emphasized process monitoring, and there has been comparatively little emphasis on post-fabrication quality assurance. Recognizing this significant inconsistency in AM technologies, many adopters have begun integrating image-recognition ML into existing models or designing new models with this technology pre-installed. A popular method of using ML to classify these defect patterns is with a convolutional neural network (CNN) [7]. CNNs are a class of artificial neural network and are structured similar to neurons in the brain, with a connectivity pattern that allows each node in the network to respond to signals only from other nodes that are relevant. Together with an automated part-scanning algorithm, an 'ex-situ' defect detection method has the potential to be integrated in place of the traditional in-situ model.

This was the motivation behind our study—how to leverage the power of programmable robotics, and continually improve a ML algorithm to automatically detect and classify defects in 3D-printed parts. Addressing this fundamental drawback in AM would allow future research pathways to consider the integration of not only the ML framework in classifying faults during the manufacturing process itself, but also to use the data to self-diagnose and improve the manufacturing process for future iterations of the model.

2 Literature Review

Significant progress has been made in introducing AM technologies to the consumer market over the past decade. The rapidly increasing userbase of 3D-printing technology, the growth and expansion of global manufacturing demand, and the push for efficiency and automation have increased the urgency for engineers and researchers to recognize and consolidate a solution for mitigating defects in printed parts. Our literature review confirmed that defect detection in AM is a problem that goes beyond small-scale production, so we present specific and general solutions, and conclude that a fundamental method of reducing 3D-printing defects is needed for the continued development and integration of AM worldwide.

2.1 Defining the Problem

It has been well recognized that defects and failure caused by 3D-printing processes are a critical barrier to the widespread adoption of AM and that it deserves serious attention. Recently in 2020, Wang et al. [8] comprehensively reviewed the state-of-the-art of ML applications in a variety of domains to study the main use-cases in the research and development of AM. Those authors state that although current ML in AM research is intensively concentrated on print parameter optimization and in-process monitoring, there is an expectation that ML research efforts will be directed to more rational manufacturing plans and automated feedback systems for AM. This is a vital step in pushing 'smart' AM into the near future. Their article highlighted the lack of reports on ML usage in AM with regard to determination of microstructure, material property prediction and topology optimization, and they conclude that further research into these areas is crucial in determining printability, the design of ML algorithms, and optimal part designs with respect to materials.

A similar state-of-the-art review performed by Oleff et al. [5] focused on the challenge of industrializing AM by analyzing current monitoring trends to identify the key weaknesses. These authors found that although the amount of research into anomaly detection evaluation was mostly for all in-situ process monitoring techniques, the emphasis was decidedly on part geometries and surface properties in terms of over- and underfill. These parameters refer to an 'infill' percentage, which simply dictates the density of the printed part. Oleff et al. noted that measurements of surface roughness and quality, as well as mechanical properties such as tensile strength, were addressed by only two of the 221 relevant publications from their research. They conclude that inspecting such material characteristics was simply not considered in any of the publications, and thus represents a significant area of further investigation.

Researchers have begun to study the implications of implementing defect detection in AM. For example, Chen et al. [3] examined the future of surface defect detection methods by comparing the key AM part inspection technologies through testing of various "traditional" detection methods, such as infrared imaging and Eddy current testing, as well as ML-based techniques such as CNN image recognition and auto-encode networks. Auto-encode networks proved to be effective at learning fault features, rather than classifying them, but required consistent input and output data dimensions. It was highlighted that ML defect detection is directly based on the characteristics of the AM field, and as such is the most sustainable program moving into the future. The authors conclude that, although more effective than traditional inspection technologies, NN-based methods of defect detection ultimately are deeply data-driven and establishing a universal model applicable at scale requires further study into each of their respective advantages and disadvantages.

Yao's findings were replicated in China by Qi et al. [9], who evaluated the effectiveness of different NN structures with the intention of optimizing the performance of AM parts. Qi et al. document the limitations in 3D-printing performance when applying numerical and analytical models to AM. Their paper reports that not only is a ML approach to AM valid due to its ability to perform complex pattern recognition without solving physical models, but also that NN are effective in CAD model design, in-situ modeling, and quality evaluation. The validity and potential of linking AM and NN technologies is brought to light, whereby they can be effectively integrated from design phases to post-treatment; however, the authors conclude that key challenges remain in the area of ML data collection and AM quality control.

2.2 Searching for Solutions

Petsiuk and Pearce [6] took a hybrid analytical approach to developing a 3D-printing defect detection method in their study aimed at supporting intelligent error-correction in AM. In their paper they present a comparison model, whereby an in-situ printing process was photographed layer-by-layer and compared with idealized reference images rendered by a physics engine, "Blender". They found that a similarity comparison did not introduce significant operational delays but rather demanded time for its virtual environment and rendering processes. A notable strength of this approach is that similarity and failure thresholds can be fine-tuned, providing both flexibility and varying sensitivity to defects during manufacturing. Emphasis is placed by the two authors on their model's ability to scale exponentially with the number of parts manufactured, needing only to render a base image set per part, as well as the model's independence of training data, presenting a significant reduction in resources required to implement the method into AM under unfavorable printing conditions, such as miscalibrated printer components, contamination or leveling discrepancies, which are prevalent control issues in mainstream AM markets.

Similarly, Paraskevoudis et al. [10] outlined a method of assessing the quality of in-situ 3D-printing using an AI-based computer vision system. Through analyzing live video of the process by utilizing a deep CNN, a larger variant of a CNN, a primary mode of AM defect was determined to be "stringing", which is caused by excess extruded material from the printer nozzle forming irregular protrusions on the printed surface, often causing issues when dimension tolerances are critical. The deep CNN model was demonstrably effective at identifying stringing from the experimental data, but proved to be unreliable when applied on external data acquired from web-based sources. This identified a critical need for continuous model improvement through training on "new" data. Those authors comment on the further applications of this form of error-detection, stating that, provided a sustainable detection model is found for other key AM failure modes, this approach may be developed to adjusting the printing process itself, whether by correcting its parameters or by terminating the process itself. Such a feature would reduce the skill ceiling needed to operate such machinery, meaning fewer engineers and more technicians in practice.

These international findings were replicated in Singapore in a 2021 study by Goh et al. [11], who explored and summarized the various types of ML techniques, alongside their current use in various AM aspects while elaborating on the current challenges being faced. Through the perspective of in-situ monitoring of 3D-printing processes, Goh et al. found the high computational cost and large-scale data acquisition for ML training to be significant challenges in practice. They also found that CNNs better capture spatial features and are hence ideal for 2D image and 3D model applications. The authors similarly stress the importance of large datasets for achieving high detection accuracy, and state that the realization of predictive modeling in digital twins for AM ultimately depends on the ML algorithm's classification accuracy, its input data quality, and its multi-task learning capabilities.

In most real-world 3D-printing applications, longevity AM as a reliable method of part production hinges on its ability to diagnose issues with print quality. A study conducted by Meng et al. [12] reviewed parameter optimization and discrepancy detection in the AM field by comparing the performance of common ML algorithms. They documented an iterative training method for an ML model wherein data from printing parameters, in-situ images and telemetry, microstructural defects and roughness, and the part's geometric deviation and mechanical strength were used to make predicative surrogate models to assist in-process optimization. This methodology is known as "active learning", whereby input–output pairs of data are formed and used to train ML models without querying new labeling data. Practically speaking, this significantly alleviates the cost, time, and human labor of conducting dedicated experiments with input labeling. The authors emphasize the gap in research with regard to active-learning ML algorithms in AM applications and conclude by stating that such an algorithm would be highly efficient in cases where a dataset is yet to be acquired.

Regardless of the defect analysis technique used, it has become clear that the best outcomes are achieved with a form of ML NN that is both accurate and scalable and able to adapt its behavior according to previous performance. Han et al. [7] considered the direct application of ML in defect classification in AM to assess its viability in the context of real-world image datasets. Through a process of localized segmentation of defects across image frames taken from in-situ print monitoring, a steady-state CNN model was established from 103 verification data sets. This model then accurately predicted defects from 101 "new" input images of 3D-printed parts. In practice, the authors' model outperformed several established detection frameworks such as "Faster RCNN" and "SSD ResNet" in terms of average precision, but lacked in terms of detection speed. The consistency of the segmentation model can be attributed to just that—the process of image instance segmentation effectively distinguishes the differences and boundaries between similar surface defects, which is an approach that can be used to help develop a similar strategy on a smaller scale.

2.3 Preparing for Tomorrow, Today

Today's global manufacturing industry continues to increase in size and pursue more time- and cost-effective means of production in AM. Unfortunately, research into "smart" AM, whereby sources of print failure are determined and used to inform better printing performance, has so far been both general and sparse. The literature reviewed by us reflected that void in knowledge. Particularly needed is a way to categorize the defects caused by erroneous printing parameters and to develop effective defect recognition in AM to address them.

3 Methods

3.1 Methodological Approach

Our goal was to evaluate the efficacy of using ML to automatically classify 3D-printing defects. We aimed to both describe the characteristic strengths and weaknesses of using such a method and gain a more in-depth understanding about its feasibility in the context of current 3D-printing applications.

We anticipated that a combination of quantitative and qualitative data would be required to achieve this aim; namely, photographic images and their derivatives such as those described by Petsiuk and Pearce [6], and Han et al. [7], to examine, decompose and categorize them as input data for the CNN model. The datasets would comprise these two data types, as the images would be parsed by a computer model and then verified by a human.

In order to maintain a control environment for testing the ML model's behavior, all preliminary data used to construct and train were primary data. From the experiments carried out by Han et al. [7] in similar CNN models, it was apparent that such a model may be very sensitive to slight deviations in input data, and so secondary data were omitted from our study but remains as a clear extension to model testing in later build iterations. Furthermore, as an act of simplifying the scope of testing, the main geometries chosen for the 3D-printed parts to be scanned were square- or rectangular-based in nature. By limiting the sample space to this particular set of shapes, the algorithm for rotating and capturing the surfaces of the part was drastically simplified. This limitation, however, did not affect the validity of what was being tested, but rather left open the door to research into converting this method into a universal model.

Finally, as a direct comparison with the ex-situ model, was created to both aid in accelerating the rate of data collection and act as a benchmark for performance and accuracy.

3.2 *Data Collection Methods*

Data were collected after establishing the CNN model in YOLOv5, whereby samples of 3D parts printed locally in the Monash Smart Manufacturing Hub Staging Lab served as a proportion of the total part sample space. The main selection criteria for these samples were their defect type and locality relative to the part's geometry, and were selected to assist in faster pattern recognition training and create a fair comparison between the in-situ and ex-situ models.

The relevant tools, equipment, and materials used to gather data were a Creality Ender 3 v2 3D printer, PLA printer filament (colored and white PLA +; 1.75 mm diameter), JAI HD-camera and Windows-based operating system (PyTorch IDE and LabelImg software).

Images were captured in a controlled environment to ensure consistency not only within the scope of the experiment itself, but also to the degree that could be replicated by another researcher or engineer (Fig. 1). This environment comprised a fixed, white-balanced background calibrated against a control in equal lighting conditions (brightness and color-temperature).

We anticipated encountering issues with object visibility when the color of the part being examined was identical to the reference background. In such cases, a unique background color was to be fixed or alternate backgrounds exchanged. Another potential issue was the base dataset not being sufficiently large. As mentioned by Chen et al. [3] and Goh et al. [11], a major limitation in mass-integration of NN-based models in AM is the sheer size of training dataset required. It is necessary to consider

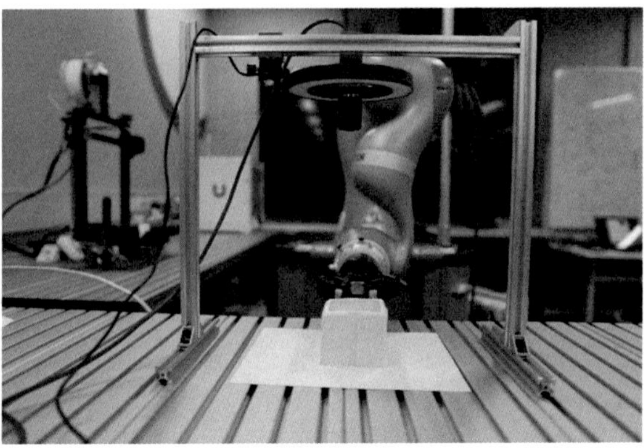

Fig. 1 Preliminary data collection setup: KUKA arm, gripper, JAI camera rig

the effect of this when evaluating the performance of ML in this application. It was only feasible for us to attain < 200 samples of data for use in training and validation.

3.2.1 KUKA Robotic Arm Algorithm

The arm was programmed to grab, move, and rotate test pieces to allow the mounted JAI camera to capture multiple isometric views of the part (Figs. 2, 3, 4 and 5). Specifically, two external isometric images of the part were captured. Based on the initial hypothesis, this method reduced the number of images needed to train the CNN for a particular defect, and thus reduced the overall duration of scanning. The advantage of having a programmable robotic arm is the ability to maneuvre the piece to expose every surface in just a few images, each captured at an isometric angle containing multiple faces.

Conditions surrounding the test rig were kept constant in terms of lighting (cool white light of 5300 K) and background (solid background sheet behind and below the sample) to limit the effect of extraneous variables.

Fig. 2 Example testing part: white hollow cube measuring 80 × 80 × 80 mm

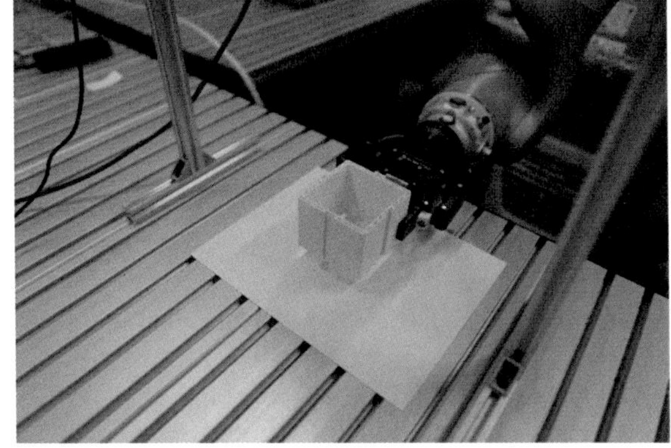

Fig. 3 Robotic arm in first position: working part is located and lifted

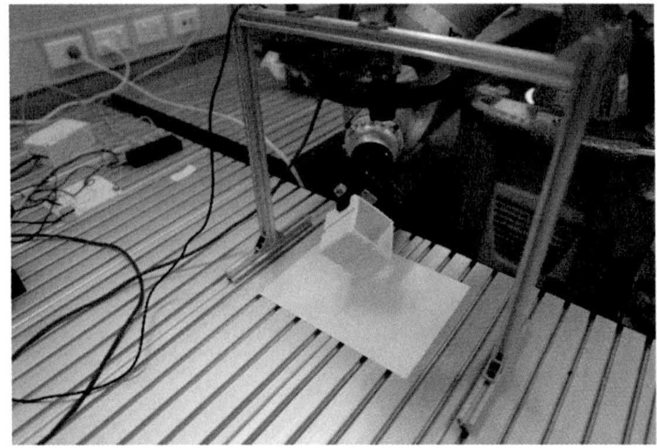

Fig. 4 Robotic arm in second position: working part is rotated to first isometric angle relative to hanging JAI camera

Fig. 5 Robotic arm in third position: working part is rotated to second isometric angle relative to hanging JAI camera

Further stages of development are: calibration pieces (homing, reference plane, improved consistency in both programming and in result collection); hanging camera mounting at isometric angle; and surrounding part background.

3.2.2 YOLO Image Classification

Using the isometric images gathered with our scanning method, the YOLOv5 model was trained under human supervision, which involved digitally labeling the input data with the LabelImg software (Fig. 6). These annotations are the mechanisms through which the CNN learns to recognize defect features.

An equivalent in-situ model provided image data that were also used as training data. Further improvements to training methodology are using a 1:9 ratio of true negative to true positive training data because 10% of image training data would **not** contain the corresponding defects and supplementing primary data with open-source secondary datasets. As a CNN model's performance improves proportionally to the volume of training input, it is expected that this will improve the overall performance.

Difficulties included larger forms of warping not being easily distinguished from curved features of the part's geometry, which restricted the effectiveness of the preliminary model to only similar cuboidal shapes. To mitigate this during labeling, larger examples of these features were bisected for better localization (Fig. 7).

Fig. 6 Example of labeling of primary defect modes using LabelImg: layer shift, stringing and warping

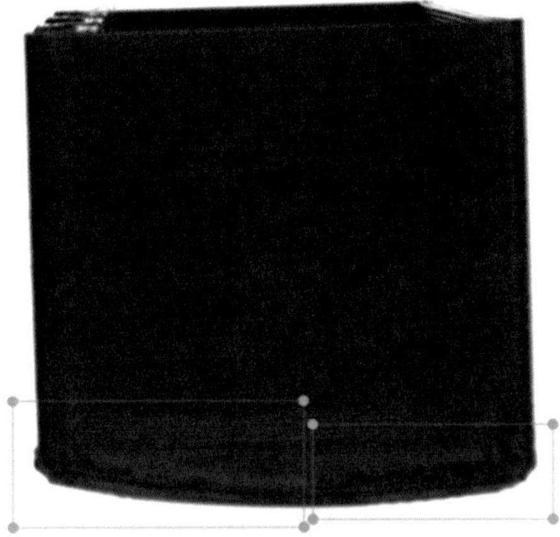

Fig. 7 Bisection method of larger examples of defects. This particular example shows dividing a large warping defect spanning the entire width of the part into two separate instances

3.3 Methods of Analysis

In both the quantitative and qualitative data analyses, outliers and anomalies were factored into the assessment of the model's functionality and reliability, but will be omitted from computational methods such as being used as training input.

The main method of quantitative analysis was the passing of image data through a CNN and recording the corresponding output. This network functions as a filtering device, using a network of calibrated nodes that respond to their associated shape or pattern. The nodes make up the decision-making mechanism of the CNN and can be adjusted during experimentation.

Qualitative analysis will comprise identification, verification and classification against the following main AM print failure categories [10]:

- spaghettification/stringing
- layer shift
- warping.

These categories can then be linked with associated mechanical errors or defects:

- under-extrusion
- over-extrusion
- nozzle blob
- poor initial layer
- poor bridging overhang
- premature detachment of print
- no extrusion
- skirt issues.

Human verification of these defect features will be used to train the CNN.

3.4 *Justifications*

An immediate limitation of a camera-based setup is detecting printing defects that are internal to the structure of the printed part. If a 3D-printed component needs to pass a quality check under the method outlined here, there would be no way of verifying the integrity of the inside surfaces and whether they fulfill certain failure mode criteria. An example of this would be a hollow cube, in which there is an interior surface that contributes to the physical properties of the part such as density, rigidity etc. Although it is possible to reveal print defects using higher energy such as X-ray [10], it would ultimately be more reliable and cost-efficient to simply detect such major print errors during the manufacturing process itself, such as was explored by Vosniakos et al. [2], Petsiuk and Pearce [6] and Oleff et al. [5].

4 Results and Discussion

4.1 *Preliminary Results and Discussion*

4.1.1 Deliverables

Our experimentation aimed to deliver a simple, working demonstration of analyzing a 3D-printed part for surface defects. The process achieved the following: (a) part retrieval from a dedicated storage location using a KUKA robotic arm, (b) an automated algorithm for part scanning using a high-definition JAI camera, (c) classification of surface defects if present and (d) percentage accuracy rating in comparison with an equivalent in-situ model.

At the time of writing, initial findings and results have only been taken from training data and experiments with the in-situ testing rig.

YOLOv5 worked consistently, but is yet to prove a reliable ML solution, noticeably because of the sheer volume of training input required.

The gathering of data will be a significant priority in the future stages of this study, with measures already taken to divide the printing load across the ex-situ and in-situ models. Further, an additional two Ender 3 v.2 rigs will be established external to the laboratory for simultaneous data acquisition.

An initial accuracy of 18.9% from the accompanying in-situ rig was due to insufficient training data with the model.

The preliminary KUKA robotic arm scanning algorithm based on isometric images was effective only for square-based prisms in a sample of existing test print pieces.

4.2 Areas of Future Work

Although this study does tackle a prominent drawback with AM technology, it is considerably limited in scope, particularly for adapting this methodology to parts of any size or shape.

4.3 Study Limitations

- The KUKA robotic arm programming is currently limited to cuboidal and rectangular-shaped parts. This is only a limitation of the scanning algorithm itself however, and once addressed by an adaptable part maneuvering strategy will present no significant challenge to training the CNN.
- The ability for this model to detect and classify minute defects or other small tolerancing issues may depend greatly on the resolution of the attached camera equipment. A solution for this is simply to upgrade to a higher resolution camera befitting for the tolerance requirements.
- Gathering large samples of verified training data is an unavoidable hurdle in assuring a high rate of accuracy for ML-based models such as YOLOv5. This will be a significant contributor to the cost and time investment for this model to be adopted at scale.
- Although defects can be quite reliably identified through the method outlined in this study, an ideal method of representing this data as a useful diagnosis for users was ultimately not investigated due to limited working constraints.
- The severity of defects was not considered. Although the type of defect may be successfully discerned, neither its degree nor magnitude can be quantified.
- The in-situ experiment running parallel to this method captured footage on different image-capturing hardware. Although this did not affect the premise of this study to a great extent, as in-situ monitoring remains limited by its fixed angle, nevertheless a closer comparison should be made using identical equipment.

References

1. Shen Z, Shang X, Zhao M, Dong X, Xiong G, Wang FY (2019) A learning-based framework for error compensation in 3D printing. IEEE Trans Cybern 49(11):4042–4050. https://doi.org/10.1109/TCYB.2019.2898553
2. Vosniakos GC, Maroulis T, Pantelis D (2007) A method for optimizing process parameters in layer-based rapid prototyping. Proc Instit Mech Eng Part B J Eng Manufact 221(8):1329–1340. https://doi.org/10.1243/09544054jem815
3. Chen Y, Peng X, Kong L, Dong G, Remani A, Leach R (2021) Defect inspection technologies for additive manufacturing. Int J Extreme Manufact 3(2). https://doi.org/10.1088/2631-7990/abe0d0

4. Yang J, Chen Y, Huang W, Li Y (2017) Survey on artificial intelligence for additive manufacturing. https://ieeexplore.ieee.org/document/8082053
5. Oleff A, Küster B, Stonis M, Overmeyer L (2021) Process monitoring for material extrusion additive manufacturing: a state-of-the-art review. Progr Addit Manufact 6(4):705–730. https://doi.org/10.1007/s40964-021-00192-4
6. Petsiuk A, Pearce JM (2022) Towards smart monitored AM: Open source in-situ layer-wise 3D printing image anomaly detection using histograms of oriented gradients and a physics-based rendering engine. Addit Manufact 52. https://doi.org/10.1016/j.addma.2022.102690
7. Han F, Liu S, Liu S, Zou J, Ai Y, Xu C (2020) Defect detection: defect classification and localization for additive manufacturing using deep learning method
8. Wang C, Tan XP, Tor SB, Lim CS (2020) Machine learning in additive manufacturing: state-of-the-art and perspectives. Addit Manufact 36. https://doi.org/10.1016/j.addma.2020.101538
9. Qi X, Chen G, Li Y, Cheng X, Li C (2019) Applying neural-network-based machine learning to additive manufacturing: current applications, challenges, and future perspectives. Engineering 5(4):721–729. https://doi.org/10.1016/j.eng.2019.04.012
10. Paraskevoudis K, Karayannis P, Koumoulos EP (2020) Real-time 3D printing remote defect detection (stringing) with computer vision and artificial intelligence. Processes 8(11). https://doi.org/10.3390/pr8111464
11. Goh GD, Sing SL, Yeong WY (2020) A review on machine learning in 3D printing: applications, potential, and challenges. Artif Intell Rev 54(1):63–94. https://doi.org/10.1007/s10462-020-09876-9
12. Meng L et al (2020) Machine learning in additive manufacturing: a review. Jom 72(6):2363–2377. https://doi.org/10.1007/s11837-020-04155-y

Electrical and Sulfate-Sensing Properties of Alkali-Activated Nanocomposites

Maliheh Davoodabadi, Marco Liebscher, Massimo Sgarzi, Leif Riemenschneider, Daniel Wolf, Silke Hampel, Gianaurelio Cuniberti, and Viktor Mechtcherine

Abstract We investigated the formation of the conductive network of carbon nanotubes (CNTs) in alkali-activated nanocomposites for sulfate-sensing applications. The matrix was a one-part blend of fly ash and ground granulated blast-furnace slag, activated by sodium silicate and water. Sodium dodecylbenzenesulfonate was used as the surfactant for dispersion of the CNTs in the aqueous media. The nanocomposites were investigated by a laboratory-developed setup to study the electrical and sensing properties of the alkali-activated material. The electrical properties (i.e., conductivity) were calculated and assessed to discover the percolation threshold of the nanocomposites. Furthermore, the sensing behavior of nanocomposites was studied upon sulfate (SO_4^{2-}) exposure by introduction of sulfuric acid ((H_2SO_4)) and magnesium sulfate ($MgSO_4$). The sensors were able to preliminarily exhibit a signal difference based on the introduced media (H_2SO_4&$MgSO_4$), CNT content

M. Davoodabadi · M. Liebscher (✉) · V. Mechtcherine
Institute of Construction Materials, Faculty of Civil Engineering, Technische Universität Dresden, Dresden, Germany
e-mail: marco.liebscher@tu-dresden.de

M. Davoodabadi · M. Sgarzi (✉) · L. Riemenschneider · G. Cuniberti
Institute for Materials Science and Max Bergmann Centre of Biomaterials, Dresden University of Technology, Dresden, Germany
e-mail: massimo.sgarzi@unive.it

Dresden Center for Nanoanalysis (DCN), Dresden University of Technology, Dresden, Germany

M. Sgarzi
Department of Molecular Sciences and Nanosystems, Ca' Foscari University of Venice, Venice, Italy

D. Wolf · S. Hampel
Leibniz Institute for Solid State and Materials Research, Dresden, Germany

G. Cuniberti
Center for Advancing Electronics Dresden (CfAED), Dresden University of Technology, Dresden, Germany

Dresden Center for Computational Materials Science (DCMS), Dresden University of Technology, Dresden, Germany

W. Duan et al. (eds.), *Nanotechnology in Construction for Circular Economy*, Lecture Notes in Civil Engineering 356,
https://doi.org/10.1007/978-981-99-3330-3_29

and H_2SO_4 volumetric quantity. The results of this research demonstrated a sensing potential of CNT alkali-activated nanocomposites and can be applied in the concrete structural health monitoring.

Keywords Aggressive ion sensing · Alkali-activated nanocomposites · Carbon nanotubes · Structural health monitoring

1 Introduction

Aggressive ions such as sulfates (SO_4^{2-}) are naturally available in the surrounding aquatic media of concrete infrastructures. Moreover, these anionic species can be liberated from the underground environments or internal sources by moisture ingress (chemical origin) or can be produced through the metabolism of microorganisms (biogenic origin). The main sources of SO_4^{2-} anions are sulfuric acid (H_2SO_4), magnesium sulfate ($MgSO_4$), and sodium sulfate (Na_2SO_4), which cause severe degradation of concrete material over long-term exposure [1–4]. An innovative idea regarding prevention of concrete material deterioration is the development of a sensor to distinguish SO_4^{2-} species coming from identical [5] or non-identical sources or having different concentrations/quantities.

Therefore, in the present study, carbon nanotube (CNT) alkali-activated sensors are proposed for SO_4^{2-} species sensing and discriminating released from H_2SO_4&$MgSO_4$, for the first time to the best of our knowledge. The sensors were fabricated from a sodium-based fly ash ground granulated blast-furnace slag (GGBS) alkali-activated material and CNTs. As the first step, the percolation threshold of the sensors was determined by measuring the electrical resistance (converted to conductivity) of CNT alkali-activated nanocomposites, incorporating different content of CNTs. To study the SO_4^{2-} sensing potential and transitional behavior of the nanocomposites from insulating to conducting mode, the sensors were fabricated to incorporate different CNT concentrations. The assessments were carried out by introducing H_2SO_4&$MgSO_4$ with identical concentration. Finally, the sensors were fabricated with the percolated CNT concentration and evaluated by introducing different volumetric quantities of H_2SO_4.

2 Methods

2.1 Materials

The applied CNTs were TUBALL™, supplied by OCSIAL Europe and their properties can be found in Davoodabadi *et al.* [5, 6]. The surfactant was a technical grade of sodium dodecylbenzenesulfonate (SDBS) produced by Merck KGaA. The utilized precursors were fly ash (Steag Power Minerals GmbH) and GGBS (Opterra GmbH); their properties are described in Davoodabadi *et al.* [7]. Sodium disilicate powder (Sikalon; Wöllner GmbH) was used to activate and geopolymerize the blend. Sulfuric acid and magnesium sulfate for the sensing measurements were diluted to 0.1 M from a stock solution of 98% ACS reagent sulfuric acid (Merck KGaA), and anhydrous magnesium sulfate powder (Merck KGaA), respectively.

2.2 Methods

The nanofluids were prepared by ultrasonication of the CNTs and SDBS in ultrapure water according to the procedures in Davoodabadi *et al.* [5–7]. The CNT concentration range spanned from 0.010 to 1.000 wt% of oxide mass and SDBS was added with the same mass as CNTs. The nanocomposites were fabricated with the formulation in Davoodabadi *et al.* [5–7]. The mixed slurries were cast into plastic molds with slot dimensions of 60 × 10 × 10 mm^3 for 24 h. After demolding, the nanocomposites were cured in the chemical laboratory ambient conditions. Thereafter, the nanocomposites were heat treated (at 105 °C for 24 h) to eliminate any negative impact of water on the CNTs' conductive network.

2.3 Characterizations

A programmed setup was used for the electrical properties and sensing measurements composed of a KEYSIGHT B2912A precision source/measure unit (SMU), a computer, and Grafana time-series database [5]. The received data in the form of the resistance (R) in Ω were converted to resistivity ($\rho = R.A.L^{-1}$) in Ω.m and conductivity ($\sigma = \rho^{-1}$) in $S.m^{-1}$ by applying the sensors' cross-sectional area (A) and length (L). For further analysis, relative resistance ($RR = 100.(R_1 - R_0).R_0^{-1}$) was used for sensor evaluations. A GeminiSEM 500 (Carl Zeiss QEC GmbH) was used for scanning electron microscopy imaging of the specimens' cross section. The FEI Tecnai F30 (ThermoFisher Scientific) was used to conduct high resolution transmission electron microscopy (HRTEM imaging) of the samples.

3 Results and Discussion

3.1 Percolation Threshold

In the available literature, the sensing properties of geopolymers exploited the electrolytic (ionic) conductance of the composites because of their ion-rich structure and pore system [8–12]. However, investigated alkali-activated composites in this study are insulators and exhibit insufficient conductive (electronic) properties to act as a sensor. The measured inherent resistances of these composites were unsteady, in the range of mega ohm (≥ 10 MΩ). Because measurements are conducted on heat-treated specimens, the electrolytic conductivity of the ion-rich framework is assumed to be negligible. Sufficient inclusion of CNTs causes the resistance range of nanocomposites to descend abruptly from mega ohm (highly insulator character) to hundreds of ohms (highly conductive behavior). This proves that the electronic conductive network of CNTs has been generated, and the nanocomposites are percolated [13].

The correlation curve of the nanocomposites' conductivity and CNTs' concentration is depicted in Fig. 1. The observed increasing trend of conductivity corresponded to the required quantity of CNTs for establishing a functional conductive network, mostly in tube-contacting mode rather than the tunneling or hopping mode [13]. This conductive network of CNTs provides the alkali-activated matrix with a percolating transition zone between 0.070 wt% and 0.200 wt% as indicated by the inset in Fig. 1. With respect to percolation theory, such curves can be fitted with power regression models [14–17]. In addition to the investigated nanocomposites, CNT Portland cement-based nanocomposites have a relatively similar percolation trend, but their documented thresholds exhibit, naturally, non-conclusive values. Some of the documented percolating transition zones of CNT Portland cement-based nanocomposites are shown in Table 1 for comparison.

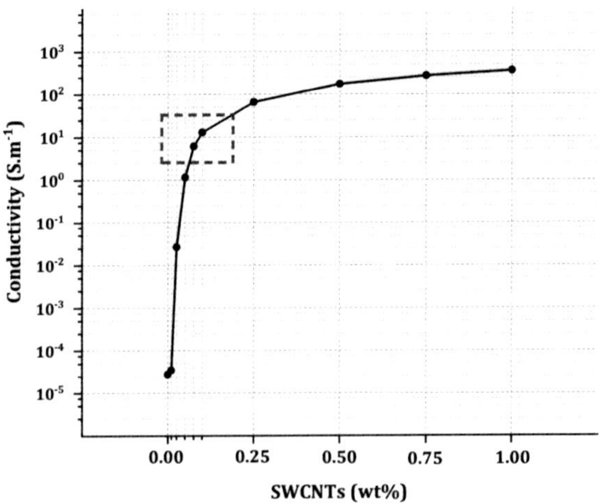

Fig. 1 Percolation diagram of CNT alkali-activated nanocomposites (concentrations (wt%) between 0 and 0.25 are 0.010, 0.025, 0.050, 0.075, & 0.100)

Table 1 Reported percolating transition zones of CNT cementitious nanocomposites

Reference	Liu et al. [24]	Danoglidis et al. [25]	Hong et al. [26]
Range	0.80–1.60 vol%	0.10–0.15 wt%	0.30–0.60 wt%

The most influential parameters on the percolation of nanocomposites are the CNTs' structure (i.e., chirality, aspect ratio, waviness, and the number of shells) and CNTs' tunneling resistance/distance. The fabrication methodology (applied surfactant, ultrasonication energy, mixing, and curing), which determines the 3D orientation, configuration, interconnection of CNTs, and geometry of agglomerates, further affects the percolating character of the nanocomposite [15, 16, 18–21]. Considering the percolation threshold as an outset, the preference of composite researchers is to maintain the nano-additives' concentration as low as possible (i.e., low or ultra-low percolation thresholds), because high concentrations of nano-additives have a destructive impact on the microstructure and consequently on the mechanical properties of the nanocomposite [14, 22, 23] (Table 1).

Based on these arguments, to achieve a firm conductive network of CNTs, in which the nanocomposite is not so overloaded as to be destroyed by the additives and not too under-loaded to be an insulator, a concentration of 0.1 wt% was considered for the sensor fabrication. In Sect. 3.2. Sulfate discrimination, the selection of CNT concentration of 0.1 wt% as the percolation threshold is further validated. In comparison, a wide concentration range of 0.05–2 wt% was reported in the literature for sensing and smart applications of the Portland cement-based nanocomposites [27–34].

3.2 Sulfate Discrimination

The CNT alkali-activated nanocomposites sensing responses upon exposure to 90 μL (30 μL in each cycle) of 0.1 M H_2SO_4& $MgSO_4$ with CNT concentration ranging from 0.0 to 1.0 wt% are illustrated in Fig. 2. The response of the sensors upon exposure to different sulfate containing regimens was explored and compared. A range of CNT concentrations were considered to confirm the influence of the generating conductive percolated network of CNTs on the sensing behavior of the nanocomposites with respect to percolation analysis.

The nanocomposites with incorporation of 0.0 and 0.010 wt% exhibited fully insulating behavior (Fig. 2a, b). The responses were highly noisy and stochastic as expected from the ultra-low conductivity of these nanocomposites; the approximate magnitude was 3E-5 S·m^{-1} as shown in Fig. 1. With CNT incorporation of 0.025 wt% and 0.050 wt%, the conductivity rose sharply to almost 3E-2 and 1 S·m^{-1}, respectively, and the responses were slightly functional but still not accurate (Fig. 2c, d). Considering the onset of the percolating area, at CNT concentration of 0.070 wt%

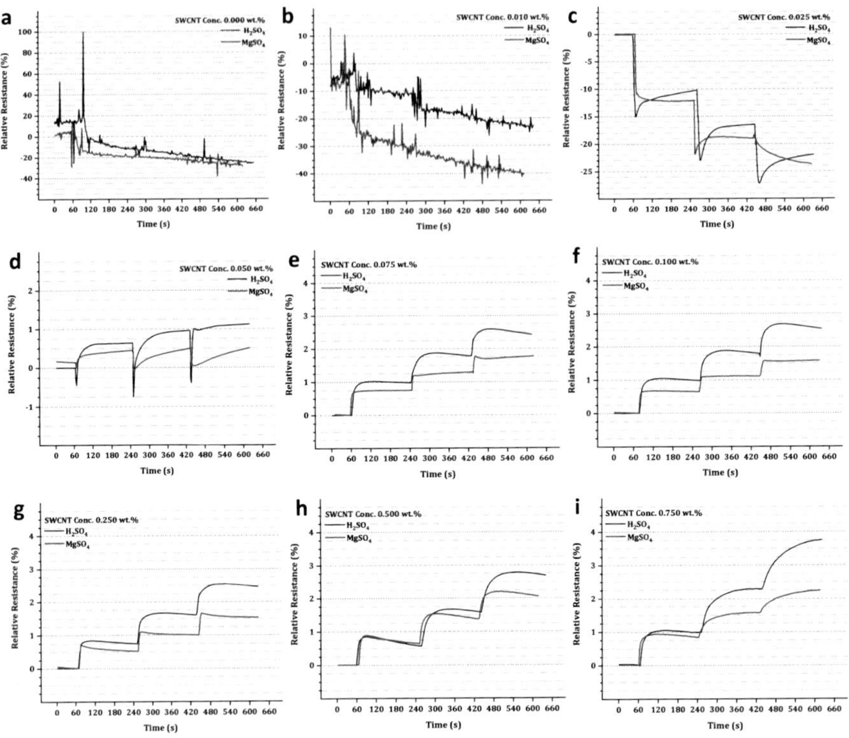

Fig. 2 The sensing responses of CNT alkali-activated nanocomposites exposed to 90 µL (30 µL in each cycle) of 0.1 M $H_2SO_4 \& MgSO_4$; CNT concentrations (wt%) in the nanocomposites are **a** 0.000; **b** 0.010; **c** 0.025; **d** 0.050; **e** 0.075; **f** 0.100; **g** 0.250; **h** 0.500; **i** 0.750

the functional response of the sensors began at 0.075 wt% with sensor conductivity of ≈6 $S \cdot m^{-1}$ (Fig. 2e). From this point onwards, the CNT signals showed a regular configuration, which corresponded to the normal behavior of pristine p-type CNTs (Fig. 2e–j). The H_2SO_4& $MgSO_4$ discrimination mechanism in the percolated area was mostly based on the difference in the signal shape and magnitude.

The correlations of maximum relative resistance and CNT concentration are plotted in Fig. 3. The main results to note were (i) the positive relationship of the relative resistance and CNT concentration, and (ii) the higher sensitivity of the nanocomposites towards H_2SO_4 exposure than $MgSO_4$. Result (ii) means the sensors can discriminate SO_4^{2-} species introduced from two different sources (H_2SO_4& $MgSO_4$), by means of relative resistance as demonstrated in Fig. 3, and by signal configuration and shape as illustrated in Fig. 2. The CNTs' signal differentiation was particularly recognizable in the percolating transition zone of the nanocomposites (between 0.075 and 0.250 wt%) shown in Fig. 2e–g.

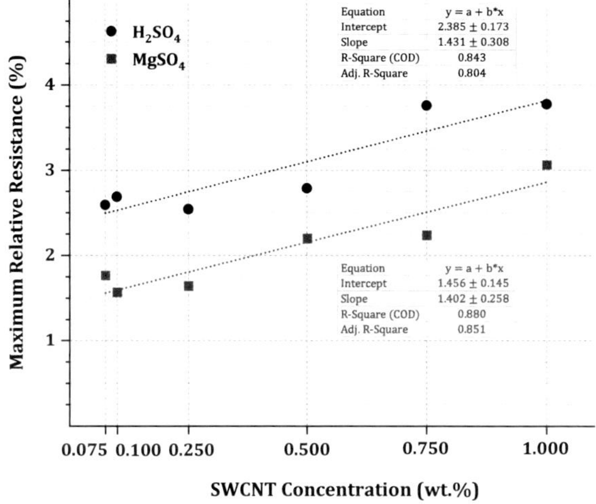

Fig. 3 Maximum relative resistance variations of CNT alkali-activated nanocomposites based on CNT concentrations upon exposure to 90 µL of 0.1 M $H_2SO_4\&MgSO_4$

3.3 *Quantity Differentiation*

The H_2SO_4 quantity differentiation potential of the sensors is shown in Fig. 4. For this purpose, the nanocomposites were fabricated with CNT inclusion of 0.1 wt% based on the percolation and sensing analyses. The same concentration regimen of H_2SO_4 (i.e., 0.1 M) was introduced in volumetric quantities of 90, 180, 270 µL (30, 60, and 90 µL, respectively in each cycle) to the sensors to evaluate the responses. The increments of relative resistance were approximately + 140% and + 60% with the volume increasing from 90 µL to 180 µL and afterwards to 270 µL, exhibiting a linear behavior.

3.4 *CNTs' Conductive Network*

The CNTs' network distribution and expansion are shown in Fig. 5. The addition of SDBS as surfactant for the dispersion of CNTs, entrained air bubbles into the alkali-activated microstructure. In addition, SBDS probably had a negative impact on the activating reactions and created a weaker microstructure. Nevertheless, the alkali-activated microstructure exhibited sufficient strength (compressive strength of 40 ± 2.48 MPa for 28-day nanocomposite) and mechanical performance. The main mechanism was the reinforcing effect of the CNTs because of their distribution (Fig. 5a), and crack covering and bridging abilities (Fig. 5b). Furthermore, this mode of interaction by CNTs has a secondary function, which is the formation of a conductive network (Fig. 5c). This network endows the nanocomposites with high

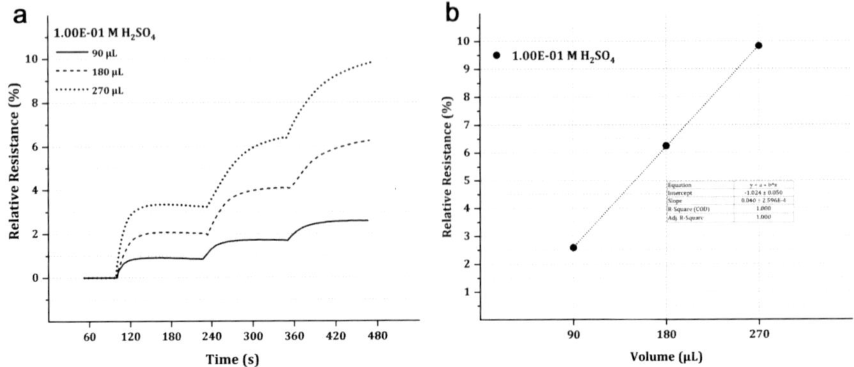

Fig. 4 Quantity differentiation of CNT alkali-activated sensors (CNT conc. 0.1 wt%). **a** Exposure to different volumes of 0.1 M H_2SO_4; **b** regression of relative resistance vs. volumetric quantity

electrical conductivity and consequently a sensing ability as explained in previously. The percolated conductive network of CNTs can be seen in the HRTEM images of Fig. 6. The network comprised overlapping CNT agglomerations, which formed directional pathways.

4 Conclusions

The present conceptual study has proposed a structural sensor for assessing the SO_4^{2-} sensing potential of CNT alkali-activated nanocomposites. The investigated discrimination criteria were CNT concentration, SO_4^{2-} bearing media (H_2SO_4vs.$MgSO_4$, and analyte volumetric quantity H_2SO_4. The obtained results can be summarized as follows.

- The percolating zone of the nanocomposites was between 0.07 and 0.20 wt% of CNT content.
- The sensors can be fabricated by incorporation of 0.1 wt% of CNT into the alkali-activated matrix material based on the percolation threshold study of the nanocomposites.
- The sensors exhibit differentiation behavior by variation of shape and magnitude of the obtained relative resistance.
- There was a linear correlation between relative resistance and CNT concentration when the sensors were exposed to 0.1 M H_2SO_4&$MgSO_4$. Similarly, the relationship between the relative resistance of the sensors and volumetric quantity of 0.1 M H_2SO_4 was linear.
- The sensors exhibited a higher magnitude of relative resistance when exposed to 0.1 M H_2SO_4 compared with 0.1 M $MgSO_4$.

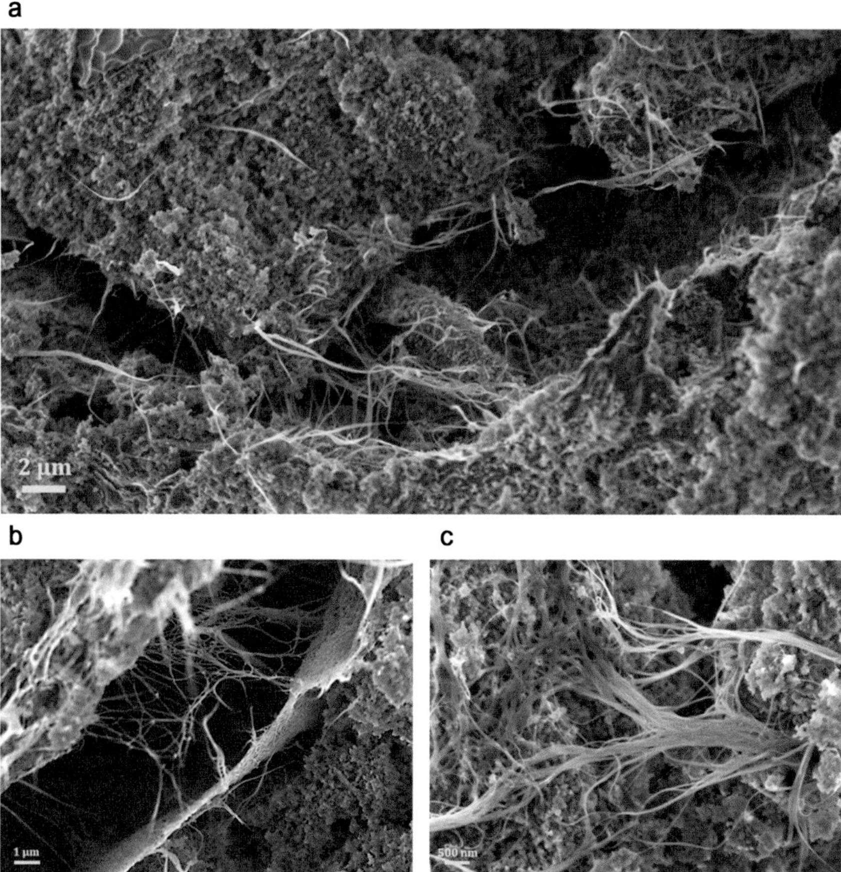

Fig. 5 The reinforcing and conductive network of CNTs in the microstructure of the alkali-activated matrix: **a** CNT distribution, **b** crack covering and bridging by CNTs, **c** CNTs' network expansion

- The percolated sensors presented a response curvilinear shape upon H_2SO_4 exposure and a rectangular shape upon $MgSO_4$ exposure.

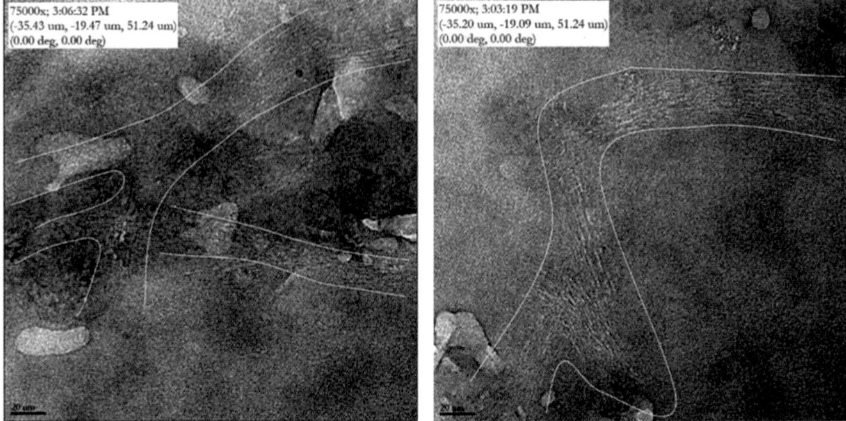

Fig. 6 Percolated conductive network of CNTs in the nanostructure of the alkali-activated matrix at the atomic scale of HRTEM. The percolated network of overlapped CNTs is shown as yellow pathways

References

1. Paul A, Rashidi M, Kim J-Y, Jacobs LJ, Kurtis KE (2022) The impact of sulfate- and sulfide-bearing sand on delayed ettringite formation. Cement Concr Compos 125:104323
2. Wang Y, Cao Y, Zhang Z, Huang J, Zhang P, Ma Y, Wang H (2022) Study of acidic degradation of alkali-activated materials using synthetic C-(N)-A-S-H and N-A-S-H gels. Compos B Eng 230:109510
3. Grengg C, Gluth GJ, Mittermayr F, Ukrainczyk N, Bertmer M, Guilherme Buzanich A, Radtke M, Leis A, Dietzel M (2021) Deterioration mechanism of alkali-activated materials in sulfuric acid and the influence of Cu: a micro-to-nano structural, elemental and stable isotopic multi-proxy study. Cem Concr Res 142:106373
4. Kobayashi M, Takahashi K, Kawabata Y (2021) Physicochemical properties of the Portland cement-based mortar exposed to deep seafloor conditions at a depth of 1680 m. Cem Concr Res 142:106335
5. Davoodabadi M, Liebscher M, Sgarzi M, Riemenschneider L, Wolf D, Hampel S, Cuniberti G, Mechtcherine V (2022) Sulphuric acid sensing of single-walled carbon nanotubes incorporated alkali activated materials. Compos Part B Eng
6. Davoodabadi M, Vareli I, Liebscher M, Tzounis L, Sgarzi M, Paipetis AS, Yang J, Cuniberti G, Mechtcherine V (2021) Thermoelectric energy harvesting from single-walled carbon nanotube alkali-activated nanocomposites produced from industrial waste materials. Nanomaterials (Basel, Switzerland) 11(5)
7. Davoodabadi M, Liebscher M, Hampel S, Sgarzi M, Rezaie AB, Wolf D, Cuniberti G, Mechtcherine V, Yang J (2021) Multi-walled carbon nanotube dispersion methodologies in alkaline media and their influence on mechanical reinforcement of alkali-activated nanocomposites. Compos B Eng 209:108559
8. Biondi L, Perry M, McAlorum J, Vlachakis C, Hamilton A, Lo G (2021) Alkali-Activated Cement Sensors for Sodium Chloride Monitoring. IEEE Sensors J 21(19):21197–21204
9. Biondi L, Perry M, McAlorum J, Vlachakis C, Hamilton A (2020) Geopolymer-based moisture sensors for reinforced concrete health monitoring. Sens Actuators B Chem 309:127775
10. McAlorum J, Perry M, Vlachakis C, Biondi L, Lavoie B (2021) Robotic spray coating of self-sensing metakaolin geopolymer for concrete monitoring. Autom Constr 121:103415

11. Vlachakis C, Perry M, Biondi L, McAlorum J (2020) 3D printed temperature-sensing repairs for concrete structures. Addit Manuf 34:101238
12. Saafi M, Gullane A, Huang B, Sadeghi H, Ye J, Sadeghi F (2018) Inherently multifunctional geopolymeric cementitious composite as electrical energy storage and self-sensing structural material. Compos Struct 201:766–778
13. Buroni FC, García-Macías E (2021) Closed-form solutions for the piezoresistivity properties of short-fiber reinforced composites with percolation-type behavior. Carbon 184:923–940
14. Khan T, Irfan MS, Ali M, Dong Y, Ramakrishna S, Umer R (2021) Insights to low electrical percolation thresholds of carbon-based polypropylene nanocomposites. Carbon 176:602–631
15. García-Macías E, D'Alessandro A, Castro- Triguero R, Pérez-Mira D, Ubertini F (2017) Micromechanics modeling of the electrical conductivity of carbon nanotube cement-matrix composites. Compos B Eng 108:451–469
16. Jang S-H, Hochstein DP, Kawashima S, Yin H (2017) Experiments and micromechanical modeling of electrical conductivity of carbon nanotube/cement composites with moisture. Cement Concr Compos 77:49–59
17. Zhang L, Ding S, Li L, Dong S, Wang D, Yu X, Han B (2018) Effect of characteristics of assembly unit of CNT/NCB composite fillers on properties of smart cement-based materials. Compos A Appl Sci Manuf 109:303–320
18. Wu D, Wei M, Li R, Xiao T, Gong S, Xiao Z, Zhu Z, Li Z (2019) A percolation network model to predict the electrical property of flexible CNT/PDMS composite films fabricated by spin coating technique. Compos Part B Eng 174:107034
19. Mahmoodi MJ, Vakilifard M (2019) CNT-volume-fraction-dependent aggregation and waviness considerations in viscoelasticity-induced damping characterization of percolated-CNT reinforced nanocomposites. Compos B Eng 172:416–435
20. Wang Y, Wang X, Cao X, Gong S, Xie Z, Li T, Wu C, Zhu Z, Li Z (2021) Effect of nano-scale Cu particles on the electrical property of CNT/polymer nanocomposites. Compos A Appl Sci Manuf 143:106325
21. Doh J, Park S-I, Yang Q, Raghavan N (2021) Uncertainty quantification of percolating electrical conductance for wavy carbon nanotube-filled polymer nanocomposites using Bayesian inference. Carbon 172:308–323
22. Wang J, Kazemi Y, Wang S, Hamidinejad M, Mahmud MB, Pötschke P, Park CB (2020) Enhancing the electrical conductivity of PP/CNT nanocomposites through crystal-induced volume exclusion effect with a slow cooling rate. Compos B Eng 183:107663
23. Kim H, Gao S, Hong S, Lee P-C, Kim YL, Ha JU, Jeoung SK, Jung YJ (2019) Multifunctional primer film made from percolation enhanced CNT/Epoxy nanocomposite and ultrathin CNT network. Compos B Eng 175:107107
24. Liu L, Xu J, Yin T, Wang Y, Chu H (2021) Improved conductivity and piezoresistive properties of Ni-CNTs cement-based composites under magnetic field. Cement Concr Compos 121:104089
25. Danoglidis PA, Konsta-Gdoutos MS, Shah SP (2019) Relationship between the carbon nanotube dispersion state, electrochemical impedance and capacitance and mechanical properties of percolative nanoreinforced OPC mortars. Carbon 145:218–228
26. Hong G, Choi S, Yoo D-Y, Oh T, Song Y, Yeon JH (2022) Moisture dependence of electrical resistivity in under-percolated cement-based composites with multi-walled carbon nanotubes. J Market Res 16:47–58
27. D'Alessandro A, Rallini M, Ubertini F, Materazzi AL, Kenny JM (2016) Investigations on scalable fabrication procedures for self-sensing carbon nanotube cement-matrix composites for SHM applications. Cement Concr Compos 65:200–213
28. Tian Z, Li Y, Zheng J, Wang S (2019) A state-of-the-art on self-sensing concrete: Materials, fabrication and properties. Compos B Eng 177:107437
29. Rao RK, Sasmal S (2021) Nanoengineered smart cement composite for electrical impedance-based monitoring of corrosion progression in structures. Cement Concr Compos:104348
30. Rao R, Sindu BS, Sasmal S (2020) Synthesis, design and piezo-resistive characteristics of cementitious smart nanocomposites with different types of functionalized MWCNTs under long cyclic loading. Cement Concr Compos 108:103517

31. Kim GM, Yang BJ, Yoon HN, Lee HK (2018) Synergistic effects of carbon nanotubes and carbon fibers on heat generation and electrical characteristics of cementitious composites. Carbon 134:283–292
32. Konsta-Gdoutos MS, Danoglidis PA, Falara MG, Nitodas SF (2017) Fresh and mechanical properties, and strain sensing of nanomodified cement mortars: the effects of MWCNT aspect ratio, density and functionalization. Cement Concr Compos 82:137–151
33. Dong W, Li W, Wang K, Han B, Sheng D, Shah SP (2020) Investigation on physicochemical and piezoresistive properties of smart MWCNT/cementitious composite exposed to elevated temperatures. Cement Concr Compos 112:103675
34. Zhan M, Pan G, Zhou F, Mi R, Shah SP (2020) In situ-grown carbon nanotubes enhanced cement-based materials with multifunctionality. Cement Concr Compos 108:103518

Advances in Characterization of Carbonation Behavior in Slag-Based Concrete Using Nanotomography

B. Mehdizadeh, K. Vessalas, B. Ben, A. Castel, S. Deilami, and H. Asadi

Abstract Exposure of concrete to the atmosphere causes absorption of CO_2 and carbonation via a chemical reaction between the CO_2 and calcium hydroxide and calcium-silicate-hydrate reaction products inside the concrete. A greater understanding of carbonation behavior and its micro- and nanoscale impacts is needed to predict and model concrete durability, cracking potential and steel depassivation behaviors. New and sophisticated techniques have emerged to analyze the microstructural behavior of concrete subjected to carbonation. High-resolution full-field X-ray imaging is providing new insights to nanoscale behavior. Full-field nano-images provide significant insight into 3D structural identification and mapping. Nanotomographic modeling of an accelerated carbonated test specimen can also provide a 3D view of the pore structure that resides inside slag-based concrete. This is critical for better understanding of the capillary porosity and pore solution behaviors of concrete in situ. We investigated the analysis of durability properties, including the carbonation behavior of slag-based concrete, by evaluating microstructural and nanotomographic identification techniques.

Keywords Accelerated carbonated test · Nanotomography · Slag-based concrete · X-ray imaging

1 Introduction

The possibility of achieving high strength and durability has made concrete a well-established structural material for use in roads, tunnels and bridge infrastructure [1–4]. Various studies have proposed alternative materials (supplementary cementitious

B. Mehdizadeh (✉) · K. Vessalas · A. Castel
School of Civil and Environmental Engineering Technology, University of Technology Sydney, NSW, Australia
e-mail: bahareh.mehdizadehmiyandehi@student.uts.edu.au

B. Ben · S. Deilami · H. Asadi
Advanced Technical Services Infrastructure and Place, Transport for NSW, NSW, Australia

W. Duan et al. (eds.), *Nanotechnology in Construction for Circular Economy*, Lecture Notes in Civil Engineering 356,
https://doi.org/10.1007/978-981-99-3330-3_30

materials (SCMs)) to partially replace cement to reduce the CO_2 emissions caused by its production [5–8]. Using SCMs such as fly ash, slag, silica fume, etc. as partial replacement of cement is effective in making concrete more cost-effective, and high performing in terms of strength and durability requirements [9–16].

The carbonation of concrete caused by the presence of CO_2 in the atmosphere ($\approx$380 ppm) acts as an environmental load and contributes significantly to the deterioration of concrete structures [17], because it causes the alkalinity of the concrete to decrease (pH 8). If the pH decreases below 12, the risk of corrosion is increased for any fixed steel components present within the concrete elements [18, 19]. Carbonation increases the porosity of the concrete, resulting in reduced compressive strength and impermeability in the carbonated zone of the concrete. The results of several experimental studies have revealed that the carbonation of slag-based concrete is highly dependent on the water/cement ratio, cement replacement ratio, curing method, and in-situ environmental conditions [20–24]. Because the carbonation process is long term in its manifestation, many researchers have used accelerated carbonation tests by adding pressurized CO_2 or increased the temperature to increase CO_2 diffusivity in the pore solution to shorten the experimental time [25–28].

Depending on the cement replacement ratio and the type of SCM, the properties of SCM-based concrete change over time as a result of both the hydration process and the carbonation behavior [29–31]. After carbonation, the total capillary porosity volume of the slag-based concrete increases, which has a negative impact on the durability of the concrete by creating a larger pore structure and thus increasing the permeability coefficients for the concrete [32–34].

To control this behavior, nondestructive techniques such as micro- and nanotomography can be used to monitor changes in the pore structure of slag-based concretes [35]. Han et al. demonstrated how these modern technologies are useful for analyzing the depth of concrete carbonation [36].

Recent advancements in micro- and nanoscale techniques give insight into forecasting and simulating concrete durability, cracking potential, and steel depassivation behaviors. We present the main advances in these techniques in investigating carbonated concrete behavior. Although several studies were reviewed there is still limited information published about the use of tomographic techniques for better understanding of the carbonation behavior of slag-based concretes.

2 Tomography Techniques

Tomography, in the broadest sense, is any technique that uses sectional views as an intermediary stage before reassembling a three-dimensional (3D) object. This characterization technique is useful for identifying the richness of the microstructure in three dimensions rather than only presenting two-dimensional projections. 3D micro- and nanoscale views are required to fully comprehend and monitor the behavior of the concrete. Materials scientists have used X-ray tomographic techniques for decades to discover the behavior of 3D microstructures [36–40]. FIB/SEM tomography (focused-ion beam scanning electron microscopy), electron tomography, X-ray micro-computed tomography (micro-CT), and X-ray nano-computed tomography (nano-CT) are nondestructive 3D imaging techniques that are useful for investigating the interior structure of a wide range of materials [34, 35, 38–41].

Tomographic techniques can give new insights into concrete deterioration by providing precise information on the changed layers of concrete affected by carbonation, corrosion, leaching or sulfate attack [42]. Carbonation as a durability concern in concrete is strongly influenced by microstructural and pore network characteristics. The porosity parameters obtained from the tomographic data can successfully indicate the internal pore network geometry and microstructural features [37]. X-ray nano- and micro-CT techniques are useful for monitoring slag-based concrete behavior subjected to carbonation [43, 44].

2.1 X-ray Micro-CT Imaging

X-ray micro-CT is a radiographic imaging technique that generates a series of cross-sectional images to identify the internal structure of materials without causing damage to the specimen [45–47]. In general, micro-CT provides a series of reconstructed images represented by a pixel, either 8- or 16-bit. Micro-CT processing transforms the 8 or 16-bit images into binary or segmented images, which are useful for examining the features of the specimen. A 3D microstructural image can be created by stacking 2D segmented images to analyze volumetric, multi-directional, and other advanced sample features. The pore structure of the concrete can affect the qualities of the concrete, including strength, durability, and permeability. Using micro-CT images can assist in discovering more information about these properties [48–51]. Although several scales are used, such as macro-, micro- and nano-CT, the spatial resolution of micro-CT can be used for microscopic CT scanning [45]. The micro-scale CT technique can scan a range from 1 to 10 μm in size, which is also a good range for investigating the size of capillary pores inside the cement paste [45, 52, 53]. For covering the whole range of cement paste pore sizes, nano-CT images can further scan a range of 10 μm to < 10 nm [48, 54]. In order to obtain high-quality nano-CT data it requires meticulous sample preparation. Measuring 3D morphology at < 10 nm can give unique information on transportation characteristics and the mechanical and durability behavior of the concrete [55].

2.2 Micro-CT and Nano-CT to Investigate Carbonated Concrete

During the hydration process, the cement paste interacts with CO_2. Carbonated cement paste contributes to steel reinforcement corrosion, causing major and long-term durability issues for concrete structures. Several studies have noted that as carbonation develops, porosity reduces due to calcite formation filling the pore microstructure [56–59]. Phenolphthalein is a well-known technique for determining carbonation depth in concrete, but it necessitates the destruction of the sample, and the findings vary based on the sampling location [60, 61]. In contrast, X-ray micro- and nano-CT are nondestructive techniques that can identify and analyze the depth of carbonation and the microstructure of the concrete during the carbonation process [62–67]. In addition, they can continually monitor the evolution of the carbonated area of the same sample at different ages [68].

Some researchers have used micro-CT to determine the microstructural development of the cement paste during the carbonation process, particularly the distribution of porosity and the effective pore width [69, 70]. In both studies, the results of average porosities revealed that the porosity reduces with additional carbonation time, which infers that calcite (calcium-bearing phase) forms during the carbonation process, and that the porosity distribution may validate the pore microstructural change (e.g., reduction in porosity caused by carbonation). Because the porous structure of concrete extends from the nano- to the macroscopic scale [71], X-ray micro- and nano-CT with good resolution are in high demand for characterizing a wide range of behavior [74]. Particle size and shape, interfacial topology, particle structure, pore structure, carbonation depth, and morphology of distinct solid phases in concrete have all been studied by these methods [72–76].

X-ray micro-CT results show that microcracks form from the surface to the inside of the cement paste after carbonation. Furthermore, the carbonated area increases in depth with increasing carbonation time. Moreover, cracks form during the carbonation process and reduce the density [77]. A new generation of laboratory-based nano-CT with high resolution [45] can provide 3D images for measuring different properties of the concrete such as durability, cracking potential, and steel depassivation behavior. Dimensional and transitional stability is necessary for generating quality data by any instrument involved in sub-micron imaging. It has been demonstrated that nano-CT with high-resolution scanning is a well-established and mature method [78] for gaining insight of the porous and hierarchical structure of the concrete at the sub-micron scale [55]. The ability to perform nano-CT scanning under ambient conditions keeps test samples in their natural form under normal and accelerated conditions [55, 71, 79–83]. Han et al. [83] reported that by using nano-CT, the solid-phase composition, pore structure, damage degradation, and nano-mechanical characteristics of the concrete can be measured at different accelerated carbonation ages. Their results confirmed that without any prior drying preparation, X-ray CT is a suitable technique for obtaining 3D images of concrete to assess the degree of microstructural damage.

2.3 *Microstructural Characteristics and Properties of Slag-Based Concrete During Carbonation*

SCMs play a critical role in controlling and improving the mechanical and durability properties of the concrete [14–16]. Han et al. [83] described how they used micro-CT to track the progression of carbonation-induced fractures and how the carbonation depth increased with exposure duration for concrete containing a slag content up to 70% of the total binder content. Figure 1 shows several cracks produced during the carbonation process, which demonstrates how micro-CT can be utilized to categorize carbonation behavior over time [81, 83]. The micro-CT results revealed that the width and length of microcracks significantly affects the carbonation behavior of concrete [83]. More CO_2 penetration causes an increase in crack length. When the fracture width is < 10 μm at 1 year, CO_2 diffusivity around the crack is nearly equal to that in the surrounding concrete. However, with fracture width > 10 μm, the CO_2 diffusivity in the concrete increases. When the crack width is > 100 μm, the CO_2 diffusivity is somewhat further increased [80, 81, 83, 84].

In another study, Han et al. [67] analyzed the carbonation depth of the cement paste with different slag addition from 0 to 70% under 0–14 days of accelerated carbonation testing (after 3 months of curing). Because of slag's cementitious and pozzolanic properties, the hydration reaction of the slag-based cement will improve the pore microstructure. Large pores will eventually transition into smaller pores with the pozzolanic reaction, which significantly reduces the CO_2 diffusion coefficient and therefore the rate of carbonation will also decrease. However, slag-based concrete generates a large amount of calcium hydroxide during the hydration process, and calcium hydroxide is an important chemical component in carbonation. The results from Han et al. demonstrated that the ideal slag addition to the binder to mitigate carbonation is < 50%. Figure 2 shows the specimen's carbonation front and depth

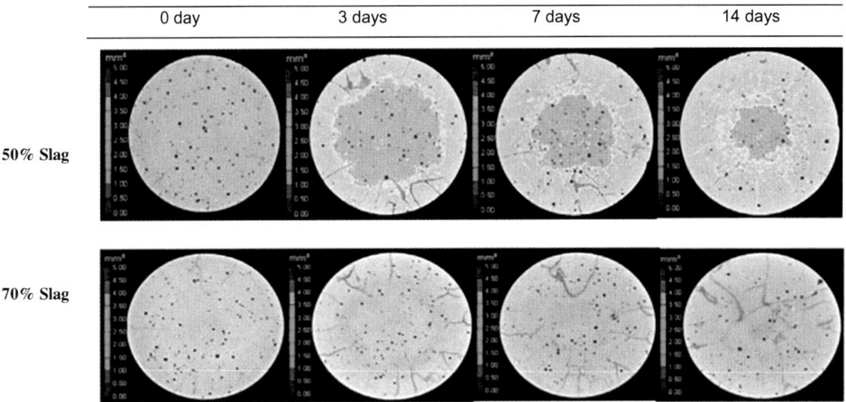

Fig. 1 Cross-sectional images of the carbonation front for 50% and 70% slag with increasing accelerated carbonation testing time, 3D voxel size 0.086 mm^3 [67, 81]

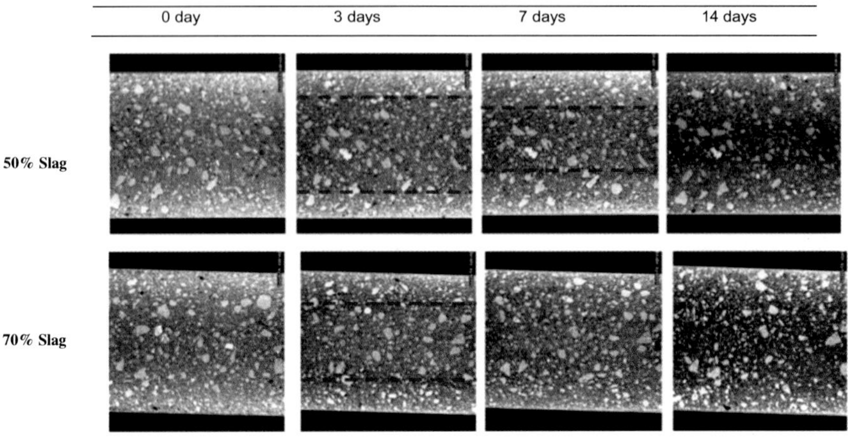

Fig. 2 Views of the carbonation depth with 50% and 70% slag during 14 days of accelerated carbonation testing, 3D voxel size 0.086 mm^3 [67]

of penetration with 50% and 70% slag during 14 days of accelerated carbonation. The carbonation front can be identified by micro-CT, which is seen to increase for 50% slag addition with the carbonation zone steadily expanding with increasing curing time. However, for 70% slag addition, the specimens are totally carbonated in only 7 days. These findings demonstrate that the ideal slag addition, which mitigates carbonation, is < 50%. Figure 3 shows that the carbonation depth estimated from micro-CT appears to be the same as with the phenolphthalein method during 14 days of accelerated carbonation without the addition of slag. These results prove that micro-CT is a reliable and appropriate technique for characterizing the carbonation depth of the concrete [67]. Table 1 is a summary of micro-and nano-CT techniques and the test conditions of different samples for identifying carbonation behavior.

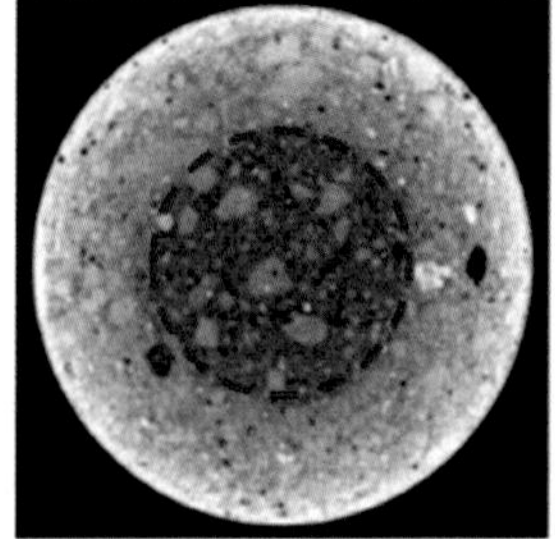

Fig. 3 Comparable results for micro-CT and phenolphthalein method for determining carbonation depth, 3D voxel size 0.086 mm^3 [67]

Table 1 Use of micro- and nano-CT with accelerated carbonation conditions

	Type of test	Sample or resolution	Test condition	Water/ cement ratio
Ji-Su Kim [45]	6 weeks of accelerated carbonation test	Resolution 65 μm Sample: 390 × 650 × 520 μm	Temperature 20 ± 2 °C, relative humidity 60 ± 5% and 5% CO_2 concentration	0.5
Dong Cui [70]	4 weeks of accelerated test	Resolution 60 μm Sample: 9 × 40 × 40 mm	Temperature 20 ± 2° C, relative humidity 70 ± 5% and 3% CO_2 concentration	0.5
Han Jiande [67]	14 weeks of accelerated carbonation tes	0.066 × 0.066 × 0.066 mm	Temperature 20 ± 2° C, relative humidity 70 ± 3% and 20% CO_2 concentration	0.53
Han Jiande [81]	12, 13, 14 and 16 weeks of accelerated carbonation test	0.066 × 0.066 × 0.066 mm	Temperature 20 ± 2 °C, relative humidity 75 ± 3%, CO_2 concentration 20 ± 3%	0.53
Keshu Wan [85]	5 and 6 weeks of the accelerated carbonation test	Pixel size of 60 μm	Temperature 20 ± 2 °C, relative humidity of 70% ± 5%, CO_2 concentration 20%	0.53

3 Conclusions

Tomographic techniques for assessing carbonation behavior in concrete are still in the early stage of development, but advances made in the previous decade have been significant. As has been demonstrated, micro-and nano-CT can efficiently and nondestructively investigate micro- and nanostructural behaviors. These techniques are suitable for analyzing micro- and nanoscale topologies and morphologies, and studying the porosity network. The advantage and benefit of using micro- and nano-CT 3D imaging to assess the pore network of the concrete is the volumetric insight into the interactions between different phases and pores. The techniques may be useful for characterizing the carbonation behavior of slag-based concretes, with r providing greater insight into accelerated carbonation testing and the impact on the pore structure. However, the need to use small-sized samples necessitates meticulous sample preparation.

Acknowledgements The authors gratefully acknowledge the University of Technology Sydney and Transport for NSW for financial support for this study.

References

1. Li XG, Lv Y, Ma BG, Chen QB, Yin XB, Jian SW (2012) Utilisation of municipal solid waste incineration bottom ash in blended cement. J Clean Prod 32:96e100
2. Chen C, Habert G, Bouzidi Y, Jullien A (2010) Environmental impact of cement production: detail of the different processes and cement plant variability evaluation. J Clean Prod 18(5):478–485. https://doi.org/10.1016/j.jclepro.2009.12.014
3. Glavind M (2009) Sustainability of cement, concrete and cement replacement materials in construction. In: Khatib (ed) Sustainability of Construction Materials. Wood Head Publishing in Materials. Cambridge, Great Abington, UK, pp 120 47
4. BM Miyandehi, A Feizbakhsh, MA Yazdi, Q Liu, J Yang, P Alipour Combined effects of metakaolin, rice husk ash, and polypropylene fiber on the engineering properties and microstructure of mortar. J Mater Civ Eng 29 (7): 04017025
5. Flower D, Sanjayan J (2007) Greenhouse gas emissions due to concrete manufacture. Int J Life Cycle Assess 12:282–288
6. Shariq M, Prasad J, Masood A (2010) Effect of GGBFS on time dependent compressive strength of concrete. Constr Build Mater 24(8): 1469–1478
7. Mehdizadeh B, Jahandari S, Vessalas K, Miraki H, Rasekh H, Samali B Fresh, mechanical, and durability properties of self-compacting mortar incorporating alumina nanoparticles and rice husk ash. Materials 14(22): 6778
8. Mehrabi P, Shariati M, Kabirifar K, Jarrah M, Rasekh H, Trung NT, Shariati A, Jahandari S (2021) Effect of pumice powder and nano-clay on the strength and permeability of fiber-reinforced pervious concrete incorporating recycled concrete aggregate. Constr Build Mater
9. Kazemi M, Hajforoush M, Khakpour Talebi P, Daneshfar M, Shokrgozar A, Jahandari S, Saberian M, Li J (2020) In-situ strength estimation of polypropylene fibre reinforced recycled aggregate concrete using Schmidt rebound hammer and point load test. J Sustain Cem-Based Mater, 1–18
10. Jahandari S, Saberian M, Tao Z, Faridfazel Mojtahedi S, Li J, Ghasemi M, Rezvani SS, Li W (2019) Effects of saturation degrees, freezing thawing, and curing on geotechnical properties of lime and lime-cement concretes. Cold Reg Sci Technol 160:242–251
11. Rasekh H, Joshaghani A, Jahandari S, Aslani F, Ghodrat M (2020) Rheology and workability of S.C.C. Woodhead Publishing Series in Civil and Structural Engineering, 31–63s
12. Mohseni E, Yazdi MA, Miyandehi BM, Zadshir M, Ranjbar MM Combined effects of metakaolin, rice husk ash, and polypropylene fiber on the engineering properties and microstructure of mortar. J Mater Civ Eng 29 (7): 04017025
13. Jung YB, Yang KH (2015) Mixture-proportioning model for low-CO2 concrete considering the type and addition level of supplementary cementitious materials. J Korea Concr Inst 27:427–434
14. Australasian (iron & steel) Slag Association, A Guide to the Use of Iron and Steel Slag in Roads, ISBN 0 9577051 58, Revision 2, 2002, Available from www. asa-inc.org.au, 27p
15. Building Research Establishment Digest 363, Sulfate and acid resistance of concrete in the ground, 1996
16. Gh SH, Miyandehi BM, Khotbehsara MM, Tarbiat M (2015) Study quality steel mill slag for use in concrete containing metakaolin. In: International Conference on advances in Engineering, Tehran, Iran.
17. IPCC, Climate Change (2007) The fourth assessment report. Cambridge University Press, Cambridge, UK
18. Yoshida N, Matsunami Y, Nagayama M, Sakai E (2010) Salt weathering in residential concrete foundations exposed to sulfate-bearing ground. J Adv Concr Technol 8(2):121–134
19. Elke G, van den Philip H, de Nele B (2013) Carbonation of slag concrete: effect of the cement replacement level and curing on the carbonation coefficient—effect of carbonation on the pore structure. Cem Concr Compos 35:39–48
20. Sisomphon K, Franke L (2007) Carbonation rates of concretes containing high volume of pozzolanic materials. Cem Concr Res 37:1647–1653

21. Monkman S, Shao Y (2010) Carbonation curing of slag-cement concrete for binding CO2 and improving performance. J Mater Civ Eng 22:296–304
22. Papadakis VG, Tsimas S (2000) Effect of supplementary cementing materials on concrete resistance against carbonation and chloride ingress. Cem Concr Res 30:291–299
23. Demis S, Papadakis VG (2012) A software-assisted comparative assessment of the effect of cement type on concrete carbonation and chloride ingress. Comput Concr 10:391–407
24. Otieno M, Ikotun J, Ballim Y (2020) Experimental investigations on the effect of concrete quality, exposure conditions and duration of initial moist curing on carbonation rate in concretes exposed to urban, inland environment. Constr Build Mater 246, Article ID 118443
25. Mo L, Zhang F, Deng M, Panesar DK (2016) Effectiveness of using CO2 pressure to enhance the carbonation of Portland cement-fly ash-MgO mortars. Cem Concr Compos 70:78–85. https://doi.org/10.1016/j.cemconcomp.2016.03.013
26. de Larrard T, Benboudjema F, Colliat JB, Torrenti JM, Deleruyelle F (2010) Concrete calcium leaching at variable temperature: experimental data and numerical model M.A.T. Marple et al. Cement and Concrete Research 156 (2022) 106760 12 inverse identification. Comput Mater Sci 49(2010): 35–45. https://doi.org/10.1016/j.commatsci.2010.04.017
27. Liu L, Ha J, Hashida T, Teramura S (2001) Development of a CO2 solidification method for recycling autoclaved lightweight concrete waste. J Mater Sci Lett 20:1791–1794
28. Bukowski JM, Berger RL (1979) Reactivity and strength development of CO2 activated non-hydraulic calcium silicates. Cem. Concr Res 9(1979): 57–68. https://doi.org/10.1016/0008-8846(79)90095-4.
29. Matalkah F, Soroushian P (2018) Carbon dioxide integration into alkali aluminosilicate cement particles for achievement of improved properties. J Clean Prod 196:1478–1485. https://doi.org/10.1016/j.jclepro.2018.06.186
30. Wang X, Zhu K, Ramli S, Xu L, Matalkah F, Soroushian P, Balachandra AM (2017) Conversion of landfilled ash into hydraulic cements under different environments. Adv Recycling Waste Manag 2:144. https://doi.org/10.4172/2475-7675.1000144
31. Zhu K, Matalkah F, Ramli S, Durkin B, Soroushian P, Balachandra AM (2018) Carbon dioxide use in beneficiation of landfilled coal ash for hazardous waste immobilization. J Environ Chem Eng 6(2018): 2055–2062. https://doi.org/10.1016/j.jece.2018.03.003.
32. Wu B, Ye G (2017) Development of porosity of cement paste blended with supplementary cementitious materials after carbonation. Constr Build Mater 145:52–61
33. Ye G (2003) Experimental study and numerical simulation of the development of the microstructure and permeability of cementitious materials: TU Delft, Delft University of Technology
34. Litvan GG, Meyer A (1986) Carbonation of granulated blast furnace slag cement concrete during twenty years of field exposure, Vol. 91. Special Publication, Symposium Paper, pp 1445–1462. concrete.org
35. Sharif A (2016) Review on advances in nanoscale microscopy in cement research, micron. Elsevier 80:45–58
36. Han J, Liang Y, Sun W, Liu W, Wang S (2015) Microstructure modification of carbonated cement paste with six kinds of modern microscopic instruments. J Mater Civ Eng 2014 27(10): 04014262. https://doi.org/10.1061/(ASCE)MT. 1943–5533.0001210
37. Provis JL, Myers RJ, White CE, Rose V, van Deventer JSJ (2012) X-ray microtomography shows pore structure and tortuosity in alkali-activated binders. Cem Concr Res 42:855–864
38. De Rosier DJ, Klug A (1968) Reconstruction of three-dimensional structures from electron micrographs. Nature 217(5124):130–134
39. Deshpande A, Bao W, Miao F, Lau CN, LeRoy BJ (2009) Spatially resolved spectroscopy of monolayer graphene on SiO2. Phys Rev B 79(20):205411
40. Frank J (2007) Electron tomography: three-dimensional imaging with the transmission electron microscope, 2nd edn. Plenum Press, New York
41. Cnudde V, Boone MN (2013) High-resolution X-ray computed tomography in geosciences: a review of the current technology and applications. Earth-Sci Rev 123:1–17

42. Takahashi H, Sugiyama T (2016) Investigation of alteration in deteriorated mortar due to water attack using non-destructive integrated CT-XRD method. In: Maekawa K, Kasuga A, Yamazaki J (Eds) Proceedings of the 11th fib International PhD Symposium in Civil Engineering. Tokyo, Japan, pp. 445–452
43. Lloyd RR, Provis JL, van Deventer JSJ (2012) Pore solution composition and alkali diffusion in inorganic polymer cement. Cem Concr Res 40
44. Provis L, Myers RJ, White CE, Rose V, van Deventer JSJ (2012) X-ray microtomography shows pore structure and tortuosity in alkali-activated binders. John, Cem Concr Res 42: 855–864
45. Chung S-Y, Kim J-S, Stephan D, Han T-S (2019) Overview of the use of micro-computed tomography (micro-CT) to investigate the relation between the material characteristics and properties of cement-based materials. Constr Build Mater 229:116843
46. du Plessis A, William P, Boshoff Y (2019) A review of X-ray computed tomography of concrete and asphalt construction materials. Constr Build Mater 119: 637−651
47. Natesaiyer K, Chan C, Sinha-Ray S, Song D, Lin CL, Miller JD, Garboczi EJ, Forster AM (2015) X-ray CT imaging and finite element computations of the elastic properties of a rigid organic foam compared to experimental measurements: insights into foam variability. J Mater Sci 50:4012–4024
48. Bossa N, Chaurand P, Vicente J, Borschneck D, Levard C, Chariol OA, Rose J (2015) Micro- and nano-X-ray computed-tomography: a step forward in the characterisation of the pore-network of a leached cement paste. Cem Concr Res 67:138–147
49. Chung S-Y, Elrahman MA, Stephan D, Kamm PH (2016) Investigation of characteristics and responses of insulating cement paste specimens with Aer solids using X-ray micro-computed tomography. Constr Build Mater 118:204–215
50. Nguyen TT, Bui HH, Ngo TD, Nguyen GD (2017) Experimental and numerical investigation of influence of air-voids on the compressive behaviour of foamed concrete. Mater Des 130:103–119
51. da Silva IB (2018) X-ray computed microtomography technique applied for cementitious materials: a review. Micron 107:1–8
52. Gallucci E, Scrivener K, Groso A, Stampanoni M, Margaritondo G (2007) 3D experimental investigation of the microstructure of cement pastes using synchrotron X-ray microtomography. Cem Concr Res 37:360–368
53. Han T-S, Zhang X, Kim J-S, Chung S-Y, Lim J-H, Linder C (2018) Area of linealpath function for describing the pore microstructures of cement paste and their relations to the mechanical properties simulated from l-CT microstructures. Cem Concr Compos 89:1–17
54. Mindess S, Young JF, Darwin D (2002) Concrete, 2nd edn. Prentice Hall, New York
55. Brisard S, Chae RS, Bihannic I, Michot L, Guttmann P, Thieme J, Schneider G, Monteiro PJ, Levitz P (2012) Morphological quantification of hierarchical geomaterials by X-ray nano-CT bridges the gap from nano to micro length scales. Am Mineral 97(2–3):480–483. https://doi.org/10.2138/am.2012.3985
56. Nagala V, Page C (1997) Effect of carbonation on pore structure and diffusional properties of hydrated cement paste. Cem Concr Res 27:995–1007
57. Johannesson B, Utgenannt P (2001) Microstructural changes caused by carbonation of cement mortar. Cem Concr Res 31: 925 391
58. Li N, Farzadnia N, Shi C (2017) Microstructural changes in alkaliactivated slag mortars induced by accelerated carbonation. Cem Concr Res 100:214–226
59. Cui D, Sun W, Banthia N (2018) Use of tomography to understand the influence of preconditioning on carbonation tests in cement-based materials. Cem Concr Compos 88:52–63
60. Han J, Sun W, Pan G, Caihui W (2012) Monitoring the evolution of accelerated carbonation of hardened cement pastes by Xray computed tomography. J. Mater. Civil Eng. 25(3):347–354
61. Han J, Sun W, Pan G (2013) Nondestructive microstructure analysis of the carbonation evolution process in hardened binder paste containing blast-furnace slag by X-ray CT. J Wuhan Univ Technique-Mater Sci Ed 28(5):955–962
62. Branch J, Epps R, Kosson D (2018) The impact of carbonation on bulk and itz porosity in microconcrete materials with fly ash replacement. Cem Concr Res 103:170–178

63. Papadakis V, Vayenas C (1989) A reaction engginering approach to the problem of concrete carbonation. AlChE J. 35:1639–1649
64. Morandeau A, Thiéry M, Dangla P (2014) Investigation of the carbonation mechanism of CH and C-S-H in terms of kinetics, microstructure changes and moisture properties. Cem Concr Res 56:153–170
65. Wan K, Xu Q, Wang Y, Pan G (2014) 3D spatial distribution of the calcium carbonate caused by carbonation of cement paste. Cem Concr Compos 45:255–263
66. Henry M, Darma I, Sugiyama T (2014) Analysis of the effect of heating and re-curing on the microstructure of high-strength concrete using X-ray CT. Constr Build Mater 67:37–46
67. Han J, Liu W, Wang S, Du D, Xu F, Li W, De Schetter G (2016) Effect of crack and ITZ and aggregate on carbonation penetration based on 3D micro X-ray CT microstructure evolution. Constr Build Mater 128:256–271
68. Konga W, Weia Y, Wang S, Chen J, Wang Y Research progress on cement-based materials by X-ray computed tomography, Chinese Society of Pavement Engineering. Int J Pavement Res Technol. Journal homepage: www.springer.com/42947
69. Kim JS, Youm KS, Lim JH, Han TS (2020) Effect of carbonation on cement paste microstructure characterised by micro-computed tomography. Constr Build Mater 263:120079
70. Cui D, Sun W, Banthia N (2018) Use of tomography to understand the influence of preconditioning on carbonation tests in cement-based materials. Cement Concr Compos 88:52–63
71. Coussy O (2010) Mechanics and physics of porous solids. John Wiley & Sons Ltd, Chichester, United Kingdom
72. Brisard S, Davy CA, Michot L, Troadec D, Levitz P (2019) Mesoscale pore structure of a high-performance concrete by coupling focused ion beam/scanning electron microscopy and small angle X-ray scattering. J Am Ceram Soc 102:2905–2923
73. Song Y, Davy CA, Troadec D, Bourbon X (2019) Pore network of cement hydrates in a high performance concrete by 3D FIB/SEM—implications for macroscopic fluid transport. Cem Concr Res 115:308–326
74. Song Y, Zhou J, Bian Z, Dai G (2019) Pore structure characterization of hardened cement paste by multiple methods. Adv Mater Sci Eng
75. Sakdinawat A, Attwood D (2010) Nanoscale X-ray imaging. Nat Photonicss 4(12): 840–848
76. Rehbein S, Heim S, Guttmann P, Werner S, Schneider G (2009) Ultrahigh-resolution soft-X-ray microscopy with zone plates in high orders of diffraction. Phys Rev Lett 103:110801
77. Han J, Sun W, Pan G, Wang C, Rong H (2012) Application of X-ray computed tomography in characterization microstructure changes of cement pastes in carbonation process. 27(2): 358
78. Bae S, Taylor R, Shapiro D, Denes P, Joseph J, Celestre R, Marchesini S, Padmore H, Tyliszczak T, Warwick T, Kilcoyne D, Levitz P, Monteiro PJM (2015) Soft X-ray ptychographic imaging and morphological quantification of calcium silicate hydrates (C–S–H). J Am Ceram Soc 98(12):4090–4095. https://doi.org/10.1111/jace.13808
79. Han J et al: Application of X-ray computed tomography in charact
80. Liu J-Z, Ba M-F, Yin-gang DU, He Z-M, Chen J-B (2016) Effects of chloride ions on carbonation rate of hardened cement paste by X-ray CT techniques. Constr Build Mater 112:619–627
81. Han J, Sun W, Pan G (2012) Analysis of different contents of blast-furnace slag effect on carbonation properties of hardened binder paste using micro-XCT technique in PRO83. Microstructural-related Durability of Cementitious Composites, RILEM Publications, pp. 228–234 ISBN: 978–2–35158–129–2, e-ISBN: 978–2–35158- 123–0.
82. Brisarda, Serdarb M, Monteiroc PJM (2020) Multiscale X-ray tomography of cementitious materials: a review Sébastien. Cem Concr Res 128:105824
83. Han J, Sun W, Pan G, Wang C, Rong H Application of X-ray computed tomography in characterization microstructure changes of cement pastes in carbonation process. 27(2): 358. HAN Jiande et al: Application of X-ray Computed Tomography in Charact.

84. Bushby AJ, P'ng KM, Young RD, Pinali C, Knupp C, Quantock AJ (2011) Imaging three-dimensional tissue architectures by focused ion beam scanning electron microscopy. Nat Protoc. 6(6): 845–858
85. Wan K, Qiong XU, Wang Y, Pan G (2014) 3D spatial distribution of the calcium carbonate caused by carbonation of cement paste. Cement Concr Compos 45:255–263

Application of Surface-Modified Nanosilica for Performance Enhancement of Asphalt Pavement

D. He, H. H. Chan, Z. H. Xiao, T. Wu, L. M. Leung, M. Sham, B. Chen, S. F. S. Lee, and C. K. K. Kwan

Abstract The fatigue performance of asphalt binder can be significantly increased by incorporating nanosilica, but due to the poor dispersion of nanosilica in asphalt, high mechanical energy is required to achieve good homogeneity. The viscosity of asphalt is nearly double when 3% nanosilica is added. Consequently, the mixing temperature of asphalt with aggregate is dozens of degrees higher to compensate for its high viscosity. To solve this problem, modification of nanosilica was introduced and the effects of its performance on asphalt are reported here. Several types of nanosilica at various dosages were modified and characterized by Fourier-transform infrared spectroscopy (FT-IR) and thermogravimetric analysis (TGA). The properties of asphalt, such as viscosity, rutting, fatigue, softening point, penetration and ductility were investigated. The viscosity of asphalt with 3% modified nanosilica was 80% lower than that of the asphalt with unmodified nanosilica and a 20 °C reduction in temperature for mixing the asphalt binder with aggregate was achieved. Moreover, the rutting and fatigue performance of asphalt with modified nanosilica were significantly enhanced. The asphalt with modified nanosilica not only had lower viscosity but also has similar performance to asphalt with unmodified nanosilica. The asphalt with modified nanosilica had excellent performance and thus good practical application potential.

Keywords Asphalt · Fatigue · Surface-modified nanosilica · Viscosity

D. He (✉) · H. H. Chan · Z. H. Xiao · T. Wu · L. M. Leung · M. Sham · B. Chen
Nano and Advanced Materials Institute Limited, Kowloon, Hong Kong
e-mail: danhe@nami.org.hk

S. F. S. Lee
Hong Kong Wah for Development Limited, Hong Kong, Hong Kong

C. K. K. Kwan
On Fat Lung Innovative Resources Ltd., Hong Kong, Hong Kong

W. Duan et al. (eds.), *Nanotechnology in Construction for Circular Economy*, Lecture Notes in Civil Engineering 356,
https://doi.org/10.1007/978-981-99-3330-3_31

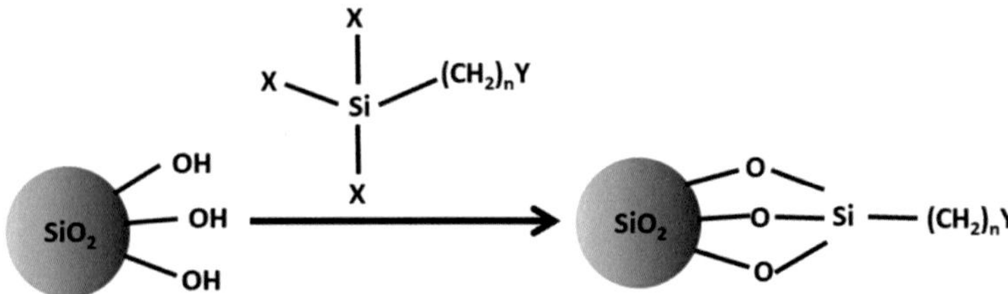

Fig. 1 Mechanism of surface-modification of nanosilica by silane coupling agents

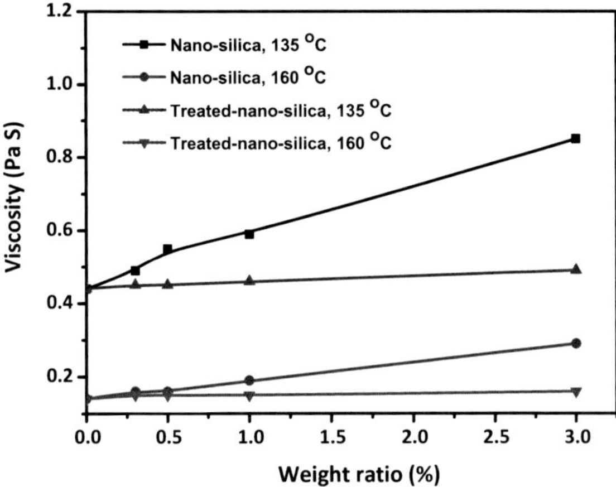

Fig. 2 Viscosity of different silica–asphalt mixes at 135 and 160 °C

Figure 1 shows the mechanism of surface-modification of nanosilica by silane coupling agents [1, 2] and Fig. 2 shows that the viscosity of asphalt was higher when either nanosilica or treated nanosilica is added [3]. The viscosity also increased with the content of silica added, regardless of its type. However, the viscosity of asphalt with modified nanosilica was comparable to that of virgin asphalt, which verified the purpose of surface modification.

The increase in viscosity caused by the nanosilica was more significant than with treated nanosilica [4, 5]. Table 1 lists the rheological properties of virgin asphalt (60/70), asphalt with nanosilica and asphalt with treated nanosilica. According to its performance grade (PG), virgin asphalt (60/70) can be classified as PG 64. The rutting parameter ($G^*/\sin\delta$, at original) increased by 12% with the addition of 1% of Treated nanosilica and the PG of asphalt binder with 1% Treated nanosilica increased by one degree (PG 64 to PG 70). The fatigue parameter ($G^* \times \sin\delta$) of asphalt binder with 1% Treated nanosilica decreased by 7% and fatigue performance was enhanced. Compared with virgin asphalt, the asphalt with modified nanosilica not only showed a significant increase in rheological properties but also had a similar viscosity [6, 7].

Table 1 Influence of treated nanosilica on the rheological properties of asphalt binder

	Viscosity[a]	Rutting parameter			Fatigue parameter	Softening point
	(135°C, Pa s)	G*/sin δ[a] (kPa)	PG[a] (°C)	G*/sin δ[b] (kPa)	G*·sin δ[c] (kPa)	(°C)
Virgin asphalt	0.44	0.89	64	3.15	3997.1	47.6
0.5% nanosilica	0.55	1.08	70	3.44	3854.3	48.3
1% nanosilica	0.59	1.1	70	3.79	2628.3	48.55
0.5% Treated nanosilica	0.45	0.99	≈70	3.24	3906.5	48.3
1% Treated nanosilica	0.46	1	70	3.3	3705	48.55

[a] Original binder; PG, performance grading, G*/sin δ ≥ 1.0 kPa
[b] Rolling Thin-Film Oven
[c] Pressure aging vessel (PAV-25 °C) residue

References

1. Karnati SR, Oldham D, Fini EH, Zhang L (2020) Application of surface-modified silica nanoparticles with dual silane coupling agents in bitumen for performance enhancement. Constr Build Mater 244:118324
2. Li F, Li H, Lai X, Wu W, Zeng X (2014) Study on the graft modification of γ-mercaptopropyltrimethoxysilane on the surface of nano-silica. Inorg Chem Ind 2014:1006–4990
3. Yao H, You Z, Li L, Lee CH, Wingard D, Yap YK, Shi X, Goh SW (2013) Rheological properties and chemical bonding of asphalt modified with nanosilica. J Mater Civ Eng 25:1619–1630
4. Bhat FS, Mir MS (2021) Rheological investigation of asphalt binder modified with nanosilica. Int J Pavement Res Technol 14:276–287
5. Abed AH, Oudah AM (2018) Rheological properties of modified asphalt binder with nanosilica and SBS. IOP Conf Ser: Mater Sci Eng 433:012031
6. Xiantao Q, Siyue Z, Xiang H, Yi J (2018) High temperature properties of high viscosity asphalt based on rheological methods. Constr Build Mater 2018(186):476–483
7. Zhang J, Yang F, Pei J, Xu S, An F (2015) Viscosity-temperature characteristics of warm mix asphalt binder with Sasobit. Constr Build Mater 2015(78):34–39

Effect of Different Additives on the Compressive Strength of Very High-Volume Fly Ash Cement Composites

R. Roychand, J. Li, M. Saberian, S. Kilmartin-Lynch, M. M. Ul Islam, M. Maghfouri, and F. Chen

Abstract The cement industry is responsible for about 5–7% of global greenhouse gas emissions and with the rapid rise in global warming, it is imperative to produce an ecofriendly alternative to Portland cement. Fly ash (FA) is an abundantly available and least utilized industrial byproduct with good pozzolanic properties that can help reduce the carbon footprint of cement composites. We investigated replacing 80% of the cement content with different blends of FA, nanosilica (NS) and silica fume (SF). Hydrated lime and a set accelerator were used to increase the pozzolanic reactivity of the blended cement composites. The portlandite released with 20% cement content was insufficient for the pozzolanic reaction of the blended cement composites containing FA and SF, requiring externally added hydrated lime. The addition of a set accelerator significantly increased the pozzolanic reaction and the resultant compressive strength, and these increased with the increasing content of the set accelerator. The replacement of SF with NS led to a remarkable increase in the pozzolanic reaction. The corresponding compressive strength of FA mixed with cement composites increased with increasing percentage composition of NS.

Keywords Cement · Compressive strength · Fly ash · Nanosilica · Set accelerator

---

R. Roychand (✉) · J. Li (✉) · M. Saberian · S. Kilmartin-Lynch · M. M. U. Islam
School of Engineering, RMIT University, Melbourne, VIC, Australia
e-mail: rajeev.roychand@rmit.edu.au

J. Li
e-mail: jie.li@rmit.edu.au

M. Maghfouri
Department of Civil Engineering, University of Ottawa, Ottawa, ON, Canada

F. Chen
Arup, Melbourne, VIC, Australia

W. Duan et al. (eds.), *Nanotechnology in Construction for Circular Economy*,
Lecture Notes in Civil Engineering 356,
https://doi.org/10.1007/978-981-99-3330-3_32

1 Introduction

An increasingly important priority worldwide is to transform waste into a valuable resource for other applications, supporting the circular economy. The cement and concrete industry is one of the major sectors that is actively contributing toward a closed loop circular economy by incorporating a range of industrial byproducts and waste materials [1–3], such as, fly ash (FA) [4, 5], slag [6, 7], waste tire rubber [8–12], recycled concrete aggregate [13, 14], organic waste [15] and medical waste [16–19]. However, the production of cement accounts for around 5–7% of global greenhouse gas emissions [20, 21]. Considering the estimated 4 billion tonnes of annual cement production by 2030, a remarkable volume of CO_2 will be generated worldwide [22, 23]. In this regard, extensive research is currently being undertaken to reduce CO_2 emissions by using industrial byproducts such as FA [24–27], slag [28, 29], silica fume (SF) [30, 31], and metakaolin [32] as supplementary cementitious materials. FA is a byproduct of coal-fired thermal power plants, and is considered an environmental pollutant and requires considerable financial and environmental input for its safe disposal. However, due to its pozzolanic properties, it can be used to substitute cement in concrete production, which not only improves the mechanical and durability properties of the concrete composite but also contributes to cutting down its carbon footprint.

Most of the worldwide FA production is categorized as ASTM Class F low calcium FA [33, 34], which has low reactivity [35]. Researchers have explored various methods of improving the pozzolanic reactivity of the FA, such as reducing the particle size of FA, adding hydrated lime (HL), sodium hydroxide, lime water, sodium silicate [12], a set accelerator (SA) [13], nano-alumina, nanosilica (NS), nanocalcium carbonate, nanoferrous oxide, nanozinc oxide and nanotitanium oxide. Although an alkaline activator is always required for the pozzolanic reaction of FA, out of all other additives, NS [36] could be the highest performing additive for significant enhancement of the mechanical properties of high-volume FA (HVFA) cement composites. The high reactivity of NS is attributed to its high surface area [37] and highly amorphous structure, which reacts with available lime to produce additional strength-forming calcium–silicate–hydrate gel counterbalancing the low early reactivity of FA. However, the key challenge in using NS in HVFA cement composites is its high cost, which hinders its application in large-scale commercial projects. An alternative to NS is SF, though with relatively lower reactivity because of its micro-sized particles compared with the nano-sized particles of NS, which delivers a considerable difference in their respective reactive surface areas and their corresponding reactivities. Therefore, we carried out a comparative analysis of the use of different micro- and nano-sized additives in HVFA cement composites at 80% replacement of the cement content.

Table 1 Chemical composition of ordinary Portland cement, fly ash, silica fume, nanosilica, and hydrated lime

Oxides	OPC (%)	FA (%)	SF (%)	NS (%)	HL (%)
SiO_2	19.8	72.5	88.2	99.5	1.4
CaO	62.7	0.5	1.2	–	74.3
Al_2O_3	4.8	22.6	1.3	–	0.7
Fe_2O_3	3.3	1.2	2.1	–	0.3
SO_3	2.8	0.3	1.7	–	–
MgO	2.2	0.3	1.6	–	0.5
Na_2O	0.2	0.1	0.1	–	0.2
K_2O	0.4	0.2	0.2	–	0.1
TiO_2	–	1.3	–	–	–

2 Methods

The materials adopted in this experimental program were ordinary Portland cement (OPC), FA, SF, NS, HL, SA, superplasticizer (SP) and potable water. The sand-to-cement ratio was kept at 2.4. The chemical compositions of the different materials are shown in Table 1, and the mineralogical compositions are presented in Fig. 1. The mix designs studied in this experimental program are provided in Table 2. Three replicates of 50 mm cube mortar samples were prepared for each mix design for 7- and 28 day compressive strength tests. The samples were cured as per AS1012.8.1:2014 until the time of testing.

3 Results and Discussion

Figure 2 shows the compressive strength values of different mix designs at 7 and 28 days of curing. It can be seen that by replacing 80% OPC with 65% FA and 15% SF in mix M1, there was a considerable reduction in compressive strength. FA has low reactivity that negatively influences strength development [38], thereby significantly reducing the 7- and 28 day compressive strength results of mix M1. Moreover, class F FA has negligible lime content, and the calcium hydroxide released by the hydration reaction of 20% OPC content was insufficient for accelerating the pozzolanic reaction. In mix M2, an additional 5% HL was added to the mix design as a percentage of the total cementitious material (CM). The increase in the alkaline activator accelerated the pozzolanic reaction, thereby bringing about 76.4% and 108.5% rise in the 7- and 28 day compressive strength results, respectively, compared with M1.

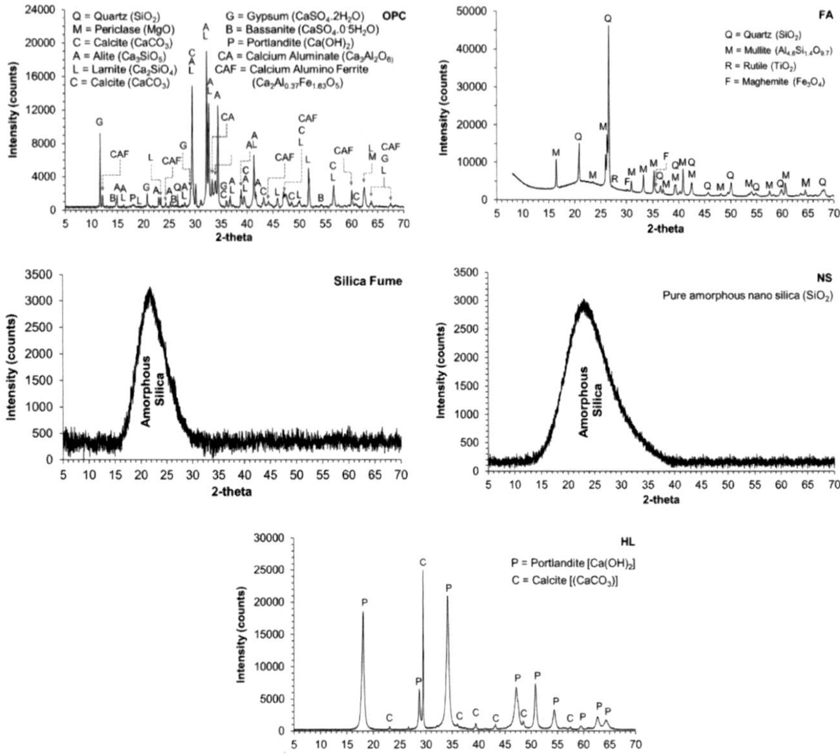

Fig. 1 Mineralogical composition of ordinary Portland cement, fly ash, silica fume, nanosilica, and hydrated lime

In mix M3, sodium thiocyanate and calcium nitrate-based SA was used at 12.5 mL/kg of CM. The addition of SA to mix M3 brought about 68% and 10% improvement in 7- and 28 day compressive strength, respectively, compared with M2. In mix M4, the quantity of the SA was doubled to 25 mL/kg of CM, which led to a considerable increase in the pozzolanic reaction, thereby increasing the 7- and 28 day compressive strength results by 11.7% and 14.1%, respectively, compared with M3. In mix M5, the water/cement ratio of the mix design was reduced from 0.3 to 0.25. Because the water–cement ratio has an inverse relationship with compressive strength, the 7- and 28 day strengths of M5 were increased by 21.4% and 7.8%, respectively.

In mix M6, SF was replaced by NS, which has a considerably higher surface area than SF and accelerates the pozzolanic reaction of NS-modified mix designs. The NS content was 4%, and the FA content was increased to 71% to keep the total cement replacement level at 80%, with the remaining 5% being the HL content. Because NS has a high surface area, replacing SF with NS significantly increased the SP requirement. Mix M6 showed a significant increase (i.e., 54.8%) in its 7 day compressive strength. At 28 days, though relatively small, it showed a 6% improvement in its compressive strength compared with M5. On increasing the NS content to 5% in

Table 2 Mix designs

Design mix	Percentage Cementitious material (CM)					S/CM (By weight)	W/CM	SA	SP
								(mL/kg of CM)	
	OPC	FA	SF	NS	HL				
C	100	–	–	–	–	2.4	0.30	–	4
M1	20	65	15	–	–	2.4	0.30	–	8
M2	20	60	15	–	5	2.4	0.30	–	8
M3	20	60	15	–	5	2.4	0.30	12.5	8
M4	20	60	15	–	5	2.4	0.30	25.0	8
M5	20	60	25	–	5	2.4	0.25	25.0	12
M6	20	71	–	4	5	2.4	0.25	25.0	20
M7	20	70	–	5	5	2.4	0.25	25.0	25
M8	20	69	–	6	5	2.4	0.25	25.0	30
M9	20	67	–	8	5	2.4	0.25	25.0	40

FA, fly ash; HL, hydrated lime; OPC; ordinary Portland cement; NS, nanosilica; S, sand; SF, silica fume; SP, superplasticizer (mL/kg of CM); W, water

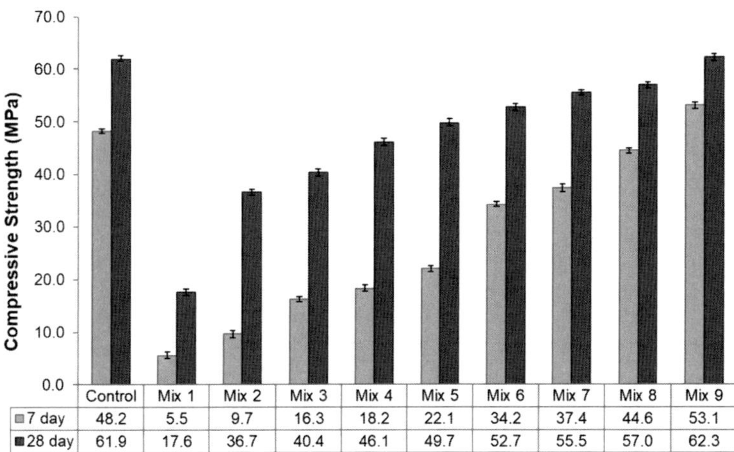

Fig. 2 Compressive strength results

mix M7, the 7- and 28 day compressive strength showed a further increase of 9.4% and 5.3%, respectively, compared with M6. In mix M8, the NS content was increased to 6%, which brought about 19.3% and 2.7% increase in the respective 7- and 28 day compressive strength results, which can be attributed to the increase in pozzolanic activity due to the addition of highly reactive NS. Mix M9 increased the NS content to 8% and reduced the FA content to 67%. The SP content was considerably increased to keep similar workability. The 7- and 28 day compressive strength increased by 19%

and 9.2%, respectively, reflecting the positive effect of nano-sized highly reactive amorphous silica on the pozzolanic activity of blended cement composites.

4 Conclusions and Recommendations for Future Work

(C1) The cementitious blend of HVFA and SF, replacing 80% of OPC content, required the inclusion of HL to increase the pozzolanic reaction of both the SF and FA.

(C2) The addition of a SA increases the pozzolanic reaction of FA and SF, which increases with increasing SA content. This increase in the pozzolanic reaction is also reflected in the increase in compressive strength.

(C3) The replacement of micro-sized SF with nano-sized highly amorphous NS significantly increases the pozzolanic reaction of the mix, resulting in significant increases in the 7- and 28 day compressive strength results as the NS content increased. However, the SP requirement increased significantly to maintain workability.

(C4) NS-blended very HVFA cement composites replacing 80% of the cement content that can give 28 day compressive strength results on par with those of a control mix containing 100% cement content.

(R1) Mix 9 had the highest results for improving the 7- and 28 day compressive strengths. It should be explored further to look for any potential of further enhancement of its mechanical properties.

(R2) Further work needs to be carried out on workability (slump), setting time, durability and techno-economic analysis to carry this work forward.

Acknowledgements The authors gratefully acknowledge the support of the RMIT X-Ray Facility and Arup.

References

1. Maghfouri M, et al (2022) Drying shrinkage properties of expanded polystyrene (EPS) lightweight aggregate concrete: a review. Case Stud Constr Mater e00919
2. Vakhshouri B, Nejadi S (2018) Review on the mixture design and mechanical properties of the lightweight concrete containing expanded polystyrene beads. Aust J Struct Eng 19(1):1–23
3. Roychand R et al (2021) Practical rubber pre-treatment approch for concrete use—an experimental study. J Compos Sci 5(6):143
4. Chalee W, Ausapanit P, Jaturapitakkul C (2010) Utilization of fly ash concrete in marine environment for long term design life analysis. Mater Des 31(3):1242–1249
5. Torii K, Kawamura M (1994) Effects of fly ash and silica fume on the resistance of mortar to sulfuric acid and sulfate attack. Cem Concr Res 24(2):361–370
6. Akinmusuru JO (1991) Potential beneficial uses of steel slag wastes for civil engineering purposes. Resour Conserv Recycl 5(1):73–80

7. Roychand R et al (2020) Recycling steel slag from municipal wastewater treatment plants into concrete applications—a step towards circular economy. Resour Conserv Recycl 152:104533
8. Youssf O et al (2019) Influence of mixing procedures, rubber treatment, and fibre additives on rubcrete performance. J Compos Sci 3(2):41
9. Abd-Elaal E-S et al (2019) Novel approach to improve crumb rubber concrete strength using thermal treatment. Constr Build Mater 229:116901
10. Youssf O et al (2020) Development of crumb rubber concrete for practical application in the residential construction sector—design and processing. Constr Build Mater 260:119813
11. Gravina RJ, et al (2021) Bond behaviour between crumb rubberized concrete and deformed steel bars. In: Structures. Elsevier
12. Youssf O et al (2022) Mechanical performance and durability of geopolymer lightweight rubber concrete. J Build Eng 45:103608
13. Hossain FZ et al (2019) Mechanical properties of recycled aggregate concrete containing crumb rubber and polypropylene fiber. Constr Build Mater 225:983–996
14. Saberian M et al (2020) An experimental study on the shear behaviour of recycled concrete aggregate incorporating recycled tyre waste. Constr Build Mater 264:120266
15. Roychand R et al (2021) Recycling biosolids as cement composites in raw, pyrolyzed and ashed forms: a waste utilisation approach to support circular economy. J Build Eng 38:102199
16. Kilmartin-Lynch S et al (2022) Application of COVID-19 single-use shredded nitrile gloves in structural concrete: case study from Australia. Sci Total Environ 812:151423
17. Kilmartin-Lynch S et al (2021) Preliminary evaluation of the feasibility of using polypropylene fibres from COVID-19 single-use face masks to improve the mechanical properties of concrete. J Clean Prod 296:126460
18. Wang G, et al (2022) Use of COVID-19 single-use face masks to improve the rutting resistance of asphalt pavement. Sci Total Environ 154118
19. Saberian M et al (2021) Repurposing of COVID-19 single-use face masks for pavements base/subbase. Sci Total Environ 769:145527
20. Benhelal E et al (2013) Global strategies and potentials to curb CO_2 emissions in cement industry. J Clean Prod 51:142–161
21. Roychand R et al (2016) Micro and nano engineered high volume ultrafine fly ash cement composite with and without additives. Int J Concr Struct Mater 10(1):113–124
22. Klee H (2004) Briefing: the cement sustainability initiative. In: Proceedings of the institution of civil engineers-engineering sustainability. Thomas Telford Ltd
23. Maghfouri M et al (2021) Impact of fly ash on time-dependent properties of agro-waste lightweight aggregate concrete. J Compos Sci 5(6):156
24. Roychand R et al (2016) High volume fly ash cement composite modified with nano silica, hydrated lime and set accelerator. Mater Struct 49(5):1997–2008
25. Khambra G, Shukla P (2021) Novel machine learning applications on fly ash based concrete: an overview. Mater Today: Proc
26. Roychand R, et al (2017) A quantitative study on the effect of nano SiO_2, nano Al_2O_3 and nano $CaCO_3$ on the physicochemical properties of very high volume fly ash cement composite. Eur J Environ Civ Eng 1–16
27. Roychand R, De Silva S, Setunge S (2018) Nanosilica modified high-volume fly ash and slag cement composite: environmentally friendly alternative to OPC. J Mater Civ Eng 30(4):04018043
28. Roychand R et al (2021) Development of zero cement composite for the protection of concrete sewage pipes from corrosion and fatbergs. Resour Conserv Recycl 164:105166
29. Amran M et al (2021) Slag uses in making an ecofriendly and sustainable concrete: a review. Constr Build Mater 272:121942
30. Roychand R (2017) Performance of micro and nano engineered high volume fly ash cement composite. RMIT University
31. Siddique R (2011) Utilization of silica fume in concrete: review of hardened properties. Resour Conserv Recycl 55(11):923–932

32. Ismail MK, Hassan AA (2016) Use of metakaolin on enhancing the mechanical properties of self-consolidating concrete containing high percentages of crumb rubber. J Clean Prod 125:282–295
33. Rangan BV (2008) Fly ash-based geopolymer concrete
34. Siddique R (2004) Performance characteristics of high-volume Class F fly ash concrete. Cem Concr Res 34(3):487–493
35. Fraay A, Bijen J, De Haan Y (1989) The reaction of fly ash in concrete a critical examination. Cem Concr Res 19(2):235–246
36. Abhilash P et al (2021) Effect of nano-silica in concrete: a review. Constr Build Mater 278:122347
37. Land G, Stephan D (2012) The influence of nano-silica on the hydration of ordinary Portland cement. J Mater Sci 47(2):1011–1017
38. Kelebopile L, Sun R, Liao J (2011) Fly ash and coal char reactivity from thermo-gravimetric (TGA) experiments. Fuel Process Technol 92(6):1178–1186

Spalling Resistance of Hybrid Polyethylene and Steel Fiber-Reinforced High-Strength Engineered Cementitious Composite

S. Rawat, Y. X. Zhang, and C. K. Lee

Abstract We analyzed the effect of elevated temperatures on the integrity of high-strength engineered cementitious composite (ECC) made with a hybrid combination of polyethylene (PE) and steel fibers. The 50 mm cube specimens were subjected to temperature ranging from 200 to 800 °C at three different heating rates: 1, 5, and 10 °C/min. Five different types of mixes with varying content of supplementary cementitious materials and fibers were evaluated. No spalling was observed at 1–5 °C/min heating rate and <400 °C. However, at a heating rate of 10 °C/min for temperature 600–800 °C, all ECC specimens with a PE fiber volume of 1.25 and 1% steel fiber spalled explosively. Moreover, cementitious matrix with silica fume was more prone to spalling at 800 °C and the use of slag or quaternary blend of slag and dolomite at an optimum content was effective in maintaining the integrity of the ECC specimens even at very high heating rates. Thus, the type of cementitious matrix is equally important to consider, as well as fiber type and content, while analyzing the spalling resistance of ECC.

Keywords Fibers · Elevated temperature · Heating rate · Spalling

1 Introduction

High-strength concrete may undergo severe spalling when exposed to elevated temperatures. The use of polypropylene (PP) fibers in concrete is the most widely recognized method of preventing explosive spalling from both the economic and technical perspective. Though existing literature validates the suitability of PP fibers in

S. Rawat (✉) · Y. X. Zhang · C. K. Lee
School of Engineering and Information Technology, The University of New South Wales, Canberra, ACT, Australia
e-mail: s.rawat@student.unsw.edu.au

Y. X. Zhang
School of Engineering, Design and Built Environment, Western Sydney University, Penrith, NSW, Australia

W. Duan et al. (eds.), *Nanotechnology in Construction for Circular Economy*, Lecture Notes in Civil Engineering 356,
https://doi.org/10.1007/978-981-99-3330-3_33

spalling prevention, their sole addition may not be very effective in post-fire strength retention. In addition, the use of PP fibers in concrete may not lead to any significant improvement in the tensile performance to prevent durability-related damage. Engineered cementitious composite (ECC) is an alternate material type with superior tensile performance at normal temperatures and has also been found promising in elevated temperature scenarios. Existing studies of the fire performance of ECC have considered individual or combined roles of steel, PP, and polyvinyl alcohol (PVA) fibers [1]. The melting of PP or PVA fibers prevents the build-up of vapor pressure and, simultaneously, steel fibers can provide the required resistance to prevent mechanical decay in the higher temperature range. However, PVA fibers are not suitable for the development of ultrahigh-performance engineered cementitious composites due to their relatively low tensile strength and hydrophilic nature [2]. Another type of low-melting fiber, polyethylene (PE) fibers, have higher strength and modulus of elasticity than PP or PVA fibers and cementitious composites with PE fibers have exhibited superior ductility, strength, and energy absorption capacity.

There has been limited work on the fire performance of PE fiber-reinforced ECC and therefore, this study is a further step towards understanding the effect of elevated temperatures on the integrity or spalling resistance of high-strength ECC made with a hybrid combination of PE and steel fibers.

2 Methods

Five different types of mixes were considered to simultaneously study the effect of fiber content and type of supplementary cementitious material (SCM) on spalling resistance. The constituents of the main mix (Mix 1) used in the study were first optimized using the Grey Taguchi method and the Taguchi method with Utility Concept to obtain optimum compressive and tensile performance. More details about the optimization methodology and other mix parameters can be found Rawat et al. [3]. In addition, four other mixes were considered to study the effect of heating rate on mixes with varying ratios of SCM (slag, dolomite, and silica fume) and steel and PE fiber content as shown in Table 1. The specimens were exposed to different temperature ranges and heating rates to analyze their effects on spalling resistance and residual compressive strength.

Fiber content is expressed as volume fraction of the mix, whereas all other constituents' ratios are expressed as weight proportion of the cement content. FA, fly ash; HRWR, high range water reducer.

Table 1 Mix proportions used in the present study

Mix ID	Binder				Sand	Water	HRWR	PE (%)	Steel (%)
	Cement	Slag	FA	Dolomite					
1	1	0.94	0.19	0.38	0.91	0.50	0.05	1.50	0.75
2	1	0.83	0.17	0.22	0.81	0.44	0.04	1.50	0.75
3	1	0.94	0.19	0.38	0.91	0.50	0.05	1.25	1
4	1	1.00	0.00	0.00	0.73	0.40	0.04	1.50	0.75
	Cement	**Silica fume**	**FA**	**Dolomite**	**Sand**	**Water**	**HRWR**	**PE (%)**	**Steel (%)**
5	1	0.11	0.00	0.00	0.44	0.22	0.02	1.50	0.75

3 Brief Discussion

Figure 1 shows the normalized compressive strength of mixes 1 and 5 after exposure to different temperatures and heating rates considered in the study. In general, the residual compressive strength decreased with increase in heating rate for all types of mixes. Nevertheless, the effect was not significant, and the difference diminished at higher heating rates. Both Mix 1 and 5 had a different cementitious matrix but the same fiber content. However, all specimens of Mix 5 spalled at 800 °C, indicating that the role of thermally better performing SCMs as adopted in Mix 1 may help in improving the spalling resistance if the fiber content is sufficient.

In addition, all the specimens of Mix 3 spalled at 600 and 800 °C at 10 °C/min. Figure 2 shows the specimens of Mix 3 spalled inside the furnace, which further confirmed the efficiency of PE fibers in mitigating spalling in ECC specimens. Mix 3 had a lower content of PE fiber, which may not be sufficient to mitigate the sudden increase in vapor pressure due to the very high heating rate.

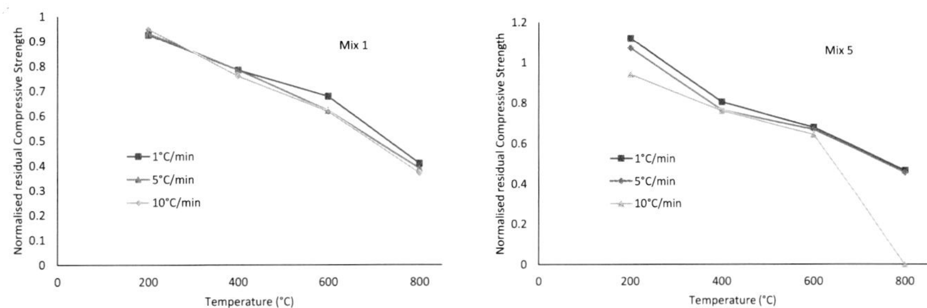

Fig. 1 Normalized residual compressive strength of Mix 1 and Mix 5 at different temperatures and heating rates

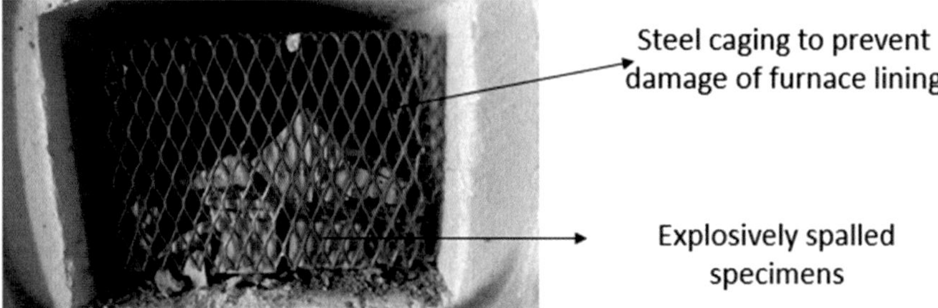

Fig. 2 Explosive spalling in Mix 3 specimens at 800 °C and 10 °C/min heating rate

4 Conclusions

We investigated the effect of heating rate on the spalling resistance of hybrid fiber-reinforced ECC and found that a mix with 1.5% PE and 0.75% steel fibers may be effective in preventing spalling even at very high heating rates. Moreover, this spalling resistance was dependent on the type of SCM used. We also observed that the use of thermally better performing SCM such as slag and dolomite can greatly improve both spalling resistance and residual compressive strength. However, a more systematic study including the role of specimen size is needed to obtain a deeper understanding of the effect of PE fiber content and the mix proportions on the spalling resistance of ECC.

References

1. Rawat S, Lee CK, Zhang YX (2021) Performance of fibre-reinforced cementitious composites at elevated temperatures: a review. Constr Build Mater 292
2. Liu JC, Tan KH (2018) Fire resistance of ultra-high performance strain hardening cementitious composite: residual mechanical properties and spalling resistance. Cem Concr Compos 89
3. Rawat S, Zhang YX, Lee CK (2022) Multi-response optimization of hybrid fibre engineered cementitious composite using Grey–Taguchi method and utility concept. Constr Build Mater 319

Roads Issues and the Social License to Operate

W. Young and M. Shackleton

Abstract Because it is agreed that the transport system needs the support and engagement of the public in its development and implementation, we explored the concept of the "social license to operate the road system" (SLORS) to show how it can assist in road policy and network implementation. We looked at 18 policy issues and identified their placement in a SLORS framework. The policy issues were categorized into 5 zones: User Advocacy Zone, Support Zone, Equilibrium Zone, Tolerance Zone and Opposition Zone. The relevant zone provides information to policy makers on the public's view of the policy. For instance: in the Advocacy Zone issues such as Driver behavior should improve, Roads must be safe for all users, the Physical quality of the road and their surface should improve, and Road travel should be more environmentally sustainable in the future are supported—and should be relatively easy to implement. In the Opposition zone People paying a toll or road charge for each trip, Private companies having a large role in planning and management of roads, Increased congestion and Increased traffic on roads in the future receive less support and thus will require considerable effort marketing their implementation. SLORS may assist in pointing the policy maker in the best direction to get the policy supported.

Keywords Policy · Roads · Social license · SLORS (Social License to Operate the Road System)

W. Young (✉)
Department of Civil Engineering, Monash University, Clayton, VIC, Australia
e-mail: bill.young@monash.edu

M. Shackleton
National Transport Research Organisation, Port Melbourne, VIC, Australia

W. Duan et al. (eds.), *Nanotechnology in Construction for Circular Economy*, Lecture Notes in Civil Engineering 356,
https://doi.org/10.1007/978-981-99-3330-3_34

1 Introduction

It is generally considered that the transport system needs the support and engagement of the public in its development and implementation. A social model approach to developing roads policy looks beyond the transport sector, beyond governments and beyond the road community to build wider acceptance of transport solutions. Here we describe at an approach for gaining insight into how the community views policies related to the performance of the road transport system: the "social license to operate the road system" (SLORS). In exploring the concept of the SLORS, we show how it can assist in the implementation of road policy and network considerations. We initially examined the (SLO) approach and how it adds to our understanding of the introduction of new products. The policy issues were then categorized into 5 zones: User Advocacy Zone, Support Zone, Equilibrium Zone, Tolerance Zone and Opposition Zone. Interpretation of the SLORS in a policy sense is outlined graphically.

2 Research into the Public's Acceptance of Road Transport Systems

At its base, road infrastructure operations require a coordinated, efficient and well-informed planning process, triple bottom line assessment and strategic asset management system. Because it is generally thought that an acceptable road system must meet the needs of the community, their view of the transport policy is an important input into these processes and, in particular, their implementation. Community consultation and people's behaviour are the main methods of collecting this information. As transport systems become more complex and invasive the general community expresses more concern about its impacts, and many transport projects and government decisions are questioned. In some cases transport projects have been stopped, delayed or not started. Public acceptance has been suggested as an important factor for the successful realization of transport plans, projects and policies, and a number of approaches have been used to quantify the performance and customer satisfaction with transport infrastructure [1]. In particular, BITRE looked at "…how customer preference might be better incorporated to improve the long term efficiency and operation of Australia's infrastructure asset" [1]. A key component of communication with the community in the roads area is transparency and a need to quantify their views. We add to the above approaches by exploring the quantification of SLORS. In particular, how can SLORS be used to assist in the implementation and operation of particular policies and the introduction of new products.

3 Consideration of SLORS

We examined how to include the public in the development of road policy decisions as part of the planning and policy development processes. The views of the community about the future of roads comprise an essential input into each of the planning processes because they are the system's end-users. This process can be assisted by the quantification of a SLORS. At the policy development level there could be varying levels of acceptance of particular issues by the community.

The survey methodology and data used in this study were collected from a series of cross-sectional questionnaires and focused group open-ended surveys over a period of 3 years. Industry and respondents were asked about the major issues and these were developed into a series of formal questionnaires. The questionnaire sample was collected using social media and a panel. The policy issues were developed over a 3-year period and changed as new issues were raised and old issues refined. Respondents were asked what "should" take place and what they think "will" take place. The "will" and "should" questions formed the basis of the quantified views using a 5-point semantic scales. Here, we only look at the stage 3 data.

There were 18 policy issues considered in stage 3 (Fig. 1) and these formed the basis for the exploration of the SLORS. Figure 2 presents the SLORS framework for consideration of these issues in a policy sense. The SLORS can be measured in terms of how the public perceives the acceptability or not of particular transport policies. That is, whether a policy issue will and should take place in road operation. The "will" provide an indication of what the respondent thinks the particular issue will be like in 30 years; "should" indicates what they think should occur. These 2 perceptions form a grid showing the implications of the SLORS: what should happen is the vertical axis and what will happen is the horizontal axis. The SLORS measure is the difference between the "should" and "will" ratings. Other relationships between the "should" and "will" ratings, such as ratios, logarithms etc. are shown by the graph, and will be explored further in the future. The vertical distance between the "should" rating and the line of equality ($\approx 45°$ line) between the "will" and "should" ratings pictorially represents the SLORS, which is the discrepancy between what the respondent thinks should take place and what will take place. This discrepancy, depending on its positive and negative value, will indicate the level of support, tolerance and opposition to the particular issue. More specifically, the line of equality (Should–Will) is the level of acceptance the community thinks these issues will take place and that they should take place at the same level at a particular point in time. Above the line of equality is where the community thinks that these policy issues should take place and that they will take place at a lower level, which is a level of advocacy and support for these issues. Below the line of equality is opposition to particular issues. These measures of the issues can be subdivided into 5 zones (Fig. 2): User Advocacy Zone, Support Zone, Equilibrium Zone, Tolerance Zone and Opposition Zone.

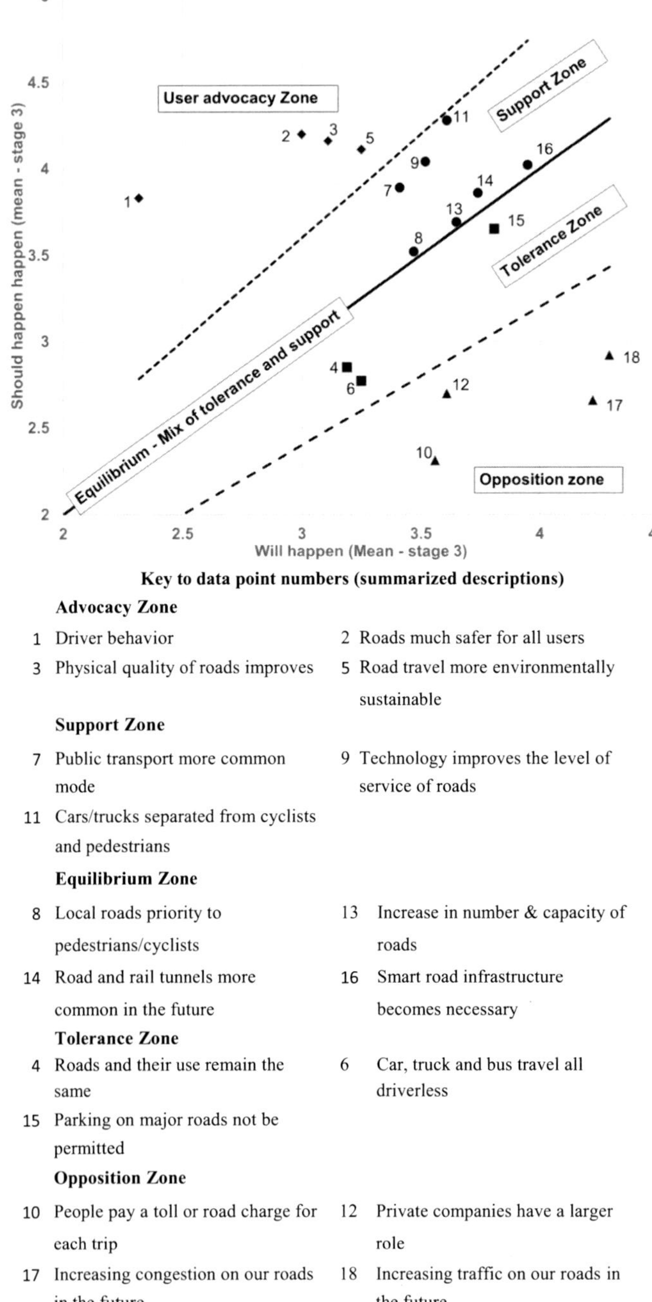

Fig. 1 Issues presented in a SLORS framework

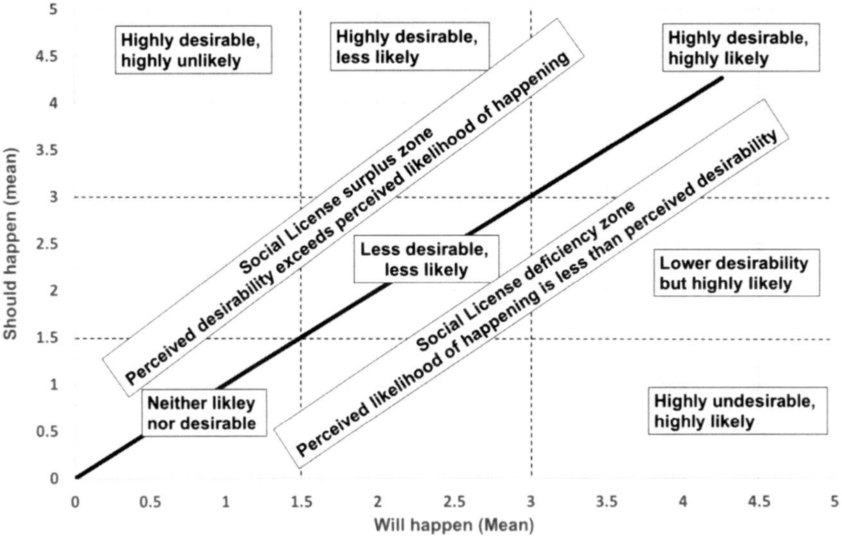

Fig. 2 Some implications of the relativity of average "should" and "will" ratings in a SLORS framework

4 The Data

The data we used was a subset of data collected for a broad study of the future of roads. Stage 3 data includes only data where all "should" and "will" ratings were given, and were collected between 24/2/20 and 24/4/20. The 18 policy issues (Fig. 1) were included in the questionnaire. The respondents were asked to answer how likely they thought that each statement described WILL occur (Fig. 1) and to what extent they agreed that what the statement described SHOULD occur (Fig. 1). These ratings formed the base for the SLORS (Fig. 1) and are discussed below.

5 Quantification of the SLORS

5.1 Respondents' View of What SHOULD Happen to Roads

The measure of the community's support for particular policies comes in many parts. One is what they think should happen and was measured using a 5-point Likert scale: Strongly Agree (5), Agree (4), Neutral (3), Disagree (2) and Strongly Disagree (1). Figure 1 presents the mean for "should" rating for each policy issue. The quantitative results of the data showed that most things were changes that people thought "should" happen, as measured by a mean equal to and above an average score of 3.00. The average rating ranged from a high of 4.22 (separation of bicyclists and pedestrians from cars and trucks) to a low of 2.34 (paying tolls and road charge)s. Overall, the "should" ratings and the percentage of people disagreeing with the issue provided a good indication of the community support for particular policy issues. This is one measure of how the community views the road system and should be taken into account when considering particular policies.

5.2 *Respondents' View of What WILL Happen to Roads*

A complementary measure to what "should" happen is the community's view of what "will" happen and was also measured on a 5-point Likert scale: Very Likely (5), Likely (4), Neutral (3), Unlikely (2) and Very Unlikely (1). The ranking of average ratings for the "will" scores indicated the preferred products. The quantitative results of the data showed that most things are changes that people think will happen. The average "will" ratings ranged from a high of 4.26 (increasing traffic) to a low of 2.37 (improving driver behaviour). The respondents' indication of what will happen is not an indication of the SLORS but taken together with the support levels for the policy issues provides a strong indication of the difference between what people want to happen and what they think will happen. This will be considered in the next section.

5.3 *The SLORS*

The "should" ratings provide one measure of the community's view of road transport policy, but do not, however, provide an indication of the community's dissatisfaction with the policies, because they understand some things will happen. The difference between the "should" and "will" ratings provides a measure of the level of discrepancy or dissatisfaction that people have with the transport policy. More specifically, the "will" ratings (Fig. 1) indicate what the respondents think the road system will be like in the future. The combination of this measure with the "should" rating (Fig. 1) gives the magnitude of dissatisfaction with what should and will take place in 30 years; that is, the SLORS. As shown in Fig. 1, the "should" minus the "will" difference for each respondent is averaged over the entire population. A positive ranking indicates that the desirability of a measure exceeds the likelihood of it happening, which is the support region in the SLORS diagram (Fig. 1). A negative indicates things that will happen but should not: The opposed region.

The SLORS diagram (Fig. 1) shows the mean "should" and "will" ratings for the data set plotted against one another, with the 45° line of equality between "should" and "will". Because the SLORS is estimated by subtracting "will" from "should", attributes with a positive rating are above the line and show where perceived desirability exceeded perceived likelihood of eventuating. For instance, it shows that although the respondents think that driver behaviour should improve (1.44), roads should be safer (1.15), and the physical quality of the roads should improve (0.99).

These are unlikely to happen despite net support. Those attributes with a negative rating are below the line of equality and show where the perceived likelihood of happening exceeds perceived desirability. Increased congestion (−1.50), increased traffic (−1.34), road charges (−1.18) and the role of private companies in planning (−0.86) fall into this category and have less support. They are likely to happen but people think they should not happen. Increased action would need to be put in place to achieve these goals.

5.4 *Interpreting the SLORS*

There are two major differences in this application of SLO from previous social license to operate applications in roads.

1. Road users have views on a wide range of policy issues from planning through to constructed infrastructure and even the behavior of users. It does not look only at 1 project as do other applications of SLO.
2. Other than in project-specific studies, users are seldom asked for their views on what they think should happen and at the same time what they think will really happen on roads. Generally, studies look at only what should happen, which is only half of the picture.

These contributions by users are potential keys to solving a number of implementation agency issues before they occur. Strategic initiatives and policy changes may meet significant stakeholder opposition, effectively preventing implementation of something that makes engineering and/or economic sense. An example is road-pricing and the concept that users should pay for the network capacity they use, and, critically, only the network capacity that they use. The net result is that custodians of the road network often face a choice—do nothing or risk a public backlash. The net result is long lead-in times for projects and changes, giving rise to perceptions that change is slow. Thus, an important part of change is the marketing of the changes to stakeholders, including road users.

The methodology described here may allow an agency that is considering a basket of policy options or initiatives to gain some insight as to the phasing strategy. They can begin with options with a high SLORS to start getting some benefits of change, while change requiring a significant shift in attitudes can be shepherded over longer periods to win road users over, or gain their license to bring about the envisaged change.

Thus, points below the equality line in Fig. 2 represent issues where "will" is greater than "should"; that is, factors where support is less than the perceived inevitability or where social license is deficient. Points above the equality line represent issues where the social license is positive; road users want the change more than they perceive it to be likely to eventuate. Points on the line represent issues where support and perceived eventuation are in balance.

This gives some idea of what the 'quick wins' might be and where significant effort may need to be put into winning the support of road users. This approach can be incorporated into the SLORS framework (Fig. 3). The idea of road users taking on an advocacy role (User Advocacy Zone) against hold outs for changes the agency wants to make would have strong appeal to the agency. It avoids the agency being accused of forcing their change through and reduces the expenditure of resources that the agency needs to effect the change. To take advantage of this, it is necessary to know at what point support tips into advocacy on the road network.

Similarly, it would be helpful to know—for those factors in the social license deficiency zone (Opposition Zone)—which are the issues where it is possible to change road user perceptions and gain social license in a reasonable amount of time, and which are candidates for really long-term efforts, or for which a rethink may be needed.

To illustrate how a finer gradation of support/opposition is introduced, 2 lines have been added to the SLORS framework in Fig. 3 to represent these tipping points: 120% (Support Zone) and 80% Opposition Zone) of the line of equality value (Equilibrium Zone = mix of tolerance and support).

By adopting such an approach to assessing social license, an agency would then have the following strategies available:

1. Supply users with materials and publicity for items in the Advocacy Zone
2. Embark on minor 'marketing' of items in the Support Zone to reduce resistance
3. Focus efforts on items in the Tolerance and Equilibrium zones in order to gain some degree of social license for them
4. Rethink the desirability of items in the Opposition Zone, form coalitions with others trying to achieve the same measures or make plans for a long process of persuasion.

Figure 3 shows the 18 policy issues identified in their SLORS zones, for the tipping points described above. Two of the Opposition Zone factors are in fact outcomes—increased traffic and congestion. Therefore, if an agency hoped to do nothing to reduce traffic or congestion that "do nothing" approach would be resisted. Conversely though, any actions taken to ameliorate outcomes in the Opposition Zone can be assumed to have the "Support", or enjoy the "Advocacy", of road users.

Put another way, the "should", "will" and SLORS ratings provide guidance on the acceptance or not of particular policy issues by the community. These need to be put into an overall policy context. Figure 4 shows the full SLORS framework and the ratings for the 18 policy issues, illustrating the policy tipping points. Issues can be divided into zones: Advocacy support (Alignment between desirability and likelihood); Issues where people show some Tolerance; and Issues where there is Opposition.

Advocacy support can be found in the areas of improvement: driver behaviour (1.44), road safety improvements for all road users (1.15), improvements in the quality of the road surface (0.99) and road transport being environmentally sustainable (0.83). **Support** is likely for policies related to: physical separation of active transport from cars and trucks (0.61), use of technology to improve level of service

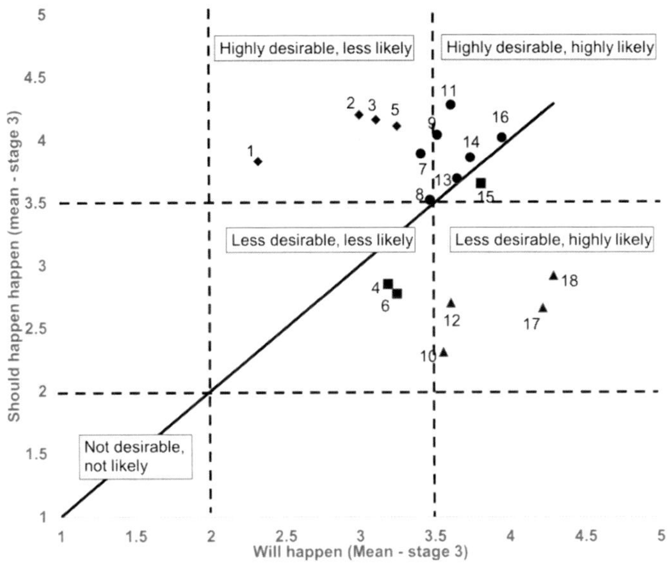

Key to data point numbers (summarized descriptions)

	Highly desirable, Highly likely		
9	Technology improves the level of service of roads	11	Cars/trucks separated from cyclists & pedestrians
13	Increase in number & capacity of roads	14	Road and rail tunnels more common in the future
15	Parking on major roads not permitted	16	Smart road infrastructure becomes necessary
	Highly desirable, less likely		
1	Driver behavior improves	2	Roads much safer for all users
3	Physical quality of roads improves	5	Road travel more environmentally sustainable
7	Public transport more common mode	8	Local roads priority to pedestrians/cyclists
	Less desirable, less likely		
4	Roads and their use remain the same	6	Car, truck and bus travel all driverless
	Less desirable, highly likely		
10	People pay a toll or road charge for each trip	12	Private companies have a larger role
17	Increasing congestion on our roads in the future	18	Increasing traffic on our roads in the future

Fig. 3 Alternate zone definitions for the SLORS framework

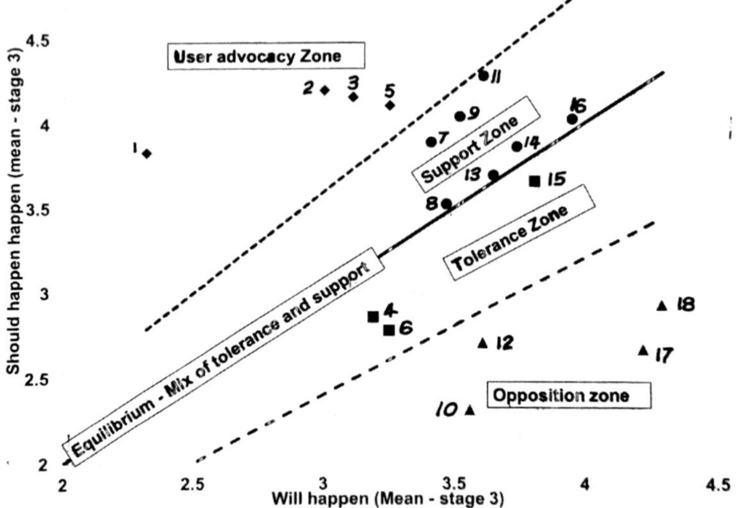

Fig. 4 A comprehensive SLORS framework

(0.49) and public transport being a more common mode choice (0.47). A mix of **tolerance and support** may be obtained for: priority given to: more tunnels for road and rail (0.12), active travel in local and shopping roads (0.05), more smart infrastructure (0.05) and increased number and capacity of roads (0.03). For these factors, the views of desirability and likelihood are similar in magnitude. The factors where the respondents may show some **tolerance** are: banning parking on major roads (−0.14), roads and their use will stay the same (−0.33), and automation of the road transport network (−0.47). Those factors where there is likely to be **opposition** are private sector being involved in planning (−0.86), paying for the use of the network (−1.18), increased traffic in the future (−1.34), and increased congestion (−1.50).

Under the SLOR zones posited, Australian road agencies can rely on user advocacy for actions and strategies to:

1. Improve driver behavior, make roads safer, better physical quality and make use of roads more environmentally sustainable.
2. Reduce traffic and congestion, by virtue of "increased traffic" and "increased congestion" being in the "opposition" zone as posited.
3. Support strategies to improve the attractiveness of public transport as a choice, increase deployment of technology (ITS/VMS) to improve levels of service on roads and to physically separate motorized traffic from cyclists and pedestrians.
4. Work on public acceptance of an all-autonomous motorized vehicle fleets, a ban on parking along major roads and any plans to allow road usage to stay the same.
5. Reconsider pricing as a strategy, or significantly change how it may be applied, or simply educate the road users on an intelligent reframing of what "pay per use" means.

6 Conclusions

We explored an approach to gaining insight into how the community views the performance of the road system: the concept of a SLORS and shows how it can assist in the implementation of policy and network considerations. The measurement of the SLORS was described and 18 policy issues were rated for whether they "should" and "will" be implemented. The SLORS is the difference between the "should" and "will" ratings. The policy issues were categorized into 5 zones: User Advocacy Zone, Support Zone, Equilibrium Zone, Tolerance Zone and Opposition Zone. The Zone into which an issue falls provides information to policy makers on the public's view of the policy. For instance: in the Advocacy Zone issues such as driver behaviour should improve, Roads must be safe for all users, the Physical quality of the road and their surface should improve, and Road travel should be more environmentally sustainable in the future are in the user Advocacy Zone. In the Opposition Zone People paying a toll or road charge, Private companies having a large role in planning and management, Increased congestion and Increased traffic on roads in the future receive less support. This information may assist in pointing policy makers in the best direction to gain community support.

Reference

1. Australian Government. Bureau of Infrastructure, Transport and Regional Economics (2017) Measuring infrastructure asset performance and customer satisfaction: a review of existing frameworks. Report 147. Department of Infrastructure and Regional Development, Canberra

An Intelligent Multi-objective Design Optimization Method for Nanographite-Based Electrically Conductive Cementitious Composites

W. Dong, Y. Huang, B. Lehane, and G. Ma

Abstract Nanographite (NG) is a promising conductive filler for producing effective electrically conductive cementitious composites for use in structural health monitoring methods. Since the acceptable mechanical strength and electrical resistivity are both required, the design of NG-based cementitious composite (NGCC) is a complicated multi-objective optimization problem. This study proposes a data-driven method to address this multi-objective design optimization (MODO) issue for NGCC using machine learning (ML) techniques and non-dominated sorting genetic algorithm (NSGA-II). Prediction models of the uniaxial compressive strength (UCS) and electrical resistivity (ER) of NGCC are established by Bayesian-tuned XGBoost with prepared datasets. Results show that they have excellent performance in predicting both properties with high R^2 (0.95 and 0.92, 0.99 and 0.98) and low mean absolute error (1.24 and 3.44, 0.15 and 0.22). The influence of critical features on NGCC's properties are quantified by ML theories, which help determine the variables to be optimized and define their constraints for the MODO. The MODO program is developed on the basis of NSGA-II. It optimizes NGCC's properties of UCS and ER simultaneously, and successfully achieves a set of Pareto solutions, which can facilitate appropriate parameters selections for the NGCC design.

Keywords Machine learning · Multi-objective design · NSGA-II · NGCC · Optimization

W. Dong · B. Lehane
Department of Civil, Environmental & Mining Engineering, University of Western Australia, Perth, WA, Australia

Y. Huang (✉) · G. Ma
Department of Civil & Transportation Engineering, Hebei University of Technology, Tianjin, China
e-mail: yimiao.huang@hebut.edu.cn

W. Duan et al. (eds.), *Nanotechnology in Construction for Circular Economy*,
Lecture Notes in Civil Engineering 356,
https://doi.org/10.1007/978-981-99-3330-3_35

1 Introduction

Nanographite (NG) is a promising conductive filler for producing highly effective electrically conductive cementitious composites. As reported by [1], the NG-based cementitious composite (NGCC) has efficient strength and conductivity as a self-sensing construction material for use in non-destructive structural health monitoring (NDSHM). There are generally three forms of NG studied in NGCC, namely graphene nanoplatelet (GNP), graphene oxide (GO), and reduced GO (rGO). Apart from the form of NG used, as well as its physical properties, it has been confirmed that other experimental factors, such as the dispersion of NG and curing condition, are linked to the strength and conductivity variations of NGCC [2, 3]. Significant experimental research has been conducted to determine the optimal values for those factors during the production of NGCC [4, 5].

The purpose of design optimization of NGCC is to approach its high strength and low electrical resistivity (ER) simultaneously, which can seem difficult to realize by experimental methods. Once, an experiment could only consider one factor with limited design values for variable-controlling analysis, making the design experience lack universality. Former experimental studies often discussed mechanical and electrical objectives separately. Nevertheless, they have provided numerous test data. We present our study that aimed to address the multi-objective design optimization (MODO) problem of NGCC using advanced data mining techniques.

The study proposed a comprehensive data-driven computing and analyzing method, which integrates algorithms of machine learning (ML) and non-dominated sorting genetic algorithm (NSGA-II). First, the uniaxial compressive strength (UCS) and ER of NGCC were modelled by Bayesian-optimized XGBoost with compiled experimental datasets. During the modelling process, the Weight and SHAP (SHapley Additive exPlanations) theories were used to interpret the ML models and determine variables that were critical for optimization. Furthermore, the established models worked as objective functions in NSGA-II to develop the MODO program. The feasibility and accuracy of the proposed MODO method are discussed and proved with a case study. Using this method, researchers can quickly find the optimal designs of NGCC that satisfy their application demands.

2 Methods

2.1 Establishment of Calculation Models

We modelled the UCS and ER of NGCC through a complex ML modelling framework, as shown in Fig. 1, to ensure the established models were accurate and reliable. The framework followed four steps. Original experimental datasets of UCS and ER were constructed after an extensive literature review. They were then processed with proper feature engineering strategies before estimator training. In Step 3 of

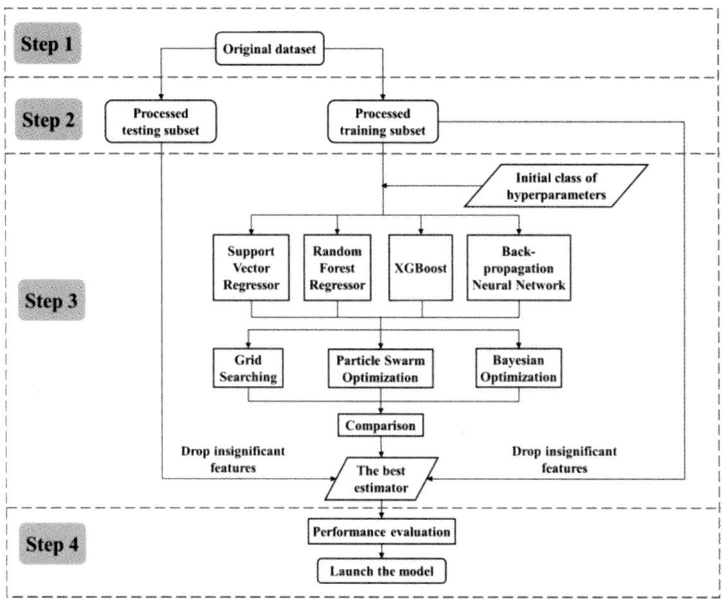

Fig. 1 Machine learning modeling framework

modelling, we applied four classic ML algorithms, namely support vector regressor (SVR), random forest regressor (RFR), XGBoost, and back-propagation neural network (BPNN), to the processed datasets. Three searching methods, comprising grid searching (GS), particle swarm optimization (PSO), and Bayesian optimization (BO), were used to tune the hyperparameters of the algorithms for achieving their best performance. At this stage, estimators based on four algorithms were compared with each other to determine the most suitable models for UCS and ER. Finally, we used SHAP theory to interpret the established models to identify critical input variables and quantify their influence.

2.2 *Development of the MODO Program*

Figure 2 shows the workflow of the developed MODO program. The structure of the program was based on the mechanism of NSGA-II. Individuals in the generated group were vectors consisting of variables to be optimized. Lower and upper boundaries should be defined for each variable. Other variables that were also considered in both datasets were fixed. Afterwards, the generated group concatenated fixed variables to form the input datasets for UCS and ER calculations using the established models. Pipelines were tools packing feature engineering methods and making the format of input datasets consistent. Individuals were ranked based on the UCS and ER results

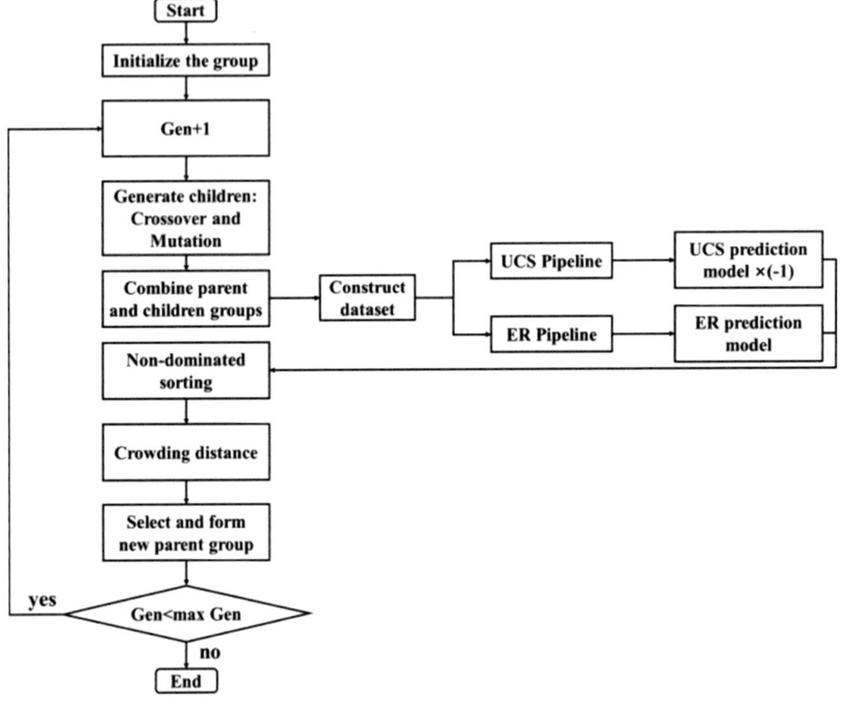

Fig. 2 Flowchart of the multi-objective design optimization (MODO) program

according to the Pareto rule. Individuals in the high Pareto ranks and with larger crowding distances were kept for creating the next generation group. At the end of the program, the final group was the Pareto set formed with optimal design solutions.

3 Results and Discussion

Bayesian-optimized XGBoost models proved the most suitable for the UCS and ER of NGCC according to the comparative results, and their hyperparameter combinations are given in Table 1. Models had minimal gaps between training and testing subsets at the end of the training, indicating no over- or under-fitting issues. Mean absolute error (MAE) and determination coefficient (R^2) were two indexes for assessing the accuracy of the established models. Small MAE values (1.24 and 3.44, 0.15 and 0.22) and high R^2 scores (0.95 and 0.92, 0.99 and 0.98) of the two models yielded satisfactory and reliable prediction abilities for the UCS and ER of NGCC. Figure 3 shows the feature importance ranking based on the SHAP interpretation results of the two models. It can be seen that the UCS and ER prediction models share almost the same feature importance ranking, where the mixing amount of NG (GC) was a

Table 1 Hyperparameter tuning results

Hyperparameter	UCS model	ER model
Estimator numbers	381	128
Learning rate	0.19	0.25
Gamma	0.31	0.74
Maximum tree depth	13	6
Subsample size	0.40	0.64

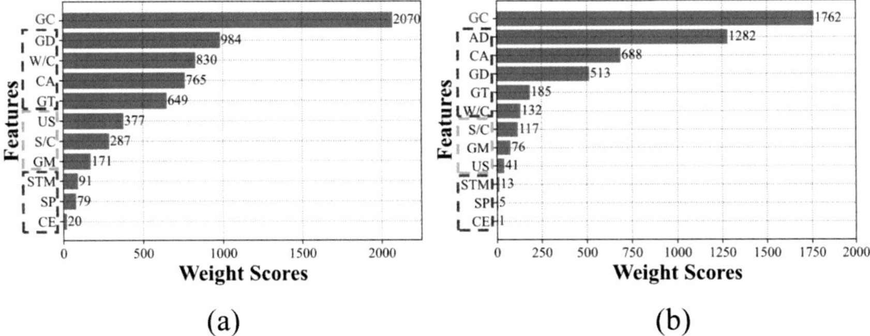

Fig. 3 Feature importance ranking: **a** uniaxial compressive strength; **b** electrical resistivity

dominant variable in both properties. The high influence could also be observed in other features: NG's physical properties of thickness and diameter (GT and GD), the water dosage and curing age.

The developed MODO program proved feasible in a case study. In the case study, GNP was selected and the curing age was set at 28 days while the other six variables were optimized. The optimization process is recorded in Fig. 4 and the program successfully converged to the final Pareto set through iteration. As listed in Table 2, optimization results indicated that all the given design solutions qualified for the NDSHM application with acceptable strength and conductivity. Higher UCS led to higher ER (lower conductivity). Additionally, optimal values could be found for some variables. For example, the ideal thickness of GNP was ≈6.36 nm. The water/cement ratio was 0.32 in most solutions. The ultrasonication process for GNP dispersion was better if it lasted for 30 min. The differences in output results between solutions were mostly due to the changes in the dosage of sand and GNP, as well as the GD. UCS and ER of NGCC became smaller by adding more GNP.

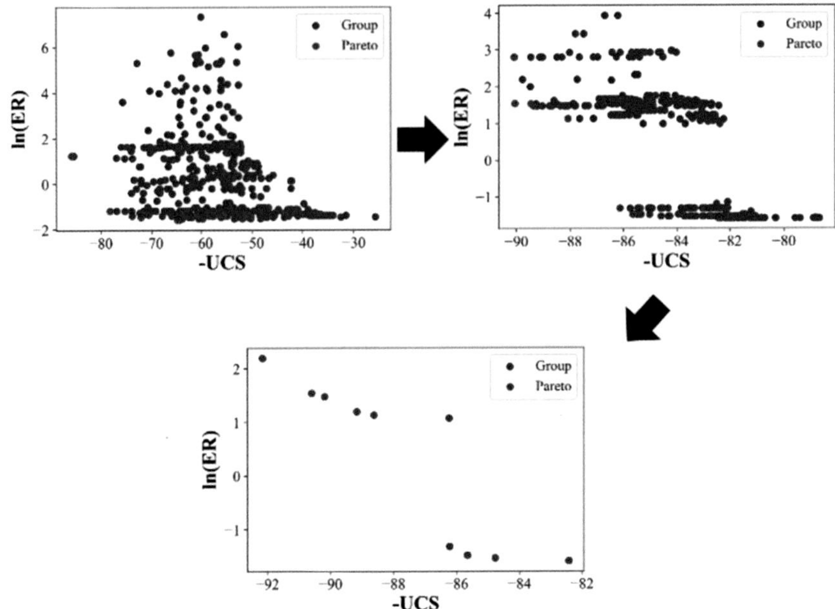

Fig. 4 Optimization process of the multi-objective design optimization (MODO) program of nanographite-based cementitious composite. (ER, electrical resistivity; UCS, uniaxial compressive strength)

Table 2 Multi-objective design optimization (MODO) design solutions for nanographite-based cementitious composite

S/C	GD	GT	GC	W/C	US	UCS	ER
0.05	21.44	5.99	0.05	0.31	0.57	92.16	8.97
0.48	21.81	6.64	0.45	0.3	0.51	90.58	4.66
0.08	22.11	6.61	0.55	0.32	0.51	90.17	4.38
1.49	21.53	6.51	0.45	0.31	0.52	89.16	3.30
1.49	24.44	6.51	0.51	0.32	0.52	88.66	3.10
2.95	24.77	6.54	0.57	0.34	0.56	87.28	2.92
2.21	23.63	6.67	0.63	0.33	0.56	86.49	2.68
0.58	21.35	6.62	2.15	0.31	0.51	86.22	0.27
0.09	21.99	5.93	4.62	0.31	0.52	85.65	0.23
0.57	21.71	5.99	4.81	0.32	0.5	84.76	0.22
0.58	21.44	5.99	7.27	0.32	0.56	82.42	0.20

(ER, electrical resistivity; GC, content of NG; GD, diameter of NG, GT, thickness of NG; NG, nanographite; S/C, sand/cement ratio; UCS, uniaxial compressive strength; US, ultrasonication; W/C, water/cement ratio)

4 Conclusions

Robust and reliable calculation models were established for the UCS and ER of NGCC by BO-tuned XGBoost. The SHAP interpretation results of the established models identified significant influential factors, including NG dosage and other variables. Moreover, we explained their quantitative influence on the properties of NGCC by analyzing their SHAP value distributions. NSGA-II was combined with the established models as objective functions to develop the MODO program of NGCC. The program proved feasible through a case study in which it successfully obtained the Pareto set of design solutions. The given solutions determined the optimal values for some variables.

Acknowledgements We appreciate the support of The University of Western Australia through the "Scholarship for International Research Fees and Ad Hoc Postgraduate Scholarship". The research work was financially supported by the ARC Research Hub for Nanoscience-Based Construction Material Manufacturing (IH150100006). Experimental data used in machine learning modelling in this study were retrieved from published literature and all data sources are acknowledged.

References

1. Qureshi TS, Panesar DK (2019) A comparison of graphene oxide, reduced graphene oxide and pure graphene: early age properties of cement composites. In: Proceedings of the International Conference on Sustainable Materials, Systems and Structures (SMSS2019) New Generation of Construction Materials. vol 1
2. Wotring EE (2015) Dispersion of graphene nanoplatelets in water with surfactant and reinforcement of mortar with graphene nanoplatelets. University of Illinois at Urbana-Champaign, Master's Theses
3. Liu J, Fu J, Yang Y, Gu C (2019) Study on dispersion, mechanical and microstructure properties of cement paste incorporating graphene sheets. Constr Build Mater 199
4. Sun S, Han B, Jiang S, Yu X, Wang Y, Li H, Ou J (2017) Nano graphite platelets-enabled piezoresistive cementitious composites for structural health monitoring. Constr Build Mater 136
5. Li X, Liu YM, Li WG, Li CY, Sanjayan JG, Duan WH, Li Z (2017) Effects of graphene oxide agglomerates on workability, hydration, microstructure and compressive strength of cement paste. Constr Build Mater 145

BY

Machine Learning-Aided Nonlinear Dynamic Analysis of Engineering Structures

Y. Feng, Q. Wang, D. Wu, and W. Gao

Abstract A machine learning (ML) technique was used to assist in the dynamic analysis of mixed geometric and material nonlinearities of real-life engineering structures. Various types of inputs of system properties were considered in the 3D dynamic geometric elastoplastic analysis, giving a series of realistic nonlinear descriptions of complex, large deformation structural behaviors. To resolve the numerical challenges of solving the mixed nonlinear problems, a newly established ML technique using a new cluster-based extended support vector regression (X-SVR) algorithm was applied. With this technique, a surrogate model can be built at each time step in the Newmark time integration process, which can then be used to predict the deflection, force and stress of the relevant structural performance at different loading time stages. To demonstrate the accuracy and efficiency of the proposed framework, practical engineering applications with linear and nonlinear properties are fully demonstrated, and the nonlinear behavior of the structure under predicted working conditions in the future was predicted and verified in numerical studies.

Keywords Engineering structures · Machine learning · Nonlinear dynamic analysis

Y. Feng (✉) · Q. Wang · W. Gao
Centre for Infrastructure Engineering and Safety, School of Civil and Environmental Engineering, The University of New South Wales, Sydney, NSW, Australia
e-mail: yuan.feng1@unsw.edu.au

D. Wu
School of Civil and Environmental Engineering, University of Technology Sydney, Sydney, NSW, Australia

W. Duan et al. (eds.), *Nanotechnology in Construction for Circular Economy*, Lecture Notes in Civil Engineering 356,
https://doi.org/10.1007/978-981-99-3330-3_36

1 Introduction

The nonlinear response of a practical structure is affected by various factors such as system properties, operational coefficients, loads and environment, as well as uncertainties in collected information and estimation models. Liu et al. investigated the random mean values of elastoplastic responses of structure using a probabilistic partial differentiation approach [1]. Feng et al. introduced a stochastic elastoplastic analysis of two-dimensional engineering structures with the aid of sampling-based machine learning algorithm [2]. In our study, the uncertain parameters were studied simultaneously within the nonlinear dynamical framework with the help of an advanced machine learning (ML) technique [3]. By using the ML algorithm, an explicit regression function can be obtained to represent the relationship between the uncertain inputs and the nonlinear responses. Subsequently, frequent, and fast nonlinear prognosis can be conducted to assess the nonlinear behavior of engineering structures during the dynamic loading process.

Here, a brief introduction to the proposed ML-aided framework for nonlinear dynamics of engineering structure is given. The two main components of the approach are briefly introduced. First, the deterministic solution to geometric–elastoplastic dynamics is presented. Then, the novel ML technique named the "extended support regression" is introduced. To demonstrate the accuracy and applicability of the proposed framework, an illustrative numerical case is incorporated to build the proposed framework and demonstrate the nonlinear response for the concerned structure.

2 Methods

2.1 *Solution to Geometric–Elastoplastic Dynamics*

For structural systems with both material and geometric nonlinearities, the plastic strain and second-order Green–Lagrange terms must be considered in the incremental strain–displacement relations as:

$$\Delta\boldsymbol{\varepsilon} = \Delta\boldsymbol{\varepsilon}_e + \Delta\boldsymbol{\varepsilon}_p + \Delta\boldsymbol{\varepsilon}_g = (\mathbf{B} + \mathbf{B}_g)\Delta\mathbf{u} \tag{1}$$

where $\Delta\boldsymbol{\varepsilon}_e$, $\Delta\boldsymbol{\varepsilon}_p$ and $\Delta\boldsymbol{\varepsilon}_g$ denote the elastic, plastic and high-order strain increments; $\mathbf{B}$ and $\mathbf{B}_g$ denote the material and geometric nonlinear strain–displacement matrixes of the deformation. In the dynamics framework without the effects of external load and damping, the equilibrium function of a nonlinear structural domain by using the principle of virtual work can be represented as:

$$\int_V \delta\boldsymbol{\varepsilon}^T \boldsymbol{\sigma} dV - \int_S \delta\mathbf{u}^T \boldsymbol{\tau} dS + \int_V \delta\mathbf{u}^T \rho\ddot{\mathbf{u}} dV = 0 \tag{2}$$

where $\ddot{\mathbf{u}}$ denotes the virtual acceleration vector; ρ denotes the material density and $\boldsymbol{\tau}$ denotes the surface traction along the boundary. By following the incremental strategy, and substituting Eq. (1) into the above virtual field, it can be re-expressed as:

$$\left(\int_V \mathbf{B}^T \mathbf{D}_{ep} \mathbf{B} dV + \int_V \mathbf{B}_g^T \mathbf{D} \mathbf{B}_g dV\right) \Delta \mathbf{u} = \int_S \boldsymbol{\Phi}^T \ \tau dS - \int_V (\mathbf{B} + \mathbf{B}_g)^T \sigma dV - \int_V \boldsymbol{\Phi}^T \rho \boldsymbol{\Phi} dV \ddot{\mathbf{u}} = 0 \tag{3}$$

Consequently, the global governing equation of the dynamic geometric-material hybrid nonlinearities problem can be represented as:

$$(\mathbf{K}_{ep} + \mathbf{K}_g)\Delta \mathbf{U} = \mathbf{F}_{ex} - \mathbf{R}_{in} - M\ddot{\mathbf{U}} \tag{4}$$

By using the Newmark time integration technique, the solution to the above equation can be acquired through:

$$\begin{cases} (a_0 \mathbf{M} + \mathbf{K}_{ep}^t + \mathbf{K}_g^t)\Delta \mathbf{U}^{t+1} = \mathbf{F}_{ex}^{t+1} - \mathbf{R}_{in}^t + (a_0 \mathbf{U}^t + a_1 \dot{\mathbf{U}}^t + a_2 \ddot{\mathbf{U}}^t)\mathbf{M} - a_0 \mathbf{M} \mathbf{U}^t \\ a_0 = 1/\left[\beta(\Delta t)^2\right],\ a_1 = 1/(\beta \Delta t), a_2 = 1/(2\beta) - 1 \end{cases} \tag{5}$$

where Δt denotes the incremental time step; $\beta = 1/4$ and $\kappa = 1/2$ are set for all the calculations based on the trapezoidal rule [4].

2.2 Extended Support Vector Regression

After thousands of simulations of deterministic nonlinear dynamic analysis based on the finite element model, the collected structural response was considered as the training dataset for the ML technique to recognize the relation between outputs and input parameters.

The adopted extended support regression was the ML technique proposed by Feng et al. [5]. It is based on the theory of classical support vector machine (SVM) and the doubly regularized SVM (DrSVM), and it offers notable practicability in handling nonlinear prediction problems. For detailed definitions of each parameter in Eq. (6), see Feng et al. [6].

$$\hat{f}_N(\mathbf{x}) = (\mathbf{p}_k - \mathbf{q}_k)^T \hat{\mathbf{k}}(\mathbf{x}) - \hat{\mathbf{e}}_k^T \hat{\mathbf{G}}_k \boldsymbol{\upsilon}_k^* \tag{6}$$

where $\mathbf{p}_k, \mathbf{q}_k$ denote two positive kernelized parameters; $\hat{\mathbf{k}}(x)$ denotes the kernel matrix; $\hat{\mathbf{e}}_k, \hat{\mathbf{G}}_k$ denote two matrix vectors; $\boldsymbol{\upsilon}_k^*$ denotes the obtained solution.

2.3 Modelling

2.3.1 Geometric–Elastoplastic Dynamics of Transmission Tower

As a numerical example, the nondeterministic nonlinear behaviors of a 3D transmission tower structure were investigated, and the geometry is presented in Fig. 1. In this case, the four supports were fixed and the magnitude of a time-varying pressure $P(t) = 8e10 \times (1 - \cos(20t))$ Pa, was applied at the tip of the tower. By establishing the surrogate system, the nonlinear responses of the tower were predicted at various time steps. The uncertainty information of the system is listed in Table 1.

By substituting the new input information into the X-SVR model, the predicted contour plots of nonlinear deflections of the tower structure at three time steps are

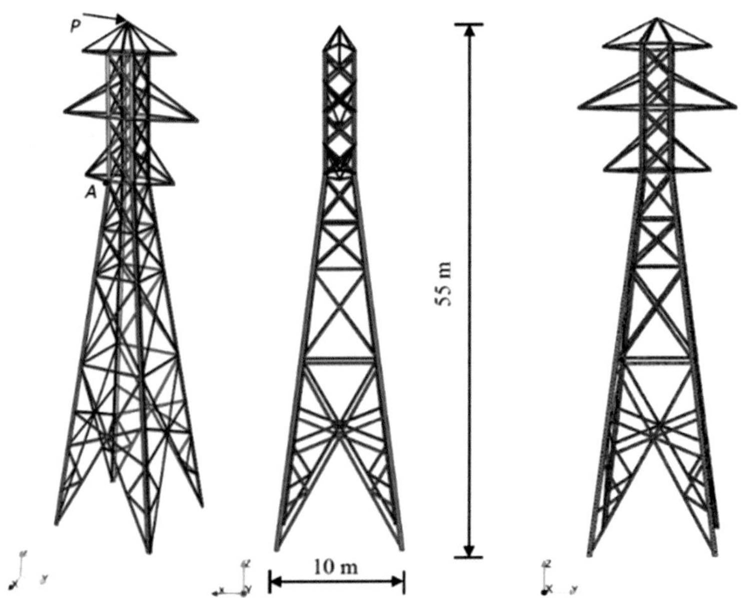

Fig. 1 Transmission tower geometry and adopted mesh condition

Table 1 Variational input data of transmission tower

Variational input	Distribution type	Mean	Standard deviation	Interval
E (GPa)	Lognormal	210	10.5	–
υ	Uniform	–	–	[0.28, 0.32]
ρ (kg/m^3)	Normal	7380	370	
σ_Y (GPa)	Beta	0.45	0.0225	–
H	Lognormal	15	0.75	–

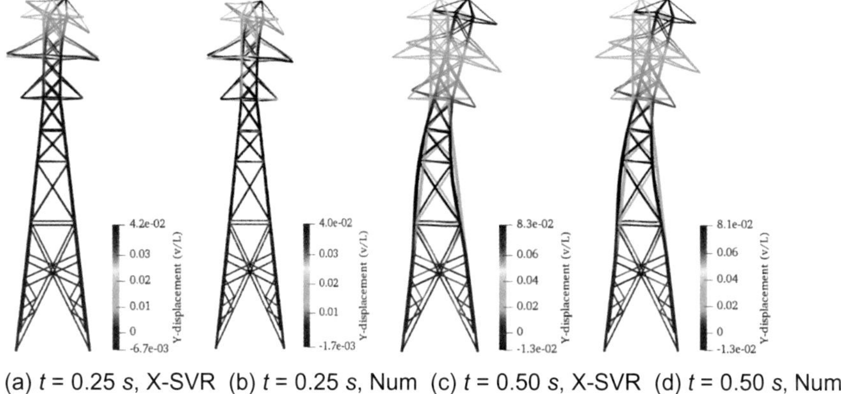

Fig. 2 X-SVR predicted and numerical simulated nonlinear deflection of a transmission tower at different times

shown in Fig. 2, as well as the deterministic numerical simulation results as verification. In Fig. 2, the predicted nonlinear displacements from X-SVR correlate well with the numerical simulation results.

3 Conclusions

In the numerical modelling example, the nonlinear response of the transmission tower predicted by the proposed method was compared with the numerical simulation result and satisfactory alignment was identified from the comparison. Consequently, the new nonlinear dynamic analysis framework aided by the ML technique for engineering structures was verified as effective for practical engineering problems. We believe that the proposed framework can improve the accuracy and efficiency of the relevant structural nonlinear dynamic's evaluation process.

Acknowledgements The numerical computations were undertaken with the assistance of resources and services from the National Computational Infrastructure (NCI), which is supported by the Australian Government.

References

1. Liu N, Tang WH, Zhou J (2002) Reliability of elasto-plastic structure using finite element method. Acta Mech Sin 18:66–81
2. Feng Y, Gao W, Wu D, Tin-Loi F (2019) Machine learning aided stochastic elastoplastic analysis. Comput Methods Appl Mech Eng 357:112576

3. Feng Y, Wang Q, Wu D, Gao W, Tin-Loi F (2020) Stochastic nonlocal damage analysis by a machine learning approach. Comput Methods Appl Mech Eng 372:113371
4. Newmark NM (1959) A method of computation for structural dynamics. J Eng Mech 85:67–94
5. Feng Y, Wang Q, Wu D, Luo Z, Chen X, Zhang T (2021) Machine learning aided phase field method for fracture mechanics. Int J Eng Sci 169:103587
6. Feng J, Liu L, Wu D, Li G, Beer M, Gao W (2019) Dynamic reliability analysis using the extended support vector regression (X-SVR). Mech Syst Signal Process 126:368–391

Insights into the Size Effect of the Dynamic Characteristics of the Perovskite Solar Cell

Q. Li, D. Wu, and W. Gao

Abstract Driven by government policy and incentives, solar power production has soared in the past decade and become a mainstay during the worldwide clean-power transition process. Among the various next-generation photovoltaic technologies, perovskite solar cells (PSCs) are the most important emerging area of research due to their outstanding power conversion efficiency and affordable scale-up operation. We adopted the nonlocal strain gradient theory and the first-order shear deformation plate theory to investigate the size-dependent free vibration behavior of PSCs. The size-dependency in the nanostructure of the PSCs was captured by coupling the nonlocal and strain gradient parameters. In accordance with the Hamilton principle, the governing equations set was derived. Subsequently, the Galerkin procedure was applied to address the dynamic characteristics analysis of PSCs with simply supported and clamped edges. Compared with the size-insensitive traditional continuum plate model, the current multiscale framework revealed a size effect on the free vibration of the PSC. Moreover, some parametric experiments were conducted to explore the impacts of scale length parameter, nonlocal parameter, and boundary conditions on the natural frequency of the PSC.

Keywords Free vibration · Nonlocal strain gradient theory · Perovskite solar cell · Size effect

Q. Li · W. Gao (✉)
Centre for Infrastructure Engineering and Safety (CIES), School of Civil and Environmental Engineering, The University of New South Wales, Sydney, NSW, Australia
e-mail: w.gao@unsw.edu.au

D. Wu (✉)
School of Civil and Environmental Engineering, University of Technology Sydney, Sydney, NSW, Australia
e-mail: di.wu-1@uts.edu.au

W. Duan et al. (eds.), *Nanotechnology in Construction for Circular Economy*, Lecture Notes in Civil Engineering 356,
https://doi.org/10.1007/978-981-99-3330-3_37

1 Introduction

The perovskite solar cell (PSC) has garned tremendous attention from the scientific community over the past few years due to outstanding optical and electronic properties [1]. This new-generation photovoltaic device rose to prominence in 2012 with an energy conversion efficiency of 9.7% and then rapidly achieved a new certified record with 25.7% in 2022 [2]. Although PSCs have proven their competitive power conversion efficiencies and the prospect of further improved performance, their structural response during their operational lifetime still lacks investigation [3]. The study of the free vibration of PSC lays a solid foundation for optimizing the structure, because it is a crucial part of analyzing the dynamic response to various loading scenarios.

For mainstream solar power generation, technologies cannot be deployed in the field without accurately estimating their structural response. Despite the significance in real-life applications, there are few studies on the size-dependent dynamic characteristics of PSCs among the broad spectrum of available scientific reports that emphasize the interest and importance of this topic. Thus, we attempted to fill this gap by using nonlocal strain gradient theory (NSGT) to reveal the size effect of the free vibration behavior of the PSC with both simply supported and clamped boundary conditions.

2 Methods

2.1 Formulations

Grounded on NSGT [4], the internal energy density potential incorporating nonlocality and higher-order strain gradient tensor were formulated as:

$$U\left(\varepsilon_{ij}, \varepsilon'_{ij}, \alpha_0, \varepsilon_{ij,m}, \varepsilon'_{ij,m}, \alpha_1\right) = \frac{1}{2}\varepsilon_{ij}C_{ijkl}\int_V \alpha_0\left(\left|x - x'\right|, e_0 a\right)\varepsilon'_{kl}dV' + \frac{1}{2}l^2\varepsilon_{ij,m}C_{ijkl}\int_V \alpha_1\left(\left|x - x'\right|, e_1 a\right)\varepsilon'_{kl}dV' \quad (1)$$

where ε_{ij} and ε'_{ij} are the strain tensors at the arbitrary point x and its neighboring point x' in V, respectively; C_{ijkl} denotes the elastic coefficients of the classical elasticity; α_0 and α_1 indicate the nonlocal attenuation function and the additional kernel function, correspondingly, which are related to the nonlocal parameters $e_0 a$ and $e_1 a$, and the distance between the considered points x and x' in V; e_0 and e_1 refer to the material constants; a is the internal characteristic length, l is a material length scale parameter which describes the higher-order strain gradient field.

The constitutive equations in NSGT herein can be described as:

$$\sigma = \int_V \alpha_0(x', \mathrm{x}, e_0 a) C_{ijkl} \varepsilon'_{kl,m} dV' \tag{2}$$

$$\sigma^{(1)} = l_0^2 \int_V \alpha_1(x', \mathrm{x}, e_1 a) C_{ijkl} \nabla \varepsilon'_{kl,m} dV' \tag{3}$$

where ∇ represents the Laplacian operator. The total stress tensor predicted by the NSGT can be then expressed as:

$$\mathrm{t} = \sigma - \nabla \sigma^{(1)} \tag{4}$$

Based on the assumption that the same nonlocal attenuation functions and parameters are taken, namely, $e_1 a = e_0 a$, then a general constitutive relation yields:

$$\left[1 - (e_0 a)^2 \nabla^2\right] t_{ij} = \left(1 - l^2 \nabla^2\right) C_{ijkl} \varepsilon_{kl} \tag{5}$$

Integrating by parts, then collecting the coefficients of $\delta u_0, \delta v_0, \delta w_0, \delta\theta_x$, and $\delta\theta_y$ to zero, the governing equations of the size-dependent PSC can be obtained. Then, corresponding to the investigated boundary conditions, the generalized displacements $\left[u_0, v_0, w_0, \theta_x, \theta_y\right]$ can be expanded in double trigonometric series. With the aid of the Galerkin method, the governing equations can be derived in the following form:

$$\mathbf{K} - \omega^2 \mathbf{M} = 0 \tag{6}$$

By solving the above eigenvalue problem, the natural frequency of the nanostructures ω_{nl} can be resolved from the smallest eigenvalue.

3 Numerical Results

In NSGT the size-dependency of the nanostructure is characterized as nonlocal parameters $e_0 a$ and material length scale parameters l_0. It should be noted that the current NSGT can degenerate into other lower-order continuum models by adjusting the related coefficients. Specifically, a pure NET model, $l_0 = 0$ is taken to eliminate the strain gradient term. In terms of the system not including stress nonlocality, $e_0 a = 0$. For the classical continuum model, herein the first-order shear deformation plate theory (FSDT), without considering the nonlocality and higher-order strain gradient, $e_0 a = l_0 = 0$.

The PSC is fabricated with a layer stack sequence of FTO/TiO_2/Perovskite/Spiro-MeOTAD/Au. The total thickness, length, and width of a PSC are $h = 1587.5$nm and $L_a = L_b = 0.2$mm, respectively. The related material properties can be found in [5]. The dimensionless parameters are introduced as $\overline{e_0 a} = \frac{e_0 a}{L_a}, \overline{l_0} = \frac{l_0}{L_a}$.

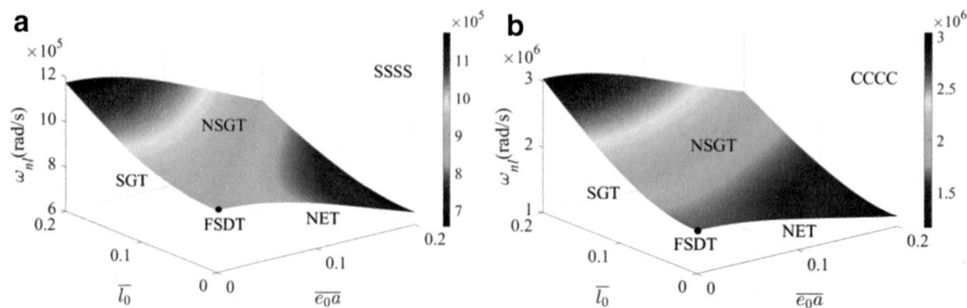

Fig. 1 Size effect on the natural frequency of **a** PSC with simply supported edges (SSSS); **b** PSC with clamped restraint (CCCC). NSGT, nonlocal strain gradient theory; PSC, perovskite solar cell

The natural frequency of the simply supported and clamped PSCs with various dimensionless nonlocal and material length parameters is shown in Fig. 1. It can be seen that, for both boundary conditions, the increase of the two size effect parameters results in distinct variations in the natural frequency of the PSC. With the attendance of the nonlocal parameter, the natural frequency of the PSC demonstrates a declined trend. However, the material length scale parameter tends to enhance ω_{nl} in the considered range. One possible explanation is that the nonlocal parameter addresses the effect of stiffness-softening, whereas the material length scale parameter addresses the effect of stiffness-hardening. In particular, the effects induced by the size-dependent coefficients become more prominent when their values approach the geometric size of the structure. Moreover, under two boundary restraints, as predicted, the clamped PSC possessed greater natural frequency than the supported plate due to the additional constraints.

4 Conclusions

In the present study, the size-dependency on the dynamic characteristics behavior of the PSC wasinvestigated based on NSGT. By introducing the nonlocal parameter and the material length scale parameter, the size-tendency of the nanostructure was captured in a complicated mix of stiffness-softening and stiffness-hardening. It was revealed that both scale parameters play crucial roles in the free vibration of the PSC. Moreover, it should be noted that the size-dependent model showed its characteristics as the scale parameters approached the dimensions of the structure.

Acknowledgements This research was supported by an Australian Government Research Training Program Scholarship and Australian Research Council projects IH150100006, IH200100010, and DP210101353.

References

1. Nayak PK, Mahesh S, Snaith HJ, Cahen D (2019) Photovoltaic solar cell technologies: analysing the state of the art. Nat Rev Mater 4:269–285. https://doi.org/10.1038/s41578-019-0097-0
2. Perovskite Solar Cells|Department of Energy n.d. https://www.energy.gov/eere/solar/perovskite-solar-cells. Accessed October 26, 2020
3. Nair S, Patel SB, Gohel JV (2020) Recent trends in efficiency-stability improvement in perovskite solar cells. Mater Today Energy 17:100449. https://doi.org/10.1016/j.mtener.2020.100449
4. Lim CW, Zhang G, Reddy JN (2015) A higher-order nonlocal elasticity and strain gradient theory and its applications in wave propagation. J Mech Phys Solids 78:298–313. https://doi.org/10.1016/j.jmps.2015.02.001
5. Li Q, Tian Y, Wu D, Gao W, Yu Y, Chen X et al (2021) The nonlinear dynamic buckling behaviour of imperfect solar cells subjected to impact load. Thin-Walled Struct 169:108317. https://doi.org/10.1016/J.TWS.2021.108317

Non-probabilistic Informed Structural Health Assessment with Virtual Modelling Technique

Q. Wang, Y. Feng, D. Wu, and W. Gao

Abstract In real-life engineering, non-probabilistic structural information is very common in many and varied disciplines. This class of information is characterized by incompleteness and imprecision, such as interval, fuzzy sets, etc. Non-probabilistic structural information can be reflected in the structural performance and cause it to fluctuate within a specific range, instead of being deterministic. Thus, without appropriate consideration of non-probabilistic information, serious or even disastrous accidents may occur. Therefore, fully estimating the structural health status using non-probabilistic information, especially detecting the lower and upper bounds of the concerned structural response, is extremely significant in uncertainty-sensitive fields. To conquer this challenge, a virtual modeling technique underpinning a structural health assessment framework is introduced. The twin extended support vector regression (T-X-SVR) approach is embedded for virtual model construction. Continuous, differentiable expression of the established virtual model allows the optimal solutions for each interval analysis to be easily achieved. Information update is another inherent feature, which enables structural health assessment to be implemented with updated conditions without rebuilding the virtual model. To demonstrate the applicability of the proposed virtual modeling technique underpinned structural health assessment framework, the non-probabilistic informed elastoplastic nonlocal damage analysis was investigated for engineering structures.

Q. Wang (✉) · Y. Feng · W. Gao
Centre for Infrastructure Engineering and Safety, School of Civil and Environmental Engineering, The University of New South Wales, Sydney, NSW, Australia
e-mail: qihan.wang@unsw.edu.au

D. Wu
School of Civil and Environmental Engineering, University of Technology Sydney, Sydney, NSW, Australia

W. Duan et al. (eds.), *Nanotechnology in Construction for Circular Economy*, Lecture Notes in Civil Engineering 356,
https://doi.org/10.1007/978-981-99-3330-3_38

1 Introduction

In the practical engineering field, structural health assessments attract more attention to meet the current higher requirement for smart engineering. Mass information is obtained via various devices (e.g., actuators, monitors, sensors, etc.), but the problem is transferring from acquiring the data to information extraction or mining of the data.

Engineering structures may contain, experience or confront multifarious information from different sources. Non-probabilistic information is one method to represent information with the features of incompleteness and imprecision, such as interval, fuzzy sets, etc. [1, 2]. Moreover, from industry practice, accumulated evidence has repeatedly revealed that stochastic- or probabilistic-based information quantification can have the dilemma that the probability distribution characteristics are challenging to be credibly determined, because of insufficient amount or poor quality of the experimental data. Thus, non-probabilistic structural information has widespread applicability in real-life industries [3].

By considering the non-probabilistic characteristics within the information of system inputs, the structural response correspondingly presents non-probabilistic features (e.g., interval, fuzzy, or imprecise). Without loss of generality, non-probabilistic information is herein considered as a fuzzy parameter. A structural health assessment involving fuzzy information was conducted by seeking the fuzzy-valued bounds or membership functions of the structural response. Through a level-cut strategy, the fuzzy problem was transformed into a series of optimization algorithms on the interval realizations.

However, in practical engineering applications, the relationship between system information and the quantity of interests is normally underpinned, sophisticated, and implicit. Directly implementing the optimization algorithms on this constitutive relationship is extremely challenging. As a single deterministic calculation could already be very time consuming, non-probabilistic uncertainty quantification with large simulations to search for the extremes would become computationally infeasible. Thus, an alternative strategy to tackle the non-probabilistic informed structural health assessment is proposed based on a supervised machine learning technique, namely twin extended support vector regression (T-X-SVR) [4]. Supreme mathematical features of T-X-SVR allow the feasibility of optimal solutions on given intervals being effectively and efficiently obtained in the established virtual model. Furthermore, the virtual model-aided health assessment has an inherent advantage of information updates without the need to reconstruct the model.

2 Methods

2.1 Non-probabilistic Information

The non-probabilistic information herein is considered as the fuzzy variable. A fuzzy variable ξ^{F} is a gradual weighting of a vector space Υ with the membership functions $\{\mu_{\xi^{\mathrm{F}}}(x) : \Upsilon \to [0,\ 1],\ \forall x \in \Re\}$. This membership function can be written as a set of ordered pairs,

$$\{(x,\ \mu(x))|x \in \Upsilon \wedge \mu(x) \in [0,\ 1]\} \tag{1}$$

The set of all fuzzy sets Υ is denoted by $\Gamma(\Upsilon)$. For numerical implementations, it is necessary to introduce $\alpha-$ levels. For a fuzzy variable ξ^{F} and $\alpha \in [0,\ 1]$, $\alpha-$ level cut can be written as,

$$\xi_{\alpha}^{\mathrm{F}} := \{x \in \Upsilon : \mu_{\xi_{\alpha}^{\mathrm{F}}} \geq \alpha\} \tag{2}$$

It is significant to note that for each given degree of truth (or membership level) α, the problem converts to an interval form. Each interval problem is conducted by seeking the lower bound (LB) and upper bound (UB) of the concerned structural response, which can be formulated as follows,

$$\begin{aligned} & \underset{\xi_{\alpha}^{\mathrm{F}}}{optimize} : X(\xi_{\alpha}^{\mathrm{F}}) \\ & s.t.,\ \xi_{\alpha}^{\mathrm{F}} \in [\underline{\xi}_{\alpha}^{\mathrm{F}},\ \overline{\xi}_{\alpha}^{\mathrm{F}}] \end{aligned} \tag{3}$$

2.2 Virtual Model Construction

To provide a more robust effective and efficient manner to tackle this engineering-simulated problem, a supervised machine learning technique, namely T-X-SVR [4] was used for the virtual model construction. The established virtual model alternatively depicts the implicit constitutive relationship between the system inputs and the concerned structural responses. T-X-SVR aims to minimize the gap between two bounds of the virtual model and the datasets from two different directions. The optimal solution for weights and bias can be effectively obtained by solving two quadratic programming problems (QPPs).

2.3 Virtual Model Aided Non-probabilistic Uncertainty Quantification

After the virtual model construction, a series of optimization programming problems for fuzzy informed engineering problems can be conducted on the virtual model instead of the original constitutive relationship. There are two main benefits in utilizing the virtual model as an alternative: (1) A mathematical function for each model possesses cheap computational costs; (2) On any specific interval of the inputs, the optimal solutions for the estimation of extremes can be effectively obtained through the derivation methods.

2.4 Numerical Investigation

A jet engine was investigated by considering fuzzy information within the material properties [5]. According to that convergence study, the virtual model was constructed by learning from 160 training samples. 1e3 Monte Carlo simulation (MCS) results were considered as the benchmark. The established virtual model had relatively high accuracy, with R-square nearly 1, and root mean squared error (*RMSE*) about 8e − 4. The Poisson's ratio ν_C and density ρ_C of the ceramic are considered as fuzzy parameters. ν_C was assumed to follow the triangular membership function, with the support of [0.33, 0.37], and top of 0.35; ρ_C followed the trapezoid membership function, with the support of [3.168, 3.232] g/cm^3, and top of [3.198, 3.202] g/cm^3. The corresponding fuzzy-valued LB and UB of the concerned structural response (i.e., critical load P (kN) of the damage analysis) was estimated through the proposed strategy, as shown in Fig. 1c.

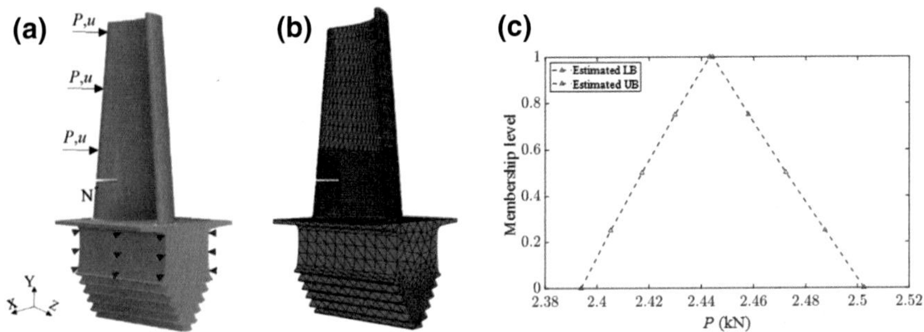

Fig. 1 **a** Numerical model, **b** adopted FEM mesh of notched blade, and **c** estimated fuzzy-valued lower bound and upper bound of the concerned structural response

3 Discussion

This extended abstract introduced a virtual model-aided non-probabilistic structural health assessment framework for engineering structures. The non-probabilistic information was considered as a fuzzy parameter. Through a level-cut strategy, the problem was formulated as a series of non-linear programming problems in the virtual model. For the virtual model construction, a supervised machine learning technique, namely T-X-SVR, has been adopted. The implicit, computational expansive constitutive relationship can be alternatively depicted as a mathematical equation, with continuous and differentiable features. Thus, the optimal solution for fuzzy-valued bounds can be effectively obtained in the virtual model. Moreover, information update can be fulfilled by importing updated information into the virtual model, instead of reconstructing the virtual model. The introduced virtual model-aided non-probabilistic informed structural health assessment has great potential applicability in real-life engineering.

Acknowledgements The work presented in this paper was supported by the Australian Research Council projects IH150100006 and IH200100010.

References

1. Wang Q, Wu D, Tin-Loin F, Gao W (2019) Machine learning aided stochastic structural free vibration analysis for functionally graded bar-type structures. Thin-Walled Struct 144:106315.1–106315.19
2. Feng Y, Wang Q, Wu D, Gao W, Tin-Loi F (2020) Stochastic nonlocal damage analysis by a machine learning approach. Comput Methods Appl Mech Eng 372:113371
3. Beer M, Ferson S, Kreinovich V (2013) Imprecise probabilities in engineering analyses. Mech Syst Signal Process 37(s 1–2):4–29
4. Wang Q, Wu D, Li G, Gao W (2021) A virtual model architecture for engineering structures with twin extended support vector regression (T-X-SVR) method. Comput Methods Appl Mech Eng 386(1):114121
5. Wang Q, Feng Y, Wu D, Yang C, Gao W (2022) Polyphase uncertainty analysis through virtual modelling technique. Mech Syst Signal Process 162(1–2):108013

BY

Modeling the Alkali–Silica Reaction and Its Impact on the Load-Carrying Capacity of Reinforced Concrete Beams

T. N. Nguyen, J. Li, V. Sirivivatnanon, and L. Sanchez

Abstract The alkali–silica reaction (ASR) is one of the most harmful distress mechanisms affecting concrete infrastructure worldwide. The reaction leads to cracking, loss of material integrity, and consequently compromises the serviceability and capacity of the affected structures. In this study, a modeling approach was proposed to simulate ASR-induced expansion considering three-dimensional stress/restraint conditions, and its impact on the structural capacity of reinforced concrete members. Both the losses in concrete mechanical properties and prestressing effects induced by the expansion under restraints are taken into account in the model. Validation of the developed model is conducted using reliable experimental datasets derived from different laboratory testings and field exposed sites. With the capability of modelling both ASR-induced expansion and its impact on structural capacity, the model provides valuable results to specify effective repair and/or mitigation strategies for concrete structures affected by ASR.

Keywords Alkali-silica reaction · Concrete deterioration · Expansion · Finite element

1 Introduction

Many concrete bridges and dam structures in Australia have been reported to be affected by various degrees of deleterious alkali–silica reaction (ASR) [1]. These affected structures require comprehensive diagnosis and prognosis protocols for assessing the current degree of damage, forecasting the potential of further deterioration, and evaluating the impact of ASR on structural capacity. Such information

T. N. Nguyen (✉) · J. Li · V. Sirivivatnanon
School of Civil and Environmental Engineering, University of Technology Sydney, Sydney, Australia
e-mail: thuc.nguyen@uts.edu.au

L. Sanchez
Department of Civil Engineering, University of Ottawa, Ottawa, Canada

W. Duan et al. (eds.), *Nanotechnology in Construction for Circular Economy*, Lecture Notes in Civil Engineering 356,
https://doi.org/10.1007/978-981-99-3330-3_39

is essential to specify efficient method(s) for remedial/rehabilitation and management procedures for the structures.

In terms of the structural implication of ASR, it is interestingly observed from several experimental studies that the load-carrying capacity of ASR-affected structures is not compromised, especially for shear capacity. An exception is evident in Swamy and AlL-Asali [2], where the affected reinforced concrete beams lost up to nearly 25% of their flexural strength at an expansion level of 0.518% measured on the beams. The studies did, however, agree that the impact on load-carrying capacity would become significant if the specimens were subjected to long periods of exposure and underwent high expansion levels. As such, besides the adverse impacts on the material performance of concrete, such as cracking and degradation of mechanical properties, there is a certain favorable effect of ASR on the capacity of structures at low expansion levels, such as a prestressing effect of restrained ASR expansion (e.g., from the reinforcement [3]). The two most important questions herein are: (1) to what extent does the prestressing effect contribute to the capacity of affected structures; and (2) is this favorable outcome maintained or decayed as the expansion increases? Therefore, investigations of the prestressing effect in relation to the expansion advancement and degradation in mechanical properties are deemed necessary.

In this study, an engineering-based finite element (FE) approach was developed to model the ASR in reinforced concrete structures, considering the impact of reinforcement restraints on anisotropic ASR expansion and the loss of concrete mechanical properties. A case study of modeling the expansion and load-carrying capacity of ASR-affected reinforced concrete beams is provided.

2 Methods

2.1 Modelling Approach

The modeling approach to assessing the impact of ASR on structural capacity is presented in Fig. 1. The approach has three main steps: (1) estimating ASR-induced expansion under non-restrained/ non-confined conditions, (2) modeling the effect of restraints/confinements in 3-dimensional expansion, and (3) assessing the impact of ASR on the load-carrying capacity.

In the first step, free ASE-induced expansion was forecasted using the thermodynamically based semi-empirical model proposed by Larive [4], then further developed by Nguyen [5], which is capable of considering the effect of aggregate reactivity, time-dependent temperature and relative humidity, concrete alkali content and alkali leaching for forecasting expansion of field concrete. More details on the semi-empirical model can be found in Nguyen [5].

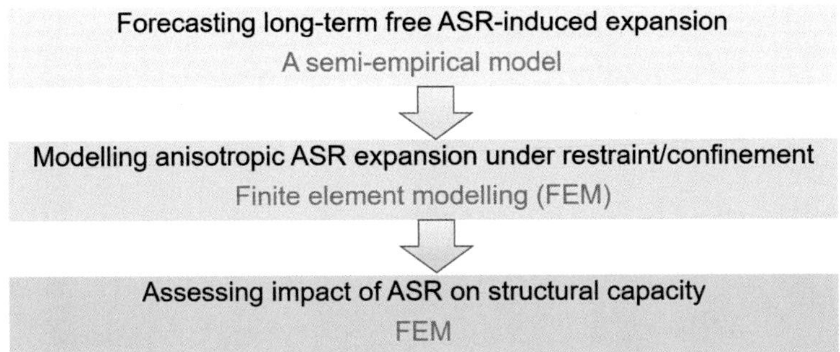

Fig. 1 Modelling of the expansion and load-carrying capacity of ASR-affected concrete members

In the second step, ASR-induced expansion in concrete under restraints/confinements, which is significantly different and far more complicated in comparison with free expansion of unrestrained concrete, was simulated in a FE model, which required a constitutive model to consider the restraints/confinements effect based on the multiaxial stresses developed in concrete elements. A general form of the incremental ASR strain tensor to be implemented in the FE analysis could be expressed as:

$$\dot{\boldsymbol{\varepsilon}}^{ASR} = \boldsymbol{E}\boldsymbol{W}\boldsymbol{E}^{T} f(\boldsymbol{\sigma})\dot{\varepsilon}_{V}^{ASR,free} \tag{1}$$

where $\dot{\varepsilon}_{V}^{ASR,free}$ is the free volumetric expansion of concrete which was calculated per the semi-empirical model presented previously, $f(\boldsymbol{\sigma})$ is expansion-stress dependent function accounting for the impact of stress state on ASR expansion, $\boldsymbol{E}$ is the eigenvectors derived from the stress tensor, and W is the weight tensor that distributes the volumetric expansion to each of three principal directions, given by:

$$\boldsymbol{W} = \begin{bmatrix} W_1 & 0 & 0 \\ 0 & W_2 & 0 \\ 0 & 0 & W_3 \end{bmatrix} \tag{2}$$

Determining the weight tensor was based on the empirical model from Gautam et al. [6], which was derived from multiaxial testing schemes of ASR-affected concrete. The weights calculated above were equivalent to the weights in three principal directions, as such the incremental ASR strain tensor is as same as in Eq. (1) to capture the ASR anisotropic behavior.

In addition, ASR causes loss of mechanical properties over time (i.e., modulus of elasticity, compressive strength and tensile strength) to various degrees. Expansion-dependent mechanical properties were implemented in the model to consider the impact of the material degradation. With all these considerations, the developed FE model was capable of assessing the impact of ASR on the structural capacity in the third step of the approach.

3 Case Study: Results and Discussion

In this section, a case study is presented for modelling of ASR-induced expansion and consequently the ASR impact on the load-carrying capacity of reinforced concrete beams tested by Fan and Hanson [7]. An overview of the test is presented followed by a modelling briefing and some selected outcomes of the model.

3.1 *Test Overview*

Fan and Hanson [7] conducted a series of tests on reinforced concrete beams (150 × 250 × 1500 mm) for ASR expansion and capacity. Two reinforced concrete beams were prepared, namely, 5R1 and 5N1 (or reactive beam and non-reactive beam, respectively), which used concrete mixtures containing reactive and non-reactive aggregates, respectively, with the same mixture proportions. They were immersed in an alkali solution at 38 °C with periodic expansion measurements for 1 year. The expansion was measured from Demec studs mounted in the beams' surfaces using a Demec dial gauge at different locations.

After 1-year immersion in an alkali solution, the beams were tested for their load-carrying capacity as shown in Fig. 2a. The load–deflection behaviors of the two beams were almost identical despite a certain reduction in mechanical properties of the concrete of 5R1 due to ASR. The behavior of the non-reactive beam can be referred to as the undamaged concrete beam in comparison with the damaged reactive beam.

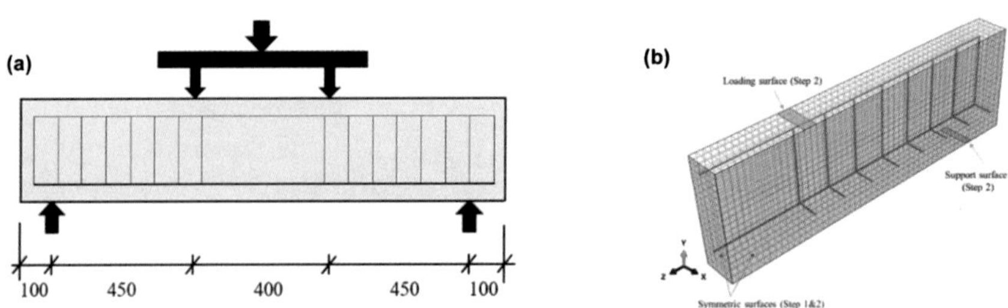

Fig. 2 Geometric and boundary conditions of the reinforced concrete tested by Fan and Hanson [7]

3.2 Modelling for ASR-Induced Expansion and Load-Carrying Capacity

Due to the symmetry of prism geometry and boundary conditions, only one-quarter of the beam was simulated utilizing symmetric boundary conditions as shown in Fig. 2b. The stress–strain behavior of the concrete defined at every 0.025% expansion level (i.e., 0%, 0.025%, 0.05%, 0.075%, 0.1%, etc.) to represent the change in the concrete's mechanical properties as expansion increased.

Distribution of average expansion (FV1) throughout the beam is shown in Fig. 3a. It shows a lower expansion in the area with both transverse and main longitudinal reinforcement at the bottom, and higher expansion on the top and at the beam-end with less reinforcement. Expansion in different locations and directions is plotted in Fig. 3b alongside the measurements. With a higher ratio of reinforcement in the longitudinal direction at the bottom, the expansion obtained at the bar level was significantly lower than at other locations. Similar to experimental observations, at the bar level, the expansion leveled off after 240 days of immersion, but kept increasing in the longitudinal direction on the top.

Load–deflection results for the 5N1 beam are shown in Fig. 4a, indicating a good agreement between the numerical and experimental results of load–deflection behavior. Figure 4b shows the predicted load–deflection curve of the reactive beam using the mean values of residual mechanical properties. First, the numerical results are comparable to the experimental in terms of capacity. The predicted ultimate loading value of the beam is $\approx$175.0 kN, and the value from test results was $\approx$177.3 kN. Similar to the test data, the numerical results showed an insignificant reduction in the capacity of the affected beam despite the reduction in mechanical properties as presented above. Second, the bending stiffness of the beam was slightly higher than the measured result despite the reduction in concrete stiffness. The observation aligned with observations from ISE [3], in which a favorable prestressing effect of restrained ASR expansion helped to increase the stiffness and capacity of several affected structures at low expansion levels.

4 Conclusions

This paper presents a modeling approach for the ASR expansion and capacity of reinforced concrete members. The approach consists of both the semi-empirical model and numerical model (i.e., FEM). The FE model could transfer the expansion modeling results such as strains, stress state, residual mechanical properties of concrete to modeling for load-carrying capacity of the affected concrete structural members. Outcomes from utilizing the proposed approach for simulation of reinforced concrete beams tested in Fan and Hanson [7] show good agreement between modeling and testing results, which indicates the capability of the model for forecasting long-term ASR-induced expansion and its impacts on structural capacity

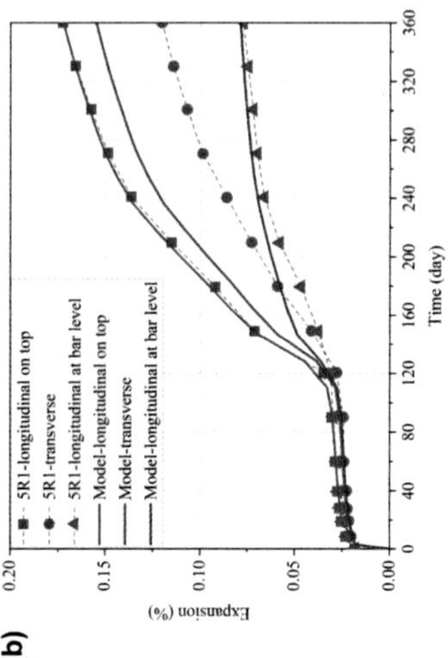

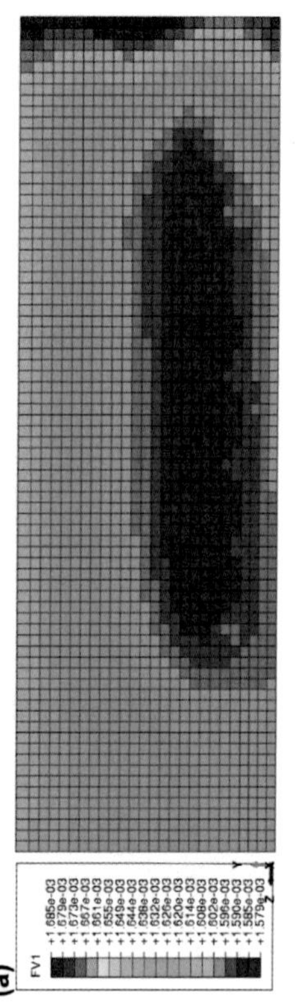

Fig. 3 Numerical and experimental ASR expansion at different locations for the reactive beam

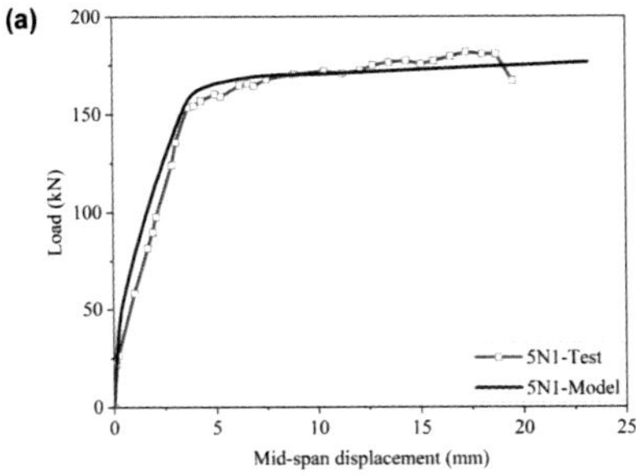

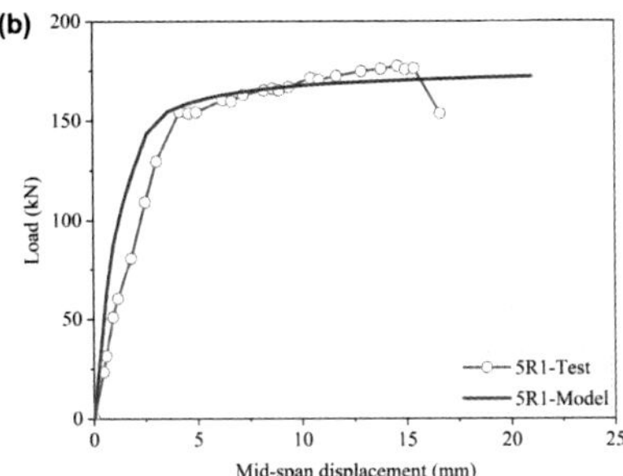

Fig. 4 Load–deflection behavior of **a** the non-reactive beam 5N1 and **b** reactive beam 5R1

of reinforced concrete structures in the field. In addition, the case study shows an insignificant impact of ASR on the load-carrying capacity at the expansion level of lower than 0.2%.

References

1. HB79 S (2015) Alkali aggregate reaction: guidelines on minimising the risk of damage to concrete structures in Australia, Cement and Concrete Association of Australia, and Standards Australia, North Sydney, NSW
2. Swamy RN, AlL-Asali M (1989) Effect of alkali-silica reaction on the structural behavior of reinforced concrete beams. Struct J 86(4):451–459
3. ISE (1992) Structural effects of alkali-aggregate reaction: technical guidance on the appraisal of existing structures. UK
4. Larive C (1997) Apports combinés de l'expérimentation et de la modélisation à la compréhension de l'alcali-réaction et de ses effets mécaniques. École Nationale des Ponts et Chaussèes, Paris
5. Nguyen TN (2021) Modelling alkali-silica reaction effects for condition assessment and capacity evaluation of reinforced concrete structures (Doctoral dissertation)
6. Gautam BP, Panesar DK, Sheikh SA, Vecchio FJ (2017) Multiaxial expansion-stress relationship for alkali silica reaction-affected concrete. ACI Mater J 114(1)
7. Fan S, Hanson JM (1998) Effect of alkali silica reaction expansion and cracking on structural behavior of reinforced concrete beams. ACI Struct J 95:498–505

Guidelines for Enzymatic Soil Stabilization

B. O'Donnell, A. Swarup, A. Sidiq, D. Robert, and S. Setunge

Abstract There are numerous manuals to guide practitioners in utilizing traditional additives in the construction of road, rail and dam construction but they fall short of specific guidance for non-standard additive-based ecofriendly and cost-effective soil stabilization. Increased attention has recently been on the use of non-standard additives for stabilizing weak soils due to environmental and cost concerns associated with traditional additives. We summarize the specific guidelines of using environmental-friendly enzymes to treat weak soils. We elaborate on the requirements and specifications for the Eko-Soil multi-enzyme product that is manufactured from water and proteins extracted from fermented exudes of plants. Specific tests (laboratory and field) and conditions required for soil stabilization using Eko-Soil enzyme are elaborated using the experience of past construction projects. The guide also elaborates enhancing the efficiency of enzymatic soil stabilization by correctly incorporating the required mixing proportions and pre-requisite condition tests. Professionals and practitioners will benefit from using novel eco-friendly sustainable stabilization techniques in the treatment of weak soils covering many applications including roads, foundations, water containment areas, landfills, working platforms and slope erosion control.

Keywords Environmental-friendly solutions · Enzymes · Guidelines · Practitioners · Soil stabilization · Weak soils

---

B. O'Donnell (✉) · A. Swarup
Centre for Pavement Excellence, Asia Pacific Ltd (CPEAP Ltd), Langwarrin, VIC, Australia
e-mail: baod1942@bigpond.com

B. O'Donnell · A. Sidiq · D. Robert · S. Setunge
School of Engineering, Royal Melbourne Institute of Technology (RMIT University), Melbourne, VIC, Australia

W. Duan et al. (eds.), *Nanotechnology in Construction for Circular Economy*, Lecture Notes in Civil Engineering 356, https://doi.org/10.1007/978-981-99-3330-3_40

1 Introduction

Worldwide, population growth at an unprecedented rate is leading to increased demand for new civil infrastructure. The American Society of Civil Engineering has estimated that private–public partnerships on roads alone have now increased to US$67 billion per year in overall investments [1]. There is also insufficient infrastructure in major countries such as India and China due to limits on expansion imposed by geographic boundaries and soil conditions [2]. In Australia, civil infrastructure is increasing due to fewer land limitations than in other major countries, but there are deficiencies in the soil conditions, leading to cracking in $\approx$50,000 residential dwellings per year [3]. Due to these serviceability concerns, there is a high demand for improving soil conditions through different techniques such as traditional method of mechanical energy (i.e., hammering, vibration or rolling) or by the application of additive-based soil stabilization.

Traditionally, lime has been used in soil stabilization and typically includes hydrated high-calcium lime, dolomite lime, monohydrated dolomite lime and calcite quicklime [4–8]. The physical mechanism of soil stabilization with the addition of lime-based products involves fluctuation of clayey particles from plate-like to rod-like, leading to a greater interlocking between particles. Hence a modified soil matrix, and effectively, as the soil turns drier, it becomes less able to absorb water [9, 10]. Also, the physiochemical progress of the additional lime content in the soil matrix potentially involves the hydration process; thus, when quicklime is added to the soil, a chemical reaction takes place between the quicklime and the available water content in the soil matrix. Subsequently, the hydrated lime reacts with alumina and silica, leading to the formation of C–S–H (calcium–silica–hydrate) and C–A–H (calcium–aluminate–hydrate), which are like Portland cement products. Hence, lime-stabilized soil results in relatively impermeable layers of soil and significant load bearing [11, 12]. Although the usage of lime-based products has positive efficacy in soil stabilization, there can also be drawbacks due to the carbonation process and sulfate attacks, which can have negative economic impacts. The addition of lime product to the soil leads to a change in the chemistry of pore water, thus increasing the pH level and affecting the charge of the clayey particles. Also, the rate of carbonation is potentially higher for soil exposed to the atmospheric environment, which places limitations on the chemical reaction and pozzolanic activities and therefore becomes less effective for the purpose of stabilization [13–15]. The addition of lime-based products leads the soil to become more susceptible to sulfate attack which also limits the pozzolanic activities, leading to less effective soil stabilization [16, 17].

Cement products are also used in soil stabilization applications and this process is well known as cement stabilization in the geotechnical engineering field. By mixing cement with the soil material, the cement particles chemically react with the available water content in the soil [18], leading to the formation of crystal particles that bond to each other and with the soil particles, achieving higher compressive strength and stability [19, 20]. Thus, using cement for soil stabilization has been considered effective, but the drawback is the excessive cost and environmental impact during the cement production process.

Fly ash (FA) is also considered as one of the most useful and sustainable additives for soil stabilization applications due to its unique characteristic of acting like a cementitious material in the soil matrix. Although FA is solely incapable of densifying the soil material, it can react chemically to achieve the cementitious compound in combination with only a small amount of activator to improve the strength of the soil. However, the implementation of FA for soil stabilization applications is limited by the water content in the soil material. Thus, in order to achieve the optimum benefit from the addition of FA, the water content in the soil matrix must be at the minimum, and then dewatering is required to maintain the optimum moisture content of the soil material [21]. On the other hand, research has shown that the sulfur content in FA potentially forms expansive soil and, in the long term, leads to reduced strength and durability of the soil [11].

In addition to the common stabilizing approaches outlined above, biotechnological products, such as enzyme-based products, are currently being used as innovative products for improving weak soils. Researchers recently reported on the Eko-Soil enzyme (Eko Enviro Services) for soil stabilization and showed sustainable benefits for the stabilization of expansive subgrades [22]. Pooni et al. [23] used the identical enzyme product from the optimum enzyme content obtained by Rintu et al. [24] and evaluated the hydraulic influence and sustainable benefits of the application to an expansive clay material. The findings revealed that the addition of enzyme resulted in an increase in California Bearing Ratio (CBR) by 58% under soaked conditions due to densification effects. They further showed that the volumetric size of the micropores in enzyme-stabilized samples was drastically decreased, in comparison with raw soil, due to the improved density obtained from enzymatic stabilization [23]. As a result of the high stabilization performance and low costs, enzymatic stabilization using Eko-Soil has been adopted in field applications for unpaved road construction [25].

Here we provide guidelines that demonstrate the methodologies of conducting subgrade soil stabilization for road pavements, typically using the enzyme-based soil stabilization technique. This guideline can be used as a tool for enhancing soil conditions by providing stronger, impervious, and cost-effective subgrade soils under rigid or flexible pavements. In addition, an impervious capping layer over reactive soils can be productive by providing a safe all-weather pavement that is dust resistant and requires minimal maintenance over the long-term life cycle. To ensure these requirements are obtained, we performed the current study to identify the steps that need to be taken by geotechnical practitioners in terms of safety instructions, laboratory investigations and evaluation of site performance in compliance with specifications.

2 Pavements

The safety and lifespan of pavement structures significantly depend on the condition of the subgrade soil, because of the significant economic impact of frequent repairing [26] if mechanical properties are weakened. Some naturally occurring soils are suitable for compaction, forming a homogeneous material that is capable of supporting commercial and residential infrastructure. For instance, weathered rock extracted from the ground is generally a suitable fill material for infrastructure, typically when used with cementitious products for stabilization applications [27]. However, there are other materials that are not suitable as fill materials because of factors such as the in-situ conditions, environmental conditions, and applied loadings during the construction timeframe. The following aspects are important to consider when selecting a fill material from the economic viewpoint.

- Clay material that has a high plasticity index (PI), and is therefore a reactive soil, must be considered under strict moisture and density control when selected as a fill material.
- After completion of compaction, there can be large particles within the soil that potentially limits trenching or drilling of piers excavation, footings, services, and driving of piles.
- Over-wet soil materials, which can be found mainly in low-lying areas, have the potential to dry out sufficiently within a shorter time during the project lifespan.
- Large-size individual graded rock fill material has limitations on the breaking mechanism during the compaction process, which leads to higher porosity within the soil that subsequently creates pathways for fine particles to migrate within the soil matrix.
- The salinity of the soil, in relation to aggressive chemical products or unwanted (polluted) soils.
- Soil carbonation leads to the occurrence of acidic products.
- The soil material has a PI between 6 and 20 with even graduation of particle size from 20 mm minus.

There are soil types that cannot be used as fill material, due to high contamination or undesirable performance of the soil material, and must be removed or transported and used elsewhere. Unsuitable fill materials may include:

- Organic soils that contain severely root-affected material and peat, where these soils are more likely to be the top soils.
- Soil materials contaminated through past site usage, which may contain toxic substances or soluble compounds harmful to water supply or agriculture.
- Soil materials containing substances that can be dissolved or leached out in the presence of moisture (e.g., gypsum), or are susceptible to volume change or loss of strength when disturbed and exposed to moisture (e.g., sandstone), unless these matters are specifically addressed in the design.

- Silts or related materials that have the deleterious engineering properties of silt.
- Soil materials with properties that are unsuitable for forming structural fill.
- Fill that contains wood, metal, plastic, boulders or other deleterious material.

Furthermore, for road pavements, it is also important that high attention be given to the shoulders abutting the pavement. Special treatments must be considered to ensure that microbial metabolism and activities of all the substrates are minimal under the permeable pavement, because they can lead to an increase in the total organic carbon content and incremental water content, which, potentially, lower the performance of the structure [28, 29].

Attention must be given to any erosion, which is the process by which the soil is worn away by water or wind and sediment is produced. Some soils are more susceptible to erosion than others, depending on their mechanical, chemical, and physical properties and the terrain [19, 30]. Effectively, erosion can be increased by several factors, including in high rainfall areas at the points where the overland flow is concentrated; where roadside activities such as vehicular traffic and maintenance practices increase the potential for erosion and sediment production or where any road construction interrupts the natural topography or drainage flows. When the runoff discharges turbid water into waterways, it can cause serious environmental harm, by reducing the sunlight penetrating the waterway [29], which can affect the growth of plant life and reduce the capacity of visual predators (e.g., fish and birds). Moreover, road dust can have a significant detrimental effect on the environment, affecting adjacent crops, waterways, buildings, vehicle amenity, aesthetics and human health by aggravating respiratory illness and road safety, through poor visibility and affecting driver behavior [31].

The designed pavement for a road mainly depends on the type of road. Rural road types are typically categorized as sealed, unsealed and stabilized. The sealed rural road has a flexible pavement that is designed, as per geotechnical recommendation, to include the addition of a top layer of bituminous concrete, asphalt or bituminous spray seal. For example, sealed roads are usually formed by excavating and preparing the existing ground (i.e., base) before placing crushed rock layers and a wearing course. For urban roads or rural roads of significance, underground drainage, footpaths, kerbing, and traffic management devices may also be considered [6, 31]. On the other hand, for unsealed rural roads, the top surface of the road has no bituminous layers but consists only of granular material (usually local gravel) or imported quarry product [32]. Within Australia, there are numerous safety, economic, social and environmental shortcomings regarding access for communities, and the extent of such shortcomings is largely dependent upon the characteristics of each individual road's construction and traffic volumes. Significant regular maintenance is required to ensure surface conditions do not change until the geometry and surface can be improved to a safe acceptable level by the construction of an all-weather sealed road [9, 33].

Soil stabilization is considered as an efficient method to ensure soil behavior is within the required shear strength, permeability and compressibility parameters. Various methods of soil stabilization have been implemented throughout the history of subgrade soil stabilization, including mechanical and chemical methods [11, 34]. Soil stabilization through mechanical methods involves changing the soil mixture by degrading and densifying the soil using compaction with heavy rollers, rammers, and vibrational equipment and may sometimes involve blasting techniques for superior stability. Mechanical stabilization methods can be costly due to the requirement for labor and specialized equipment, so soil stabilization using chemical additives is becoming more common, using academic and practical engineering applications to ensure soil stabilization and densification are obtained by mixing minerals or biological additives [35]. In addition, chemical additives have the potential to reduce the timeframe of the construction by using available construction equipment, which is beneficial from the economic aspect. Stabilized road pavements are constructed with one or more layers/courses mixed with an additive to bind the pavement material [36, 37]. The preferred option for a conventional pavement is using the in-situ subgrade soils and gravels as a subgrade. This is especially important for saving natural resources, particularly where deficiencies in the existing in-situ gravels or clays can be rectified by importing more suitable gravels or clays and mixing them with a stabilizer additive to construct an unsealed pavement that is an environmental friendly, cost-effective, impervious and strong road [38].

3 Construction Commencement

This section covers the specific working steps that need to be taken before commencing the earthworks and soil stabilization process.

3.1 Construction Site Fencing

In the initial stage of construction, it is highly important that fencing around the perimeter of the construction site is installed before any earthwork commences. Fencing installation is one of the safest methods for identifying the boundaries of the working zone because permission to enter the construction site is then only given to authorized users, thereby preventing the public from entering the site and disturbing the work performance while ensuring their protection.

3.2 Drainage, Erosion and Sedimentation Control

Before conducting the stabilization works, precautions must be taken to ensure the earthworks will not cause siltation or erosion of adjoining lands, streams or watercourses. Drainage, erosion and sedimentation controls should be installed before the natural surface is disturbed. Sedimentation basins, stream diversion or other works may be appropriate in some environments or topographies. Careful planning is required to ensure both erosion and sedimentation controls are effective by minimizing the area of disturbance and through progressive revegetation or redevelopment of the site. Wherever water may tend to accumulate, provision for temporary drainage should be made by the contractor and care should be taken to guard against scour during any part of the construction. All temporary provisions for drainage should be installed to the satisfaction of the superintendent before stabilization and/or pavement materials are placed. The cost of temporary drainage, unless otherwise directed to be retained for use as catch or shoulder drains, is the responsibility of the contractor. The location of each drainage line is determined on geotechnical advice and with the approval of the superintendent.

3.3 Site Clearing

The site must be cleared (to the minimum extent required for the work) of all trees, stumps and other materials unsuitable for incorporation in the works. The roots of all trees and debris, such as old foundations, and buried pipelines are removed to sufficient depth to prevent any inconvenience during subsequent excavation or foundation work. The resulting excavations should be backfilled and compacted to the same standard as required for subsequent filling operations. Disposal of cleared combustible material may have to be off-site if clean air or bushfire regulations prevent on-site burning.

3.4 Stripping

The area in which fill is to be placed and the area from which the cut is to be removed are stripped of all vegetation and of such soils that may be unsuitable for incorporation into fills, subject to density, moisture or other specified controls. Topsoil may need to be stripped either as unsuitable material or as required for subsequent revegetation. Extreme care needs to be considered to ensure that materials that will inhibit or prevent the satisfactory placement of subsequent fill layers are not allowed to remain in the foundations of the fills. Geotechnical assessment of the depth and quality of topsoil or vegetable cover of the underlying soils and of the quality and depth of the proposed fill may obviate the need for such stripping in some circumstances. All stripped materials should be deposited in temporary stockpiles or permanent dumps in locations available for subsequent re-use if required and where there is no possibility of the material being unintentionally covered by or incorporated in the earthworks.

3.5 Slope Preparation

Where a fill abuts sloping ground, benches should be cut progressively with each lift as appropriate. It is unlikely that slopes flatter than 8:1 (horizontal to vertical) gradient will require benching. The benches should be shaped to provide free drainage. The boundary of cut-and-fill areas requires special consideration. All topsoil and other compressible materials should be stripped prior to benching into the natural material of the cut zone. The depth of the cut can vary depending upon the natural slope of the ground, the nature and proposed end use of the fill and the equipment being used.

3.6 Foundation Preparation

The ground surface exposed after stripping should be shaped to assist drainage and be compacted to the same requirements as for the overlying layers of fill. The surface exposed upon completion of excavation works may also require preparation prior to the fill placement proceeding. This will typically be the case when the subsequent fill to be placed is for pavement construction or the base material of a project. In such circumstances, it is necessary to loosen the exposed excavation surface to a certain depth (depending on the soil conditions), then moisture-condition and compact this loosened material. The depth to which this loosening is carried out should not exceed that of the compacted soil layer above it. The degree of compaction achieved should be consistent with the required subsequent filling operations unless design advice has been obtained. In such cases a working platform generally of granular material, end-dumped and spread in sufficient depth to allow the passage of earthmoving equipment with minimal surface deflection, can provide a suitable foundation for subsequent filling. Localized springs or seepages in the foundation area, detected during site investigation for the work, should be noted and considered in the design. If such problems are not detected before the work progresses, it is critical they be assessed so that measures such as subsoil or rock rubble drains can be designed for incorporation in the works.

4 Laboratory Test Investigations

This section elaborates on the required laboratory tests for evaluating the soil condition and required stabilization measures prior to construction.

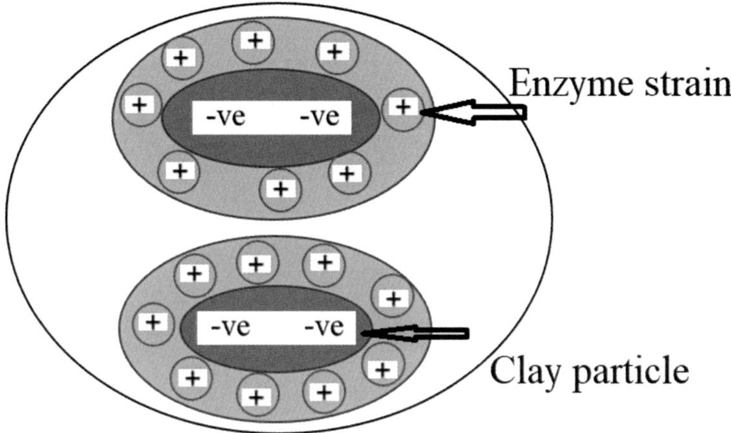

Fig. 1 Representation of the clay particle entrapped by the enzyme strains

4.1 Enzyme Stabilization

An enzyme is a biological macromolecule that catalyzes biological reactions, found in ribozymes (catalytically active RNA molecules) and some proteins (protein enzymes that have the ability to catalyze specific biochemical reactions), which are capable of initiating a biochemical reaction. The process involves changes (both formation and breakage) in chemical bonds. This method creates an interaction between the enzyme and bacterial strains within the soil product and is a replication of the natural construction of termite mounds [23]. Enzyme soil stabilization is based on reducing surface tension in soil particles through an ionic reaction, hence incrementing the soil compaction conditions. After the absorbed water is reduced through the compaction efforts, the soil particles agglomerate and as a result of the relative movement between particles, the surface area is reduced and less absorbed water can be held, which in turn reduces the swelling capacity. Figure 1 shows clay particles when entrapped with the enzyme strains.

4.2 Laboratory Results Illustrating the Enzymatic Soil Stabilization Mechanism

The behavior of the enzyme has been established through various studies. Laboratory results have indicated that the addition of enzyme causes water content reduction, as shown by different critical tests including FTIR (Fourier-transform infrared spectroscopy), SEM (scanning electron microscope) and microtomography (μ-CT). The addition of the enzyme to the soil has shown a marginal reduction in the intensity of the interlamellar water region due to the reduced affinity by the enzyme product

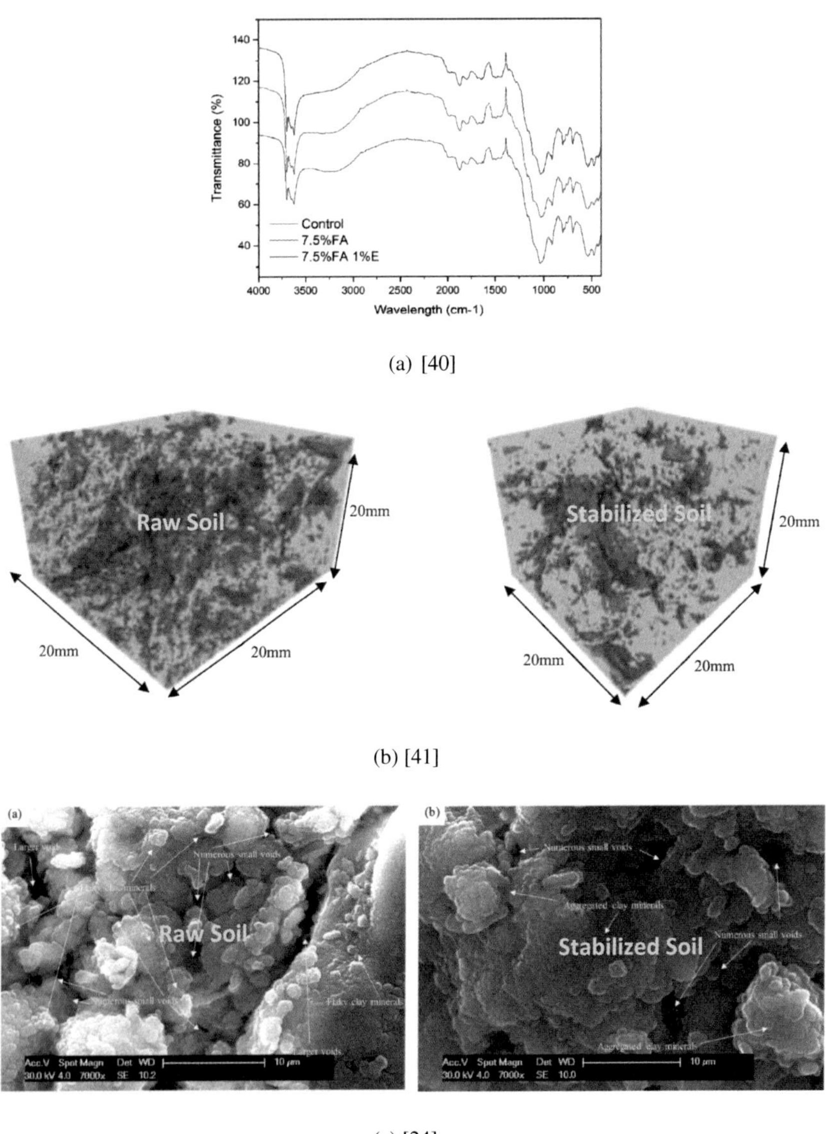

(a) [40]

(b) [41]

(c) [24]

Fig. 2 Laboratory results for the mechanism of enzyme-treated soil. **a** FTIR evaluating the enzyme base material in the soil and densification determination; **b** μ-CT analysis of the compactness of treated soil compared with control soil material; **c** SEM images demonstrating the lower void content and clay aggregation with the addition of enzyme product (Right) compared with control soil material (Left)

(Fig. 2) [23, 39]. Porosity analysis and clay microstructure via SEM images indicate that enzyme-stabilized soil samples potentially show reduced permeability and increased mechanical strength as ingress of water is restricted and density is enhanced through clay aggregation [3]. Pooni et al. [40] showed by μ-CT analysis that the addition of enzyme can reduce the porosity from 2.67 to 1.44% (i.e., 46.07% reduction in pore volume). Effectively, the enzyme mechanism is increasing the density with decreased affinity for water.

4.3 Testing Techniques

Subject to the scale of the project, difficult conditions may be expected, and it is not envisaged to relax the test frequencies specified herein; in some cases, more frequent testing may be required. These testing frequencies relate to acceptance on a 'not one to fail' basis.

In order to obtain optimized performance of the stabilized road, it is recommended to estimate the performance of the stabilized soil through a comprehensive test plan in the laboratory prior to the field application. This is mainly due to the performance of the stabilized soil (i.e., treated road pavement), which is governed by the in-situ soil type and its condition. Figure 3 shows the recommended laboratory tests that can be conducted in the application of enzymes to stabilize pavements. The proposed tests will facilitate determination of the suitability of in-situ soil in stabilizing the pavement, as well as obtaining the optimal amounts of enzymes that will result in

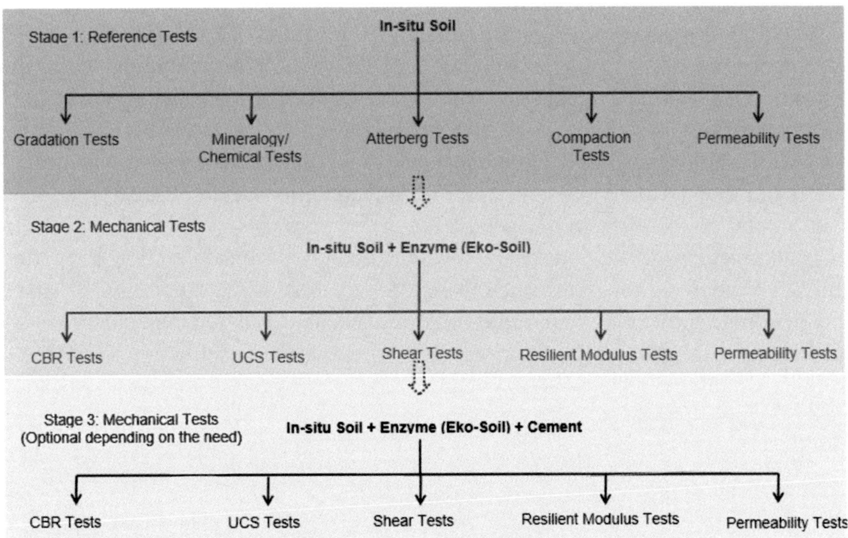

Fig. 3 Typical soil stabilization using different testing techniques at different stages

the expected road performance. It is to be noted that oven-drying is not allowed in any of the tests prescribed in this plan. The soil needs to be air-dried, or lime could be used as a drying agent prior to testing.

The proposed test methods can be conducted in three stages. In stage 1, initial tests are proposed to obtain the description and physical (and chemical) properties of the raw soils (possibly in-situ/natural soil). These tests include gradation, mineralogy, chemical, Atterberg, compaction and permeability tests. Having obtained the description and properties of the raw soils, samples are prepared at maximum dry density and optimum moisture content and left to cure for 72 h before mechanical testing (Stage 2 in Fig. 3). Should the natural soils be over-saturated, then add 3% lime mixture to the soil, and allowed to dry for 3 days, repeat the procedure until the natural soils are dried. This procedure can be used in the field to obtain the desired result. Furthermore, every mechanical test can be repeated at 4 and 7 days after curing to characterize the time-dependent strength gain of stabilized soils. Such mechanical tests will be helpful to assess the performance of the pavement material in accordance with relevant road standards. In addition to the tests noted in Stage 2, a permeability test of the stabilized samples is also recommended to determine whether the stabilization has improved the permeability of the mix. If the material is shown to be unsuitable, the strength of the stabilized mix could be improved with the addition of cement. The proposed series of mechanical tests on the stable mix with cement and enzymes will be able to ascertain the suitability of in-situ soil as a pavement material.

In order to proceed with testing using the additive/s for stabilization, initially samples should be taken from the field for each different soil type that is observed. In the case of a road, samples should be taken every 300 m, or less if there appears to be inconsistency within the sampled area. Moisture content within the sample bore is also a good measure for indicating a change in soil characteristics. In a remote area, geotechnical maps will assist in the location of samples if required. The initial tests are conducted to evaluate the mechanical strength of the soils. Figure 3 shows the types of tests that need to be conducted when enzyme is used as an additive for stabilization. Soil stabilization enhances an in-situ soil to support loads well in excess of those normally possible under natural conditions. Geotechnical analysis of soil samples collected from prospective stabilized sites must be carried out.

In all practical applications where moisture enters a pavement section, the permeability of the designed pavement mix must be tested in the laboratory to ensure specification requirements are either met or exceeded. It is recommended that the soil or soil with additives being considered for use at the construction site is tested to ensure that it conforms with the ideal specifications when using Eko-Soil. These specifications include the following.

1. Minimum 18% non-granular cohesive fines passing the 0.075 mm screen or a PI $\geq$ 6%. These fines will react best with Eko-Soil.
   Test method: ASTM D-11, D-422 [41], AS1289.3.6.3 or similar particle size analysis.
2. PI of 6–15 is the ideal soil condition when using Eko-Soil.

Test method: ASTM D-4318 [42] Atterberg Limits AS1289.3.9.1 [43].

3. pH 4.5–8.5 is the most satisfactory. Soils with low pH may be amended with crushed limestone. Alkaline soils may be treated with a cheap acid such as sodium acid sulfate. Eko-Soil is classified as a naturally produced weak acid with pH 4–5.

Rintu et al. [44] used Eko-Soil to evaluate the effectiveness of different contents of Eko-Soil enzyme on fine-grained field soil. The authors found that the optimum enzyme stabilization is achieved with a diluted additive of 1% to the weight of soil in a diluted mass ratio in water of 1:500. The performance at an optimum dosage showed an increase in CBR value for the tested soil by 500% in comparison with the control and a reduction in water demand and improved dry density.

5 Field Procedure to be Followed

To assess the quality of materials and workmanship provided on a project, regular inspections and testing are required in the field at suitable time intervals. However, site construction should not rely on test results alone; where good supervision is essential for inspection measures such as test rolling. Such inspections should be carried out by experienced and knowledgeable users in earthworks.

Having conducted laboratory tests as outlined in Sect. 4, careful measures should be put in place in the field to apply the stabilization. Enzyme stabilization will prove less than effective if the following minimum procedures are not followed.

- Ensure a soil analysis is undertaken on the subject soils and proposed additives before treatment begins.
- During initial preparation, it may be necessary to add further plain water (Fig. 4a) to bring the soil close to optimum moisture level for best compaction, or to aerate and allow it to dry. Proper engineering supervision is essential (Fig. 4b).
- Add the appropriate quantity of enzyme additive to the water truck after the truck is filled with water; as the quantity of enzyme depends on the in-situ soil type and its condition. The recommended tests proposed in Fig. 3 will facilitate the designer determining such optimized amounts of enzyme.
- After the soil/gravel and any other cementitious additives is mixed, the enzyme-treated water is dispensed by the tanker and the material thoroughly mixed using a purpose-built roto-mill (Fig. 4c). In most cases, adequate mixing can be achieved using a single pass.
- The mixed material should then be spread and shaped before compaction by rolling in maximum layers of 250 mm. For fill depths, refer to engineering plans.
- The rollers must make enough passes to ensure adequate compaction is taking place. A vibrating pad foot roller is most effective for this application. On the final passes, when the tire tracks no longer show, the vibrator should be turned off to prevent excessive surface cracking caused by rapid drying. The top surface should be rolled until it shows a uniform sealed appearance. The final stages of

Fig. 4 **a** For an optimum moisture content mix water lightly through the broken soil. **b** Ripping and shaping an existing road base. **c** Adding stabilizer using a mixing machine and water tanker in tandem. **d** Compaction by 16-tonne vibrating roller

rolling should be performed by a 16-tonne smooth drum roller (Fig. 4d), followed by a pneumatic (rubber) roller to assist in drainage and preventing ponding of water on the surface.

The ideal curing time for a 250-mm pavement depth would normally be 72 h. Light traffic may be permitted, as soon as tyre tracks are not visible from the surface. Light rain or high humidity will increase curing time. Application of enzyme is not to be undertaken during rain unless otherwise approved by the superintendent. Once field stabilization is completed, tests (Fig. 5) can be performed to ascertain field efficiency.

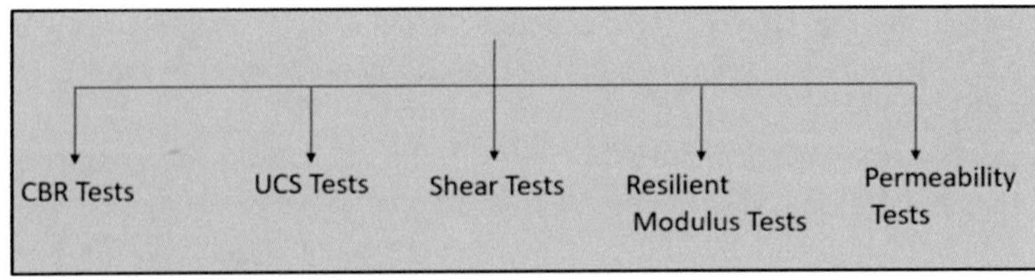

Fig. 5 Different testing techniques that can be conducted on the basis of field samples. CBR, California bearing ratio; UCS, unconfined compressive strength

5.1 Unconfined Compressive Strength

Unconfined compressive strength (UCS) is an effective test in ensuring pavement strength is within the range of 0.5 and 1.5 MPa or if the UCS value exceeds such a range the pavement is classified as a rigid type and therefore subjected to cracking. To assess the quality of the material through mechanical testing methods for areas of a minimum of 1500 m^2, such as subdivisions, large industrial lots, road embankments, etc., the following requirements must be considered in corresponding different testing locations within the field.

Not less than:

- 1 test per layer or 250 mm thickness per material type per 2500 m^2 or
- 1 test per 500 m^3 distributed reasonably evenly throughout full depth and area or
- 3 tests per visit, whichever requires the most tests.
- Confined operations: 1 test per 2 layers per 50 m^2.

To ensure that enzyme and cementitious modified materials (clays mixed with local gravels, cement, and enzyme) are achieving desired strengths in the order of 0.5–1.5 MPa (at 7 days), it is recommended that a minimum of two molded samples of these materials be re-compacted in the laboratory and allowed to cure for 7 days for the purpose of obtaining UCS values.

5.2 Deflection Test

Field deflection testing can be used to evaluate the strength of the soil. This test is highly recommended when working with large platforms or pavements (Table 1).

5.3 Plasticity Index

The PI needs to be 6–15 for a minimum of 18% non-granular cohesive fines passing the 75-micron sieve.

Table 1 Results of deflection testing of both treated and untreated pavement sections: Harvey Norman/Ikea site, Springvale, November 2008

Pavement section	Characteristic deflection (mm)	Tolerable deflection (mm)
In-situ stabilized material as found (chainage 0–0.210 km)	1.486	1.5
Unstabilized existing subgrade material as found (chainage 0–0.220 km)	2.157	1.5

5.4 *Maximum Density/Optimum Moisture*

This test will determine the amount of enzyme required to obtain maximum results. The test method follows ASTM D-1557 [45] modified proctor. Typical enzyme rate (based on Eko-Soil) from field experience can be 1–1.5 L of the enzyme to 30 m^3 of compacted pavement in general. Moisture content to achieve maximum compaction should be 1–2% below optimum. In the field, the moisture content is determined by hand squeezing of the mixed material. If it crumbles, then add more water and retest. If it has an excess of water, allow drying.

5.5 *Bearing Strength or CBR*

The bearing strength, or CBR, is an effective test for determining the bearing strength of the soil. However, the laboratory CBR may not conform to or replicate the field bearing strength because the compacted CBR samples must be allowed air dry for 72 h before submerging in water. CBR tests should be undertaken on cementitious modified materials for each 2nd day's production or every 2500 m^3 whichever is the lesser. Desired CBR values of cementitious modified materials should be >15%.

5.6 *Permeability*

With the addition of enzymes, the reduction in moisture directly affects the design of the structural section of the pavement. Reductions of up to 100-fold are achieved. The test method can be performed in accordance with ASTM D–5084 [46]. The representative values of relative permeability of the different types of soils are shown in Table 2. To ensure that the cementitious materials maintain the desired permeability of $<5 \times 10^{-8}$ m/s, preferably $<5 \times 10^{-9}$ m/s, permeability testing should be undertaken on cementitious modified materials for every 4th day's production or every 5000 m^3 whichever is lesser. The permeability for the enzyme treated soil is approximately 10^{-8}–10^{-11} m/s.

Table 2 Representative soil permeability for various soil types (obtained from AS1547 [47])

Soil	Permeability coefficient
Sand	10^{-5} m/s
Silt	10^{-6} m/s
Clay	$<10^{-7}$ m/s

5.7 *Field Density*

Methods for the determination of field dry density are as described in Sects. 5.8–5.11 below.

5.8 *Direct Density Test*

In order to perform the direct density test, various standards describe specific methods of conducting the test and evaluating the analysis including: AS1289.5.3.1 [48], AS1289.5.3.2 [49], AS1289.5.8.1 [50] and AS1289.5.3.5 [51].

5.9 *Indirect Density Test*

This test provides an empirical measure of achieved density by measuring another engineering property, principally shear strength, and may be used to further validate density; however, the method of direct testing will govern acceptance.

5.9.1 Establishment of a Reference Density for Calculation of Relative Compaction

To permit relative compaction to be calculated, it is necessary to establish a laboratory reference density. Procedures for establishing such reference densities have been developed empirically over many years and standardized with test procedures of AS1289.5.1.1 [52], AS1289.5.2.1 [53], AS1289.5.7.1 [54], and AS1289.5.5.1 [55].

5.9.2 Sample Selection for Reference Density

For routine "compaction" testing, the sample for determination of laboratory reference density should comprise either the material recovered from the field density determination, (see AS1289.5.3.1 [48]) or from the volume of material considered in the field density.

For cement-modified stabilized materials, including enzyme, the reference density may vary with time but the laboratory compaction should still be carried out on material that has been mixed and compacted by on-site purpose-built machinery. The density tested and re-compacted in the laboratory must be conducted as soon as practicable to ensure minimum curing has occurred. For granular materials, including pavement base and sub-base materials that have been manufactured from a hard rock source under controlled conditions, consideration may be given to providing an assigned value as further discussed within the procedures of AS1289.5.4.2 [56].

6 Construction Applications

This section describes various applications of enzymatic soil stabilization.

6.1 Flood Mitigation

Enzyme-stabilized pavements in the field are constructed in a similar manner to conventional stabilization methods, using water tankers, motor graders with rippers, and a 16-tonne vibrating steel roller. Additives should be verified for weight/square meter to ensure that the quantity of additive meets specifications. For speed of construction and to ensure a homogeneous mix, self-propelling and towing mixing machines are recommended. Both types of machines can be fitted with a computer water feed system and in a recent project that we were involved in, 7000 m^2 of 300-mm deep stabilization were completed in an 8-h day. Should rock or cement be required as part of the geotechnical design as an addition to the pavement mix, this should be placed and mixed before the stabilization process.

Cross fall should be at least 3% and side drains should be cut into the pavement (Fig. 6). Just as the pavement is losing its plasticity, a skim of one-size crushed rock (10 mm) should be rolled into the top layer to provide a skid-resistant surface. Light traffic can be allowed to traverse the pavement as soon as the tyre marks of the stabilization equipment disappear. Full curing is achieved in 72 h after which the road may be opened to all traffic.

Stabilized pavements subject to inundation must be tested in the laboratory to ensure permeability either achieves or exceeds specification. Enzymes, or an enzyme blended with up to 3% cement, in both laboratory and field testing, can achieve the required specification but will have an advantage over clay pavements due to their

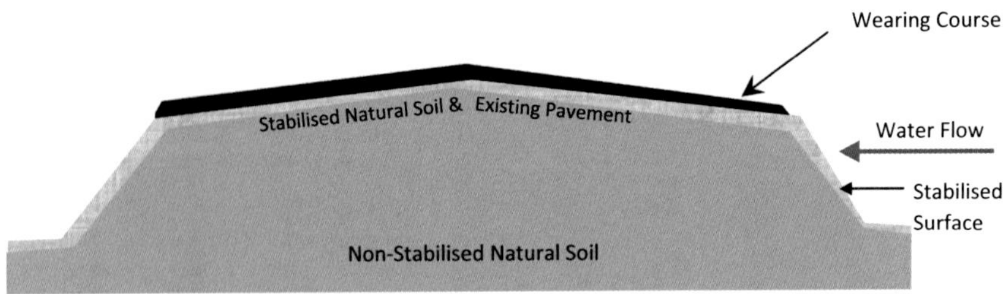

Fig. 6 Typical cross-section of road pavement subject to flooding

residual tensile strength, which prevents cracking of the pavement once the water recedes; hence, stabilized pavements must envelop the road or dam surfaces.

6.2 Embankments and Slopes

Enzymatic soil stabilization can be used successfully in the stabilization of embankments and slopes. Preliminary work is presently underway on problems with some highly plastic soils (e.g., colloid clays that are highly expansive). These soils contain fatty clays and fine gravel or sands and silts with a PI ranging from 45 to 80.

In making calculations for enzymatic stabilization application for these types of soils, the ratio of cohesive fines is different from the standard gradation specification. Normal gradation is calculated on 18–30% cohesive fines. Therefore, correlating these expansive clay results in a greater number of cohesive materials being present. The enzymatic composition works on the molecules within these expansive clays but not the other materials present. The standard application of 1 L per 30 m^3 of compacted pavement must be increased to compensate for the increased amounts of cohesive fines. The standard rate for high amounts of expansive soil is calculated using the same ratio as for a standard application. As an example, the rate of application for soils containing 48% cohesive fines would be 1 L per 15 m^3 of compacted stabilized slope pavement. However, this has to be determined from adequate tests as explained previously.

Recommended thicknesses for slopes are:

- Up to 30 degrees, 150 mm (Fig. 7)
- 30–50 degrees, 150–225 mm

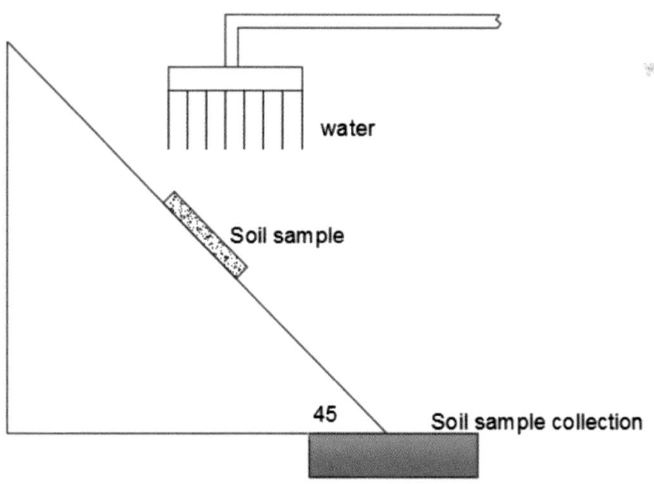

Fig. 7 Erosion simulation equipment used in modified version of the tests [57]

- 50–65 degrees, 250–350 mm
- ≥65 degrees must be used in conjunction with soil nails.

Compaction has always been difficult in slope construction. In recent years, new mechanical attachments have been introduced for compaction of trenches, slopes, and embankments. Compaction load is applied by the backhoe. A filtration membrane, mostly likely sand, must be placed between the slope and the pavement. Weep holes are placed in the pavement at locations as directed by a civil engineer experienced in such methods. Recent compaction equipment introduced into slope stabilization enhances the ability to compact slopes at much steeper angles than was previously possible.

6.3 Soil Nailing

Soil nails have been used for many years in slope stabilization. The normal soil nail is constructed using concrete and steel reinforcement, which can be an expensive additional cost in embankment stabilization projects.

It has been observed that concrete soil nails will catch water on their upper surface, allowing water to penetrate around the soil nail, loosening adjacent soil and aggregate materials and resulting in loss of support of surrounding materials.

Enzymatic-stabilized soil nails are economic alternatives. As with concrete soil nails, they can be pre-manufactured or constructed on-site, and similarly, it is recommended that enzymatic soil nails be reinforced with vinyl-coated reinforcement rods, which should be ≈30% below the bottom of the soil nail for anchoring into the in-situ material. Enzymatic soil nails are recommended to be pre-formed and pressed into pre-bored holes that are slightly undersized. It is also recommended that enzymatic soil nails be placed on the slope after being dampened with a 1:10,000 mist of enzymatic composition and water.

An alternating pattern of rows should be used. The spacing of the nails and the rows should be approximately 1.5 times the diameter of the soil nails. A civil engineer experienced in slope stabilization should design the use of soil nails. Construction and replacement of soil nails should be performed in favorable construction climatic conditions by avoiding freezing and wet weather conditions.

6.4 Enzymatic Composition Blocks (Hollow Blocks/Cement Blocks) and Bricks

The construction of cemented crushed shell blocks can be achieved with the aid of enzymatic composition. A reduction of ≈5% in the use of water will be realized and a reduction of mold breakage rate of between 35 and 50% can be achieved. This is an economical saving for any manufacturing process.

Enzymatic-stabilized soils have been used to construct solid construction blocks designed for the construction of interior walls and low retaining walls. They were manufactured using small, fractured gravels and cohesive fines within standard gradation specifications, and have proved to be exceptionally strong, durable, and easily constructed with a hand-operated lever-style press and a steel mold.

More efficient methods are used to manufacture enzymatic-stabilized blocks/bricks, where facilities are available. Hydraulic power units have the capacity to produce 6000–8000 blocks per shift per machine.

Mortar for joining the enzymatic-stabilized blocks and bricks should be a combination of the makeup of the blocks/bricks (without the larger aggregate) and applied as normal mortar to all connecting surfaces. Curing times of the mortar joints will be in the order of 24 h. The utilization of enzyme is currently under investigation for unfired brick fabrication.

6.5 *Working Platform*

A typical example of soil stabilization by utilizing enzyme product in the field was undertaken on an area of 65,000 m^2 for a whitegoods site in Australia (Fig. 8). The pavement was 150 mm of in-situ fill material and 150 mm of recycled material recovered from demolished site buildings. The testing was conducted in the laboratory and the field as recommended here. The compaction density of the subgrade achieved results >100% while the strength of the subgrade, in field testing on this site achieved CBRs of ≈80% and permeability of 10^{-14} m/s.

The process for working platforms also applies to roads, large industrial/commercial building sites, and most civil infrastructure sites. They are useful for subgrade improvement of over-reactive clays to stabilize the subgrade moisture and to limit differential movement in the subgrade. Figure 9 shows the comparisons of road pavement based on unstabilized and stabilized base in the same locality.

6.6 *Reclaimed Materials*

Testing performed for the Hong Kong University recommended that the sludge retrieved from the Hong Kong Harbour be used as a working platform for the Hong Kong Housing Department. The material was delivered to the site and dried by using 3% lime mixed into the sludge and left to dry for 3 days. A pavement mixture consisting of 150-mm of dried reclaimed material, 150 mm of 19-mm of recycled concrete, 3% cement (due to the variation of soil consistency) and 1% of enzyme stabilizer produced a CBR of 80%.

Fig. 8 Using enzyme as an additive for soil stabilization of a construction site consisting of 65,000 m^2

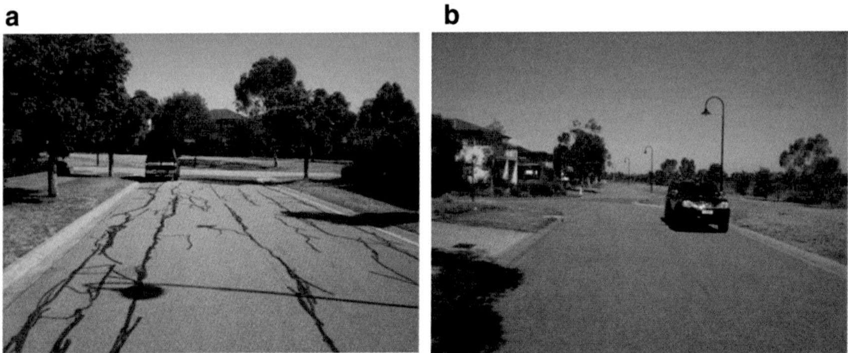

Fig. 9 **a** Cracked seal: base not stabilized. **b** Good seal: enzyme-stabilized base

6.7 Water Containment

Ponds were constructed at Warrnambool Airfield. The method of construction was to build a base with an enzyme additive, build up the pavement mix for the walls then stabilize the existing base. The stabilized material was then moved from the top of the pond walls to the base on the 1 in-3 slope using vibrating rollers.

7 Conclusions

Based on the procedures and criteria from different standards and guidelines, the utilization of additives for soil stabilization is more effective when compared with the mechanical methods traditionally used for soil stabilization. Soil stabilization through enzymatic bonding, although highly dependent on the soil material, requires soil materials to biochemically react with the additives in order to obtain an effective stabilization. An excellent understating of the topography and geology of the construction site is imperative.

Consequently, different mechanical and chemical composition tests must be undertaken on the soil types available from the site for each and every combination of chemical additives that will potentially be used for stabilization of those materials. Specifically, when using enzyme products, the soil properties need to be carefully evaluated before any stabilization commences. Once the laboratory results are evaluated, fieldwork can be performed based on the optimum additive content determined. Subsequently, to maintain the safety of infrastructure, comprehensive construction management of the site must be considered before conducting any construction work. Thus, monitoring the field test in accordance with available guidelines, standards, and contractor management protocols is equally important and the field tests are undertaken regularly throughout the lifespan of the construction work to ensure that construction is performed using the highest quality standards and with minimum geotechnical issues.

Stabilization is a science, so a suitably qualified engineer must design and sign off the pavement stabilization construction to ensure the pavement has met or exceeded standards of permeability, strength and density. It is to be noted that the information provided in this document is for guidance only and should not be used without required tests and suitable evaluations as detailed herein.

References

1. Levy S (2008) Public-private partnerships in infrastructure. Leadership Manage Eng
2. Saleh S et al (2019) Improving the strength of weak soil using polyurethane grouts: a review. Constr Build Mater 202:738–752
3. Considine M (1984) Soils shrink, trees drink and houses crack. J ECOS Magazine 41:13–15
4. Baldovino JJ et al (2021) Strength, durability, and microstructure of geopolymers based on recycled-glass powder waste and dolomitic lime for soil stabilization. Constr Build Mater 271:121874
5. Clare K, Cruchley A (1957) Laboratory experiments in the stabilization of clays with hydrated lime. Geotechnique 7(2):97–111
6. Hoang T et al (2019) Sand and silty-sand soil stabilization using bacterial enzyme–induced calcite precipitation (BEICP). Can Geotech J 56(6):808–822
7. Davidson D et al., Comparison of various commercial limes for soil stabilization. HRB Bull 335
8. Sidiq A et al (2023) Investigation of enzyme–based soil stabilization in field application. J Mater Civ Eng 35(5):04023086

9. Keller I (2011) Improvement of weak soils by the deep soil mixing method. Keller Bronchure 23–30
10. Sherwood P (1993) Soil stabilization with cement and lime
11. Afrin H (2017) A review on different types soil stabilization techniques. J Int J Transp Eng Technol 3(2):19–24
12. Babu N, Poulose E (2018) Effect of lime on soil properties: a review. Microbiology 5(11)
13. Bagoniza S et al. (1987) Carbonation of stabilised mixtures 29–48
14. Jawad IT et al. (2014) Soil stabilization using lime: advantages, disadvantages and proposing a potential alternative 8(4):510–520
15. Vitale E et al. (2021) Effects of carbonation on chemo-mechanical behaviour of lime-treated soils 80(3):2687–2700
16. Chaibeddra S, Kharchi F (2014) Study of sulphate attack on earth stabilized blocks
17. Rajasekaran G, Narasimha Rao SJMG (2005) Sulphate attack in lime-treated marine clay. Mar Georesour Geotechnol 23(1–2):93–116
18. Pham TA, Koseki J, Dias D (2021) Optimum material ratio for improving the performance of cement-mixed soils. Transp Geotech 28:100544
19. Firoozi AA et al. (2017) Fundamentals of soil stabilization. 8(1):1–16
20. Kennedy TW et al. (1987) An evaluation of lime and cement stabilization (1119)
21. Cetin D et al. (2014) Dewatering performance of fiber-reinforced fly ash slurry. In: New frontiers in geotechnical engineering, pp 98–107
22. Taha MR et al. (2013) Recent experimental studies in soil stabilization with bio-enzymes–a 18:3881–3894
23. Pooni J et al (2021) Mechanism of enzyme stabilization for expansive soils using mechanical and microstructural investigation. Int J Geomech 21(10):04021191
24. Renjith R et al (2020) Optimization of enzyme-based soil stabilization. J Mater Civil Eng 32(5):04020091
25. Renjith R et al. (2017) Enzyme based soil stabilization for unpaved road construction. In: MATEC Web of Conferences. EDP Sciences
26. Beeghly JH (2003) Recent experiences with lime-fly ash stabilization of pavement subgrade soils, base and recycled asphalt. In: Proceedings of the International Ash Utilization Symposium, University of Kentucky, Lexingston, USA
27. Saunders M, Fookes PJEG (1970) A review of the relationship of rock weathering and climate and its significance to foundation engineering 4(4):289–325
28. Fu B et al. (2010) A review of surface erosion and sediment delivery models for unsealed roads 25(1):1–14
29. Drapper D, Tomlinson R, Williams PJJOEE (2000) Pollutant concentrations in road runoff: Southeast Queensland case study 126(4):313–320
30. Toy TJ, Foster GR, Renard KG (2002) Soil erosion: processes, prediction, measurement, and control. Wiley
31. Gourley C et al. (2001) Cost effective designs for low volume sealed roads in tropical and sub tropical countries. In: XIVth International Road Federation Meeting, Paris. Citeseer
32. Ismail MSN, Ghani ANA (2017) An overview of road damages due to flooding: case study in Kedah state, Malaysia. In: AIP Conference Proceedings. AIP Publishing LLC
33. Andrews R, Sharp K (2010) A protocol for conducting field trials for best value management of unsealed roads. In: ARRB Conference, 24th, 2010ARRB
34. Weiss PT et al. (2019) Permeable pavement in northern North American urban areas: research review and knowledge gaps 20(2):143–162
35. Puppala AJ, Wattanasanticharoen E, Hoyos LRJTRR (2003) Ranking of four chemical and mechanical stabilization methods to treat low-volume road subgrades in Texas 1819(1):63–71
36. Paige-Green PJTRR (1999) Materials for sealed low-volume roads 1652(1):163–171
37. Consoli NC et al. (2019) Use of sustainable binders in soil stabilization 31(2):06018023
38. Ramaji AEJJOASR (2012) A review on the soil stabilization using low-cost methods 8(4):2193–2196

39. Karami H et al (2021) Use of secondary additives in fly ash based soil stabilization for soft subgrades. Transp Geotech 29:100585
40. Pooni J et al. (2019) Durability of enzyme stabilized expansive soil in road pavements subjected to moisture degradation. J Transp Geotech 21:100255
41. D422 A (1998) Standard test method for particle-size analysis of soils
42. D-4318 A, Standard test methods for liquid limit, plastic limit, and plasticity index of soils
43. 1289.3.9.1 A (2002) Method 3.9.1: Soil classification tests—determination of the cone liquid limit of a soil. Methods of testing soils for engineering purposes
44. Renjith R et al. (2020) Optimization of enzyme-based soil stabilization 32(5):04020091
45. D1557–12 A (2021) Standard test methods for laboratory compaction characteristics of soil using modified effort (56,000 ft-lbf/ft3 (2,700 kN-m/m3))
46. D5084 A, Standard test methods for measurement of hydraulic conductivity of saturated porous materials using a flexible wall permeameter
47. Standard A (1994) Disposal systems for effluent from domestic premises
48. 1289.5.3.1 A (2004) Determination of the field density of a soil—sand replacement method using a sand-cone pouring apparatus. Methods of testing soils for engineering purposes soil compaction and density tests
49. 1289.5.3.2 A (2004) Determination of the field density of a soil—sand replacement method using a sand pouring can, with or without a volume displacer. Methods of testing soils for engineering purposes Soil compaction and density tests
50. 1289.5.8.1 A (2007) Determination of field density and field moisture content of a soil using a nuclear surface moisture-density gauge—direct transmission mode. Methods of testing soils for engineering purposes Soil compaction and density tests
51. 1289.5.3.5 A (1997) Determination of the field dry density of a soil—water replacement method. Methods of testing soils for engineering purposes Soil compaction and density tests
52. 1289.5.1.1 A (2017) Determination of the dry density/moisture content relation of a soil using standard compactive effort. Methods of testing soils for engineering purposes
53. 1289.5.2.1 A (2003) Soil compaction and density tests—determination of the dry density/moisture content relation of a soil using modified compactive effort. Methods of testing soils for engineering purposes
54. 1289.5.7.1 A (2006) Method 5.7.1: soil compaction and density tests—compaction control test—Hilf density ratio and Hilf moisture variation (rapid method) 2006. Methods of testing soils for engineering purposes
55. 1289.5.5.1 A (R2016) Determination of the minimum and maximum dry density of a cohesionless material. Methods of testing soils for engineering purposes—soil compaction and density tests 1998
56. 1289.5.4.2 A (2007) Compaction control test—assignment of maximum dry density and optimum moisture content values. Methods of testing soils for engineering purposes soil compaction and density tests
57. Liu J et al (2012) Effect of polyurethane on the stability of sand–clay mixtures. Bull Eng Geol Env 71(3):537–544

Deterioration Modeling of Concrete Bridges and Potential Nanotechnology Application

H. Tran and S. Setunge

Abstract Management of aging concrete bridges with limited resources can be a challenge for state authorities. Deterioration modeling of concrete bridges at the component level is essential to optimize maintenance actions and ensure the safety and serviceability of the bridge network. In this study we examined the Level 2 visual inspection data of a concrete bridge's components collected over 4–5 inspection cycles with the objective of predicting deterioration of components and the bridge's life cycle. With the increasing application of nanotechnology to increase the mechanical properties and durability of concrete material for bridge structures, the deterioration of nano-based concrete could be significantly different from conventional concrete. A range of deterioration prediction methods, including deterministic models and stochastic models, were examined to understand the validity of the different methods in predicting the deterioration of bridge components made of conventional and nano-based materials. A case study with a demonstration on a concrete open girder was investigated with regard to linear regression models and the stochastic Markov deterioration model. The outcomes can be used to support future study on the performance of conventional and nano-based concrete materials and their lifecycles in the asset management of bridges.

Keywords Bridges · Deterioration · Nanotechnology · Markov model

1 Introduction

Bridges are important assets of the transport infrastructure network that play a crucial role in ensuring the well-being of populations and the economy. Concrete is the popular material used in bridges, but after many years of service, concrete bridge structures are aging and deteriorating, which becomes a hazard to traffic users in the

H. Tran (✉) · S. Setunge
Civil and Infrastructure Engineering, School of Engineering, RMIT University, Melbourne, VIC, Australia
e-mail: huu.tran@rmit.edu.au

W. Duan et al. (eds.), *Nanotechnology in Construction for Circular Economy*, Lecture Notes in Civil Engineering 356,
https://doi.org/10.1007/978-981-99-3330-3_41

event of structural failure or falling apart. The causes of concrete deterioration have been intensively studied, such as the carbonation of concrete and chloride attack [1, 2]. Contributing factors to deterioration include traffic volume, exposure to corrosive soil and airborne chloride near the coastline and acidic gases (including airborne carbon dioxide, nitrous and sulfurous oxides) in urban area. The deterioration process can be single, such as concrete creep, or combined processes, such as stress corrosion cracking in steel, and coupled with random damage events such as flooding and earthquake. The search for better concrete materials against natural and man-made hazards is ongoing. Recent advances in nanotechnology and its application in concrete bridge construction can help not only ensure safety and serviceability throughout service life but also cost-effective maintenance, rehabilitation and replacement (MRR).

Nano-concrete has been developed in the past decade to improve key characteristics of normal concrete such as tensile and compressive strength, anticorrosion and durability [3, 4]. The efficiency of the nano-concrete depends on its density, which can be maximized by minimizing the particle gaps. The optimal particle gap within the concrete mass can be achieved by incorporating a homogeneous gradient of fine and coarse particles in the mixture. In this regard, nanomaterials such as nano-silica, nano-graphite platelets, carbon nanotubes, graphene, nano-titanium dioxide and nano clay have been used to reinforce cementitious composites (cement paste, mortar, and concrete) [5]. They can be highly effective because, with their extremely small size, nanomaterials can fill the voids between cement and silica fume particles, leading to higher level of compaction and generating a denser binding matrix. This high level of compaction of concrete particles can significantly improve both the durability and mechanical properties of the nano-concrete. Several studies have reported significantly improved properties of the nano-concrete. For example, the use of 0.02% graphene oxide in ultra-high performance concrete can increase its strength characteristics, such as compressive, tensile, and flexural strength, up to 197%, 160%, and 184%, respectively [5]. Kancharla et al. reported that replacing 0.5 and 1.0% cement with nanosilica showed good improvement in bending strength of 7.8 and 15.7%, respectively, in the crushing stage and slight improvement in bending strength of 0.42 and 1.26%, respectively, in the failure stage [6]. Mostafa et al. found that nano glass waste can increase the bending strength of ultra-high performance concrete up to 1.5-fold if it is added at 1% [7].

With the increasing use of nano-concrete, it is useful to compare deterioration rates between traditional concrete and nano-concrete over service life. Such a comparison is useful for decision making on the use of nano-concrete in bridge engineering and to provide better knowledge for managing the maintenance and life cycle costs for current and future use of the nano-concrete. One method is to use deterioration models to compare the deterioration rates. Deterioration models can be divided into knowledge-based, data-driven and mechanism-based models. Few bridge management agencies use knowledge and experience to determine future deterioration from inspected defects [8]. The data-driven models use inspection data to predict future deterioration [9–11]. The mechanism-based model is based on deterioration mechanisms such as corrosion of reinforcing steels [1, 12]. The physical models are considered more advanced and accurate but they require intensive and detailed data, which can be costly and difficult to obtain. Therefore, they were not selected for this study.

The deterministic linear model is well known for its ease of use and implementation, but is criticized for failure to capture the uncertainty of the deterioration process, which was addressed by the stochastic Markov model in this study.

We aimed to investigate the deterioration rate of traditional concrete as the benchmark for future study of deterioration of nano-concrete used for bridge components such as bridge girders and the bridge deck. For traditional concrete, visual inspection data of bridge components using traditional concrete are available and collected from the bridge agency for deterioration modeling. However, such data are not available for the nano-concrete. Therefore, scenario analysis was conducted for the deterioration rate of the nano-concrete based on its reported performance in public literature to assess its potential economic benefit. The outcomes of this study will demonstrate the benefit of condition monitoring and data collection for deterioration modeling traditional concrete for expanding the application of nano-concrete in bridge engineering.

2 Case Study

VicRoads is the registered business name of the Roads Corporation in the State of Victoria, Australia (www.vicroads.com.au). It is a Victorian statutory authority established under the *Transport Act 1983* and continued in the *Transport Integration Act 2010*. One of VicRoads' core services is to plan, develop and manage the arterial road network, including roads, bridges, culverts and traffic signs. The bridge management of VicRoads is supported through its computerized database, which contains basic data of 6207 bridge structures including bridges, culverts and tunnel with their attributes such as number of spans, length and width. The oldest structure was built in 1899. The database also stores 26,1324 records of Level 2 inspections as of 2017. The first recorded inspections in the computerized database began in 1995.

The VicRoads's inspection practice is published in an open-access inspection manual, which basically has three levels of inspection [13]. Level 1 is considered a screening inspection with a maximum interval of 6 months. Level 2 is a routine inspection with a typical interval of 2 years and Level 3 is an in-depth inspection for special cases. The VicRoads' inspection manual describes the breakdown of bridge structures into superstructures and substructures and into structural elements such as piles, decks and bearings. Each bridge element is coded by its function and one of five types of material (i.e., timber, steel, in-situ concrete, precast concrete and others); for example, the 2C mean open girder/stringer with the concrete material. The Level 2 inspection provides q condition rating for individual elements based on their inspected defects as per inspection guidelines. The condition rating is a 1–4 scale, with 1 being good and 4 being worst. The condition state percentage distribution is used to provide inspection reports. For example, an inspection report of [90% 10% 0% 0%] of 2C means 90% of the open girders in condition 1, 10% in condition 2 and 0% in conditions 3 and 4. The case study dataset of Level 2 inspection data from 1997 to 2017 was used for deterioration modeling of bridge components in this study.

A collaborative research project between VicRoads and RMIT University with support from the ARC Nanocom Hub was established to develop deterioration models using Level 2 inspection data for bridge structures. The outcomes of the collaborative project could provide justification for future state funding of bridge management and also provide support for more effective and efficient asset management programs.

2.1 Deterioration Models

2.1.1 Linear Regression Model

Equation (1) shows the linear relation between bridge condition (output) with age (input) [9]. More input factors can be added into Eq. (1) in a similar manner.

$$Y = a_0 + a_1 * \text{age} + \varepsilon \tag{1}$$

where a_0 is a constant coefficient, a_1 is slope coefficient of age (year) and ε is the measurement error assumed to have zero mean and independent constant variance.

The interpretation of Eq. (1) is that (a) for the case of 1 input and 1 output, the relation is a straight line; (b) for the case of $\geq$2 inputs, the relation is a plane and hyperplane; (c) a linear relation means a constant deterioration rate over time; (d) measurement error means that for the same values of the inputs, every time the output is measured or assessed, the output value might not be the same due to equipment sensitivity or human subjectivity; and (e) deterioration increases with age, meaning the unchanged condition is not valid.

For calibrating or estimating the model coefficients, the least square method is often used to minimize the error term between the observed output and model output. The predictive performance of the linear model is often assessed using the coefficient of determination R2 between observed values and model outputs.

2.1.2 Markov Model

The Markov model is based on the stochastic theory of Markov chain [14], which describes a system that can be in one of several defined condition states at any time, and it can stay still or move to another state at each time step with some transition probabilities over time. This theory is well suited for observation of Level 2 condition data of bridge elements because the snapshot inspection reveals the condition state at the time of inspection and condition state movement over time can be captured with the transition probabilities.

The most important property of the Markov chain is that the probability of movement depends only on the current condition regardless of the history of movement, called memoryless property. This means that (a) the currently known condition can represent the accumulated deterioration up to the current time and (b) the future condition does not depend on how long it stays in previous conditions. This property is well suited for long-life assets such as concrete. The opposite is usage-based assets such as light bulbs and machinery in which their future condition depends on how long they have been used in the past.

The Markov model is mathematically expressed as a matrix M of transition probabilities:

$$M = \begin{bmatrix} P_{11} & P_{12} & P_{13} & P_{14} \\ P_{21} & P_{22} & P_{23} & P_{24} \\ P_{31} & P_{32} & P_{33} & P_{34} \\ P_{41} & P_{42} & P_{43} & P_{44} \end{bmatrix} \tag{2}$$

where P_{ij} is the probability of moving from condition state i to condition state j over a unit time step. In this study, P_{ij} was assumed to be 0 if $i > j$, meaning improved condition was not modeled due to the lack of maintenance data.

The memoryless property and the theory of total probability can be used to predict the probability vector $[P1\ P2\ P3\ P4]$ at any future time T, given the known current condition with certainty or probabilities $[C1\ C2\ C3\ C4]$ as shown in the Chapman–Kolmogorov equation [15]:

$$[P1\ P2\ P3\ P4] = [C1\ C2\ C3\ C4] * M^T \tag{3}$$

Equation (3) is a matrix multiplication of the current condition with the transition matrix powered to future time T. To utilize Eq. (3), the transition matrix M of Eq. (2) needs to be estimated. Among several calibration/estimation techniques, we used the proven Bayesian Markov chain Monte Carlo simulation technique [16] to estimate the transition matrix M of the Markov model. In brief, the Bayesian technique transforms the unknown elements P_{ij} of the transition matrix M into a multivariate distribution by using observed data, Eq. (3) and the Bayesian theory. The Markov chain Monte Carlo simulation is then used to generate sampling data of the elements, which become the estimator of the unknown elements (see [16]).

The predictive performance of the Markov model is commonly validated using the Chi-square test on a separate dataset that is not used in the calibration of the Markov model [15, 16]. The validation dataset is often randomly selected from 15 to 20% of the entire dataset.

3 Results

The deterioration models were applied to the case study and the results are demonstrated for a concrete open girder.

3.1 Model Fitness and Effect of Long-Term Data Collection

The effect of long-term and short-term data collection was investigated using the linear and Markov models. Table 1 shows that the Markov model passed the fitness test for both datasets. The negative values for R2 of the linear model on both datasets indicate the poor fitness of the linear model and it should not be used. The poor fitness of the linear model can be seen in Fig. 1 where the observed data have high uncertainty as shown by the unpatterned scattering. Instead of trying to fit an impossible line through the observed percentage in condition 1 with age, the Markov model takes a different approach by capturing the percentage changes between condition states. This shows the good generality of the Markov model in such cases.

Despite the unacceptable fitness of the linear models, Fig. 1 can still be used to illustrate the significant effect of long-term data collection. This effect showed the predicted deterioration rate is sensitive to the range of calibration data. The predicted deterioration rate for long-term data collection (Fig. 1a) has a mild slope as compared with the steep slope of the predicted deterioration rate for short-term data collection (Fig. 1b).

The calibrated transition matrix for an old bridge is shown as a demonstration in Eqs. (4) and (5) for long-term and short-term data collection. Figure 2a, b shows the effect of long-term data collection on the Markov model. It appears that removal of the first inspection accelerated the deterioration. For example, at the age of 100 years, the percentages in condition 1 were 45 and 18% between the long-term and short-term data. Similarly, the percentage of 20 and 30% in condition 4 can be observed between the long-term and short-term data, implying a 15% difference.

Table 1 Results of fitness test for Markov and linear models for percentage prediction on condition 1

Data	Markov model	Linear model
	Chi-square test	R2 value
Long-term	55.8 < 85.9	−0.544
Short-term	7.8 < 67.5	−0.247

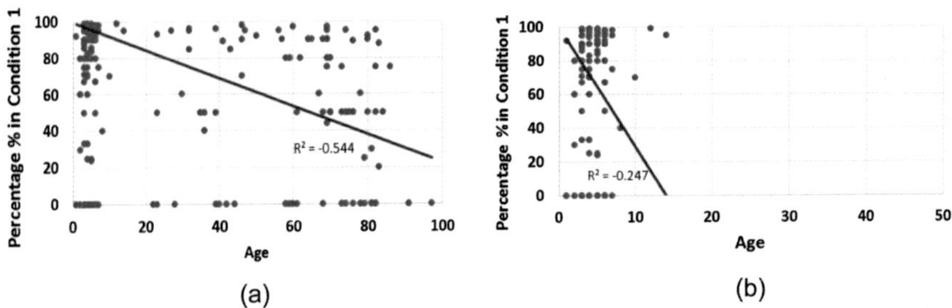

Fig. 1 Linear model fitted with long-term (**a**) and short-term (**b**) data for percentage in condition 1

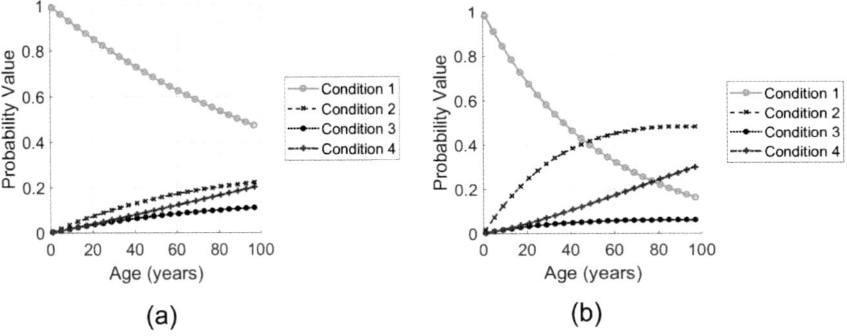

Fig. 2 Markov model using long-term data (**a**) and short-term data (**b**)

$$M(\text{all}) = \begin{bmatrix} 0.9934 & 0.0052 & 0.0008 & 0.0006 \\ 0 & 0.9984 & 0.0009 & 0.0007 \\ 0 & 0 & 0.9972 & 0.0028 \\ 0 & 0 & 0 & 1 \end{bmatrix} \quad (4)$$

$$M(\text{second}) = \begin{bmatrix} 0.9894 & 0.0082 & 0.0018 & 0.0006 \\ 0 & 0.9889 & 0.0104 & 0.0007 \\ 0 & 0 & 0.9991 & 0.0009 \\ 0 & 0 & 0 & 1 \end{bmatrix} \quad (5)$$

3.2 *Potential Benefit of Nano-concrete*

The deterioration rate of traditional concrete was estimated using visual inspection data. There were 3000 bridge structures in the case study. If it is assumed that at current year 2022, all bridge structures in failure condition 5 have already been repaired or replaced, then the budget for proactive asset management over the next 10 years (selected as an example) can be estimated as: (a) unit cost of major repair or replacement AUD 100,000 per bridge girder for its average length of 30 m. The penalty cost is assumed being equal to unit cost of inspection; (b) Markov deterioration model predicts the increase by 4.0% of 3000 bridge structures (i.e. ≈120 structures) that will have its girder in failure condition 5 and require major repair or replacement. The replacement budget is therefore AUD 12 million (which is 120 × $100,000); and (c) if nano-concrete is used instead of traditional concrete, the rate of deterioration could be 20% lower because of its better strength and durability, resulting in 4%*0.8 = 3.2*3000 = 96 girders that require major repair or replacement. The budget cost is AUD 9.6 million dollars and the saving is AUD 2.4 million over 10 years.

It should be noted that the cost figures are hypothetical values that were used in this study only for demonstration of methodology.

4 Discussion

This study was limited to a case of concrete material with a demonstration of an open girder 2C. However, the methodology can be extended to all bridge elements with different materials. The effect of contributing factors such as traffic volume and exposure condition will be investigated in further studies.

This study revealed the weakness of the linear model and thereby supports the use of the stochastic Markov model in modelling deterioration of bridge elements. Despite this advantage, the Markov model still shows deterioration in the first few years, which is considered unrealistic as found in a previous study [17]. Sobanjo and Thompson claimed that new structures are expected to stay in condition 1 for the first 10–30 years before transitioning into a deteriorated condition. If all structures follow this pattern, then the Markov model should be applied to the data after transition start. If some new structures start to move to condition 2 after a few years due to unknown causes such as uncertainty of material batch, inconsistent construction practice or damage events, then the Markov model can reflect that. Therefore, it is not the shortfall of the Markov model but how valid data are compiled for the deterioration models. The comparison between the Markov model and other stochastic models such as the Gamma process model [11] will be conducted in further works.

5 Conclusions

Nanomaterials including nano-silica, nano-alumina and nano-titanium oxide are used in current research into developing nano-concrete. The addition of nanomaterials to concrete can be viewed as having a similar effect to that with micro-based materials such as metakolin and silica fume. The effect comes from pore refinement and thereby increases the strength and durability of concrete. The long-term performance of nano-concrete used in bridge structures is still unknown. We used deterioration models to estimate the deterioration rate of traditional concrete and conducted a scenario analysis to understand the potential benefit of the nano-concrete. This can significantly improve the efficiency of MRR management of bridge networks. We also compared the deterministic linear model against the stochastic Markov model. The demonstration on a concrete open girder showed that the Markov model is more suitable than the linear model. Further investigations will be conducted to study the effect of contributing factors, and to compare with other stochastic models.

References

1. Jamali A, Angst U, Adey B, Elsener B (2013) Modeling of corrosion-induced concrete cover cracking: a critical analysis. Constr Build Mater 42:225–237
2. Voyiadjis GZ (2022) Handbook of damage mechanics: nano to macro scale for materials and structures. Springer International Publishing, Cham
3. Norhasri MSM, Hamidah MS, Fadzil AM (2017) Applications of using nano material in concrete: a review. Constr Build Mater 133:91–97
4. Silvestre J, Silvestre N, De Brito J (2016) Review on concrete nanotechnology. Eur J Environ Civ Eng 20:455–485
5. Ahmad F, Jamal A, Iqbal M, Alqurashi M, Almoshaogeh M, Al-Ahmadi HM, Hussein E (2021) Performance evaluation of cementitious composites incorporating nano graphite platelets as additive carbon material. Materials 15:290
6. Kancharla R, Maddumala VR, Prasanna TVN, Pullagura L, Mukiri RR, Prakash MV (2021) Flexural behavior performance of reinforced concrete slabs mixed with nano- and microsilica. J Nanomater 2021:1–11
7. Mostafa SA, Faried AS, Farghali AA, El-Deeb MM, Tawfik TA, Majer S, Abd Elrahman M (2020) Influence of nanoparticles from waste materials on mechanical properties, durability and microstructure of UHPC. Materials 13:4530
8. Safia M, Sundquista H, Karoumia R, Racutanu G (2013) Development of the Swedish bridge management system by upgrading and expanding the use of LCC. Struct Infrastruct Eng 9:1240–1250
9. Agrawal A, Kawaguchi A, Chen Z (2010) Deterioration rates of typical bridge elements in New York. J Bridg Eng 15:419–429
10. Bu G (2013) Development of an integrated deterioration method for long-term bridge performance prediction. PhD thesis, Griffith University
11. Edirisinghe R, Setunge S, Zhang G (2013) Application of gamma process for building deterioration prediction. J Perform Constr Facil 27:763–773
12. Lu C, Jin W, Liu R (2011) Reinforcement corrosion-induced cover cracking and its time prediction for reinforced concrete structures. Corros Sci 53:1337–1347
13. Principal Bridge Engineers (2014) Road structures inspection manual. VicRoads, Melbourne
14. Madanat S, Mishalani R, Ibrahim WHW (1995) Estimation of infrastructure transition probabilities from condition rating data. J Infrastruct Syst, ASCE 1:120–125
15. Tran HD (2007) Investigation of deterioration models for stormwater pipe systems. PhD thesis, Victoria University
16. Micevski T, Kuczera G, Coombes P (2002) Markov model for storm water pipe deterioration. J Infrastruct Syst 8:49–56
17. Sobanjo JO, Thompson PD (2011) Enhancement of the FDOT's project level and network-level bridge management analysis tools final report. Florida State University

Transfer and Substrate Effects on 2D Materials for Their Sensing and Energy Applications in Civil Engineering

Q. Zhang, C. Zheng, K. Sagoe-Crentsil, and W. Duan

Abstract The recent emergence of two-dimensional (2D) materials such as graphene and transition metal dichalcogenides (TMDs) of the family (Mo, W)(S, Se)$_2$ has attracted interest from a broad range of engineering applications, including advanced sensing and energy harvesting and conservation, because of their distinctive properties. However, it is critical important to achieve intact delamination and transfer of these atomically thin materials, as well as to understand the effects of the target substrates on their optical and electronic properties. Therefore, we developed and compared techniques for transferring as-grown WS_2 crystals to arbitrary substrates. Polystyrene-assisted wet transfer can realize improved preservation of monolayer WS_2 crystals than the commonly used poly(methyl methacrylate) (PMMA)-assisted wet transfer method, due to minimal chemical etching involved in the 2D material delamination process. The intercalation of alkali ions in the PMMA-based transfer method induces chemical doping over the transferred 2D crystals, leading to the formation of trions. Moreover, the edges of the crystals on hydrophilic substrates, such as sapphire or SiO_2/Si, are subject to ambient water intercalation, which locally affects the photoluminescence behavior of the monolayer WS_2 by doping and changing of the dielectric environment. This non-uniform optical behavior is absent when the crystal is transferred onto a hydrophobic substrate through which ambient water cannot penetrate. These results have important implications for the choice of target substrate and transfer method adopted for 2D TMD-based applications such as next-generation strain sensing, photodetectors, gas sensing, bio sensing, solar energy harvesting and radiative cooling in which uniform behavior of the channel material is required.

Keywords Two-dimensional materials · Transfer · Photoluminescence · Sensing · Dielectric screening

Q. Zhang (✉) · K. Sagoe-Crentsil · W. Duan
Department of Civil Engineering, Monash University, Clayton, VIC, Australia
e-mail: zqhmonashcsu@gmail.com

C. Zheng
Department of Physics, School of Science, Key Laboratory for Quantum Materials of Zhejiang Province, Westlake University, Hangzhou, Zhejiang Province, China

W. Duan et al. (eds.), *Nanotechnology in Construction for Circular Economy*,
Lecture Notes in Civil Engineering 356,
https://doi.org/10.1007/978-981-99-3330-3_42

1 Introduction

Because of their distinctive optical, electrical and mechanical properties, two-dimensional (2D) materials such as graphene and transition metal dichalcogenides (TMDs) are promising candidates for the high performance electronics and opto-electronics needed for next-generation sensing and energy applications [1–3]. For example, monolayer WS_2 in the TMDs family has been reported as a promising strain sensing that are integrated into infrastructure such as building windows [5]. To realize their ultimate utilization in diverse applications, a feasible non-defective and quality-preserving transfer technique is essential to create functional devices or van der Waals heterostructures [6, 7]. The polymer-assisted wet transfer technique using poly(methyl methacrylate) (PMMA) is widely used because of its better applicability and material conservation [4, 8, 9]. However, this method usually involves soaking the sample in heated alkaline solution, resulting in chemical doping as well as a certain level of damage to the transferred materials [4]. To overcome this drawback, hydrophobic polystyrene (PS) has been adopted as the supporting polymer layer for the Chemical Vapor Deposition (CVD)-grown WS_2 single crystal transfer process, which enables less exposure of the materials to the alkaline solution [10]. Nevertheless, there is still a lack of detailed characterization of the crystals before and after transfer to fully understand the doping effect of the transfer process.

Furthermore, because of their atomic thickness and reduced dielectric screening, the properties of 2D materials are greatly influenced by the underlying substrate [11]. For 2D TMDs, the reduced Coulomb screening gives rise to high exciton binding energies and quasiparticles of trions, which are charged excitons. Trions have been observed in doped monolayer TMDs [12]. The characteristics of excitons and trions, which significantly affect electronic transport and optical transitions in the monolayer TMDs, are found to be strongly coupled with the dielectric environment [13], as well as doping effects [14] induced by the underlying substrate. In this regard, non-uniform photoluminescence (PL) emissions of CVD-grown TMDs crystals on dielectric substrates are widely observed [4, 15–18]. However, the underlying physics have not been clearly demonstrated and further investigation is required to understand the mechanism of this non-uniform PL behavior.

In this study, the distinctive effects of the substrate and the influence of doping from the transfer process on the optical properties of CVD-grown monolayer WS_2 triangular crystals were demonstrated. Optical and morphological characterizations indicated that the non-uniformity of the monolayer WS_2 PL behavior is highly likely related to the intercalation of a layer of ambient water molecules between the WS_2 crystal and the underlying hydrophilic substrate, where the intercalated water changes the dielectric environment and the doping state. Comparison was made between the commonly used PMMA transfer and the modified PS transfer methods in transferring single WS_2 crystals from as-grown sapphire substrates to target SiO_2/Si substrates. Detailed morphology mappings showed that PS transfer maintained better preservation of the crystals in transfer using PMMA, due to the avoidance of soaking in alkaline solution. Optical characterizations of crystals before and after PS transfer indicated that the residual alkali metal ions facilitated the formation of trions in monolayer WS_2 under ambient conditions, leading to variations in PL behavior. In contrast to the transferred WS_2 crystal on hydrophilic SiO_2/Si substrates, where non-uniformity of PL along the edges remained, the transferred WS_2 on hydrophobic PS presented uniform PL emission across the crystal, supporting the theory that water intercalation is the source of the inhomogeneous PL behavior on hydrophilic dielectric substrates. After removal of alkali dopants by annealing at 100 °C in an argon gas environment, the trion peak diminished, with only the exciton peak contributing to the PL emission, indicating that the trion peak formation closely correlated with the electron doping caused by alkali metal ions.

2 Methods

2.1 CVD Growth of Monolayer WS_2 Crystals

The WS_2 crystals were grown by CVD on substrates of sapphire [Al_2O_3 (0001)] using WO_3 and S powders as precursors. The detailed CVD setup for monolayer single WS_2 crystal growth can be found in Zhang et al. [4].

2.2 Polymer-Assisted Wet Transfer Using PMMA and PS

For PMMA wet transfer, the CVD-grown monolayer WS_2 on the sapphire substrate was spin-coated at 3000 rpm for 60 s by PMMA (A4) and soft baked at 80 °C for 3 min. Next, the spin-coated sample was soaked in 2 mol/L KOH solution and heated to 100 °C on a hot plate for 1 h. Before soaking, the edges of the PMMA thin film on the sample were scratched with a scalpel to facilitate penetration of alkali between the polymer film and the substrate. After soaking, the PMMA film was separated from the as-grown sapphire substrate at a deionized (DI) water surface with the help of surface tension. The separated PMMA film floated on the surface of the DI water with the WS_2 crystals attached. The film was then fished out by the target substrate from underneath (WS_2 side). The PMMA film was removed by soaking in acetone and isopropanol (IPA) in turn.

For the PS transfer, PS (M_w ~ 192,000) in toluene solution (50 mg/mL) was used to spin-coat as-grown WS_2 on a sapphire sample. The steps of the PS transfer process were the same as those introduced in Xu et al. [10]. Similarly, the PS film was fished out by SiO_2/Si from underneath (WS_2 side) and the polymer was washed off by acetone and IPA. A further cleaning step of soaking the sample in PG remover was used for thorough removal of the PS. To transfer WS_2 onto a hydrophobic PS surface, a rigid substrate of SiO_2/Si was used to attach the detached PS film from the top, leaving the WS_2 sitting on the PS surface.

2.3 Optical and AFM Characterizations

Raman and PL characterizations were performed with a confocal microscope system (WITec alpha 300R) with ×50 objective lens and 532 nm laser excitation. A weak laser power of 50 μm and a short integration time of 1 s was adopted. The PL intensity mappings were obtained by totaling the PL intensity from 1.9 to 2.1 eV. The Raman and PL spectra were collected using a 600 line mm^{-1} grating. The atomic force microscopy (AFM) measurements were carried out on the Bruker Dimension Icon in tapping mode.

3 Results and Discussion

Characterizations of a representative as-grown monolayer WS_2 crystal by CVD on an atomically flat sapphire substrate are presented in Fig. 1. The optical image in Fig. 1a shows a typical triangular single WS_2 crystal ~5 μm in size with uniform contrast. The AFM morphological mappings of the crystal (Fig. 1b, c) revealed that the grown WS_2 followed the atomic steps of the surface of the bare sapphire with step heights of 0.2 nm, indicating uniform height and atomic flatness. However, it was observed that the heights of the edges were slightly raised by ~0.37 nm, as shown in Fig. 1c, g. This was inferred to be raised by a single layer of water molecules from the ambient environment that intercalated between the sapphire and the monolayer WS_2 crystal. This intercalated water layer can induce localized strong dielectric screening to the monolayer WS_2 because water has a much higher dielectric constant value than sapphire [19].

This finding was further supported by the optical features reflected by Raman intensity, PL intensity and PL peak position mappings for this WS_2 crystal on

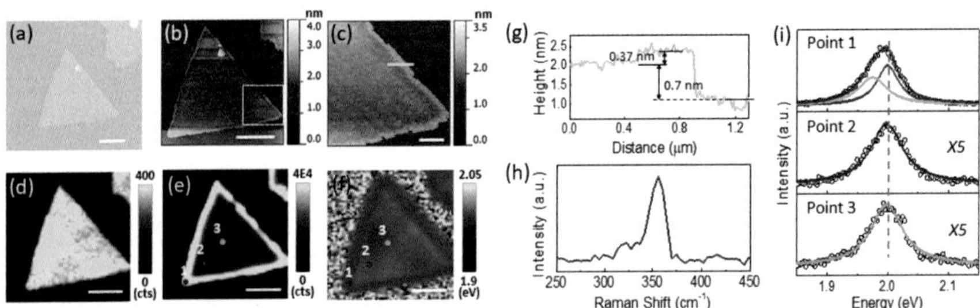

Fig. 1 Characterizations of a CVD as-grown monolayer WS_2 crystal. **a** Optical image; **b** AFM height scan; **c** zoomed-in AFM mapping of the region shown in the white square in (**b**); **d–f** Raman intensity, PL intensity and PL peak position mappings; **g** height profile along the yellow line in (**c**); **h** Raman spectrum collected in (**d**); **i** PL spectra taken from points 1–3 indicated in (**e**, **f**). Scale bars: 5 μm in (**a**, **b**, **d–f**); 1 μm in (**c**)

sapphire. A Raman spectrum collected from the crystal (Fig. 1h) shows a distinctive in-plane E' Raman peak of monolayer WS_2 at 350 cm^{-1}. Although uniform Raman signals were detected across the crystal (Fig. 1d), distinctive non-uniform PL behavior was observed (Fig. 1e, f) where the edges of the crystal exhibited stronger PL emissions and redshifted peak positions. The regions of edges with varied PL behavior were consistent with the regions where the height was raised by water intercalation, indicating that the trapped water had a notable influence on the optical properties of the monolayer crystal.

To further understand the inconsistent PL behavior, three PL spectra were collected at points 1–3 from the edge to the center of the crystal as indicated in Fig. 1e, f and shown in Fig. 1i. Information regarding the three spectra is summarized in Table 1. Lorentzian fitting indicated that the PL spectra at points 2 and 3 featured a single peak each at 1.998 eV, coinciding well with the exciton peak of the CVD WS_2 monolayers on sapphire [20]. For the PL spectrum collected at point 1 located at the edge of the crystal, however, the peak position was redshifted. Lorentzian fitting revealed two distinctive peaks of an exciton (X^0, magenta curve) at 1.999 eV and a trion (X^-, cyan curve) at 1.975 eV, with an energy difference of 24 meV. This value was in good agreement with the negative trion binding energy of the monolayer WS_2 [21]. The exciton to trion intensity ratio was 1.4, agreeing well with the value for the monolayer WS_2 with 10 V positive gate voltage [21], indicating n-type doping in the water-intercalated regions.

Besides the change in peak composition and the shift in peak position, the integrated PL emission at the edge with water trapped underneath (point 1) also had fivefold higher intensity than the inner region of the crystal (points 2 and 3). This was likely due to the high dielectric constant (κ) of the water trapped underneath, which led to prolonged lifetimes of the quasiparticles and increased recombination efficiency of the excitons and trions [22, 23].

Next, the influence of the wet transfer method on the properties of the transfer monolayer WS_2 was investigated. First, the morphological changes of the crystal transferred by processes using different polymers were studied. The transfer steps with the two different polymers (i.e., PMMA and PS) are schematically illustrated in Fig. 2a. As can be seen, compared with PMMA transfer, the PS film could be directly

Table 1 Spectral information for WS_2 crystal on sapphire (corresponding to Fig. 1i)

	Trion			Exciton			Trion binding energy (meV)	Exciton to trion intensity ratio
	Peak position (eV)	Peak intensity (a.u.)	Peak width (meV)	Peak position (eV)	Peak intensity (a.u.)	Peak width (meV)		
Point 1	1.975	208	58	1.999	322	44	24	1.4
Point 2				1.998	97	65		
Point 3				1.998	100	64		

separated from the sapphire at the KOH solution surface with minimal exposure to the alkaline environment. The effect of the alkali exposure is reflected in the morphological characterization of the transferred crystals presented in Fig. 2b–e. Obvious damage can be observed along the edges of the PMMA-transferred crystal, with jagged shapes evident by the optical contrast (Fig. 2b) and AFM scan (Fig. 2d). This damage to the edges was presumed to be caused by KOH etching from the long soaking at elevated temperature [10]. Specifically, the raised edges of the crystals by ambient water intercalation prior to the transfer process caused them to be more easily subjected to etching by the alkali. In contrast, the morphology of the WS_2 crystal transferred by PS indicated good preservation, including that at the edges (Fig. 2c, e).

However, PS cannot easily be removed by acetone and IPA washing, so polymer residue on the crystals can be observed. Therefore, PG remover, an NMP-based solvent stripper, was further used to thoroughly remove the PS residue. As shown in Fig. 3, the polymer residue was cleanly removed with edges well preserved. These results suggested that PS transfer was superior to PMMA transfer, with better preservation of the morphology of the transferred 2D crystals.

The change in the optical properties of the WS_2 crystal after the PS transfer process was investigated for the same crystal characterized in Fig. 1. To avoid confusion, the crystal before PG remover washing is here referred to as the "as-transferred crystal" and the crystal after PG remover washing is denoted as the "PG washed crystal". Figure 3a–d shows the optical images and PL mappings of the as-transferred crystal. As can be seen clearly, non-uniform behavior of PL remains after the crystal was

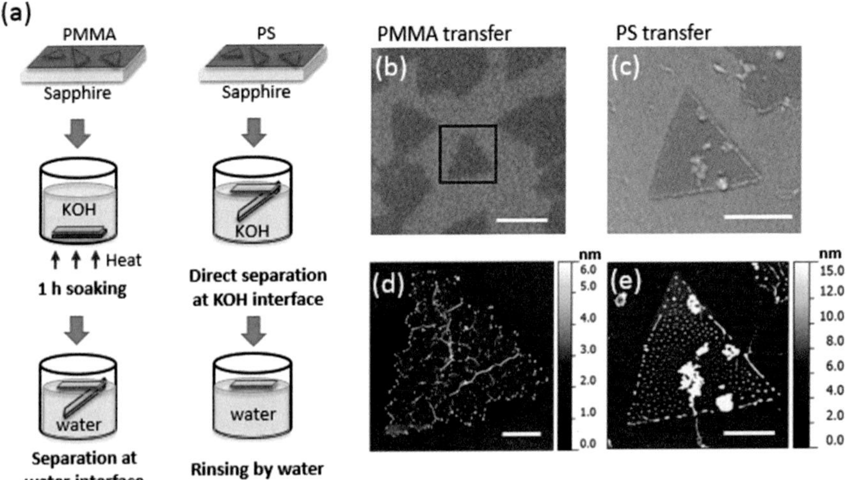

Fig. 2 **a** Schematic of PMMA- and PS-based wet transfer methods. **b**, **c** Optical images of PMMA- and PS-transferred WS_2 crystals on SiO_2/Si; **d** AFM mapping of the PMMA-transferred crystal shown in the square in (**b**); **e** AFM mapping of the PS-transferred crystal in (**c**). Scale bars: 10 min (**b**, **c**); 2 min (**d**); 5 min (**e**)

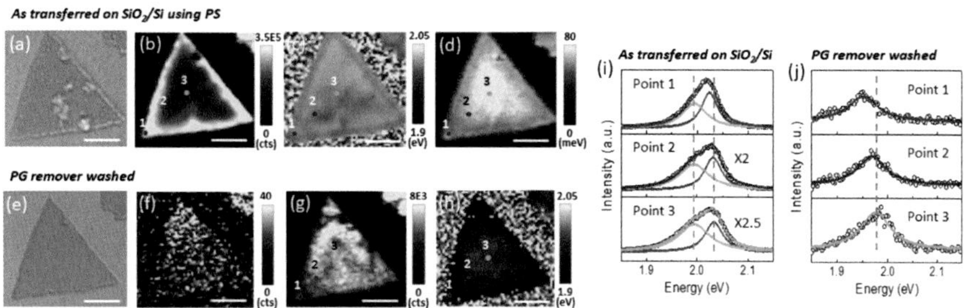

Fig. 3 Optical characterization of the PS-transferred WS_2 crystal onto SiO_2/Si. **a** Optical image, **b** PL intensity mapping, **c** PL peak position mapping and **d** PL width mapping of the crystal as transferred onto SiO_2/Si. **e** Optical image, **f** Raman intensity mapping, **g** PL intensity mapping and **h** PL peak position mapping of the crystal transferred onto SiO_2/Si and soaked in PG remover. **i** PL spectra taken at points 1–3 in (**b–d**). **j** PL spectra taken at points 1–3 in (**g**, **h**). Scale bars: 5 μm in (**a–h**)

Table 2 Trion and exciton peak information in spectra for WS_2 crystal transferred onto SiO_2/Si by PS transfer (corresponding to Fig. 3i)

	Trion			Exciton			Trion binding energy (meV)	Exciton to trion intensity ratio
	Peak position (eV)	Peak intensity (a.u.)	Peak width (meV)	Peak position (eV)	Peak intensity (a.u.)	Peak width (meV)		
Point 1	1.995	791	58	2.024	1143	38	29	1.4
Point 2	1.995	421	72	2.031	532	42	36	1.2
Point 3	1.995	352	75	2.033	382	42	38	1.1

transferred onto SiO_2/Si, with the edges exhibiting stronger PL intensity, redshifted peak position and narrowed peak width. Because SiO_2 has a hydrophilic surface and a very small dielectric constant of 3.9 compared with water, it was deduced that this PL non-uniformity was induced by water intercalation. It has been reported that the trapped water can be removed by annealing at high source-drain bias or prolonged baking in air [24, 25]. Figure 3i and Table 2 shows three PL spectra collected at points 1–3 from the edge to the center. Unlike the spectra taken before the transfer, all three PL spectra for the as-transferred crystal feature trion emissions comparable to excitons. This strongly suggested that the WS_2 crystal was subjected to doping after the transfer process. The doping was presumed to have been induced by the alkali metal ions of K^+ from the transfer process acting as electron donors. The notable PL peak position blueshift compared with before transfer (~25–35 meV) can be explained by the release of strain induced by high temperature CVD growth after the PS transfer process [20].

After PG washing, the crystal was free of polymer residue, as indicated by the optical image in Fig. 3e. However, the optical characterizations suggested severe degradation of the optical properties of the crystal, reflected in the weak signals in Raman and PL intensity mappings in Fig. 3f–h. The PL spectra fittings in Fig. 3j show a weak single peak at the low energy of 1.950–1.975 eV for each spectrum of points 1–3, consistent with a defect-bound exciton emission derived from defects across the crystal [26]. Further study is required to address the negative impact of PG remover on the transferred WS_2.

To further study the doping induced by alkaline solution, a CVD as-grown monolayer WS_2 crystal was transferred onto a hydrophobic PS surface, using the PS wet transfer to eliminate the influence of intercalated water. As can be seen from the PL intensity, peak position and peak width mappings in Fig. 4a–c, the as-transferred WS_2 crystal on PS exhibited uniform contrast over the entire crystal, with no variation along the edges. This observation strongly supported the theory that the non-uniform PL behavior along the edges stems from ambient water intercalation between the WS_2 crystal and the underlying hydrophilic substrate. The PL spectra collected at points 1–3 on the as-transferred crystal on PS are shown in Fig. 4g, with details presented in Table 3. The features of the three spectra are nearly identical, implying consistent optical behavior over the crystal without trapped water along the edges. The negative trion emission across the crystal with a binding energy of 28 meV was strongly indicative of n-type doping of the alkali metal ions after the PS transfer process.

Ar annealing of the crystal at 100 °C for 1 h was then performed to confirm the effect of alkali doping. The optical mappings of the crystal after Ar annealing are presented in Fig. 4d–f, with the PL spectra at points 1–3 shown in Fig. 4h. At elevated temperatures, impurities, including the alkali dopants, could be detached from the crystal and exhausted by Ar gas flow. Thus, after Ar annealing, it can be seen in Fig. 4h that only a single peak was required to fit the PL spectrum for each point. The peak positions are at ~2.013 eV, corresponding to exciton emission energy. The suppression of trion peaks and the decrease of the PL intensity indicated that electron doping of the crystal was significantly diminished by removal of residue alkali during the annealing process. The exciton peak position was blueshifted compared with

Table 3 Trion and exciton peak information in spectra of WS_2 on PS by PS transfer (corresponding to Fig. 4g)

	Trion			Exciton			Trion binding energy (meV)	Exciton to trion intensity ratio
	peak position (eV)	Peak intensity (a.u.)	Peak width (meV)	Peak position (eV)	Peak intensity (a.u.)	Peak width (meV)		
Point 1	1.976	127	55	2.003	223	37	27	1.8
Point 2	1.975	119	58	2.003	249	37	28	2.1
Point 3	1.974	137	55	2.002	214	39	28	1.6

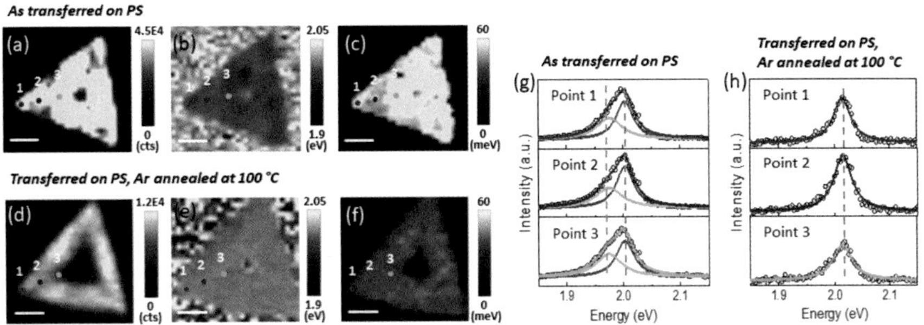

Fig. 4 Optical characterization of a WS_2 crystal transferred onto hydrophobic PS using the PS wet transfer method. **a** PL intensity, **b** PL peak position and **c** PL width mappings of the crystal as transferred onto PS. **d** PL intensity mapping, **e** PL peak position and **f** PL width mappings of the crystal transferred onto PS and annealed by Ar at 100 °C. **g** PL spectra taken at points 1–3 in (**a–c**). **j** PL spectra taken at points 1–3 in (**d–f**). Scale bars: 2 min (**a–f**)

that of the crystal as transferred onto PS (2.003 eV) as a result of reduced electron doping, which also led to decreased intensity and narrowed width of the PL peaks after annealing.

4 Conclusions

We investigated the effects of polymer-assisted wet transfer process and type of substrate on the optical properties of CVD-grown monolayer WS_2 crystals. We found that transfer using PS instead of the traditionally adopted PMMA led to better preservation of the morphology of the transferred WS_2 crystals due to minimized exposure to alkali solution. The electron doping induced by the alkali metal ions from the transfer processes caused trion formation as well as intensified PL emission across the transferred crystal. Annealing by Ar at 100 °C can efficiently remove alkali residue on the transferred crystals. We also demonstrated that ambient water intercalation between the edges of the monolayer crystal and the underlying hydrophilic substrates induces variations in PL due to doping and dielectric screening effects. The non-uniform PL behavior of the WS_2 crystals disappeared when the crystals were transferred onto a hydrophobic PS substrate. These results contribute to the understanding of trion formation in atomically thin TMDs via chemical doping and non-uniform PL behavior stemming from the dielectric screening effect of water intercalation when the material is deposited on hydrophilic substrates. This work has implications for the provision of consistent crystal behavior of 2D TMDs for their applications in sensing and energy harvesting and conservation devices.

References

1. Khan K et al (2020) Recent developments in emerging two-dimensional materials and their applications. J Mater Chem C 8(2):387–440
2. Gant P et al (2019) A strain tunable single-layer MoS_2 photodetector. Mater Today 27:8–13
3. Mahmoudi T, Wang Y, Hahn Y-B (2018) Graphene and its derivatives for solar cells application. Nano Energy 47:51–65
4. Zhang Q et al (2016) Strain relaxation of monolayer WS_2 on plastic substrate. Adv Func Mater 26(47):8707–8714
5. Oliva N et al (2017) Van der Waals MoS_2/VO_2 heterostructure junction with tunable rectifier behavior and efficient photoresponse. Sci Rep 7(1):1–8
6. Novoselov K, et al (2016) 2D materials and van der Waals heterostructures. Science 353(6298):aac9439
7. Liu Y et al (2016) Van der Waals heterostructures and devices. Nat Rev Mater 1(9):1–17
8. Chiu M-H et al (2014) Spectroscopic signatures for interlayer coupling in MoS_2–WSe_2 van der Waals stacking. ACS Nano 8(9):9649–9656
9. Cun H et al (2019) Wafer-scale MOCVD growth of monolayer MoS_2 on sapphire and SiO_2. Nano Res 12(10):2646–2652
10. Xu Z-Q et al (2015) Synthesis and transfer of large-area monolayer WS_2 crystals: moving toward the recyclable use of sapphire substrates. ACS Nano 9(6):6178–6187
11. Zeng L et al (2013) Remote phonon and impurity screening effect of substrate and gate dielectric on electron dynamics in single layer MoS_2. Appl Phys Lett 103(11):113505
12. Godde T et al (2016) Exciton and trion dynamics in atomically thin $MoSe_2$ and WSe_2: effect of localization. Phys Rev B 94(16):165301
13. Sercombe D et al (2013) Optical investigation of the natural electron doping in thin MoS_2 films deposited on dielectric substrates. Sci Rep 3(1):1–6
14. Buscema M et al (2014) The effect of the substrate on the Raman and photoluminescence emission of single-layer MoS_2. Nano Res 7(4):561–571
15. Cong C et al (2014) Synthesis and optical properties of large-area single-crystalline 2D semiconductor WS_2 monolayer from chemical vapor deposition. Adv Opt Mater 2(2):131–136
16. Gutiérrez HR et al (2013) Extraordinary room-temperature photoluminescence in triangular WS_2 monolayers. Nano Lett 13(8):3447–3454
17. Peimyoo N et al (2013) Nonblinking, intense two-dimensional light emitter: monolayer WS_2 triangles. ACS Nano 7(12):10985–10994
18. Kim MS et al (2016) Biexciton emission from edges and grain boundaries of triangular WS_2 monolayers. ACS Nano 10(2):2399–2405
19. Malmberg C, Maryott A (1956) Dielectric constant of water from 0 to 100 C. J Res Natl Bur Stand 56(1):1–8
20. McCreary KM et al (2016) The effect of preparation conditions on Raman and photoluminescence of monolayer WS_2. Sci Rep 6(1):1–10
21. Zhu B, Chen X, Cui X (2015) Exciton binding energy of monolayer WS_2. Sci Rep 5(1):1–5
22. Sivalertporn K et al (2012) Direct and indirect excitons in semiconductor coupled quantum wells in an applied electric field. Phys Rev B 85(4):045207
23. Lin Y et al (2014) Dielectric screening of excitons and trions in single-layer MoS_2. Nano Lett 14(10):5569–5576
24. Shu J et al (2017) Influence of water vapor on the electronic property of MoS_2 field effect transistors. Nanotechnology 28(20):204003
25. Zhang C et al (2012) Electrical transport properties of individual WS_2 nanotubes and their dependence on water and oxygen absorption. Appl Phys Lett 101(11):113112
26. Plechinger G, et al (2015) Identification of excitons, trions and biexcitons in single-layer WS_2. Physica Status Solidi (RRL)—Rapid Res Lett 9(8):457–461

Experimental and Numerical Studies on the In-Plane Shear Behavior of PVC-Encased Concrete Walls

Kamyar Kildashti and Bijan Samali

Abstract The effective application of lightweight stay-in-place concrete forms for casting shear walls subjected to wind and seismic loading is of particular concern to practitioners. Insufficient technical data available for new kinds of wall systems, such as Polyvinyl Chloride (PVC) form walls, hinder their implementation in construction practice. To that end, an effective experimental and numerical campaign was launched at Western Sydney University to investigate the structural performance of PVC form walls when subjected to in-plane shear loading. A set of push-out specimens was designated to conduct monotonic in-plane shear tests until failure. All failure phenomena, capping strengths, and ductility capacities were monitored. Test results indicated that the embedded PVC latticed webs could efficiently protect the concrete web from sudden crushing and improve ductility capacity and failure pattern of the specimens. Nonlinear finite element analysis on test specimens was also conducted and good correlation with experiment results was achieved.

Keywords Stay-in-place forms · Squat concrete walls · Finite element analysis

1 Introduction

The common construction technique of stay-in-place (SIP) concrete forms delivers low installation costs and high speed of construction. A novel SIP form, cast with high-strength fabrics, was developed by Li and Yin [1] and its composite action with inner concrete was studied. Both finite element (FE) analyses and experimental lab works were conducted by various researchers to explore the performance of SIP encasements under mechanical and thermal loading conditions [2–4]. The application of cold-formed sections in the off-site manufacturing of SIP forms was studied by Wu et al. [5]. The so-called Insulated Concrete Forms (ICF) were extensively employed

K. Kildashti (✉) · B. Samali
Centre for Infrastructure Engineering, School of Engineering, Design, and Built Environment, Western Sydney University, Penrith, NSW, Australia
e-mail: k.kildashti@westernsydney.edu.au

W. Duan et al. (eds.), *Nanotechnology in Construction for Circular Economy*, Lecture Notes in Civil Engineering 356,
https://doi.org/10.1007/978-981-99-3330-3_43

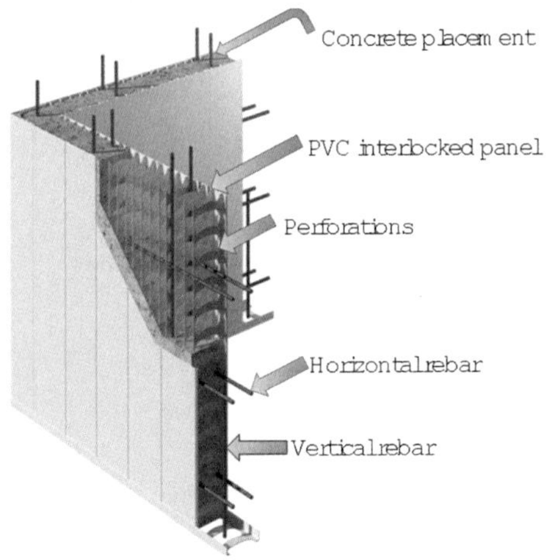

Fig. 1 Geometry of PVC form

to cast either squat or slender walls and their structural composite action with the concrete core were mostly studied under monotonic and cyclic loading regimes [6–10]. A novel stackable Polyvinyl Chloride (PVC) SIP form was introduced to cast concrete walls even with complex geometries [11–13], as shown in Fig. 1. As the PVC form engages with the inner concrete, the new design represents a departure from traditional solid concrete core in responding to in-plane shear loads. Therefore, the in-plane shear behavior of PVC form walls needs to be explored through experimental and numerical studies. In this paper, we present in-plane shear behavior of PVC form walls using experimental tests on push-out specimens as well as FE analysis.

2 Methods and Results

2.1 Test Outline

A total of three PVC form wall push-out specimens were tested to failure to acquire a better understanding of the in-plane shear capacity and ductility with the influence of PVC encasement. Further, one push-out wall specimen was cast using the traditional form to be enabled for direct comparison with PVC form walls. The geometry of specimens along with the test parameters are, respectively, shown in Fig. 2 and in Table 1. These tests were performed to investigate the effects of vertical and horizontal reinforcement ratio, wall thickness, and the PVC encasement. Constants were concrete compressive strength and steel yield stress. Concrete compressive strength and steel tensile yield stress were, respectively, 40 and 550 MPa. A 10,000 kN servo hydraulic machine was employed to apply vertical force parallel to the shear surface. Figure 3 Shows photographs of PVC and traditional form walls secured in the testing rig. Linear Potentiometers (LPs) were employed to measure vertical

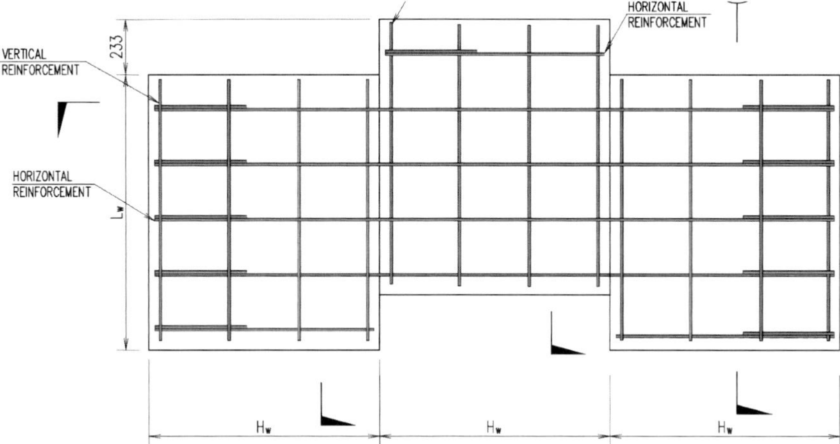

Fig. 2 Geometry of specimens

Table 1 Push-out test specimens

Specimen name	Length L_w (mm)	Height H_w (mm)	Wall thickness t_w (mm)	PVC encasement	Horizontal reinforcement	Vertical reinforcement
F16-RW156S	1165	1000	150	Yes	N12-350 mm	N12-250 mm
F17-RW200D			195	Yes	N12-233 mm	2N12-250 mm
F18-RW200D			195	Yes	N16-233 mm	2N12-250 mm
F19-STW200D			195	No	N16-233 mm	2N16-250 mm

displacement that occurred along the shear plane. The location of LPs at the top and bottom face of specimens is shown in Fig. 4.

2.2 *Test Results*

Qualitative observations of the response of PVC form specimens were recorded during the tests. Most of the observations were visual including crack initiation, growth, and propagation, as well as excessive deformation through PVC encasements. Figure 5 shows the failure shapes of PVC form and traditional form concrete specimens. In Fig. 5a, it is identified while the slip occurred due to concrete failure along shear plane, the excessive deflection in PVC encasement took place due to outward dilatation of concrete. As seen in Fig. 5b, the concrete crushing near the bottom corner close to the shear plane is mostly attributed to substantial slip along the shear plane. In Fig. 5c, inclined shear cracking near the shear plane as well

(a) PVC form specimen

(b) Traditional form specimen

Fig. 3 Specimens in testing rig

(a) Bottom face (b) Top Face

Fig. 4 Instrumentation of specimens

as crushing and spalling of the concrete is associated with failure at shear plane. More specifically, this failure occurred abruptly accompanied with explosive sound. Figure 6 depicts force-slip relationships pertinent to different specimens. As seen, specimens cast with PVC forms demonstrate substantially high ductility compared to one cast with the traditional form. The major reason is the confining action of PVC encasement to protect concrete core against dominating premature spalling off. Further, the higher in-plane shear strength of the traditional form specimen is attributed to the thickness and amount of vertical and horizontal reinforcement.

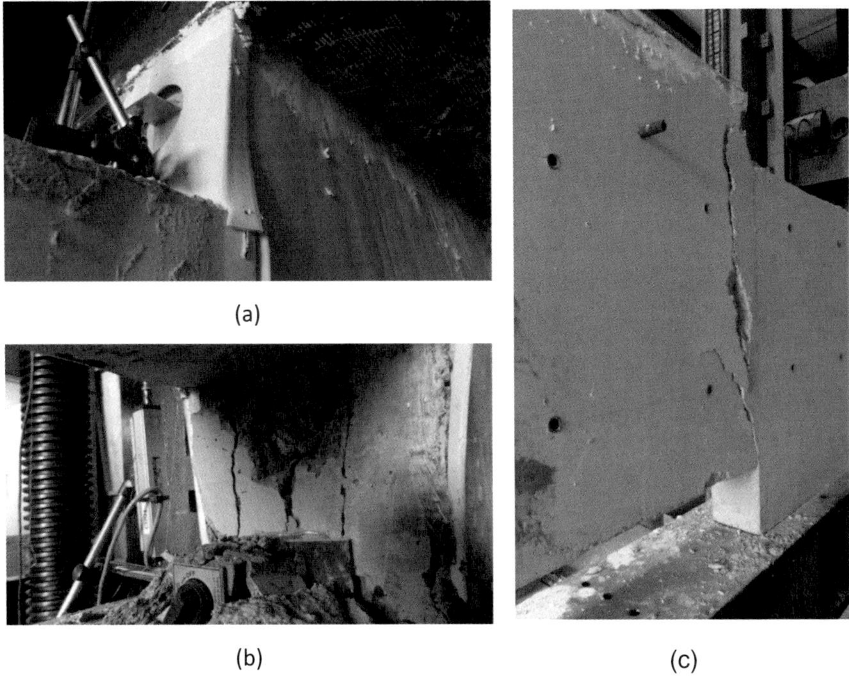

Fig. 5 Phenomena of failure **a** excessive deformation of PVC encasement in a PVC form specimen **b** concrete crushing at the bottom face in a PVC form specimen **c** diagonal concrete crushing in traditional form specimen

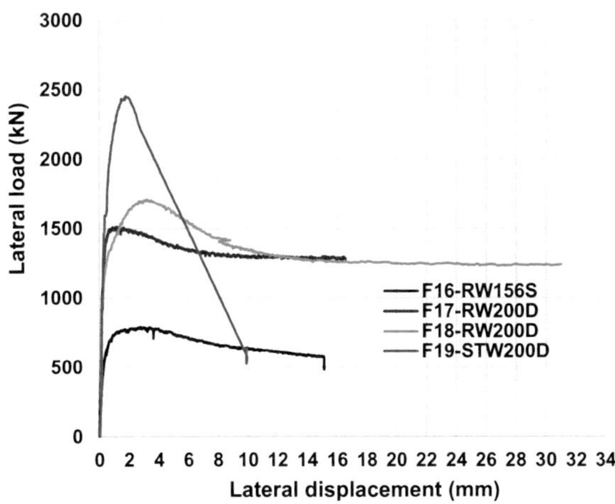

Fig. 6 Force-slip behavior of different specimens

2.3 FE Simulation

In this study, the Abaqus program [14] was employed for FE analyses. The proper selection of constitutive material model played an important role in verifying FE models against experiments. Three built-in material models including *Concrete Damaged Plasticity* (CDP), *Crushable Foam* (CF) [20], and J_2 Plasticity were used, respectively, to model concrete, PVC encasement, and horizontal/vertical reinforcements. The ascending and descending branches of uniaxial stress–strain curves in compression pertinent to CDP model were assumed as follows [15]:

$$\sigma_c = (1 - H[|\varepsilon_c| - |\varepsilon_{c1}|])\frac{\kappa.\eta - \eta^2}{1 + (\kappa - 2).\eta} f_{cm} + H[|\varepsilon_c| - |\varepsilon_{c1}|]\left[\frac{2 + \gamma_c f_{cm}\varepsilon_{c1}}{2 f_{cm}} + \gamma_c \varepsilon_c + \frac{\gamma_c \varepsilon_c^2}{2\varepsilon_{c1}}\right]^{-1} \quad (1)$$

where

σ_c = compression stress in (MPa)

ε_c = compression strain

$$\eta = \varepsilon_c / \varepsilon_{c1}$$

$$\kappa = E_{ci} / E_{c1}$$

γ_c = constant crushing energy [16]

$$\gamma_c = \frac{\pi^2 f_{cm}\varepsilon_{c1}}{2\left[g_{cl}^* - \frac{1}{2} f_{cm}\left(\varepsilon_{c1}(1 - \beta_c) + \beta_c \frac{f_{cm}}{E_c}\right)\right]}$$

g_{cl}^* = volume specific localised crushing energy

ε_{c1} = strain at maximum compressive stress

E_{ci} = tangent modulus = $10^4 f_{cm}^{1/3}$ in (MPa)

f_{cm} = characteristic compressive strength

E_{c1} = secant modulus from the origin to the peak compressive stress

H = Heaviside function

The concrete stress–strain relation in uniaxial tension consisted of a linear phase up to tensile strength, followed by a nonlinear strain softening phase that depended on the specimen geometry. The stress-crack opening relation can be expressed as [17, 18]

$$\sigma_{ct}(w) = \left\{ \left[1 + \left(\frac{c_1 w}{w_c} \right)^3 \right] \exp\left(-\frac{c_2 w}{w_c} \right) - \frac{w}{w_c}(1 + c_1^3) \exp(-c_2) \right\} f_{ctm}$$
$$c_1 = 3, \quad c_2 = 6.93 \tag{2}$$

where

σ_{ct} = tensile stress in (MPa)

$w = l_t \varepsilon_{ct}^{ck} = l_t(\varepsilon_{ct} - \sigma_{ct}/E_{ci})$ = crack opening in (mm)

$w_c = 5G_F/f_{ctm}$ = crack opening when $\sigma_{ct} = 0$ in (mm)

$G_F = 0.073 f_{cm}^{0.18}$ = fracture energy in (N/mm)

$f_{ctm} = 0.3 f_{cm}^{2/3}$ = tensile strength in (MPa)

ε_{ct} = tensile strain

l_t = characteristic length in FE modelling

ε_{ct}^{ck} = cracking strain

List of parameters used for CDP model calibration was presented in [19]. To capture nonlinear behavior of the PVC encasement, parameters of the CF model were calibrated against test results as reported in [11]. To establish yield surface for J_2 Plasticity constitutive law, steel Young's modulus, Poisson's ratio, yield stress, ultimate stress, and ultimate strain was, respectively, assumed as 200 GPa, 0.3, 550 MPa, 650 MPa, and 0.05. Figure 7 shows the comparison between failure phenomena obtained from the experiment and FE simulations. The equivalent plastic strain was chosen from FE analysis to represent localization of damage. As seen, the FE simulation reasonably replicates the failure modes triggered and evolved across the PVC encasement and concrete core. As shown in Fig. 8, there is a good correlation between experimental results and those obtained from FE analysis.

3 Conclusion

Push-out experimental lab tests were conducted to explore in-plane shear capacity of PVC form concrete walls under monotonic loading. Further, a numerical FE model was established to predict the shear behavior of the walls. Experimental observations revealed gentler failure phenomena for PVC form walls compared to the one cast using the traditional form. It was also concluded that PVC encasement provided confinement pressure against concrete outward dilation and therefore enhanced ductility capacity was observed. Further, the established FE model showed correlative results with those obtained from experiments.

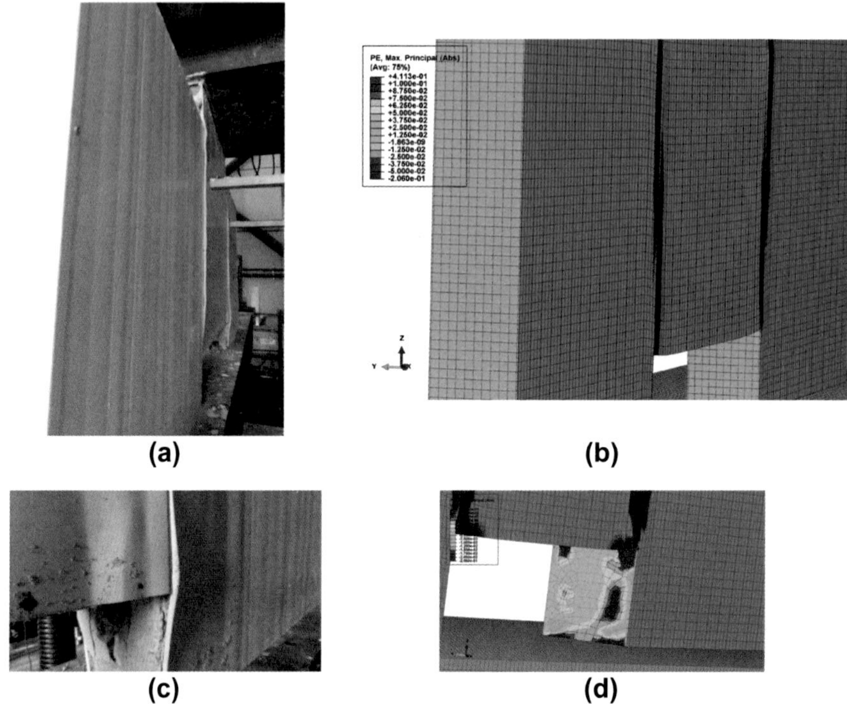

Fig. 7 **a** A photo of excessive deformation on PVC encasement **b** localized plastic deformation on PVC encasement obtained from the FE **c** a phot of concrete crushing at the bottom end **d** damage evolution at the bottom end obtained from FE

Fig. 8 Comparison between FE and test results

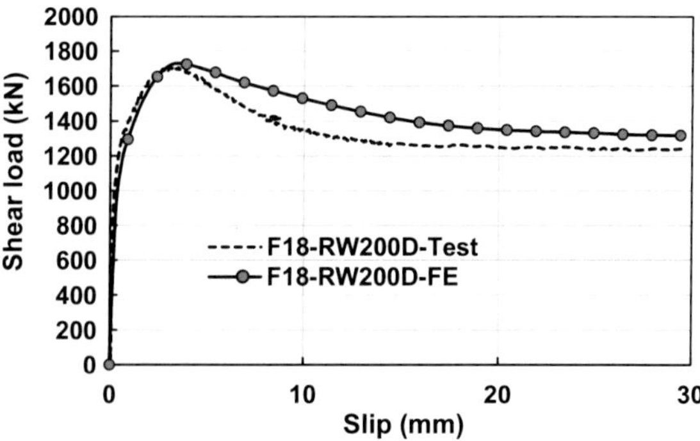

References

1. Li S, Yin S (2021) Research on the mechanical properties of assembled TRC permanent formwork composite columns. Eng Struct 247:113105
2. Li M et al (2018) Modelling of heat transfer through permanent formwork panels exposed to high temperatures. Constr Build Mater 185:166–174
3. Kim GB, Pilakoutas K, Waldron P (2008) Development of thin FRP reinforced GFRC permanent formwork systems. Constr Build Mater 22(11):2250–2259
4. Bruno R, et al (2022) A novel stay-in-place formwork for vertical walls in residential nZEB developed for the Mediterranean climate: hygrothermal, energy, comfort and economic analyses. J Build Eng 45
5. Wu YT et al (2017) Seismic performance of reinforced concrete squat walls with embedded cold-formed and thin walled steel truss. Eng Struct 132:714–732
6. Tang BZ, et al (2019) Seismic performance of RC frames with EPSC latticed concrete infill walls. Eng Struct 197
7. Dusicka P, Kay T (2011) In-plane lateral cyclic behavior of insulated concrete form grid walls. J Struct Eng-ASCE 137(10):1075–1084
8. Zhou Z-Y, Cao W (2016) Experimental study on seismic performance of low-rise recycled aggregate concrete shear wall with single-layer reinforcement. Adv Struct Eng 20(10):1493–1511
9. Lopez A et al (2021) Experimental study of in-plane flexural behavior of screen-grid insulated concrete form rectangular and T-shaped walls. Eng Struct 247:113128
10. Karamlou A, Kabir MZ (2012) Experimental study of L-shaped slender R-ICF shear walls under cyclic lateral loading. Eng Struct 36:134–146
11. Kildashti K, Samali B, Malik A (2020) Experimental and numerical studies on the comparison between stay-in-place- and conventionally-formed reinforced concrete columns under concentric loading. Constr Build Mater 258:119631
12. Kildashti K, et al (2021) Computational simulation of eccentrically loaded reinforced concrete walls formed with modular thin-walled permanent formwork system. J Build Eng 36
13. Wahab N, Soudki KA (2013) Flexural behavior of PVC stay-in-place formed RC walls. Constr Build Mater 48:830–839
14. Simulia DS (2018) Abaqus analysis user's manual. Dassault Syst., Pawtucket, USA
15. Fib model code for concrete structures. Materials 74–150 (2010)
16. Kratzig WB, Polling R (2004) An elasto-plastic damage model for reinforced concrete with minimum number of material parameters. Comput Struct 82(15–16):1201–1215
17. Hordijk DD (1992) Tensile and tensile fatigue behaviour of concrete; experiments, modelling and analyses
18. Birtel V, Mark P (2006) Parameterised finite element modelling of RC beam shear failure. In: ABAQUS users' conference
19. Kildashti K, Nash S, Samali B (2022) In-plane lateral behaviour of PVC modular concrete form squat walls: experimental and numerical study. J Build Eng 52
20. Deshpande VS, Fleck NA (2000) Isotropic constitutive models for metallic foams. J Mech Phys Solids 48(6–7):1253–1283

Recycled Glass-Based Capping Layer for Foundations in Expansive Soils

H. Karami, D. Robert, S. Costa, J. Li, S. Setunge, and S. Venkatesan

Abstract Construction on weak expansive soils is challenging due to their low bearing capacity and high-volume susceptibility under moisture fluctuation. The uplift pressure from expansive clay can induce significant swelling pressure on foundations, but on the other hand, shrinkage of clay can result in substantial foundation settlement during dry seasons. This differential movement of the foundation can distress the superstructure of a building, resulting in serviceable and ultimate limit state failures. The current approach to dealing with foundations in expansive clay soils is to construct a rigid slab that can withstand the anticipated movement or a pier-type approach using engineering design principles or a normal-type slab based on the fill being placed under controlled conditions. We introduce a capping layer under the foundation to control moisture fluctuation and increase the bearing capacity of the foundation using a recycled glass-based stabilization approach. A prototype foundation was constructed in the laboratory using an optimum stabilization mix design that was derived using glass and other sustainable additives. Slab movements and soil conditions were monitored over a 6 month period under simulated dry/wetting moisture fluctuations and operational loads. Results revealed that the performance of the foundation under controlled conditions was outperformed by the capping layered foundation during service loadings including seasonal moisture fluctuation. The outcomes from this research will have a significant impact on improving foundation performance in expansive soils, as well as proposing a sustainable foundation construction process using recycled glass waste.

Keywords Capping layer · Expansive soils · Foundations · Glass waste · Prototypes · Recycling

---

H. Karami · D. Robert (✉) · J. Li · S. Setunge · S. Venkatesan
Civil Engineering Department, School of Engineering, RMIT University, Melbourne, VIC, Australia
e-mail: dilan.robert@rmit.edu.au

S. Costa
School of Engineering, Faculty of Science, Engineering and Built Environment, Deakin University, Waurn Ponds, VIC, Australia

W. Duan et al. (eds.), *Nanotechnology in Construction for Circular Economy*, Lecture Notes in Civil Engineering 356,
https://doi.org/10.1007/978-981-99-3330-3_44

1 Introduction

Expansive soils are a significant hazard for lightweight building foundations and highway pavements, because they experience changes in volume due to seasonal moisture fluctuation, swelling during wet seasons and shrinking during dry periods. If a pavement rests on expansive soil, longitudinal surface cracks may occur because of the seasonal volume change of the subgrade expansive soil [1]. Infrastructure damage due to expansive soils is commonly reported in many countries, such as Australia, Canada, England, China, India, and the USAs. Consequently, the need for change in conventional foundation construction systems is becoming a very important requirement to construct more sustainable houses and buildings with low maintenance costs over their lifetime.

Soil stabilization is an effective way to enhance the durability, mechanical characteristics and to reduce or eliminate the amount of volume change in expansive soils. Using lime and cement for soil stabilization are less cost effective and not environmentally friendly strategies due to the use of energy, resources and carbon footprint produced during the manufacturing process. Moreover, investigations revealed that cyclic wetting and drying cause arresting volume change behavior to be lost after the first wet–dry cycle, and consequently swelling potential increases after each cycle due to the formation of expansive material such as ettringite in calcium-based stabilized soils [2].

Much attention has been given to reducing the utilization of natural resources in cement and other traditional construction materials. For example, the construction sector generated roughly 92.5 million tonnes of asphalt concrete and 4.6 million tons of cement in 2013 and 2015, respectively [3, 4], necessitating alternative construction materials to minimize environmental impact and to preserve natural resources. Hence, the main goal of research by both scientists and engineers has been reducing the demand for natural resources as well as to minimizing the disposal of wastes such as glass [5–7]. Significant research has been performed into the use of waste glass in construction. For example, cullet has been tested as aggregate in the construction sector ranging from concrete and cementitious materials to roadway and asphalt construction. In addition, waste glass can be utilized in the manufacture of ceramic-based products [8].

The mechanical behavior of clay soils can be significantly improved by adding glass powder to the raw soil [9]. Adding fly ash and cement together with recycled glass powder increases the shear strength and CBR of soil [10–12]. Individually, glass powder can increase the strength of cement-stabilized expansive soil and decrease the plasticity index of the soil mixture. Furthermore, the addition of glass powder increases dry density, CBR and UCS with a reduction in optimum moisture content [13]. However, the effect of crushed glass on soaked CBR can be less pronounced compared with unsoaked conditions. In terms of other additives, Phanikumar showed that the addition of fly ash to expansive soils can modify the pore orientation of the soil while significantly improving its compaction behavior [14].

Previous studies identified the optimum mix design using waste glass and secondary additives, proposing specifications for a novel capping layer that can be applied to minimize the impact of soft subgrades on foundations [15]. A series of mechanical tests (UCS, standard compaction test and direct shear test), hydraulic conductivity test (permeability) and microscopy test (X-ray diffraction, scanning electron and porosity test) were carried out to explore the optimum combination of glass wastes and secondary additives for field trials in expansive soils.

The aim of current study is to evaluate the performance of a capping layer by applying it in large-scale tests as verification of the proposed novel capping layered foundation for buildings and roads. A laboratory experiment was carried out to investigate the performance of a foundation slab placed on an expansive soil. The proposed capping layer was placed between the foundation and weak subgrade clays to evaluate the foundation's performance under environmental and operational loads. Performance of the foundation and soil conditions were carefully monitored across a period of time under operational loads for verification of the proposed capping layered foundation system.

2 Methods

2.1 *Materials*

2.1.1 Soil Characteristics

Soil was obtained from a land excavation site in Melbourne, Australia. The soil was categorized as fine-grained (CL) [16] with a high degree of expansion. All the physical properties of the soil are shown in Figs. 1, 2 and Table 1.

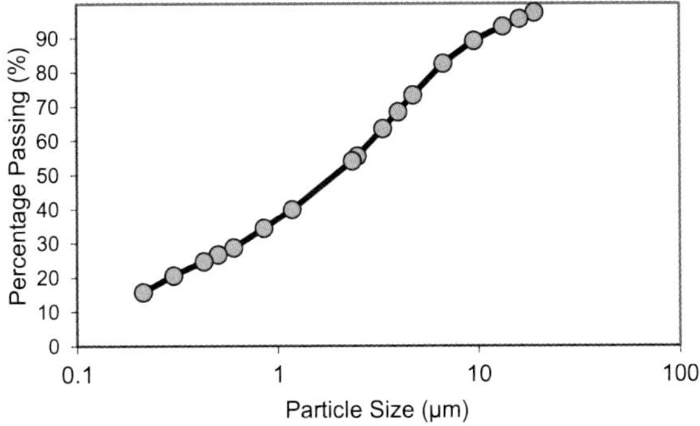

Fig. 1 Particle size distribution of soil

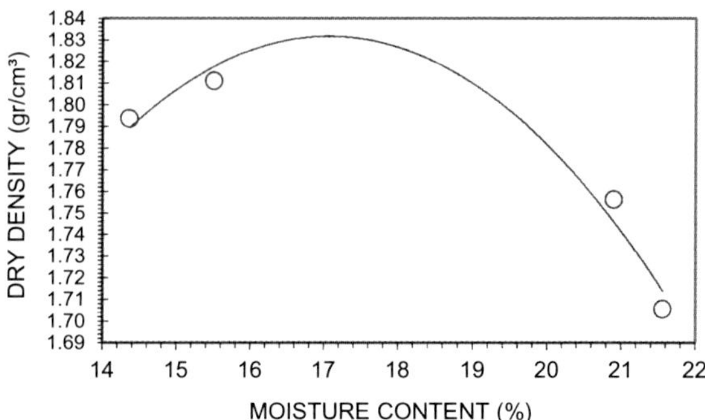

Fig. 2 Soil standard compaction test result

Table 1 Summary of soil characteristics

Property	MDD g/cm^3	OMC %	LL %	PL %	PI
Value	1.832	17.10	36.9	20.3	16.6

2.1.2 Crushed Recycled Glass Powder

Crushed glass behaves like natural rock and is totally inert and non-biodegradable. Glass powder (GP) was used in this study. Particle size distribution and samples are shown in Figs. 3 and 4 respectively.

2.1.3 Fly Ash

Class F fly ash from a local supplier was the primary additive used in the capping layer. For all types of fly ash, the particle size distribution is similar to that of silt

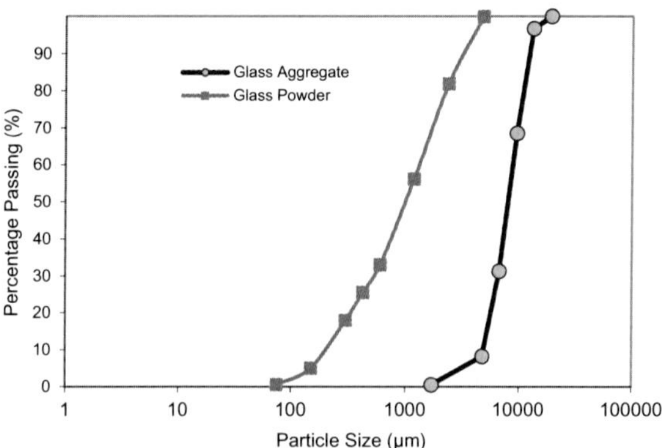

Fig. 3 Particle size distribution of glass powder and glass aggregate [15]

Fig. 4 Glass powder used in the study [15]

(Fig. 5). Typical physical and mechanical properties for fly ash are shown in Table 2. Other characteristics of the fly ash including chemical composition can be found in Karami et al. [17].

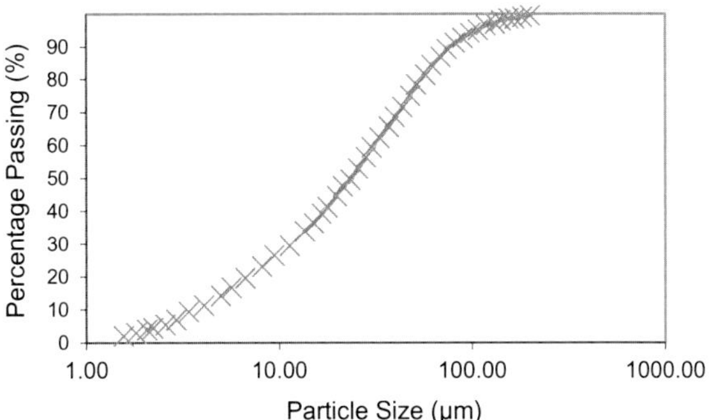

Fig. 5 Fly ash particle size distribution [17]

Table 2 Physical properties of fly ash [14]

Property	Value
Class	F or low lime fly ash
Specific gravity	2.14
Liquid limit	43
Plastic limit	Non-plastic
Optimum moisture (%)	34
Max. dry density (gr/cm^3)	1.1

Table 3 Physical and chemical properties of hydrated lime [14]

Property	Value	Property	Value
Specific gravity (g/cm^3)	2.05	Calcium hydroxide $Ca(OH)_2$ (%)	82
Density (kg/m^3)	510	Chloride (%)	0.01
Normal consistency (%)	43.50	Silicon dioxide (SiO_2) (%)	2.5
Initial setting time (min)	165	Aluminum (Al_2O_3)/iron (FeO) (%)	3.5
Final setting time (h)	46.25	Water (H_2O) (%)	0.6
Fineness (% by weight on 300 μm sieve)	2.65	Arsenic (%)	0.0004
Soundness (Le Chatelier's expansion mm) (mm)	1.8	Lead (%)	0.0001
Compressive strength (14 days) (N/mm^2)	1.45	Sulfate (SO_3) (%)	0.9
Compressive strength (28 days) (N/mm^2)	2.18	Magnesium oxide (MgO) (%)	3.5

2.1.4 Lime

Hydrated lime, which is the most concentrated form of lime, was used as a secondary additive. Lime decreases the liquid limit of soil mixtures, resulting a reduction in plasticity index. Typical physical and engineering properties of hydrated lime are similar to what was used in this study are shown in Table 3.

2.2 *Experimental Set-Up*

2.2.1 Stage 1: Control Test

Preparation of the soil required an extremely thorough and careful process to obtain as homogeneous material as possible. In preparation for the test, the natural soil was first sequentially sieved with a 25 mm sieve. A large-scale soil box (L: 1900 mm, W: 750 mm, H: 1000 mm) was built to enable the control of hydraulic and mechanical processes of the soil by changing the water table in the soil. A drain pipe and a valve system were installed at the bottom of the box (i.e., at a depth of 1000 mm) to allow drainage and to control the water table inside the box. Sieved soil was mixed with water in a concrete mixer until a homogenous mixture with optimum moisture content was obtained. The soil was then placed in the test box in 4 layers and compacted in 100 mm rises (Table 4). Figure 6 shows the details of the test box setup. A geotextile sheet was placed over the bottom surface of the box, a 100 mm thick drainage layer made of gravel was placed on top and finally a filter paper before the soil. The top soil surface at the top was left exposed to atmospheric conditions to facilitate evaporation and infiltration. The experiment began late December 2022 during the summer, and continued for 6 months.

Table 4 Soil layer characteristics and thickness for stages 1 and 2

Stage 1			
Layer	**Depth (mm)**	**Material**	**Color**
A	650–820	Reinforced concrete slab	
B	100–500	Raw Soil	
C	0–100	Gravel	

Stage 2			
Layer	**Depth (mm)**	**Material**	**Colour**
A	650–820	Reinforced concrete slab	
B	500–650	Capping layer	
C	100–500	Raw soil	
	0–100	Gravel	

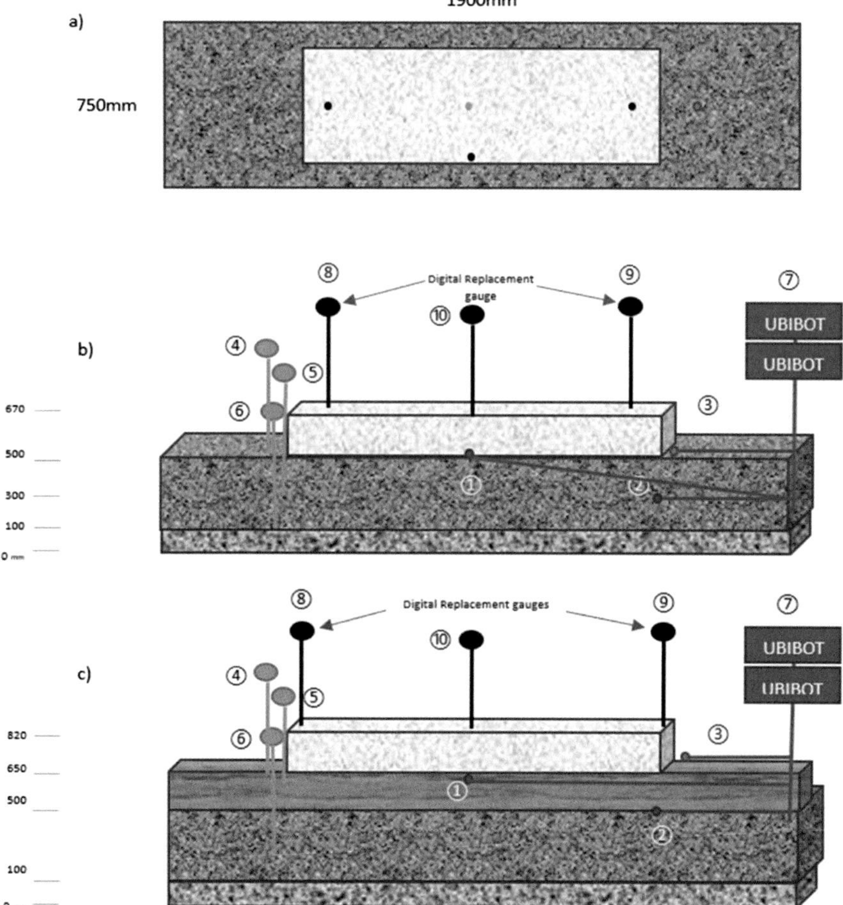

Fig. 6 Proposed sensor locations for prototype test: **a** top view, **b** stage 1, **c** stage 2

Table 5 Instruments

No.	Description
1	UBIBOT soil moisture probe
2	UBIBOT soil moisture probe
3	Humidity and temperature probe
4	Surface moisture meter
5	Middle moisture meter
6	Bottom moisture meter
7	UBIBOT soil moisture monitoring station
8	Edge movement gauge
9	Edge movement gauge
10	Central movement gauge

Table 6 Load calculations

Item	Standard residential load (kPa)	Actual applied load to the soil (kN)
Dead load	3.35	1.38
Live load	2	0.82
Total	**5.35**	**2.2**

As shown in Fig. 6 and Table 5, six moisture probes and three digital movement sensors were installed in different locations. Two temperature/humidity sensors were also installed in the test box to record both the soil and room temperatures. Data were recorded in the controller at a frequency of 1 h and transferred to a computer daily. In addition, weekly manual readings of the movement were taken. A sprinkler system was constructed above the sandbox to simulate rain events similar to field conditions. Water spray was based on the water table depth at specific times and amounts. Water percentage was obtained from the swelling test results for raw soil according to Australian Standard (AS 1289.7.1.1-2003). The amount of water was calculated based on the raw soil amount in the box (Table 7).

For Stage 1, a concrete slab scaled down to L: 750 mm, W: 550 mm, and H: 100 mm was placed on top of the compacted soil to apply the dead loads during the test period. The size of the concrete block was scaled down from an actual designed slab for a residential building following AS 2780. In this test, the dead load

Table 7 Water added to the sample

Water percentage (%)	Time (weeks)	Amount (L)
2.20	W2	22.36
1.91	W4	19.41
3.15	W6	32.01

was added to the slab for increasing the thickness of the slab from 100 to 170 mm for a residential-purpose building. The live load was applied to the soil by specific weights at the end of wetting and drying periods. The standard load amounts and the calculated applied load based on the slab area are shown in the Table 6.

2.2.2 Stage 2: Test with Capping Layer

In Stage 2, all steps in Stage 1 were repeated, but a capping layer was constructed on top of the raw soil (Fig. 6c). The design mix of the capping layer was based on previous studies [15]. Materials were mixed with soil in percentages of 25% WG powder, 7.5% fly ash and 3% lime by weight compared with raw soil. To prepare the stabilized soil, all ingredients were mixed properly in the dry state in a mechanical mixer. Water was added to the mixture up to the optimum moisture content according to the Standard Compaction test and well mixed. Similar to Stage 1, the soil (subgrade soil) was uniformly compacted in the box. The size of the installed capping layer was same as the raw soil in length and width to cover the whole soil surface area. Thus, the size of the capping layer was L: 1900 mm, W: 750 mm, and H: 100 mm.

3 Results and Discussion

3.1 Slab Settlement

At the beginning of both stages of the prototype test, the slab started settling down for the first few days after it was placed on the soil surface, due to the weight of both the slab and the capping layer on the raw soil, which caused a secondary compaction process during those first few days. The results show that settlement in Stage 2 was greater than in Stage 1 at early stages. The maximum settlement in Stage 1 was 0.06 mm and for Stage 2, it was 0.10 mm after week 1 for the control and stabilized capping layer tests respectively (Fig. 7). The larger settlement in Stage 2 resulted from elastic compression of the capping layer, which consisted of coarser materials than raw soil. However, this increment in initial settlement was negligible and did not imply a reduction in bearing capacity.

It is important to note that the subsequent slab movement due to swelling was reduced when the capping layer was introduced. The difference in maximum slab movement between the two stages was 0.2 mm. More importantly, the rate of increase in slab movement was less with the capping layer in place. In the control test, slab movement reached 0.4 mm in less than 5 weeks, whereas it took nearly 9 weeks to move by the same amount under stabilized conditions. Therefore, it can be concluded that the capping layer was effective in mitigating the magnitude and the rate of the vertical slab movement.

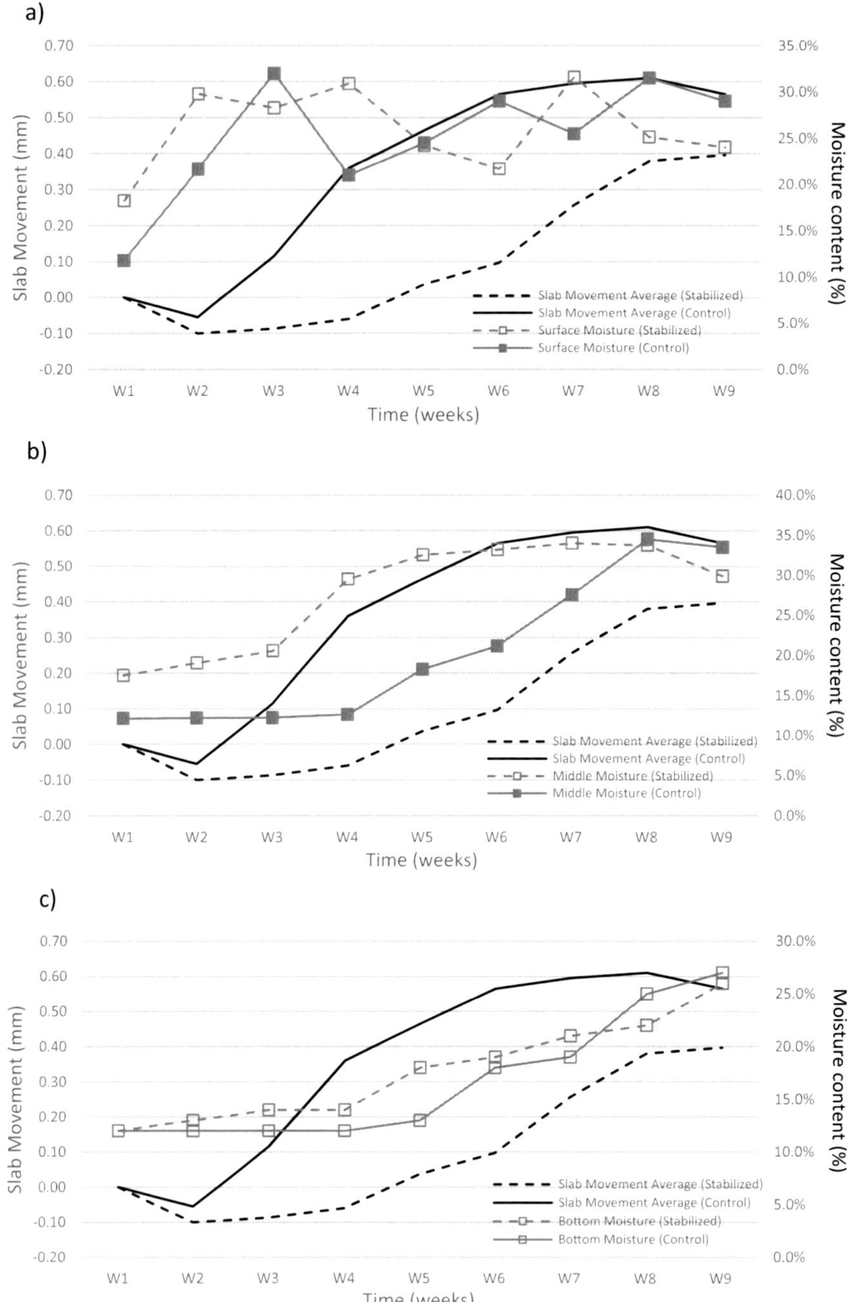

Fig. 7 Slab movement and moisture profiles for stage 1 and stage 2 during the wetting period from week 1–9. Horizontal profiles: **a** at soil surface level, **b** 200 mm below the soil surface level, and **c** at the base of the soil layer for Stage 1 and 100 mm above the base for stage 2

3.2 Expansive Soil Movement (Swelling)

The soil started swelling and increasing in volume as expected for any expansive soil. During the wetting period, the soil in the control test swelled by 0.6 mm on average (maximum 0.72 mm for individual gauges). In Stage 2, with the capping layer in place, the slab moved by only 0.397 mm (maximum 0.61 mm for individual gauges) (Fig. 7a–c). As shown in Fig. 7, the swelling in Stage 2 of the test with the stabilized capping layer was lower than with the control soil during the wetting period.

Because the capping layer was more permeable than the control soil, it assisted water to drain away faster and remove it from the slab and the expansive soil underneath. Moreover, the load applied to the soil from the weight of the capping layer helped to arrest or reduce the swelling of the expansive soil.

3.3 Moisture Distribution

Water was sprayed on top of the soil with the same pattern (time, amount, and size of spray nozzle) for both stages of the test. As shown in Fig. 7a, the capping layer tended to be mostly dryer than the raw soil in the control test, because of the higher permeability of the capping layer (Fig. 8). In the control test, water pooled on the surface or remained at a shallow depth (based on visual observations) after a wetting event whereas more water infiltrated the soil through the capping layer in Stage 2.

This was evident from the moisture variation at 200 mm from the surface (Fig. 7b). Moisture content at the raw soil-capping layer interface was higher than the moisture content recorded at 200 mm within the raw soil in the control test (Fig. 9). This was due to the rapid flow of water through the capping layer. However, the capping layer

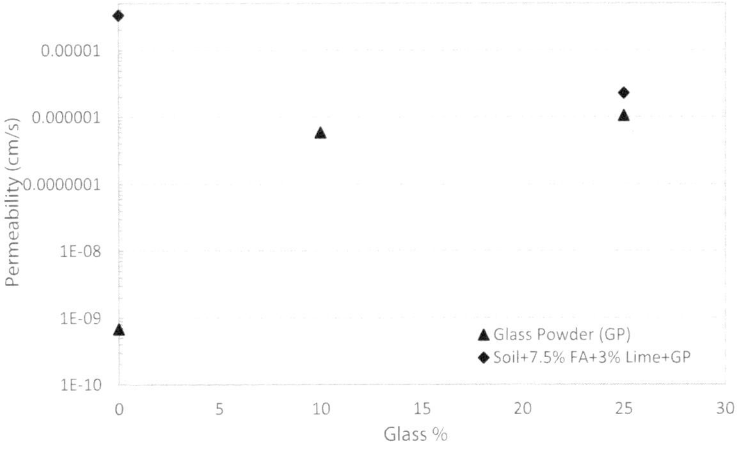

Fig. 8 Permeability results for soil–glass powder and soil–fly ash–lime–glass powder mix [17]

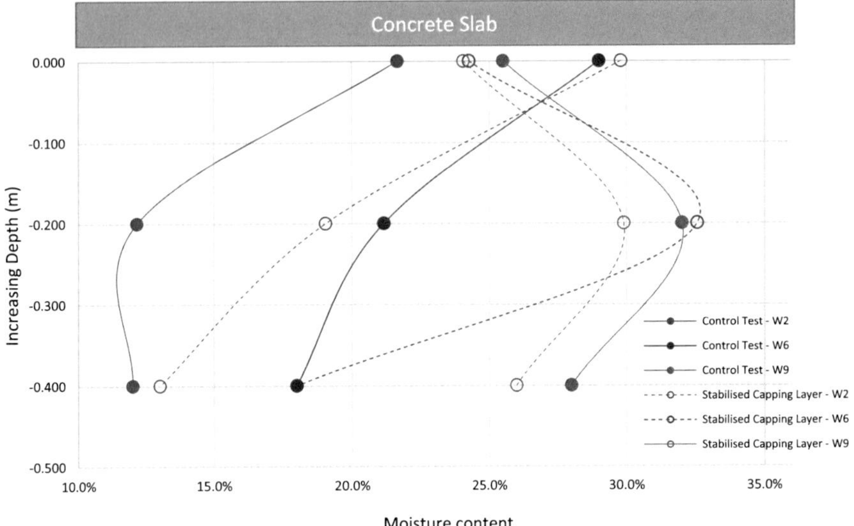

Fig. 9 Variation of water content in control soil and stabilized capping layer in the wet season of the test: vertical profile

was effective in delaying water reaching the raw soil layer. Without the capping layer, the moisture content at the surface of raw soil was >30% in less than 3 weeks. When the capping layer was in place, it took almost 5 weeks for the water content at the top of the raw soil layer (interface between the raw soil and capping layer) to reach >30%. Consequently, the raw soil layer remained drier for a longer period, which will reduce the vertical movement of the slab.

The moisture variation shown in Fig. 7c corresponded to the condition at the base of the control test and at 400 mm depth below the surface for Stage 2. Interestingly, the difference in moisture content at these locations was narrow, indicating the raw soil layer remained drier in Stage 2 throughout the wetting period. Thus, it was evident that the capping layer kept the water away from the expansive soil underneath.

4 Conclusions

This research investigated a novel sustainable foundation system that could mitigate the adverse influence of expansive soil conditions for construction. The proposed approach is based on a capping layer constructed from recycled glass waste and other sustainable additives and placed directly above the existing expansive soil. A prototype foundation was constructed in the laboratory using an optimum stabilization mix design derived from a detailed investigation of mechanical and hydraulic characteristics. The slab movement and soil conditions were monitored over 6 months under simulated dry/wet moisture fluctuations and operational loads. Results showed that the foundation's performance on the novel capping layer was significantly productive compared with control conditions. The higher efficiency of the new capping layer system was evidenced by 35% reduction in slab displacement during wet season simulation, which is a significant improvement in foundation serviceability on expansive soils. The outcome from this research will have a significant impact on minimizing infrastructure maintenance costs, which are a heavy burden for asset managers in any country. The proposed capping layer, which incorporates waste materials (32.5% of total), is a waste management strategy that minimizes the serviceability concerns of civil infrastructure while introducing a value-added benefit for waste materials. It is to be noted that the results provided in this document are based on the referred tests/materials conditions and should not be used in field applications without appropriate verification.

Acknowledgements This research was funded by the RMIT postgraduate scholarship and the Australian Research Council Industrial Transformation Research Hub (grant no. IH150100006, IH200100010).

References

1. Khan MA, Wang JX, Sarker D (2020) development of analytic method for computing expansive soil-induced stresses in highway pavement. Int J Geomech 20:04019160
2. Al-Rawas AA, Hago A, Al-Sarmi H (2005) Effect of lime, cement and Sarooj (artificial pozzolan) on the swelling potential of an expansive soil from Oman. Build Environ 40:681–687
3. Du H, Tan KH (2017) Properties of high volume glass powder concrete. Cem Concr Compos 75:22–29
4. Adaway M, Wang Y (2015) Recycled glass as a partial replacement for fine aggregate in structural concrete—effects on compressive strength. Electron J Struct Eng 14:116–122
5. Mohajerani A, Tanriverdi Y, Nguyen BT, Wong KK, Dissanayake HN, Johnson L, Whitfield D, Thomson G, Alqattan E, Rezaei A (2017) Physico-mechanical properties of asphalt concrete incorporated with encapsulated cigarette butts. Constr Build Mater 153:69–80
6. Mohajerani A, Vajna J, Cheung THH, Kurmus H, Arulrajah A, Horpibulsuk S (2017) Practical recycling applications of crushed waste glass in construction materials: a review. Constr Build Mater 156:443–467
7. Ukwatta A, Mohajerani A, Eshtiaghi N, Setunge S (2016) Variation in physical and mechanical properties of fired-clay bricks incorporating ETP biosolids. J Clean Prod 119:76–85
8. Majdinasab A, Yuan Q (2019) Post-consumer cullet and potential engineering applications in North America. Resour Conserv Recycl 147:1–9
9. Bilondi MP, Toufigh MM, Toufigh V (2018) Experimental investigation of using a recycled glass powder-based geopolymer to improve the mechanical behavior of clay soils. Constr Build Mater 170:302–313
10. Turgut P (2013) Fly ash block containing limestone and glass powder wastes. KSCE J Civ Eng 17:1425–1431
11. Baldovino JDJA, Izzo RLDS, Silva ÉRD, Rose JL (2020) Sustainable use of recycled-glass powder in soil stabilization. J Mater Civ Eng 32:04020080
12. Olufowobi J, Ogundoju A, Michael B, Aderinlewo O (2014) Clay soil stabilisation using powdered glass. J Eng Sci Technol 9:541–558

13. Ikara I, Kundiri A, Mohammed A (2015) Effects of waste glass (WG) on the strength characteristics of cement stabilized expansive soil. Am J Eng Res (AJER) 4:33–41
14. Phanikumar RSS (2007) Volume change behavior of fly ash-stabilized clays. Mater Civ Eng 19:8
15. Karami H, Sidiq A, Costa S, Robert D, Li J., Setunge S (2022). Hydro-mechanical and microstructural properties of an expansive soil stabilized with waste glass and other additives. Under Review
16. ASTM-D2487 (2017) Standard practice for classification of soils for engineering purposes (unified soil classification system), vol 17e1. ASTM International, West Conshohocken, PA
17. Karami H, Pooni J, Robert D, Costa S, Li J, Setunge S (2021) Use of secondary additives in fly ash based soil stabilization for soft subgrades. Transp Geotech 100585

Submicroscopic Evaluation Studies to Minimize Delayed Ettringite Formation in Concrete for a Sustainable Industry and Circular Economy

Yogesh Kumar Ramu, Paul Stephen Thomas, Kirk Vessalas, and Vute Sirivivatnanon

Abstract The high cost of maintenance, repair and retrofitting of concrete infrastructure to keep these structures durable and serviceable is not sustainable, so the design process needs to consider all aspects of deterioration mechanism/s that can potentially occur in a concrete structure. The ideal solution should contribute to sustainability by enhancing the durability of concrete elements and supporting a circular economy. We studied delayed ettringite formation (DEF), a potential deterioration mechanism, including mitigation measures, in various heat-cured cementitious systems. The results showed that continuously connected pore/crack paths at the submicroscopic level favor the transportation of DEF-causing ions in heat-cured systems. DEF increases the chance of developing cracks, which is a durability concern. To mitigate DEF, fly ash produced from an Australian bituminous coal-burning power station was incorporated in the binder to support the circular economy concept. Changes in heat-cured cementitious systems were evaluated using expansion, electrical resistivity, dynamic modulus, and microstructural studies. The pozzolanicity of fly ash was found to greatly enhance the formation of denser calcium-silica-hydrate, which in turn restricted the transportation of DEF-causing ions at the submicron level, leading to less DEF occurrence and enhancement of the durability and sustainability of concrete in field structures.

Keywords Circular economy · Delayed ettringite formation · Fly ash · Precast concrete · Sulfate attack

---

Y. K. Ramu (✉) · P. S. Thomas · K. Vessalas · V. Sirivivatnanon
School of Civil and Environmental Engineering, University of Technology Sydney, Ultimo, NSW, Australia
e-mail: yogeshaccet@gmail.com; yogesh.ramu@uts.edu.au

W. Duan et al. (eds.), *Nanotechnology in Construction for Circular Economy*,
Lecture Notes in Civil Engineering 356,
https://doi.org/10.1007/978-981-99-3330-3_45

1 Introduction

Modern-day concrete infrastructure requires routine maintenance, repair and/or retrofit to ensure structures achieve their intended service life without compromising durability and sustainability. Unfortunately, the modernization of the construction industry solved a problem but created another. The precast concrete industry produces offsite concrete members to enable faster construction throughput rates for infrastructure megaprojects, but deleterious/expansive delayed ettringite formation (DEF), which has been linked to crack formation, is a significant concern for this industry [1–5] Elevated temperature curing (>70 °C) in precast concrete production is believed to be primarily responsible for inhibiting or decomposing early ettringite and, at a later stage, recrystallizing ettringite results in expansion and cracking [6]. Cracking in concrete causes permeability, which increases the risk of infiltration of external harmful agents (e.g., sulfates and chlorides) that negatively impact concrete durability. However, despite the potential risk, Australia lacks research in DEF, creating uncertainty within the industry. In this study we addressed the issue by evaluating Australian cements and their susceptibility to DEF under heat-cured conditions. The influence of sulfate and alkali spiking of these cements on DEF was also systematically studied and as mitigation of expansive/deleterious DEF, the incorporation of fly ash (FA) to enhance the durability, and thereby the sustainability of concrete, was also studied.

2 Materials and Methods

Two Australian general-purpose cements conforming to AS 3972:2010 [7] and locally sourced Gladstone FA, complying with AS 3972:2010 [8], were used. The composition of the binders was determined by X-ray fluorescence analysis (Table 1). The specific surface area of Cement 1 (C1) and Cement 2 (C2) was 395 m^2/kg and 420 m^2/kg, respectively. The Na_2O_{eq} content of both cements (C1: 0.5%; C2: 0.35%) complied with the Australian Technical Infrastructure Committee's specification ATIC SP-43 [9]. C1 contained 7.5% limestone as mineral addition and C2 was free from mineral addition. Modified Bogue's phase composition analysis indicated that C2 had a higher tri-calcium aluminate, C_3A (11.15%) content than C1 (7.48%). Sand sourced from Rockhampton, Queensland, was used to prepare the mortar prisms. Distilled water was used to prepare all mortar and paste specimens used in the study. Calcium sulfate and sodium hydroxide were used to chemically modify the cements.

Four different cement systems were used for mortar prisms prepared with cement:sand:water mass ratio of 1:3:0.45. Mixing of mortars was carried out according to AS2350.12-2006 [10]. Mortars were cast in steel molds (complying with AS 2350.11-2006 [11]) of size 40 × 40 × 160 mm, incorporating steel expansion studs. Three prisms per mix were prepared for the expansion studies. Cement paste

Table 1 Chemical composition of binders by oxide analysis

Oxide %	CaO	SiO_2	Al_2O_3	Fe_2O_3	SO_3	MgO	K_2O	Na_2O	TiO_2	LOI	Total
Cement 1 (C1)	64.2	19.7	4.8	3.1	2.4	0.9	0.4	0.2	0.2	4.1	99.9
Cement 2 (C2)	65.7	22.8	4.4	0.2	3.0	0.7	0.3	0.1	0.2	2.2	99.6
Fly ash	3.9	55.5	24.4	6.9	0.2	1.2	0.9	0.7	1.0	2.5	97.2

specimens were also prepared for characterization studies. To investigate the impact of different cement systems and heat curing on DEF, three cementitious systems and heat-curing regimes were designed: (1) cements without any additional sulfates (C1, C2); (2) cement systems with 4% sulfate and 1% alkali added (C1SN, C2-N); and (3) cement systems with 25% FA, 4% sulfate and 1% alkali added (C1SNF, C2SNF). These cement systems were used to make mortar specimens that were subjected to two different regimens (Fig. 1): ambient temperature (23 °C) cured in a humidity chamber (relative humidity (RH) 90%) for 24 h (C1-A, C2-A); heat-curing in a programmable laboratory oven. For the heat-curing process, the specimens were pre-cured at 30 °C for 4 h, followed by a temperature ramp to 90 °C, which was maintained for 12 h before natural cooling to 30 °C (≈6 h) (C1-H, C2-H, C1SN-H, C2SN-H, C1SNF-H, C2SNF-H). In both curing regimes, specimens were demolded 24 h after casting and transferred to a lime-saturated water bath for further curing. Changes in the length of the specimens were monitored in accordance with AS 2350.13-2006 [12], using a digital length comparator with an accuracy of 0.001 mm. The frequency of length change monitoring was every 7 days for the first 90 days, then every 30 days up to 600 days and after that, every 90 days.

For the characterization studies, the hydration of the cement pastes was arrested with solvent replacement methods at the respective ages. Hydration-arrested samples were kept in a desiccator with silica gel (as the drying agent) under laboratory-controlled conditions (23 ± 2 °C). The stored samples were tested for phase changes using Thermo gravimetric analysis. Thermal analysis was conducted with a Netzsch Jupiter F5 STA instrument using helium flowing at 40 mL/min on 20 ± 0.3 mg cement powder over a temperature range of 40–1000 °C. Dynamic modulus was calculated by the ultrasonic pulse velocity test method. Bulk electrical resistivity was evaluated by conducting electrochemical impedance spectroscopy on mortars and the results are provided in a previous study [13].

3 Results and Discussion

Figures 2 and 3 show the linear expansion trends of the cement mortar systems made with the two cements, which were exposed to ambient (23 °C) and heat-cured (90 °C) cycles as described. It can be seen that the ambient-cured mortars

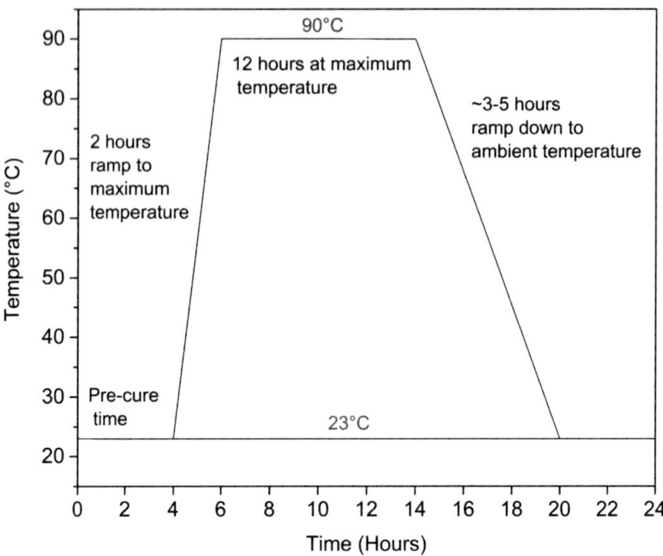

Fig. 1 Representation of ambient and heat curing regimens for cement mortars

(C1-A and C2-A) showed no expansion. Cement mortars C1-H and C2-H (in as received condition) satisfied Australian cement standards [7] and specifications [9] by showing no expansion, despite being cured at 90 °C, likely due to the lesser volume precipitation of DEF typically observed in low sulfate cements [14].

In contrast, the chemically modified cement mortars (C1SN-H, C2SN-H) containing higher levels of sulfate (4%) and alkali (1%) showed significant DEF expansion, aligning with other research studies [15, 16]. Of the chemically modified cements, C2SN-H expanded more than C1SN-H, likely because of the higher C_3A content of C2 (11.15%) compared with C1 (7.5%), which translates to higher DEF and corresponding expansion. Furthermore, from the 6 month (0.5 year) expansion data, the rate of expansion of the C2SN-H mortar was higher than the C1SN-H, also likely due to the higher quantity of C_3A, which is known to accelerate hydration reactions.

Although C1-H and C2-H did not show any significant DEF expansion, they may have formed porous hydrates because of the accelerated heat-curing regimen. The data in Table 2 support this hypothesis as the dynamic modulus and bulk resistivity values were less than those for the ambient-cured systems. This finding infers poor quality microstructure despite no observable expansion. For the highly expansive mortars, C1SN-H and C2SN-H, dynamic modulus and bulk resistivity values were found to be lower than for C1-H and C2-H, which indicated that other than the formation of porous hydrates, expansive DEF may be causing microcracks and further deteriorating the microstructure. Hence whether heat-cured cementitious systems undergo expansive DEF or not, there will be durability issues. DEF only further

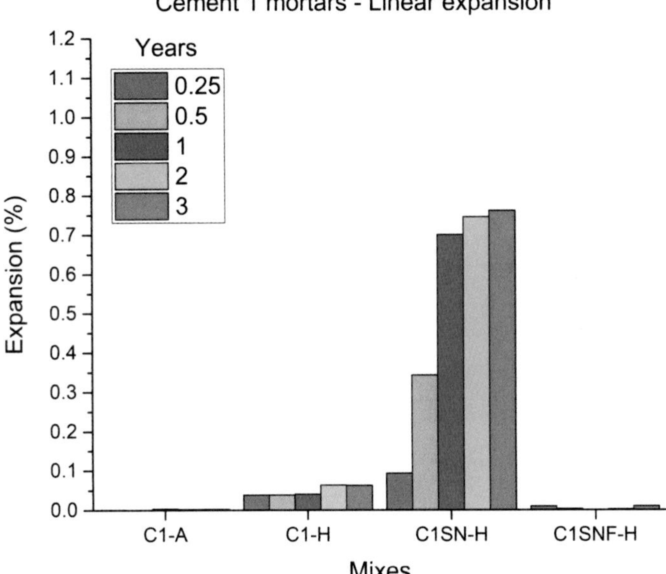

Fig. 2 Linear expansion of cement 1 mortars over 3 years

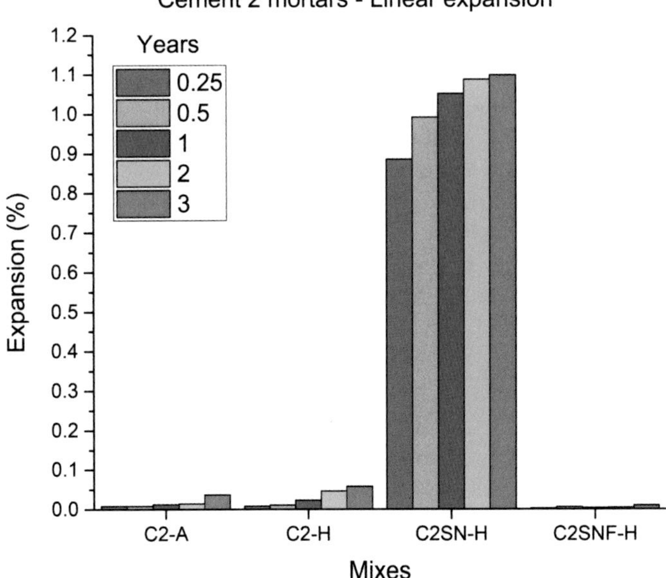

Fig. 3 Linear expansion of cement 2 mortars over 3 years

Table 2 Later-age mechanical characteristics of cement mortars

Mix	Physical characteristics at 1 year		
	Expansion (%)	Dynamic modulus (GPa)	Bulk resistivity (Ω)
C1-A	0.006	46	30
C2-A	0.015	37	22
C1-H	0.067	25	9
C2-H	0.049	26	12
C1SN-H	0.729	21	7
C2SN-H	1.058	18	7
C1SNF-H	0.001	350	51
C2SNF-H	0.003	309	40

complicates the durability issue. Therefore, the solution for DEF mitigation must also address mitigation of porous hydrates forming in these cementitious systems.

FA was included in this study to investigate its ability to mitigate expansive DEF and restrict the formation of porous hydrates. Although there has been significant research undertaken in understanding mitigation of DEF using FA [17–19], this study is unique because we tested the ability of FA to mitigate DEF in high-sulfate and high-alkali cements. As shown in Figs. 2 and 3, complete elimination of expansive DEF occurred with additional FA in both C1SNF-H, and C2SNF-H mortars despite the cement being spiked with 4% sulfate and 1% alkali. This result could be partially due to a change in the dissolution behavior of the cement, thereby reducing the overall content of aluminum and sulfate ions available in the pore solution. Thus, less ettringite would precipitate in the cementitious system. From the data in Table 2, the FA addition led to significant improvement in the dynamic modulus and bulk resistivity values, contributing to a denser and high-quality microstructure. The denser microstructure is likely due to the pozzolanicity of FA, which converts porous (10 μm–10 nm) portlandite to denser (10–0.5 nm) calcium–silica–hydrate (C–S–H). For comparison, the portlandite (calcium hydrate (CH)) content of the heat-cured cementitious systems was also studied, as shown in Fig. 4. It is evident that the portlandite content in C1SN-F was the lowest, which represents that the greater reaction of CH with reactive silica in the FA formed additional C–S–H, thus making the microstructure denser, in turn restricting the transportation of DEF-causing ions at the submicron level and eliminating any increase in deleterious DEF occurrence.

Thus, it can be deduced that incorporating FA in heat-cured cementitious systems eliminates DEF expansion and improves the microstructure by densification through the production of more gel pores and less capillary pores due to the formation of additional C–S–H. Based on our results, the following microstructural pattern (Fig. 5) is hypothesized from the circuit model proposed by Guangling [20]. The model considers concrete/mortar as a circuit when a potential difference is applied to the sample. The free transportation of ions in the pore solution is controlled by the action of continuous conductive paths. In contrast, blocked passages by the arrangement

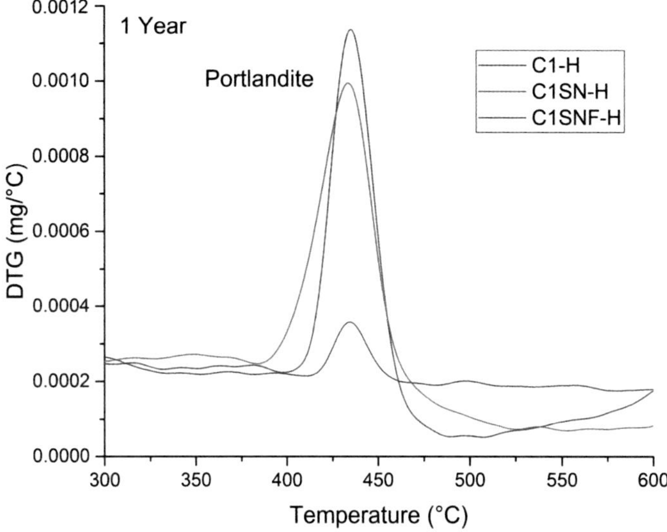

Fig. 4 Portlandite consumption of heat-cured mortars with and without fly ash

of a series of gel particles prevent the transfer of pore solution ions and are termed "insulator paths". If some gel particles exist in the conductive paths, then points of discontinuation lead to discontinuous conductive paths. For further understanding of the concrete/mortar circuit models, please refer to Guangling [20] and Ramu et al. [13].

Figure 5a shows that the ambient-cured mortars (C1-A and C2-A) may have more insulator paths than conductive paths due to the absence of heat-induced accelerated curing, as reflected in their higher dynamic modulus and bulk resistivity values shown in Table 2. However, for the heat-cured mortars (C1-H, C2-H, C1SN-H, and C2SN-H), because of accelerated curing at 90 °C, microcracks may develop, which leads to the gel particles not being tightly arranged and giving rise to more conductive paths as represented in Fig. 5b. This scenario correlates well with the significant reduction in dynamic modulus and bulk resistivity values noted in Table 2, which creates a favorable situation whereby various ions such as Na, K, S, Ca, and Al are readily transported. Although this scenario is the same for all heat-cured mortars in our study, the C1SN-H and C2SN-H mortars contained more sulfates than C1-H and C2-H. Therefore, greater transportation and deposition of sulfur ions at different locations would increase the likelihood of DEF. Higher expansion of DEF causes expansion and microcracks, which justifies the observation from Table 2 for further reduction in dynamic modulus and bulk resistivity values.

With the incorporation of FA, greater C–S–H formation is inferred by the consumption of (CH) portlandite (Fig. 4) and this creates a pore blocking effect that leads to tightly packed gel particles creating more insulator paths, as represented in Fig. 5c. More tightly packed gel particles reduces the likelihood of ion

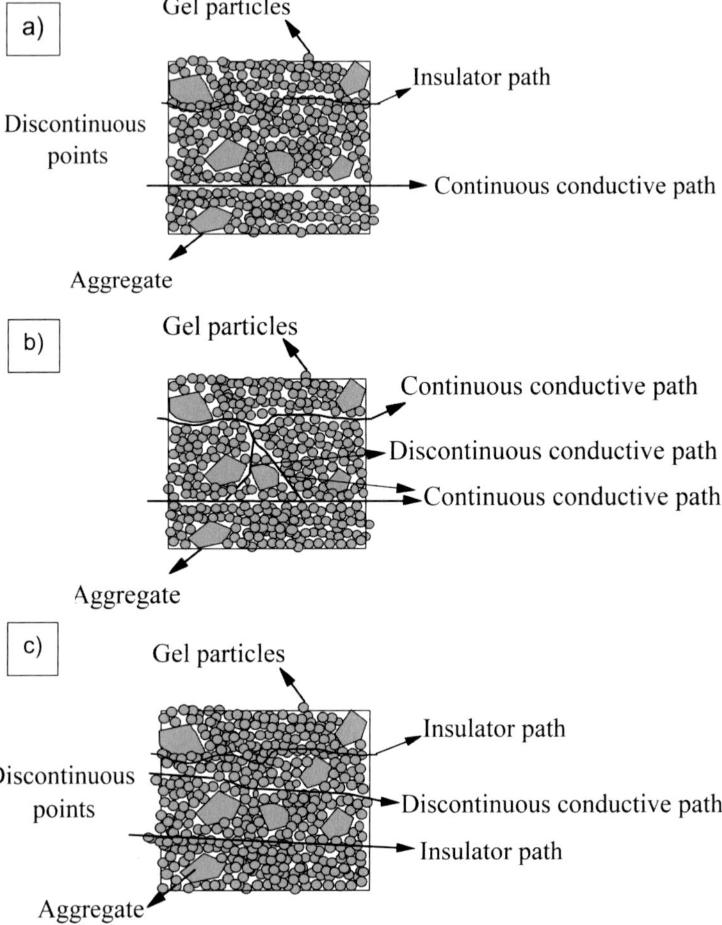

Fig. 5 Schematic representation of different mortars' microstructure (not to scale): **a** cement mortar cured at ambient temperature (25 °C) [20]; **b** cement mortar cured at 90 °C; **c** cement mortar with fly ash cured at 90 °C

transportation, thereby reducing ettringite precipitation and eliminating deleterious DEF. Thus, the incorporation of FA in heat-cured mortars has two significant benefits: (1) in highly DEF-prone cementitious systems, it can eliminate harmful expansive DEF; and (2) in non-DEF-prone cementitious systems, it can be helpful in creating a denser microstructure. Both these benefits contribute to more durable and sustainable concrete.

4 Conclusions

From this research work, the following conclusions can be made.

(1) Cements satisfying the requirements of AS 3972:2010 [7] and ATIC SP-43 [9] for sulfate and alkali content are not prone to expansive DEF even if the concrete is heat-cured. However, due to elevated temperature curing, porous hydrates form, which leads to poor microstructural quality and eventual reduction in the durability performance of the concrete.

(2) Chemically modified cement (sulfate 4%, alkali 1%) was prone to expansive DEF when heat-cured. A reduction in physical characteristics such as dynamic modulus and bulk resistivity was observed. Thus, if such cements are used in concrete that is heat-cured, there is a high likelihood of durability issues arising from the combined action of expansive DEF and permeability, leading to the detriment of concrete service life.

(3) The incorporation of FA eliminated expansive DEF even in chemically modified cement mortars (sulfate 4%, alkali 1%). Furthermore, the consumption of portlandite to form additional C–S–H resulted in a much denser microstructure, as inferred from the high values for dynamic modulus and bulk resistivity. Thus, the addition of FA undoubtedly contributes to increasing the durability of heat-cured (precast) concrete, thereby also contributing to enhanced sustainability.

(4) The conclusions presented in this study are limited to the two Australian general-purpose cements used in the research and mortar specimens. This study needs to be extended to a range of cements/clinkers and concrete specimens.

Acknowledgements We acknowledge the financial support provided by Holcim Australia and the award of a UTS international research scholarship for carrying out this research project. The Construction Materials Research Laboratory, IIT Madras, is gratefully acknowledged for laboratory support to do part of the work presented in this report.

References

1. Adamopoulou E, Pipilikaki P, Katsiotis MS, Chaniotakis M, Katsioti M (2011) How sulfates and increased temperature affect delayed ettringite formation (DEF) in white cement mortars. Constr Build Mater (Elsevier Ltd) 25(8):3583–3590. https://doi.org/10.1016/j.conbuildmat.2011.03.051
2. Menéndez E (2002) Cracking and sulphate attack in field concrete in Spain. In: International RILEM workshop on internal sulfate attack and delayed ettringite formation, September, 4–6
3. Stark J, Bollmann K (2000) Delayed ettringite formation in concrete. Nord Concr Res 1–25
4. Tosun K (2006) Effect of SO_3 content and fineness on the rate of delayed ettringite formation in heat cured Portland cement mortars. Cement Concr Compos 28:761–772. https://doi.org/10.1016/j.cemconcomp.2006.06.003
5. Yang R, Sharp JH (2001) Hydration characteristics of Portland cement after heat curing: II, degree of hydration of the anhydrous cement phases. J Am Ceram Soc 84(3):608–614. https://doi.org/10.1111/j.1151-2916.2001.tb00707.x

6. Taylor HFW, Famy C, Scrivener KL (2001) Delayed ettringite formation. Cem Concr Res 31:683–693
7. Australian Standard (2010) General purpose and blended cements (AS 3972-2010)
8. Australian Standard (2016) Supplementary cementitious materials part 1: fly ash, AS 3582.1
9. Australian Technical Infrastructure Committee (2017) Cementitious materials for concrete (ATIC SP-43)
10. Australian Standard (2006) Methods of testing Portland, blended and masonry cements method 11: compressive strength (AS 2350.11:2006)
11. Australian Standard (2006) Methods of testing Portland, blended and masonry cements method 12: preparation of a standard mortar and moulding of specimens (AS 2350.12-2006)
12. Australian Standard (2006) Methods of testing Portland, blended and masonry cements method 13: determination of drying shrinkage of cement mortars (AS 2350.13-2006)
13. Ramu YK, Sirivivatnanon V, Thomas P, Dhandapani Y, Vessalas K (2021) Evaluating the impact of curing temperature in delayed ettringite formation using electrochemical impedance spectroscopy. Constr Build Mater 282:122726. https://doi.org/10.1016/j.conbuildmat.2021.122726
14. Ramu YK, Thomas PS, Sirivivatnanon V, Thomas PS (2022) Non-expansive delayed ettringite formation in low sulphate and low alkali cement mortars. Aust J Civ Eng 1–12. https://doi.org/10.1080/14488353.2022.2075077
15. Famy C (1999) Expansion of heat-cured mortars. University of London. https://books.google.com.au/books?id=aN7xtgAACAAJ
16. Zhang Z, Olek J, Diamond S (2002) Studies on delayed ettringite formation in heat-cured mortars II. Characteristics of cement that may be susceptible to DEF. Cem Concr Res 32:1737–1742
17. Amine Y, Leklou N, Amiri O (2016) Effect of supplementary cementitious materials (SCM) on delayed ettringite formation in heat cured concretes. In: International conference on materials and energy 2015, ICOME 15, 19–22 May 2015, Tetouan, Morocco; International conference on materials and energy 2016, ICOME 16, pp 565–570. https://doi.org/10.1016/j.egypro.2017.11.254
18. Ramlochan T, Thomas MDA, Hooton RD (2004) The effect of pozzolans and slag on the expansion of mortars cured at elevated temperature: part II: microstructural and microchemical investigations. Cem Concr Res 34(8):1341–1356. https://doi.org/10.1016/j.cemconres.2003.12.026
19. Silva AS, Soares D, Matos L, Salta MM, Divet L, Pavoine A, Candeias AE, Mirão J (2010) Influence of mineral additions in the inhibition of delayed ettringite formation in cement based materials—a microstructural characterization. Mater Sci Forum 636–637:1272–1279. https://doi.org/10.4028/www.scientific.net/MSF.636-637.1272
20. Guangling S (2000) Equivalent circuit model for AC electrochemical impedance spectroscopy of concrete. Cem Concr Res 30(11):1723–1730. https://doi.org/10.1016/S0008-8846(00)00400-2

A Novel Concrete Mix Design Methodology

D. Kumar, M. Alam, and J. Sanjayan

Abstract Concrete mix design is the methodology for mixing binder, aggregate and water to achieve required physical, mechanical, and thermal properties. In particular, the physical properties depend on the volume fraction of each element in the concrete recipe. In this study we considered cement mortar, complying with ASTM C105, as the reference concrete with cement as the binder and silica sand as the aggregate. The reference mortar was denser with high thermal conductivity and compressive strength at given rheological properties. A denser concrete presents difficulty in material handling and imposes a safety risk, and high thermal conductivity increases building energy consumption. Therefore, lightweight concrete (LWC) has been developed by replacing silica sand with porous materials. LWC includes cement as the binder, with silica sand and other porous materials as the primary and binary fillers. The mass of the filler materials is determined by their particle density and volume fraction. LWC has low thermal mass, thereby exacerbating the summertime overheating and peak cooling demand of buildings. Therefore, there is a need to design a LWC with high thermal mass by incorporating phase change materials (PCM), which are mainly incorporated as tertiary filler. Here, we propose a novel concrete mix design methodology to incorporate PCM composite as a partial replacement of the porous material without changing binding materials.

Keywords Capric acid · Cement mortar · Expanded perlite · Mix design methodology · Silica aerogel

---

D. Kumar · M. Alam (✉) · J. Sanjayan
Centre for Sustainable Infrastructure and Digital Construction, Department of Civil and Construction Engineering, Swinburne University of Technology, Hawthorn, VIC, Australia
e-mail: malam@swin.edu.au

D. Kumar
Department of Mechanical Engineering, Mehran University of Engineering and Technology, (SZAB Campus), Khairpur Mir's, Sindh, Pakistan

W. Duan et al. (eds.), *Nanotechnology in Construction for Circular Economy*,
Lecture Notes in Civil Engineering 356,
https://doi.org/10.1007/978-981-99-3330-3_46

1 Introduction

Energy use in buildings is increasing due to population growth, changes in lifestyle, and urbanization, and it accounts for 30–40% of the total energy use in many countries [1]. In buildings, the heating, ventilation, and air conditioning (HVAC) system is responsible for 42–68% of the total energy use in building to maintain acceptable levels of thermal comfort [2, 3]. Energy consumption by the HVAC system depends on internal heat gain due occupancy, lighting and appliances, and heat transfer through the building envelope (50–60% of HVAC system energy consumption). The latter is reduced by using various insulating materials to reduce building energy consumption by 20–80% [4]. However, inappropriate use of insulation materials accounts for 30% of heat loss in buildings [5]. Traditional insulation materials such as extruded polystyrene boards, expanded polystyrene panels, glass felt, and polyurethane foam have lower resistance, compressive strength, and poor hydrophobicity [4].

Lightweight concrete (LWC) was developed using silica aerogel (SA) as an alternative insulating material with comparable thermal conductivity (0.026–0.37 W/m–K) to foam and autoclave aerated concrete, high compressive strength, and high fire resistance [6]. Ibrahim et al. [7] reported that the thermal performance of SA-based LWC materials was better than that of insulation. Gao et al. [8] developed an aerogel-incorporated concrete by replacing 10–60 vol% of sand with SA, having thermal conductivity and compressive strength of 0.26 W/m–K and 8.3 MPa, respectively. Fickler et al. [9] used SA granules instead of SA particles to develop LWC with minimum thermal conductivity and compressive strength of 0.16 W/m–K and 3.0 MPa, respectively. Suman et al. [10] developed ultra-LWC using expanded glass particles, SA particles and prefabricated plastic bubbles to replace the weight percentage of cement. They found a 42% decrease in compressive strength at 15% decrease in cement content with SA particles. The thermal conductivity of the cement mortar was reduced by 60, 53 and 48% by replacing 20 wt% of fine aggregates with SA, expanded perlite and vermiculite, respectively. They achieved the same thermal performance to the reference coating material by reducing its thickness by 60% with the addition of SA [11]. Li et al. [12] used alkali-activated binder instead of cement because of its low thermal conductivity. They developed an ultra-LWC with thermal conductivity equivalent to insulating materials (0.0434 W/m–K) with minimum compressive strength of 0.38 MPa at 30 °C. The volumetric heat capacity and thermal conductivity of the SA-LWC was lower than those of the reference concrete by 0.17–0.227 J/m^3-K and 1.22–1.27 W/m–K, respectively. The thermal mass of LWC is lower than that of normal concrete because of its low density and high porosity [13].

Some researchers have integrated form-stable phase change material (FSPCM) into expanded perlite [14], expanded clay [15] and hollow ceramsite [16]-based LWCs to develop low thermal conductivity and high thermal storage composites. They first developed LWC composites by volumetrically replacing sand with lightweight fillers, the mass of which was calculated using a particle density approach.

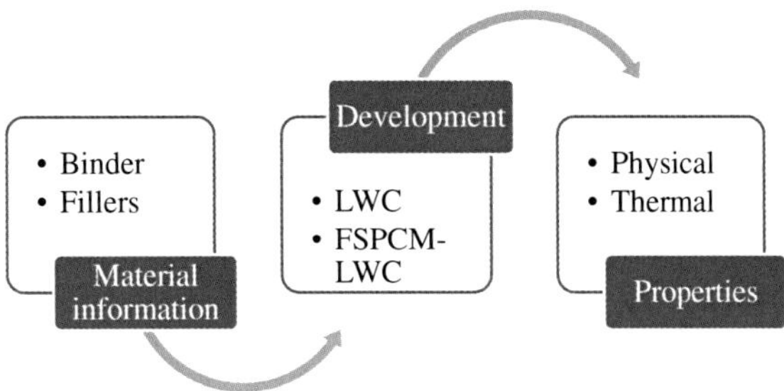

Fig. 1 Research methodology. Note: FSPCM, form-stable phase change material; LWC, lightweight concrete

The FSPCM composites were added as a mass fraction of dry mixture [14] and total aggregate [15, 16]. In these mixing and trial methods [17], the mass of binder and of aggregate was changed to those of the integrated FSPCM composite, resulting in a significant loss of materials to achieve desirable thermophysical properties. Han et al. [18] collected data regarding concrete laboratory waste for the period of 2011–17. They found that a cylindrical specimen was annually producing 50 m^3 (61%) concrete waste, followed by 18% of tensile specimen, 15% cubes and remaining pavement block and mortar specimen. Annually, the concrete and cement waste was 80 m^3 and 20 m^3, respectively, with CO_2 emissions >15 m^3. Therefore, there is need to develop a scientific methodology to design concrete mixtures for achieving desirable thermophysical properties and performance without wasting materials.

In this study we aimed to develop a novel concrete mix methodology using the particle density of lightweight filler and FSPCM (Fig. 1). This study considered SA granules as the filler. The FSPCM is made up of capric acid and hydrophobic expanded perlite [19]. We also investigated the thermophysical properties of the LWC and FSPCM-LWC composites by considering bulk density, compressive strength, thermal conductivity, and latent heat storage.

2 Methods

2.1 Materials

Ordinary Portland Cement (OPC; purchased from Bunnings, Australia) was used as the binder in accordance with AS3972. Silica sand was selected as the aggregate for the reference concrete in compliance with ASTM C105. SA granules (SAG; purchased from Enersen, France) was the lightweight aggregate for the LWC composite. Finally, CAHEP was used to develop the FSPCM-LWC composite, as described previously [19, 20] (Table 1).

Table 1 Properties of different aggregates

Property	Units	Sand	CAHEP	SAG
Surface	–	Hydrophilic	Hydrophobic	Hydrophobic
Particle Density	kg/m^3	2657	600	120
Particle size	Å	327,376	35	69
Thermal conductivity	W/m.K	NA	0.38	0.024
Latent heat	kJ/kg	–	95	–

CAHEP, capric acid/hydrophobic expanded perlite; SAG, silica aerogel granules

2.2 Development of LWC and FSPCM-LWC Composite

Our proposed novel concrete mix design methodology for developing LWC and FSPCM-LWC composites is shown in Fig. 2. The reference concrete was according to ASTM C105 regarding cement, water and sand as a binder, activator, and primary filler, respectively. At a given quantity of binder, the sand-to-cement and water-to-cement ratios were 2.75 and 0.485, respectively. LWC was developed by volumetrically replacing the sand particles (primary filler) with SAG (secondary filler). The complete replacement of sand with SAG makes SAG the primary filler due to the presence of only one filler. The primary filler is replaced by a secondary filler from top to bottom at a given mass of cement and sand. The secondary filler (SAG) is volumetrically replaced by a tertiary filler (CAHEP) to develop the FSPCM-LWC composite without changing the primary filler (sand) from left to right. When the secondary filler (SAG) is completely replaced by CAHEP, the developed concrete has sand as the primary filler and CAHEP as the secondary filler.

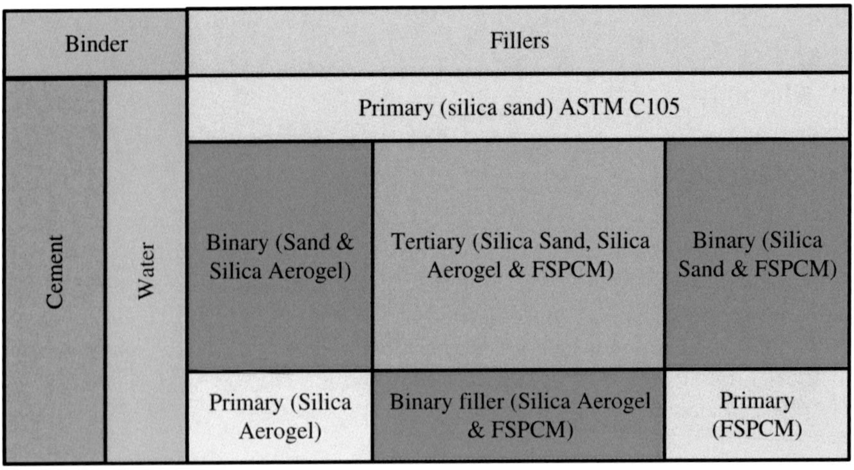

Fig. 2 Novel concrete mix design methodology. PCM, phase change material

The reference concrete composite was developed using a water to OPC ratio of 0.485 and silica sand to OPC ratio of 2.73, in accordance with ASTM C105. The LWC composite was developed by volumetrically replacing silica sand with SAG at 20–80%. The mass of sand and SAG was calculated using Eq. 1 and Eq. 2, respectively. To develop the CAHEP-LWC composite, the mass of SAG was volumetrically replaced by CAHEP at 20–80% without changing the mass of sand. The mass of CAHEP was calculated using Eq. 3.

$$m_{Sand} = \rho_{sand} V_{sand} \tag{1}$$

$$m_{SAG} = \frac{\rho_{SAG} m_{sand} (V_{sand} - V_{SAG})}{\rho_{sand} V_{sand}} \tag{2}$$

$$m_{CAHEP} = \frac{\rho_{CAHEP} m_{SAG} (V_{SAG} - V_{CAHEP})}{\rho_{SAG} V_{SAG}} \tag{3}$$

where, m, ρ and V show the mass (kg), density (kg/m^3) and volume (m^3) of sand, SAG and CAHEP composites, respectively, to develop the reference concrete, LWC composite, and FSPCM-LWC composite. The workability of the LWC and FSPCM-LWC composites was kept the same as that of the reference concrete by using a superplasticiser. The mass of OPC, water, and the calculated mass of fillers are given in Table 2.

Table 2 Mix design recipes

Notation	Binder	Activator	Filler			Superplasticizer
	OPC	Water	Sand	SAG	CAHEP	
	kg	kg	kg	kg	kg	%
RC	520	252	1430	0	0	–
LWC1	520	252	1144	13	0	–
CAHEP-LWC1	520	252	1144	10	13	–
CAHEP-LWC2	520	252	1144	8	26	–
CAHEP-LWC3	520	252	1144	5	39	–
CAHEP-LWC4	520	252	1144	3	52	–
LWC2	520	252	172	52	0	–
CAHEP-LWC5	520	252	172	41	52	1.2
CAHEP-LWC6	520	252	172	31	103	3.7
CAHEP-LWC7	520	252	172	21	155	5.0
CAHEP-LWC8	520	252	172	10	207	6.0

CAHEP, capric acid/hydrophobic expanded perlite; LWC, lightweight concrete; RC, reference concrete; SAG, silica aerogel granules

Fig. 3 Techno-test machine **a** and test specimen **b**

2.3 Properties of LWC and FSPCM-LWC Composites

2.3.1 Physical Properties

Cubic specimens were cast using our mix design recipes in 50 × 50 × 50 mm metallic molds, in accordance with ASTM C1009. Three specimens were cast for each mix design for precision and accuracy of results. They were demolded after 24-h curing in an environmental chamber at temperature and relative humidity of 23 °C and 90%, respectively. The demolded specimens were water cured until the test date. The mass of the cubes was measured by a simple balance with accuracy of 0.1 g. The techno-test machine with accuracy of 0.1 kN was used to measure the compressive strength of the test specimens, as shown in Fig. 3.

2.3.2 Thermal Properties

Thermal properties include thermal conductivity and latent heat storage. Thermal conductivity of the developed cementitious composites was measured using a transient line source (TLS-100), complying with ASTM D5334, as shown in Fig. 4. A 50-mm diameter cylinder of length 120 mm was cast and demolded after 24-h curing in an environmental chamber. The samples were air dried in the environmental chamber at temperature and relative humidity of 23 °C and 50%, respectively, until the test date. The TLS-100 probe was inserted into the test specimen and kept for 15 min to achieve thermal equilibrium between the specimen and probe surface. Final measurements by executing the test were obtained using a digital display meter.

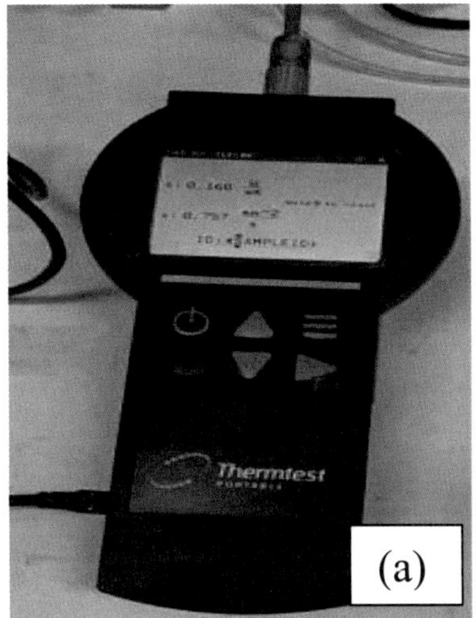

Fig. 4 Thermal conductivity meter **a** and sensor **b**

Differential scanning calorimetry provided inaccurate data due to the smaller sample size of 5–20 mg. The mass fraction of each material in the developed composites could not be accurately balanced in such a small sample. Consequently, the latent heat storage of the FSPCM-LWC composite ($h_{FSPCM-LWC}$) was calculated using Eq. 4 [20].

$$h_{FSPCM-LWC} = \frac{m_{FSPCM} h_{FSPCM}}{m_w + m_{OPC} + m_S + m_{FSPCM}} \tag{4}$$

where, $m_w, m_{OPC}, m_S\ and\ m_{FSPCM}$ are the mass of water, OPC, sand and FSPCM, respectively, and h_{FSPCM} is the latent heat storage of the FSPCM composite.

3 Results and Discussion

Figure 5 shows the density of the LWC and CAHEP-LWC composites. The density of the reference concrete was 2226 kg/m^3 and that of the composites was 9 and 40%, respectively, lower than the reference concrete due to replacement of 20 and 80% volume of silica sand with SAG, because the particle density of SAG is 20-fold lower than that of silica sand. The effect of CAHEP on density depends on the presence of SAG. For instance, the density of LWC with 20 vol% of SAG was only 3%, but the density of LWC with 80 vol% of SAG was maximally increased by 30% with the addition of CAHEP. Thus, the density of LWC increases dramatically at higher volume fractions of the lightweight filler. LWC must contain the highest proportion of lightweight fillers as given binding materials.

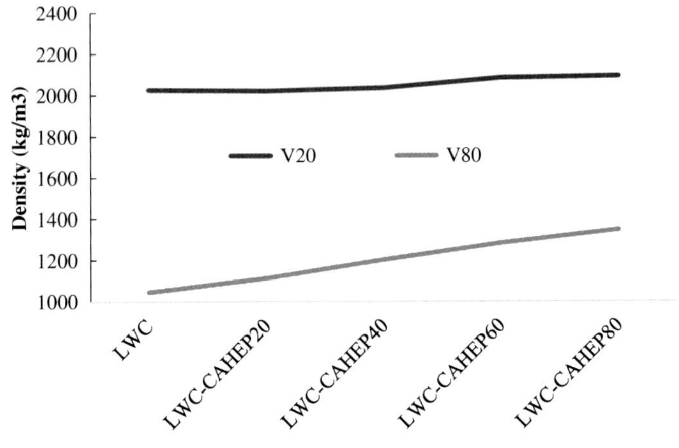

Fig. 5 Effect of capric acid/hydrophobic expanded perlite (CAHEP) on bulk density of lightweight concrete (LWC)

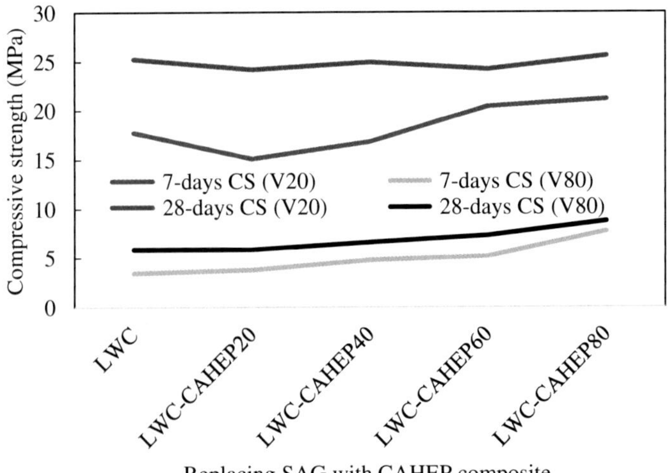

Fig. 6 Effect of capric acid/hydrophobic expanded perlite (CAHEP) on compressive strength of lightweight concrete (LWC)

Figure 6 exhibits the compressive strength of LWC and CAHEP-LWC. The compressive strength of 31 and 46 MPa were measured for the reference concrete after 7 and 28 days, respectively, of water curing. The addition of 20 vol% and 80 vol% of SAG as a partial replacement of sand reduced compressive strength by 56% and 81%, respectively. The decrement percentage of density was smaller than the compressive strength even at the same proportion of SAG because SAG have smaller particle

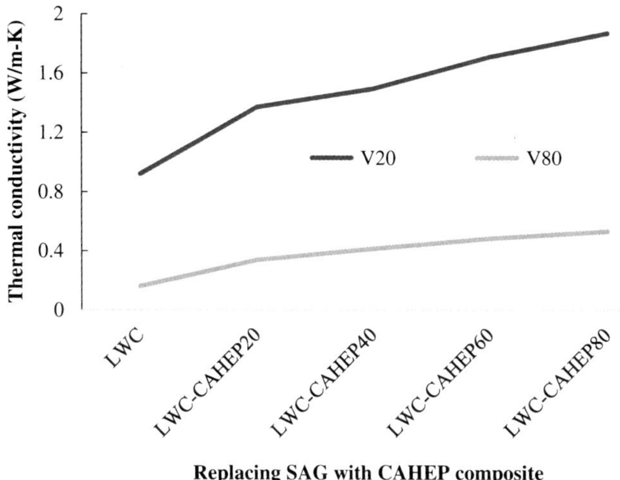

Fig. 7 Effect of capric acid/hydrophobic expanded perlite (CAHEP) on thermal conductivity of lightweight concrete (LWC)

density and more open porous structure, making SAG more fragile and breaking at very small loading. Moreover, SAG have an amorphous shape, promoting heterogeneous porosity, and resulting in lower compressive strength. The addition of CAHEP increased the compressive strength of the LWC by filling the large pores of the LWC. The compressive strength hardly changed with the addition of CAHEP to the LWC with 20 vol% of SAG. At higher volume fractions of SAG, the compressive strength of LWC doubled with the addition of CAHEP. The developed LWC and LWC-CAHEP composites both had higher than the minimum compressive strength (4.14 MPa) required for nonload-bearing structural material as required by ASTM C129–17 [15].

Thermal conductivity and theoretical latent heat storage of LWC and CAHEP-LWC are shown in Figs. 7 and 8, respectively. Thermal conductivity of the reference concrete was 2.27 W/m–K, which decreased by 60 and 93% by adding 20 vol% and 80 vol% of SAG, respectively. The effect of SAG on thermal conductivity was higher than on density and compressive strength because of the 100-fold lower thermal conductivity of SAG (0.01–0.02 W/m–K [5, 18]) compared with sand, and the creation of macro-porosity due to the heterogeneous porous structure of SAG. The addition of CAHEP almost increased thermal conductivity of the LWC from 0.92 to 1.87 W/m–K and from 0.16 to 0.532 W/m–K at 20 vol% and 80 vol% of SAG, respectively. The addition of CAHEP dramatically increased the thermal conductivity of the LWC due to the higher thermal conductivity of CAHEP (0.38 W/m–K) and its smaller particle size which filled the macro-porosity of the LWC. The latent heat storage increased linearly with CAHEP proportion. The reference concrete and

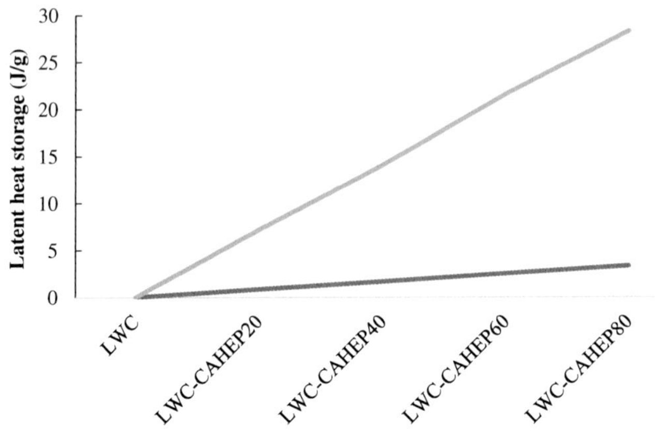

Fig. 8 Effect of capric acid/hydrophobic expanded perlite (CAHEP) on latent heat storage of lightweight concrete (LWC)

LWC stored heat directly, whereas the CAHEP-LWC composite stored latent heat by changing the material's phase. Heat storage increased by 3.3 kJ/kg and 28 kJ/kg at 20 vol% and 80 vol%, respectively, of SAG replaced by CAHEP (Fig. 8).

4 Conclusions

We proposed a new concrete mix design methodology for developing tertiary filler cementitious composite to meet structural and thermal properties. We considered three fillers—silica sand, silica aerogel granules and CAHEP composites—with different particle densities. Silica sand was used for high compressive strength of cementitious composites, but the addition of SAG decreased the thermal conductivity of the reference concrete to an acceptable level for non-load bearing application in buildings. The developed LWC with the lowest thermal conductivity had low thermal mass due to the high porosity of SAG. To increase the thermal mass of the LWC, CAHEP partially replaced the volume fraction of SAG. The mass of the LWC-CAHEP composite was determined by a particle density approach instead of the hit and trail mixing methodology. The particle density-based mix design methodology revealed that the high fraction, denser filler increased compressive strength, thermal conductivity, and density, whereas the porous material reduced thermophysical properties. Thermophysical properties are a function of particle density, surface morphology and porous structure. The use of a denser filler increases compressive strength at given binding materials. A hydrophobic surface results in lower compressive strength due to less affinity with cement paste. Finally, an amorphous porous structure promotes macro-porosity, reducing the strength of the concrete. Future studies should consider the structure and surface morphology of fillers to investigate optimal mass fractions of silica sand, SAG and CAHEP in thermally enhanced LWC composites.

The selection of the CAHEP-LWC composite depends on thermal conductivity and latent heat storage. Both properties are essential for energy-efficient building design. We found that increasing the thermal storage of LWC and increasing its thermal conductivity was undesirable due to heat transfer. Therefore, there is a need to conduct a sensitivity analysis to investigate the optimum thermal conductivity and latent heat storage of heat resistive and storage panels to design energy efficient buildings.

Acknowledgements This research was funded by the Higher Education Commission (HEC), Pakistan and Swinburne University of Technology, Australia with grant no. 5-1/HRD/HESTPI/(Batch-VI)/6021/2018/HEC.

References

1. Kumar D, Zou PXW, Memon RA, Alam MM, Sanjayan JG, Kumar S (2020) Life cycle cost analysis of building wall and insulation materials. J Building Phys 43(5):428–455
2. Kumar D, Ali I, Hakeem M, Junejo A, Harijan K (2019) LCC optimization of different insulation materials and energy sources used in HVAC duct applications. Arab J Sci Eng 44(6):5679–5696. https://doi.org/10.1007/s13369-018-3689-x
3. Pérez-Lombard L, Ortiz J, Pout C (2008) A review on buildings energy consumption information. Energy Build 40(3):394–398. https://doi.org/10.1016/j.enbuild.2007.03.007
4. Kumar D, Alam M, Zou PXW, Sanjayan JG, Memon RA (2020) Comparative analysis of building insulation material properties and performance. Renew Sustain Energy Rev 131:110038. https://doi.org/10.1016/j.rser.2020.110038
5. Goulouti K, Castro JD, Keller T (2016) Aramid/glass fiber-reinforced thermal break – thermal and structural performance. Comp Struct 136:113–123. https://doi.org/10.1016/j.compstruct.2015.10.001
6. Kumar D, Alam M, Sanjayan JG (2021) Retrofitting building envelope using phase change materials and aerogel render for adaptation to extreme heatwave: a multi-objective analysis considering heat stress, energy, environment, and cost. Sustainability 13(19). https://doi.org/10.3390/su131910716
7. Ibrahim M, Biwole PH, Wurtz E, Achard P (2014) A study on the thermal performance of exterior walls covered with a recently patented silica-aerogel-based insulating coating. Build Environ 81:112–122. https://doi.org/10.1016/j.buildenv.2014.06.017
8. Gao T, Jelle BP, Gustavsen A, Jacobsen S (2014) Aerogel-incorporated concrete: an experimental study. Constr Build Mater 52:130–136. https://doi.org/10.1016/j.conbuildmat.2013.10.100
9. Fickler S, Milow B, Ratke L, Schnellenbach-Held M, Welsch T (2015) Development of high performance aerogel concrete. Energy Procedia 78:406–411. https://doi.org/10.1016/j.egypro.2015.11.684
10. Adhikary SK, Rudžionis Ž, Vaičiukynienė D (2020) Development of flowable ultra-lightweight concrete using expanded glass aggregate, silica aerogel, and prefabricated plastic bubbles. J Build Eng 31. https://doi.org/10.1016/j.jobe.2020.101399
11. Bergmann Becker PF, Effting C, Schackow A (2022) Lightweight thermal insulating coating mortars with aerogel, EPS, and vermiculite for energy conservation in buildings. Cement Concrete Comp 125. https://doi.org/10.1016/j.cemconcomp.2021.104283

12. Li X, Cui D, Zhao Y, Qiu R, Cui X, Wang K (2022) Preparation of high-performance thermal insulation composite material from alkali-activated binders, foam, hollow glass microspheres and aerogel. Const Build Mater 346. https://doi.org/10.1016/j.conbuildmat.2022.128493
13. Strzałkowski J, Garbalińska H (2022) The dynamic thermal properties of aerogel-incorporated concretes. Const Build Mater 340. https://doi.org/10.1016/j.conbuildmat.2022.127706
14. Guardia C, Barluenga G, Palomar I, Diarce G (2019) Thermal enhanced cement-lime mortars with phase change materials (PCM), lightweight aggregate and cellulose fibers. Constr Build Mater 221:586–594. https://doi.org/10.1016/j.conbuildmat.2019.06.098
15. Mokhtari S, Madhkhan M (2022) The performance effect of PEG-silica fume as shape-stabilized phase change materials on mechanical and thermal properties of lightweight concrete panels. Case Stud Const Mater 17. https://doi.org/10.1016/j.cscm.2022.e01298
16. Wang F, Zheng W, Qiao Z, Qi Y, Chen Z, Li H (2022) Study of the structural-functional lightweight concrete containing novel hollow ceramsite compounded with paraffin. Const Build Mater 342. https://doi.org/10.1016/j.conbuildmat.2022.127954
17. Adhikary SK, Rudžionis Ž, Tučkutė S (2022) Characterization of novel lightweight self-compacting cement composites with incorporated expanded glass, aerogel, zeolite and fly ash. Case Stud Const Mater 16. https://doi.org/10.1016/j.cscm.2022.e00879
18. Han AL, Setiawan H, Hajek P (2019) Laboratory concrete specimens waste, a case study on life cycle assessment. IOP Conf Ser Earth Environ Sci 290:012015
19. Kumar D, Alam M, Sanjayan J (2022) An energy-efficient form-stable phase change materials synthesis method to enhance thermal storage and prevent acidification of cementitious composite. Constr Build Mater 348:128697. https://doi.org/10.1016/j.conbuildmat.2022.128697
20. Kumar D, Alam M, Sanjayan J, Harris M (2023) Comparative analysis of form-stable phase change material integrated concrete panels for building envelopes. Case Stud Const Mater 18. https://doi.org/10.1016/j.cscm.2022.e01737

Characterization of the Nano- and Microscale Deterioration Mechanism of the Alkali–Silica Reaction in Concrete Using Neutron and X-ray Scattering Techniques: A Review

E. Nsiah-Baafi, M. J. Tapas, K. Vessalas, P. Thomas, and V. Sirivivatnanon

Abstract Alkali–silica reaction (ASR) is one of the most recognized chemical reactions that lead to the deterioration and premature failure of concrete. The severity of ASR is largely dependent on the expansive nature of the reaction product (ASR gel). As such, it is important to expound the developed knowledge on the formation, structure, composition, and swelling mechanism of ASR gel, to provide a greater understanding of ASR deterioration and to facilitate the development of more reliable prediction and mitigation methods. We present a summary of existing methods for assessing ASR and the state-of-the-art techniques that use neutron and X-ray scattering methods to characterize the nano- and microstructural properties of concrete and elucidate the potential transport dynamics of reactants that determine the mechanism and extent of ASR.

Keywords Alkali–silica reaction · Characterization · Microstructure · Nanostructure

1 Introduction

The alkali–silica reaction (ASR) is a deleterious chemical reaction that occurs in concrete when certain reactive silica phases in aggregates react with alkali ions (Na, K) from the pore solution [1–3]. Through decades of research the mechanism of ASR is relatively well documented, but controversy remains regarding the reported reaction sequence leading to the formation of ASR gel [4, 5] and the expansion

E. Nsiah-Baafi (✉) · M. J. Tapas · V. Sirivivatnanon
Innovation Factory, UTS-Boral Centre for Sustainable Building, Sydney, NSW, Australia
e-mail: elsie.nsiah-baafi@uts.edu.au

K. Vessalas · P. Thomas · V. Sirivivatnanon
School of Civil and Environmental Engineering, University of Technology Sydney (UTS), Sydney, NSW, Australia

W. Duan et al. (eds.), *Nanotechnology in Construction for Circular Economy*, Lecture Notes in Civil Engineering 356,
https://doi.org/10.1007/978-981-99-3330-3_47

mechanism of the gel to form cracks in concrete [6]. Undoubtedly, ASR begins with the dissolution of reactive silica in the alkaline pore solution of concrete and the dissolution rate depends on several factors including the degree of alkalinity of the pore solution, the type and particle size of the silica mineral present, temperature, and the presence of other cations such as Li^{+}, Al^{3+} and Ca^{2+} in the pore solution [1, 2]. Following silica dissolution, Ca^{2+} and alkali cations in the pore solution react with the dissolved silica to form a C-(Na, K)-S–H reaction product known as ASR gel, as well as other calcium-rich hydrates [2]. The composition and structure of the ASR gel may vary in the same concrete system, and in one concrete system from the other, depending on the location of the reaction site, the type and amount of silica and cations at the reaction site, and the age and curing conditions of the concrete. The contention in the sequence of ASR gel formation stems from an earlier proposal described by Hou et al. [4]. By comparing ASR gels in laboratory specimens to gels in field structures, those authors concluded that the continuous formation of a calcium-rich C-S-H product occurs when calcium is locally available at the reaction site [4]. This initial C-S-H is typically dense and acts as a physical barrier that isolates the reaction sites in the concrete structure [7]. Upon depletion of calcium at the localized sites, the concentration of silicon increases until a low calcium and high alkali C-(Na,K)-S-H ASR gel is formed at silicon saturation. Although this sequence of ASR gel formation has been supported by other studies [4, 8, 9], it is worth noting that the results were obtained from batch experiments using model reactant methods; thus, they may not be representative of actual concrete systems. Furthermore, a recently reported study [10] demonstrated that ASR gel may first form in cracks on the aggregate surface and around the aggregate, then penetrate towards the inside of the aggregate, which suggests that a C-S-H physical barrier may not be evident in the sequence of ASR gel formation. Moreover, once the ASR gel is formed, the mechanism by which it expands upon moisture absorption is still a topic of discussion [6, 11]. It is, however, well recognized that the addition of supplementary cementitious materials (SCMs) such as fly ash and slag effectively mitigates ASR in concrete [1, 12–14]. With the current depletion of these conventional SCMs, several studies are emerging to discover and optimize potential alternative SCMs and techniques for mitigating ASR. Understanding the dynamics of ASR gel formation, the transport of the gel through the concrete structure, the expansion mechanism of the gel and the effects of its expansion at the micro-and nanoscale is ultimately the key to developing effective mitigation against deleterious ASR. Currently, a number of techniques exist for assessing the structure and composition of ASR gel. In this paper we recap some of the reported studies on ASR characterization methods and present the state-of-the-art techniques that use neutron and X-ray scattering to identify the micro- and nanostructure of concrete to characterize additional features and propagation of ASR in concrete.

2 Traditional Characterization Techniques for Assessing ASR in Cement-Based Materials

Microscopic techniques such as petrography and scanning electron microscopy (SEM) are by far the most commonly used to identify ASR gel in concrete and other cement-based materials [15–18]. These methods can be used in conjunction with other techniques such as electron dispersive spectroscopy (EDS) [17], electron backscattered diffraction [19], Raman spectroscopy [20], nuclear microwave resonance (NMR) [21] and nonlinear impact resonance acoustic spectroscopy [22] to identify the morphology, composition and effect of ASR and its reaction products on concrete. A typical example of petrographic characterization of an ASR-affected mortar and concrete is presented in Fig. 1. Other less commonly used techniques that have shown proficiency in providing information on the nanostructure of cement-based materials and ASR are scanning transmission electron microscopy (STEM) [23, 24], atomic force microscopy [25] and scanning confocal microscopy [26].

Studies carried out by Leeman et al. [10, 27] to characterize the structure and composition of ASR gel formed in concrete using a combination of SEM–EDS, Raman spectroscopy and ^{29}Si NMR demonstrated that both crystalline and amorphous ASR products form in concrete aggregates. To follow the sequence of ASR gel formation and structure of the ASR products at the nanoscale, the studied concrete mixes were doped with $CsNO_3$ and KNO_3 tracers [10]. The authors established that ASR gel first forms in pre-existing pores or cracks in aggregates close to the cement paste and is generally amorphous in structure. The initial cracks that result from the expansion of the formed gel are mostly empty; therefore, secondary ASR products, identified as crystalline ASR gel, begin to fill the cracks while progressing from the aggregate–cement interface to the aggregate's interior. Further characterization in that study revealed that both the amorphous and crystalline ASR products have a structure dominated by Q^3-sites; however, there is a difference in their composition

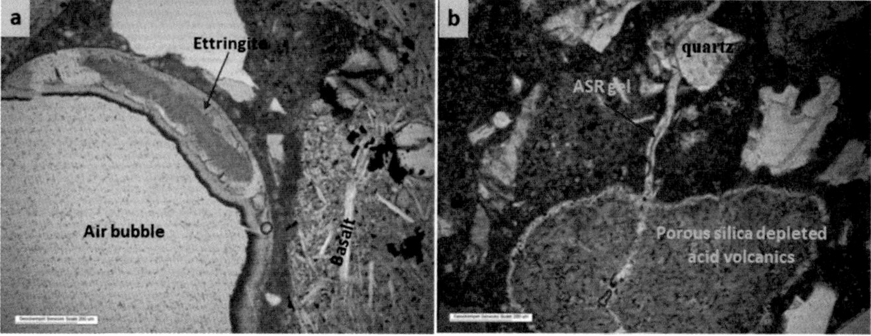

Fig. 1 Petrographic images of **a** concrete specimen showing ASR gel lining an air bubble that is thinly outlined by low birefringent ettringite and **b** ASR-filled crack passing through an acid volcanic fragment and a silica-depleted (porous) acid volcanic fragment

such that the amorphous product has a higher Na/K ratio. Additionally, ASR products in concrete samples cured at 38 °C exhibited a similar structure to ASR gel in field concretes, whereas at temperatures above 50 °C, a K-shylkovite structure was observed [27]. A similar observation has been reported in other studies [28, 29]. The difference in the ASR gel structure with temperature potentially contributes to the expansion capacity of the gel. This outcome supports that temperature is a significant factor to consider when selecting a suitable accelerated test method for assessing ASR, such as the accelerated concrete prism test and accelerated mortar bar test, and rationalizes the differences observed in the laboratory and field reactivity predictions of some aggregates.

Although these characterization methods are effective in assessing the microstructure and composition of ASR and other cement reaction products, they are generally destructive methods that depend on proper representative sampling from bulk material and somewhat rigorous sample preparation, which may influence the outcome of the characterization studies. For example, there are reports in the literature that cite instances where the sectioning and polishing of suspected ASR-affected samples for SEM analysis may have resulted in dislodgement of the ASR gel from cracks [30, 31]. Similarly, to obtain TEM lamellae for nanostructural characterization, cement-based samples are usually milled down to a low micron thickness (<3 μm) to allow the penetration of electron beams [24]. This potentially affects the structure of the formed product. The destructive nature of these methods also inhibits time-lapse characterization and continuous or in-situ monitoring of ASR development and crack propagation in the same region of the bulk sample. This is a major drawback to improving insight on the transport of reactants in concrete and identifying features of ASR that can contribute to the development of novel mitigation strategies.

3 X-ray and Neutron Scattering Techniques

Cement-based materials are characteristically porous. Pores play a key role in the durability and mechanical performance of concrete. For one, they act as a conduit, thus determining the extent of permeation of chemical agents and triggers of deterioration mechanisms. For example, during ASR, moisture containing alkali and other solutes may ingress from the service environment of the concrete structure or be transported from one region to the other within the concrete system through pores. Furthermore, as ASR gel takes up water and expands to form cracks, the propagation of the cracks provides a channel for the spread of less rigid, high-expansive ASR gel [32] through the concrete, promoting ASR. These cracks also become potential sites for the repolymerization and crystallization of new ASR products. However, it is worth noting that pores may be closed (air bubble) or open with a network of micro- and nanosized distribution. These features of the pore system are critical in understanding the influence and extent of porosity on the durability and strength of a concrete structure. For instance, during salt attack, the crystallization pressure of salts in the concrete will vary with pore size. Similarly, the mechanism of drying shrinkage in the concrete is dependent on pore size and relative humidity such that in larger pores with higher relative humidity, capillary pressure is the driving force for drying shrinkage [33, 34]. Considering that the nature of the pore system in concrete has a major influence on its durability properties, destructive characterization techniques that generally sample thin sections from bulk concrete material may not provide precise information on pore features, including size, volume fraction, distribution, and network.

In the past decade, there has been a significant increase in the use of non-destructive neutron and X-ray scattering techniques, such as ultra-/small-angle neutron scattering (USANS, SANS), small angle X-ray scattering, X-ray computed tomography (X-ray CT) and neutron tomography, to characterize the micro- and nanostructure of the concrete and particularly to elucidate the chemo-poromechanics of ASR reactants through time-lapse damage evolution monitoring [35–37]. USANS and SANS use the elastic scattering of neutrons passing through a sample (neutron diffraction) to study the atomic structure of the bulk material and determine structural inhomogeneity at the mesoscopic scale length, typically ranging from 1 to 300 nm [38]. This technique is similar to X-ray diffraction as both principles obey the Beer Lambert law [39]. However, neutrons are unaffected by electrons, therefore when encountering matter, they penetrate to interact with the atomic nuclei whereas X-rays intermingle with the electron cloud around the atom [40]. As such, neutrons are relatively more sensitive to atoms with low atomic number such as hydrogen. This explicates the proficient use of neutron scattering techniques to characterize ASR gel (C-S-(Na,K)-H) and other cement hydrated reaction products, as well as the transport of reactant through the pores in the pore solution of concrete structures.

In tomography, neutron and X-ray beams passing through a sample are attenuated according to the sample's composition and geometry. A series of transmission images (tomographs) representing slices of the sample at several rotation angles are generated. These tomographs can be superimposed to form a 3D representation of the bulk sample showing the surface and internal features in respective volumetric locations [35, 41]. Typically, incident neutrons will provide a high imaging contrast for hydrates in concrete and a good transmittance of metals (e.g., steel reinforcements), whereas X-rays display a high contrast for metals and an adequate transmittance for other light element materials. Therefore, the information obtained from both scattering techniques is complementary. In characterizing concrete structures, the combined use of neutron and X-ray diffraction for imaging has proven to be very efficient for investigating the pore system, propagation of cracks, the presence and dynamics of reaction products, and monitoring of concrete reinforcement materials. For example, the neutron imaging facility at the Paul Scherrer Institute in Switzerland (NEUTRA) has in the past 7 years shown a development in imaging techniques by mounting an X-ray tube before the initial collimator to enable characterization of samples with X-rays and neutrons under the same geometric conditions and the use of an identical detection system [42]. A typical application of this bimodal approach to investigate the internal structure of stainless steel fiber-reinforced concrete is presented in Fig. 2.

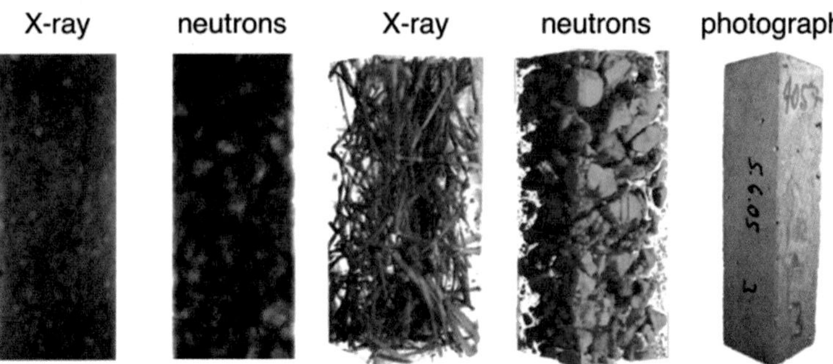

Fig. 2 Neutron and X-ray tomography of a concrete sample [42]

In Fig. 2, it can be seen that the neutron tomographic data set provides information on the segregation of aggregates and pore network in the concrete, whereas the X-ray tomographic results show detailed information on the steel reinforcement. In another study, the authors reconstructed a 3D microstructure of concrete using finite modelling by stacking 2D slices of the segmented concrete's microstructure obtained from both neutron and X-ray CT [41]. The reconstructed 3D microstructure, shown in Fig. 3, revealed the distribution and volume fraction of pores, aggregate and paste binder, as well as the sizes of the respective constituents.

Recently, the inclusion of nanomaterials in construction such the addition of carbon nanotubes as reinforcement in concrete to increase fire resistance, reduce porosity and improve strength properties, and the addition of novel nano- and micro-sized SCMs to improve strength and reduce ASR in concrete have become the focus of emerging research [43, 44]. The application of nondestructive neutron and X-ray techniques to understand how these new materials alter the microstructure of concrete to deliver the desired properties and performance is undoubtedly crucial in accomplishing such ground-breaking innovations.

Fig. 3 Segmented aggregate and void phases obtained from **a** X-ray computed tomography (CT), **b** neutron CT, and **c** combined CT [41]

4 Conclusions

The characterization of ASR in concrete using microscopic and spectroscopic techniques have to date provided useful information that has contributed to the understanding of the mechanism of ASR and the development of mitigation measures. However, due to the destructive nature of these conventional techniques, crucial microstructural features that influence the dynamics, progression, and extent of ASR with age under different conditions are unobserved. To obtain these characteristics, non-destructive X-ray and neutron scattering techniques have shown excellent results, providing 3D volumetric insights into the morphology, interfacial bonding and interaction between constituents, and the pore system, which has a key effect on the durability of concrete. Nonetheless, X-ray and neutron scattering techniques usually do not provide information on the composition of the material. With advancements in the engineered materials used in construction, the ultimate characterization tool kit that combines microscopy and spectroscopy with neutron and X-ray scattering techniques will provide the best approach to understanding and optimizing the performance of concrete for modern structures.

References

1. Thomas MDA, Fournier B, Folliard KJ (2013) Alkali-aggregate reactivity (AAR) facts book. http://www.fhwa.dot.gov/pavement/concrete/asr/pubs/hif13019.pdf
2. Rajabipour F, Giannini E, Dunant C, Ideker JH, Thomas MD (2015) Alkali–silica reaction: current understanding of the reaction mechanisms and the knowledge gaps. Cem Concr Res 76:130–146
3. Thomas M, Fournier B, Folliard K, Ideker J, Shehata M (2006) Test methods for evaluating preventive measures for controlling expansion due to alkali–silica reaction in concrete. Cem Concr Res 36(10):1842–1856
4. Hou X, Struble LJ, Kirkpatrick RJ (2004) Formation of ASR gel and the roles of CSH and portlandite. Cem Concr Res 34(9):1683–1696
5. Hou X, Kirkpatrick RJ, Struble LJ, Monteiro PJ (2005) Structural investigations of alkali silicate gels. J Am Ceram Soc 88(4):943–949
6. Nsiah-Baafi E, Vessalas K, Thomas P, Sirivivatnanon V (2019) Mitigating alkali silica reactions in the absence of SCMs: a review of empirical studies. In: FIB 2018-proceedings for the 2018 fib congress: better, smarter, stronger
7. Kim T, Olek J (2014) Chemical sequence and kinetics of alkali-silica reaction part I. Exp J Am Ceram Soc 97(7):2195–2203
8. Leemann A, Le Saout G, Winnefeld F, Rentsch D, Lothenbach B (2011) Alkali–silica reaction: the influence of calcium on silica dissolution and the formation of reaction products. J Am Ceram Soc 94(4):1243–1249
9. Mahanama D, De Silva P, Kim T, Castel A, Khan M (2019) Evaluating effect of GGBFS in alkali–silica reaction in geopolymer mortar with accelerated mortar bar test. J Mater Civ Eng 31(8):04019167
10. Leemann A, Münch B (2019) The addition of caesium to concrete with alkali-silica reaction: implications on product identification and recognition of the reaction sequence. Cem Concr Res 120:27–35
11. McGowan J (1952) Studies in cement-aggregate reaction XX, the correlation between crack development and expansion of mortar. Aust J Appl Sci 3:228–232

12. Tapas M, Vessalas K, Thomas P, Sirivivatnanon V, Kidd P (2019) Mechanistic role of supplementary cementitious materials (SCMs) in alkali-silica reaction (ASR) mitigation, concrete in practice-progress through knowledge
13. Duchesne J, Bérubé M-A (2001) Long-term effectiveness of supplementary cementing materials against alkali–silica reaction. Cem Concr Res 31(7):1057–1063
14. Thomas M (2011) The effect of supplementary cementing materials on alkali-silica reaction: a review. Cem Concr Res 41(12):1224–1231
15. Fernandes I (2009) Composition of alkali–silica reaction products at different locations within concrete structures. Mater Charact 60(7):655–668
16. Rivard P, Ollivier J-P, Ballivy G (2002) Characterization of the ASR rim: application to the potsdam sandstone. Cem Concr Res 32(8):1259–1267
17. Fernandes I, Silva AS, Gomes JP, de Castro AT, Noronha F, dos Anjos Ribeiro M (2013) Characterization of deleterious expansive reactions in fagilde dam. Metallogr Microstruct Anal 2(5):299–312
18. ASTM C856 (2018) Standard practice for petrographic examination of hardened concrete. ASTM International, West Conshohocken, PA
19. Rößler C, Möser B, Giebson C, Ludwig H-M (2017) Application of electron backscatter diffraction to evaluate the ASR risk of concrete aggregates. Cem Concr Res 95:47–55
20. Balachandran C, Muñoz J, Arnold T (2017) Characterization of alkali silica reaction gels using Raman spectroscopy. Cem Concr Res 92:66–74
21. Peyvandi A, Harsini I, Holmes D, Balachandra A, Soroushian P (2016) Characterization of ASR in concrete by 29Si MAS NMR spectroscopy. J Mater Civ Eng 28:4015096
22. Rashidi M, Knapp MC, Hashemi A, Kim J-Y, Donnell KM, Zoughi R, Jacobs LJ, Kurtis KE (2016) Detecting alkali-silica reaction: a multi-physics approach. Cem Concr Comp 73:123
23. Chaunsali P, Peethamparan S (2013) Influence of the composition of cement kiln dust on its interaction with fly ash and slag. Cem Concr Res 54:106–113
24. Boehm-Courjault E, Barbotin S, Leemann A, Scrivener K (2020) Microstructure, crystallinity and composition of alkali-silica reaction products in concrete determined by transmission electron microscopy. Cem Concr Res 130:105988
25. Liu Q, Tong T, Liu S, Yang D, Yu Q (2014) Investigation of using hybrid recycled powder from demolished concrete solids and clay bricks as a pozzolanic supplement for cement. Constr Build Mater 73:754–763
26. Collins C, Ideker J, Kurtis K (2004) Laser scanning confocal microscopy for in situ monitoring of alkali–silica reaction. J Microsc 213(2):149–157
27. Leemann A, Shi Z, Lindgård J (2020) Characterization of amorphous and crystalline ASR products formed in concrete aggregates. Cem Concr Res 137:106190
28. Shi Z, Geng G, Leemann A, Lothenbach B (2019) Synthesis, characterization, and water uptake property of alkali-silica reaction products. Cem Concr Res 121:58–71
29. Shi Z, Park S, Lothenbach B, Leemann A (2020) Formation of shlykovite and ASR-P1 in concrete under accelerated alkali-silica reaction at 60 and 80 C. Cem Concr Res 137:106213
30. Kawabata Y, Dunant C, Yamada K, Scrivener K (2019) Impact of temperature on expansive behavior of concrete with a highly reactive andesite due to the alkali–silica reaction. Cem Concr Res 125:105888
31. Chen J, Jayapalan AR, Kim J-Y, Kurtis KE, Jacobs LJ (2009) Nonlinear wave modulation spectroscopy method for ultra-accelerated alkali-silica reaction assessment. ACI Mater J 106(4):340
32. Gholizadeh-Vayghan A, Rajabipour F (2017) The influence of alkali–silica reaction (ASR) gel composition on its hydrophilic properties and free swelling in contact with water vapor. Cem Concr Res 94:49–58
33. Scherer GW (2004) Stress from crystallization of salt. Cem Concr Res 34(9):1613–1624
34. Sasano H, Maruyama I, Nakamura A, Yamamoto Y, Teshigawara M (2018) Impact of drying on structural performance of reinforced concrete shear walls. J Adv Concr Technol 16(5):210–232
35. Oesch T, Weise F, Meinel D, Gollwitzer C (2019) Quantitative in-situ analysis of water transport in concrete completed using x-ray computed tomography. Transp Porous Media 127(2):371–389

36. Kong W, Wei Y, Wang S, Chen J, Wang Y (2020) Research progress on cement-based materials by X-ray computed tomography. Int J Pavem Res Technol 13(4):366–375
37. Yang S, Cui H, Poon CS (2018) Assessment of in-situ alkali-silica reaction (ASR) development of glass aggregate concrete prepared with dry-mix and conventional wet-mix methods by X-ray computed micro-tomography. Cement Concr Compos 90:266–276
38. Mühlbauer S, Honecker D, Périgo ÉA, Bergner F, Disch S, Heinemann A, Erokhin S, Berkov D, Leighton C, Eskildsen MR (2019) Magnetic small-angle neutron scattering. Rev Mod Phys 91(1):015004
39. Lehmann EH, Mannes D, Kaestner AP, Hovind J, Trtik P, Strobl M (2021) The XTRA option at the NEUTRA facility—more than 10 years of bi-modal neutron and X-ray imaging at PSI. Appl Sci 11(9):3825
40. Jiang M Neutron Scattering for Experimental Research
41. Kim H-T, Szilágyi V, Kis Z, Szentmiklósi L, Glinicki MA, Park K (2021) Reconstruction of concrete microstructure using complementarity of X-ray and neutron tomography. Cem Concr Res 148:106540
42. Kuhne G (2006) Nutron imaging facilities at the paul scherrer institut and their application for non-destructive testing of abrasive water jet nozzle's. In: ECNDT 2006 proceedings
43. Carriço A, Bogas J, Hawreen A, Guedes M (2018) Durability of multi-walled carbon nanotube reinforced concrete. Constr Build Mater 164:121–133
44. Olafusi OS, Sadiku ER, Snyman J, Ndambuki JM, Kupolati WK (2019) Application of nanotechnology in concrete and supplementary cementitious materials: a review for sustainable construction. SN Appl Sci 1(6):580

Life Cycle Assessment of the Environmental Impacts of Virgin Concrete Replacement by CO_2 Concrete in a Residential Building

M. Ma, Y. Zhou, V. W. Y. Tam, and K. N. Le

1 Introduction

Concrete is one of the most consumed materials in construction, with 25 billion tons produced globally per year [1]. However, it is considered as the most non-sustainable material. The acquisition of virgin aggregate consumes a considerable amount of energy and emits a large amount of greenhouse gasses [2]. Recycled aggregate from construction and demolition (C&D) waste could be a viable substitution in concrete production, both avoiding landfills and conserving natural resources [3]. Recycled aggregate could replace part or all virgin aggregate in concrete and the product is referred to as "recycled concrete" [1]. Although recycled concrete containing recycled aggregate is considered as comparable to virgin concrete, it is not widely accepted by the industry, because of uncertainty about material performance [4]. Recycled concrete has lower mechanical properties and higher shrinkage and creep than virgin aggregate with the same mix design [1]. In order to improve the properties of recycled concrete, CO_2 gas is injected into the recycled aggregate and the CO_2-treated aggregate is mixed into concrete as normal. This new concrete is known as "CO_2 concrete" and rivals the virgin concrete in its mechanical and durability qualities.

Environmental performance of the concrete has attracted increasing attention from academics [5]. Marinkovic et al. [1] summarized two research focuses on sustainable solutions for concrete production: (1) using recycled aggregate to partly or entirely replace virgin aggregate, and (2) replacing cement with cementitious materials [1]. Life cycle assessment is a commonly used tool to evaluate the environmental impact of a product [5]. Specifically, Xing et al. [6] compared the environmental benefits of

M. Ma · Y. Zhou · V. W. Y. Tam (✉) · K. N. Le
School of Design, Engineering and Built Environment, Western Sydney University, Penrith, NSW, Australia
e-mail: vivianwytam@gmail.com

W. Duan et al. (eds.), *Nanotechnology in Construction for Circular Economy*, Lecture Notes in Civil Engineering 356,
https://doi.org/10.1007/978-981-99-3330-3_48

virgin concrete, recycled concrete and CO_2 concrete, and found that CO_2 concrete was the best-performing product for greenhouse gas reduction, because the CO_2 was retained in the recycled aggregate during the carbonation process [6]. Residential buildings use a wide range of resources in their construction, including a great amount of concrete [7]. The environmental performance of a residential building using CO_2 concrete as a partial replacement for virgin concrete remains unknown, so our aim was to conduct a lifecycle assessment to evaluate the environmental impact of CO_2 concrete as a replacement in a residential building. Specifically, virgin concrete replaced by 0, 30, 50, 75 and 100%.

2 Methods

To fulfil the aim of this study, a building information modeling (BIM) and life cycle assessment integration program was used to conduct the life cycle assessment of a building in five scenarios where 0, 25, 50, 75 and 100%, respectively, of the virgin concrete was replaced by CO_2 concrete. The life cycle assessment was conducted according to the ISO 14040 framework, which provides a standard process of four phases, namely, goal and scope definition, life cycle inventory analysis, life cycle assessment analysis and life cycle interpretation phases. In this study, the process started with creating a BIM of a residential building as the goal and scope definition phase. The lifespan of the building was assumed to be 50 years. The analysis accounted for the full cradle-to-grave life cycle of the building studied across all stages, including material manufacturing, transportation, building construction, maintenance and replacement, and eventual end of life. In the life cycle inventory analysis phase, the bill of quantities for each building component was extracted from the BIM, and the life cycle inventory data of each component was retrieved from the GaBi 2018 databases, the Australian life cycle inventory database, and a literature review. The quantities of building components and their corresponding environmental impact coefficient were recorded in a spreadsheet. The life cycle environmental impact of the building in five scenarios was assessed by multiplying the quantities of building components by the corresponding environmental impact coefficient. In the life cycle interpretation phase, the building's environmental impact was expressed as global warming potential (reported in kg CO_2eq) based on the Traci 2.1 method [8].

3 Results and Discussion

The life cycle assessment results of the building in the five scenarios are presented in Table 1.

The results showed that replacing virgin concrete with CO_2 concrete in a building could greatly reduce its carbon emissions. By increasing the proportion of CO_2

Table 1 Life cycle environmental impact of a building in five scenarios

	Scenario 1: 0% CO_2 concrete	Scenario 2: 30% CO_2 concrete + 70% virgin concrete	Scenario 3: 50% CO_2 concrete + 50% virgin concrete	Scenario 44: 75% CO_2 concrete + 25% virgin concrete	Scenario 5: 100% CO_2 concrete
Global warming (t CO_2eq/m^2) (declined ratio compared with scenario 1)	275.830 (0)	271.33 (1.6%)	268.36 (2.7%)	264.632 (4.1%)	260.848 (5.4%)

concrete in a building, its carbon emission decreases over its life cycle. As much as 5.4% of the CO_2eq/m^2 can be reduced when 100% of the virgin concrete is replaced by CO_2 concrete.

This study evaluated the life cycle environmental performance of a residential building using CO_2 concrete as a replacement for virgin concrete. The results suggested that the application of CO_2 concrete in the building sector will bring great benefits in terms of environmental performance. However, the mechanical and durability qualities of CO_2 concrete have been considered in this study. In future work, more emphasis should be put on the mechanical and durability qualities of CO_2 concrete for application in the building sector.

References

1. Marinkovic S, Dragas J, Ignjatovic I, Tosic N (2017) Environmental assessment of green concretes for structural use. J Clean Prod 154:633–649. https://doi.org/10.1016/j.jclepro.2017.04.015
2. Kisku N, Joshi H, Ansari M, Panda SK, Nayak S, Dutta SC (2017) A critical review and assessment for usage of recycled aggregate as sustainable construction material. Constr Build Mater 131:721–740. https://doi.org/10.1016/j.conbuildmat.2016.11.029
3. Tam VWY, Soomro M, Evangelista ACJ (2021) Quality improvement of recycled concrete aggregate by removal of residual mortar: A comprehensive review of approaches adopted. Constr Build Mater 288:123066. https://doi.org/10.1016/j.conbuildmat.2021.123066
4. Makul, N. 2020, 'Cost-benefit analysis of the production of ready-mixed high-performance concrete made with recycled concrete aggregate: A case study in Thailand', *Heliyon*, vol. 6, no. 6, pp. e04135-e, DOI https://doi.org/10.1016/j.heliyon.2020.e04135.
5. Serres N, Braymand S, Feugeas F (2016) Environmental evaluation of concrete made from recycled concrete aggregate implementing life cycle assessment. Journal of Building Engineering 5:24–33. https://doi.org/10.1016/j.jobe.2015.11.004
6. Xing W, Tam VWY, Le KN, Butera A, Hao JL, Wang J (2022) Effects of mix design and functional unit on life cycle assessment of recycled aggregate concrete: evidence from CO2 concrete. Constr Build Mater 348. https://doi.org/10.1016/j.conbuildmat.2022.128712

7. Islam H, Jollands M, Setunge S (2015) Life cycle assessment and life cycle cost implication of residential buildings—A review. Renew Sustain Energy Rev 42:129–140. https://doi.org/10.1016/j.rser.2014.10.006
8. Ryberg M, Vieira MDM, Zgola M, Bare J, Rosenbaum RK (2014) Updated US and Canadian normalization factors for TRACI 2.1. Clean Technol Environ Policy 16(2):329–339. https://doi.org/10.1007/s10098-013-0629-z

Economic Impacts of Environmentally Friendly Blocks: The Case of Nu-Rock Blocks

V. W. Y. Tam, K. N. Le, I. M. C. S. Illankoon, C. N. N. Tran, D. Rahme, and L. Liu

Abstract There are numerous industry byproducts that have negative environmental impacts. Pond ash accumulated from coal power plants is one such byproduct that creates major environmental and social issues, especially with regard to decommissioning a coal power plant. Using pond ash in the block manufacturing process is a promising solution proposed by Nu-Rock. This research study evaluated the economic impact throughout the life cycle of Nu-Rock blocks. Nu-Rock block production use a technology called "Nu-creeting" in ash dams to prevent dust generation followed by the block manufacturing process using pond ash as a raw material. Nu-Rock technology can process approximately 250,000 tonnes of ash per annum and manufacture the equivalent of up to 330,000 tonnes volume of traditional building materials. This manufacturing process already generates jobs and pays tax and royalty fees to local governments, which is an added advantage. The total operating cost for 1 tonne of Nu-Rock blocks amounts to approximately AUD48, and the cost of a Nu-Rock block is AUD1.50–2.40, which is within the range of common bricks. Although there is a considerable initial cost to this process, it derives significant economic benefits in terms of manufacturing blocks using industrial byproducts, job creation and even tax revenue. Apart from these economic benefits, the Nu-Rock block manufacturing process generates environmental benefits through the reuse of pond ash from decommissioned coal power plants.

V. W. Y. Tam (✉) · K. N. Le · L. Liu
School of Engineering Design and Built Environment, Western Sydney University, Sydney, NSW, Australia
e-mail: vivianwytam@gmail.com

I. M. C. S. Illankoon
School of Architecture and Built Environment, University of Newcastle, Sydney, NSW, Australia

C. N. N. Tran
Ho Chi Minh City University of Technology, Ho Chi Minh, Vietnam

D. Rahme
Nu-Rock Australia Pty Ltd, Sydney, NSW, Australia

W. Duan et al. (eds.), *Nanotechnology in Construction for Circular Economy*, Lecture Notes in Civil Engineering 356,
https://doi.org/10.1007/978-981-99-3330-3_49

Keywords Coal power plants · Economic issues · Industrial byproducts · Nu-Rock

1 Introduction

The manufacturing process of most conventional building materials poses various environmental issues, and the brick and block manufacturing process is no exception. To name a few of these issues, when manufacturing burnt clay bricks there is depletion of fertile topsoil and emissions of CO_2 to the atmosphere [1]. Conventional bricks are produced from clay with high-temperature kiln firing or from ordinary Portland cement concrete, and thus contain high embodied energy and have a large carbon footprint [2]. To protect the clay resource and with the vision of developing ecofriendly building materials, certain cities in China have banned the use of clay bricks [3]. Brick manufacturing with higher resource efficiency and less carbon footprint is the priority. Industrial byproducts such as fly ash/pond ash are alternatives in brick and block manufacturing process. Fly ash is a fine industrial waste byproduct often found after coal combustion for power generation.

There are research studies of using byproducts in the brick manufacturing process. Cicek and Tanriverdi [4] conducted research on using fly ash from thermal coal power plants for brick manufacturing in Turkey. Several research studies have used various percentages of fly ash in the process of manufacturing bricks; for example, Wang et al. [5] introduced a technology to use up to 50% of fly ash in brick manufacturing. Using pond ash as an alternative to sand is another viable alternative in the block manufacturing process [6]. Similarly, there are many research studies on using fly ash as a replacement material in both brick and block manufacturing [7, 8].

Although research has been conducted on ecofriendly brick/blocks using industrial waste, the commercial production of such bricks is still very limited [2]. Slow acceptance of waste materials-based bricks by industry and the public is one of the challenges faced when commercializing these innovative products [2]. One of the main reasons for this limited use in the commercial space is economic considerations, because incorporating these byproducts is deemed to be costlier compared with conventional materials.

Most research studies rarely discuss the economic aspects from initial investment, construction cost, operational cost and recycling or reuse opportunities. Although these studies used industrial waste, the environmental impacts were rarely quantified using valid parameters. Therefore, this research study conducted a whole of life-cycle cost analysis of Nu-Rock blocks.

1.1 Nu-Rock Blocks: Properties and Manufacturing Process

In Australia, 54% of electricity is generated from coal combustion [9]. Thermal coal power plants generate fly ash and bottom ash, and these are mixed with water, resulting in a slurry that is carried to ash ponds [10]. When the Port Augusta power plant closed, it created a huge pile of pond ash covering hectares of land, giving rise to many environmental and social issues [11]. Nu-Rock technology uses this pond ash to manufacture a range of building materials.

Nu-Rock Technology Pty. Ltd. is a privately owned company providing solutions for environmental and economic issues related to the growing stockpiles of industrial byproducts occurring worldwide. Nu-Rock produces a range of products, including common or rendered bricks and blocks, masonry bricks, blocks, pipes, pavers, tiles, sheeted products and any other shaped building products [12]. Our study focused on the manufacturing of three types of Nu-Rock blocks, namely Mighty block (90 × 245 × 470 mm), 200 series (190 × 190 × 390 mm), and 100 series (90 × 190 × 390 mm). The Nu-Rock Technology has a Proprietary IP. The recycling process usually converts 100% of the pond ash in abandoned power plants into building blocks and brick products. The product has undergone rigorous testing including (but not limited) AS3700D Flexural strength, AS1191 Acoustic fire and AS1530.4 Fire resistance.

The first step in the Nu-Rock block manufacturing process is called Nu-creeting, which prevents dust generation and builds "modules" to process the pond ash. Nu-Rock can process 250,000 tonnes of ash per annum and manufacture the equivalent of up to 330,000 tonnes of traditional building materials using the pond ash as an input [13]. Each module in Nu-Rock can process approximately 0.25 million tonnes of ash per annum and each module can produce a specific type of product. Nu-Rock Blocks require only 3% of the embodied energy required to produce traditional concrete blocks and the Nu-Rock manufacturing process produces zero waste [14]. According to Tam et al. [11], there are great benefits in using this technology to remediate ash ponds.

2 Methods

The primary objective of this research study was to conduct a life-cycle cost analysis of the entire Nu-Rock block manufacturing process: block manufacturing, transportation, operational stage of the building and the demolition. However, here we present the results for initial stages up to construction.

2.1 Scope of the Analysis

The scope of the analysis was the entire process of blocks from cradle to cradle, calculated in four stages:

Stage 1—Initial stage including setting up the plant.
Stage 2—Raw material extraction, manufacturing, transportation to construction site.
Stage 3—Construction, operation, and maintenance.
Stage 4—Demolition/re-use phase.

The data are for stages 1 and 2 up to construction. The calculations included sensible assumptions and limitations, which are given with the relevant calculations.

2.2 Life-Cycle Cost

Key parameters of a life-cycle analysis are determined to a large extent by its purpose and objectives [15]. The main parameters of this project included costs, period of analysis, method of economic evaluation, extent of environmental input and sensitivity analysis. Each of these parameters is discussed in detail.

Costs included those incurred by Nu-Rock Technology Pty Ltd and by the end-users. The life-cycle cost calculation followed ISO15686-5:2017: Building and construction assets—service life planning—Part 5: Life cycle costing" as a guideline. When a decision required including any cost or income in the analysis, the ISO standard was followed. However, additional notes illustrate if any exception was made. All the assumptions (if any) relevant to each calculation are given with it. This project selected an analysis period of 60 years to reflect the anticipated total lifespan of Nu-Rock Technology Pty Ltd.

All costs incurred within the life cycle must be captured and discounted into present day values to calculate the life-cycle cost. Net present value (NPV) was the economic evaluation method used for life-cycle cost calculation, as shown in Eq. 1 (adapted from Dell'Isola and Kirk [15])..

$$NPV(i, N) = \sum_{t=0}^{N} \frac{R_t}{(1+i)^t} \tag{1}$$

In Eq. 1, i denotes the discount rate; t denotes the time of cash flow; R_t denotes the net cash flow, and N is the total number of periods. The discount rate considers the time value of money and the associated risk. The return on equity (RoE) of Nu-Rock Technology Pty Ltd was used as the discounting rate in the life-cycle cost calculation to reflect the capital used by the company (RoE was provided by Nu-Rock).

2.2.1 Life-Cycle Cost Analysis

The initial investment cost for Nu-Rock was AUD12,000,000, which included land cost, specialized design costs, construction cost, cost for initial approvals, electricity connection charges, water connection charge, cost of machinery, cost of specialized equipment and other professional fees. The site establishment cost of AUD200,000 was not included in the investment cost. Therefore, the total initial investment was AUD12,200,000 including site establishment.

Life- cycle cost calculations were based on the following sensible assumptions.

- Transportation costs included loading and unloading and bulk discounts for blocks were not considered. Investment costs provided by Nu-Rock were for the Mt. Piper plant.
- On-site labor costs include one site manager, four factory staff, including two for lift drivers and accounts manager. Salary and associated costs were provided by Nu-Rock.
- Repair and maintenance costs were 5% of the plant costs. The initial investment cost of AUD12,000,000.00 was taken as the plant costs.
- Distribution cost included AUD30 per tonne as provided by Nu-Rock.
- Operational costs included other miscellaneous operations and energy costs. The cost per kWh was considered to be 66 cents/kWh.

Table 1 summarizes the life-cycle costs for Nu-Rock during the raw material extraction and manufacturing stage. The life-cycle cost for the 60-year period at 15% discount rate per tonne of blocks was AUD321.

Table 1 Life-cycle costs for Nu-Rock during the raw material extraction and manufacturing stage

Description	Cost (AUD)
Factory costs	
Site manager	174,750
Outsourced factory operations	360,000
Accounts Manager/plant administration	79,220
Miscellaneous costs	300,000
Selling and distribution costs	
Advertising	220,000
Distribution costs	167,893
Other costs	
Repairs and maintenance cost p.a	600,000
Other operational costs	40,032
Total cost per annum	**1,941,895**
Total cost per tonne of blocks	**48**
Life-cycle costs for 60 years 15% discount rate per tonne of blocks	**321**

Table 2 Construction cost for Nu-Rock block types

	Type of block		
	Mighty block	200 series	100 series
Size	90 × 245 × 470 mm	190 × 190 × 390 mm	90 × 190 × 390 mm
No. of blocks per m^2	8	12.5	12.5
Cost per block (AUD)	2.40	1.80	1.50
Cost of blocks per m^2 (AUD)	19.20	22.50	18.75
Other costs including mortar, tools etc. (AUD)	Labor 4.20/block	Labor 4.20/block	Labor 3.20/block
Total cost of construction (per m^2)	**52.80**	**75.00**	**58.50**
Total cost of construction (per block)	**6.6**	**6**	**4.68**

The next phase was the construction, operations and maintenance stage. This cost calculation included a wall construction using Nu-Rock blocks. "Wall" was assumed as a face brick without any finishing. The size and further details of the three types of Nu-Rock block are given in Table 2.

Stages 3 and 4 are not discussed here but are expected to derive savings by using Nu-Rock.

The selling price of a Nu-Rock block varies between AUD 1.50 and 2.40 which is within the range of common bricks. When setting up the factory Nu-Rock incurred an initial cost of AUD12,200,000. The cost attributed during material extraction was AUD48 per tonne of Nu-Roc blocks (refer to Table 1). During this phase Nu-Rock blocks absorb industrial byproducts such as waste from coal-fired power stations, steel mills, non-ferrous smelters and alumina smelters. This is a non-quantifiable benefit of using Nu-Rock blocks. According to Tam et al., the Nu-Rock manufacturing plant provides almost AUD20 million worth of jobs [11], which is a significant social benefit.

3 Conclusions

The life-cycle cost impacts of Nu-Rock blocks for the first two stages of the manufacturing process, starting from setting up the factory up to the actual construction, were calculated. Nu-Rock technology uses pond ash to manufacture blocks. Ash ponds in discontinued power plants pose serious environmental and social threats. According to our life-cycle calculations, the cost of setting up the factory was ≈ AUD12.2 million. The life-cycle cost for the 60-year period at 15% discount rate per tonne of blocks was AUD321. However, it is interesting to note that the cost of Nu-Rock blocks was within the range of conventional bricks. Although there are many research studies of environmentally friendly materials using industrial byproducts, the commercialization of these products is very slow. It is necessary to conduct similar economic and environmental impact analyses of these products to ensure end-users of their importance. Although this research study was limited to the first two stages, we are planning to extend the study to include stage 3 and 4 in the life-cycle cost calculation.

References

1. Waheed A, Azam R, Riaz MR, Zawam M (2021) Mechanical and durability properties of fly-ash cement sand composite bricks: an alternative to conventional burnt clay bricks. Innov Infrastruct Solut 7(1):24. https://doi.org/10.1007/s41062-021-00630-w
2. Zhang L (2013) Production of bricks from waste materials—a review. Constr Build Mater 47:643–655. https://doi.org/10.1016/j.conbuildmat.2013.05.043
3. Chen Y, Zhang Y, Chen T, Zhao Y, Bao S (2011) Preparation of eco-friendly construction bricks from hematite tailings. Constr Build Mater 25(4):2107–2111. https://doi.org/10.1016/j.conbuildmat.2010.11.025
4. Cicek T, Tanrıverdi M (2007) Lime based steam autoclaved fly ash bricks. Constr Build Mater 21(6):1295–1300. https://doi.org/10.1016/j.conbuildmat.2006.01.005
5. Wang L, Sun H, Sun Z, Ma E (2016) New technology and application of brick making with coal fly ash. J Mater Cycles Waste Manage 18(4):763–770. https://doi.org/10.1007/s10163-015-0368-9
6. Wahab AA, Arshad MF, Ahmad Z, Ridzuan ARM, Ibrahim MHW (2018) Potential of bottom ash as sand replacement material to produce sand cement brick. Int J Integr Eng 10(8)
7. Naganathan S, Mohamed AYO, Mustapha KN (2015) Performance of bricks made using fly ash and bottom ash. Constr Build Mater 96:576–580. https://doi.org/10.1016/j.conbuildmat.2015.08.068
8. Sahu S, Ravi Teja PR, Sarkar P, Davis R (2019) Variability in the compressive strength of fly ash bricks. J Mater Civ Eng 31(2):06018024
9. Department of Industry Science Energy and Resources (2021) Australian electricity generation–fuel mix. https://www.energy.gov.au/data/australian-electricity-generation-fuel-mix
10. Ghosh P, Goel S (2017) Leaching behaviour of pond ash. In: Goel S (ed) Advances in solid and hazardous waste management. Springer International Publishing, Cham, pp 171–204
11. Tam VWY, Rahme D, Illankoon IMCS, Le KN, Yu J (2021) Effective remediation strategies for ash dam sites in coal power plants. Proc Instit Civil Eng Eng Sustain 174(2):94–105. https://doi.org/10.1680/jensu.20.0005212Nu-Rock. Technology Pty Ltd. (2017). Nu-Rock technology. http://www.nu-rock.com/
12. Nu-Rock Technology Pty Ltd. (2017) Nu-Rock technology. http://www.nu-rock.com/
13. Allen M (2018) Nu-Rock: the product which rose from the ashes. Western Advocate. https://www.westernadvocate.com.au/story/5658840/product-rises-from-ashes/
14. Nu-Rock (2020) Nu Rock–greener, smarter, stronger. https://nu-rock.com/
15. Davis and Langdon (2007) Life cycle costing (LCC) as a contribution to sustainable construction: a common methodology

Life Cycle Assessment (LCA) of Recycled Concrete Incorporating Recycled Aggregate and Nanomaterials

W. Xing, V. Tam, K. Le, J. L. Hao, J. Wang, and P. Yang

1 Extended Abstract

Nanotechnology is widely used in construction to advance high performance and durability materials, as represented by concrete, in which nanomaterials such as nanosilica (SiO_2), nano-alumina (Al_2O_3), and nanotitanium oxide (TiO_2) are used to replace cement as a partial binder. Suitable amounts of adding these nanomaterials are beneficial with regard to the mechanical, economic, and environmental aspects of concrete. For example, the pores in the cement matrix can be filled by nano-SiO_2, which has been demonstrated to enhance by $\approx$10% the comprehensive strength and by 25% flexural strength of the concrete [1]. Nano-TiO_2 concrete shows outstanding effective self-cleaning because TiO_2 nanoparticles can trigger a photocatalytic degradation of gas emissions (e.g. NOx, CO and VOCs), reducing the environmental impact of such concretes in a promising way [2].

To seek better performance of the concrete with the least amount of virgin material and address the global issues of construction and demolition waste and emissions, some researchers have introduced both recycled aggregate and nanomaterials into concrete, but environmental evaluations of such concrete products are scarce. Therefore, we assessed a broad range of environmental impacts of recycled aggregate concrete in a comprehensive literature review of sustainable concretes containing recycled aggregate and nanomaterials to establish an experimental database. Only concretes with recycled aggregate and nano-SiO_2 were included after the review

W. Xing · V. Tam (✉) · K. Le · J. Wang · P. Yang
School of Engineering, Design and Built Environment, Western Sydney University, Sydney, NSW, Australia
e-mail: vivianwytam@gmail.com

J. L. Hao
Department of Civil Engineering, Xi'an Jiaotong-Liverpool University, Sydney, China

W. Duan et al. (eds.), *Nanotechnology in Construction for Circular Economy*,
Lecture Notes in Civil Engineering 356,
https://doi.org/10.1007/978-981-99-3330-3_50

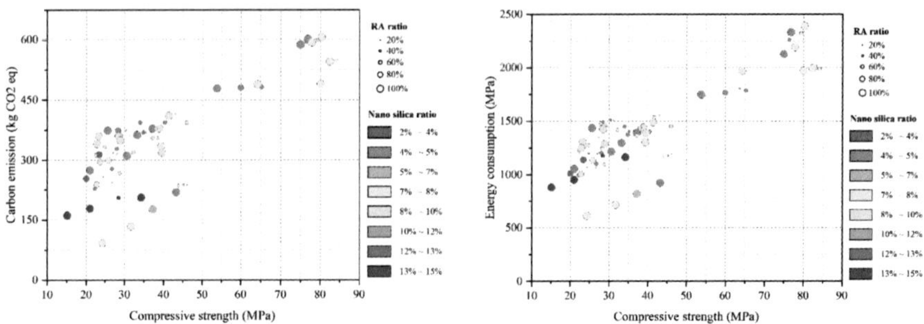

Fig. 1 Carbon emissions and energy consumption of concretes

process. These concrete products were evaluated by life cycle assessment (LCA), considering both volume and compressive strength.

The LCA model fitting sustainable concretes incorporating recycled aggregate and nanomaterials was modified from the framework of Xing et al. [3], based on the normal conditions and current technology applied to the Australian concrete industry. The functional unit was 1 m^3 of ready-mix concrete, considering its 28-day compressive strength, and the system boundary was from cradle to gate. The LCA methodology was CML 2002, but focused primarily on carbon emissions and energy consumption.

The literature review process resulted in 92 mix designs containing recycled aggregate and nano-SiO_2 and their carbon emissions and energy consumption were quantified by the LCA model, as shown in Fig. 1. Generally speaking, the incorporation ratio of nano-SiO_2 was limited to 15% whereas virgin aggregate could be fully replaced by the recycled aggregate. The carbon emission and energy consumption of the concretes examined were 91.6–607.1 kg CO_2 eq., and 611.6–2453.6 MJ, respectively. With the development of compressive strength, both the carbon emissions and energy consumption of the concrete increase. When other supplementary cementitious materials are simultaneously used in the concrete mix, the concrete products perform much better.

Compared with the effect of nano-SiO_2 on the strength and environmental impact of the concrete, substituting recycled aggregate was less effective. We therefore suggest introducing nanomaterials into recycled aggregate concrete for higher potential regarding the sustainability of the concrete.

Our study quantified the effects of nanomaterials and recycled aggregate to the overall environmental impacts of designed concrete products and estimated the ecological performance based on the mix design and compressive strength.

References

1. Qing Y, Zenan Z, Deyu K, Rongshen C (2007) Influence of nano-SiO2 addition on properties of hardened cement paste as compared with silica fume. Constr Build Mater 21(3):539–545
2. Zhu W, Bartos PJ, Porro A (2004) Application of nanotechnology in construction. Mater Struct 37(9):649–658
3. Xing W, Tam VW, Le KN, Butera A, Hao JL, Wang J (2022) Effects of mix design and functional unit on life cycle assessment of recycled aggregate concrete: Evidence from CO2 concrete. Constr Build Mater 348:128712

Analysis of the Compressive Strength of CO_2 Concrete While Eliminating Overshadowing Concrete Variables

V. W. Y. Tam, A. Butera, K. N. Le, and L. Liu

1 Extended Abstract

Although other technologies for the improvement of recycled aggregate, such as removal of adhered cement, improvement of aggregate by methods including nanomaterials, sequential mixing and precoating, the ideology of sequestration of CO_2 into cementitious materials provides great potential for both the enhancement and utilisation of recycled concrete materials and sequestration of CO_2. However, a great deal of study has focused on the complete carbonation of recycled aggregate without consideration of a practical timeframe for use with ready-mix concrete. Furthermore, the carbonation of recycled aggregate concrete does provide strength increases that can be overshadowed by other concrete mix design variables such as the water to cement (w/c) ratio and aggregate replacement percentage. Consequently, here, we focus on practical carbonation timeframes with a maximum of 2 h where the carbonation reaction is most prevalent, and keep other major concrete variables consistent to ensure the effect of carbonation can clearly be observed. The injection of CO_2 into recycled aggregate sequesters the undesirable gas while reducing the porosity and water absorption of the recycled aggregate. When CO_2 is injected into recycled aggregate, a chemical reaction occurs, converting the CO_2 into a mineral, calcium carbonate. The converted calcium carbonate fills air voids within the aggregate and consequently improves its mechanical performance. The carbonated aggregate concrete is also known as CO_2 concrete. Many researchers have investigated this mechanism, but on many occasions the true effect of carbonation variables (e.g., pressure and duration) on the compressive strength of the concrete cannot be observed due to other concrete variables, such as the w/c ratio and aggregate replacement percentage, which can

V. W. Y. Tam (✉) · A. Butera · K. N. Le · L. Liu
School of Engineering, Design and Built Environment, Western Sydney University, Penrith, NSW, Australia
e-mail: vivianwytam@gmail.com

W. Duan et al. (eds.), *Nanotechnology in Construction for Circular Economy*, Lecture Notes in Civil Engineering 356,
https://doi.org/10.1007/978-981-99-3330-3_51

overshadow the performance compressive increase provided by the carbonated recycled aggregate. In our study, we provided a constant w/c ratio and separate recycled aggregate percentages so that the effect of the carbonation variables could be truly observed. The longer carbonation duration of 120 min surpasses the lesser durations of 60 and 30 min. Furthermore, generally the lower carbonation pressure of 25 kPa provided greater results over the 75 and 200 kPa treated samples. Our results show that practical carbonation variables can help to increase the mechanical properties of recycled aggregate concrete. Furthermore, without other mix design variables a longer carbonation duration is of great importance to improving the properties of recycled aggregate concrete. The carbonation of recycled aggregate can help to make the greener product mainstream.

Harmonic Vibration of Inclined Porous Nanocomposite Beams

D. Chen and L. Zhang

Abstract This work investigated the linear harmonic vibration responses of inclined beams featured by closed-cell porous geometries where the bulk matrix materials were reinforced by graphene platelets as nanofillers. Graded and uniform porosity distributions combined with different nanofiller dispersion patterns were applied in the establishment of the constitutive relations, in order to identify their effects on beam behavior under various harmonic loading conditions. The inclined beam model comprised of multiple layers and its displacement field was constructed using Timoshenko theory. Forced vibration analysis was conducted to predict the time histories of mid-span deflections, considering varying geometrical and material characterizations. The findings may provide insights into the development of advanced inclined nanocomposite structural components under periodic excitations.

Keywords Functionally graded porosity · Graphene platelets · Harmonic vibration · Inclined beams

1 Introduction

The inclined beam problem has attracted many researchers and engineers due to its application potential in various fields (bridges and skytrain rail [1], for instance). The axial force induced by the inclined angle leads to a different deformation pattern compared with horizontal beams [2]. Meanwhile, in order to achieve enhanced performance, inclined structures with novel material compositions have been proposed, such as functionally graded (FG) inclined pipes [3] and inclined FG sandwich beams [4]. The recent development in this field using graded distributions of porosity and graphene platelets (GPLs) is demonstrated to be promising. Chen et al. [5] pointed

D. Chen (✉) · L. Zhang
Department of Infrastructure Engineering, University of Melbourne, VIC, Australia
e-mail: da.chen1@unimelb.edu.au

D. Chen
School of Civil Engineering, University of Queensland, St. Lucia, QLD, Australia

W. Duan et al. (eds.), *Nanotechnology in Construction for Circular Economy*,
Lecture Notes in Civil Engineering 356,
https://doi.org/10.1007/978-981-99-3330-3_52

out that strategic arrangements of internal pore size/density and GPL weight fractions significantly benefit the low-velocity impact properties of inclined beams under various impulses. The combination of FG porosities and graphene nanofillers can boost the mechanical performance of lightweight structural components [6, 7].

In this work, we aimed to further reveal the behavior of this novel porous nanocomposite by focusing on the responses of corresponding inclined beams under harmonic excitations, which represent steady wind, unbalanced rotating machine force, or vehicle loadings, etc. The theoretical formulations were first briefed and validated, then the mid-span deflections of fully clamped beams with changing inclined angles, porosity coefficients, GPL weight fractions and slenderness ratios were examined, considering the typical harmonic force sitting on the top mid-span surface.

2 Formulation Briefing

Figure 1 is a schematic illustration of the examined FG beam, of which the elastic properties varied along the height direction with $E(z) = E^*[1 - e_0\alpha(z)]$ for Young's modulus, $G(z) = E(z)/[2(1 + v(z))]$ for shear modulus, and $\rho(z) = \rho^*[1 - e^*\alpha(z)]$ for mass density, where Young's modulus of non-porous nanocomposites $E^* = \phi(E_{Al}, \Delta_{GPL})$ were determined using the Halpin–Tsai micromechanics model, e_0 reads the porosity coefficient related to internal pore size/density, $\alpha(z) = \cos(\pi z/h)$ and $\alpha(z) =$ constant correspond to the graded and uniform porosities, respectively, and $\rho(z)$ and $G(z)$ root in the closed-cell morphologies. The GPL weight fraction was $\Delta_{GPL} = \zeta_1[1 - \cos(\pi z/h)]$ for non-uniform dispersions and $\Delta_{GPL} = \zeta_2$ for the uniform one, while ζ_1 and ζ_2 were related to material distributions. The mass density and Poisson's ratio of non-porous nanocomposites were computed via the rule of the mixture with the corresponding values of the matrix materials (ρ_{Al}, v_{Al}) and GPLs (ρ_{GPL}, v_{GPL}). Note that the FG variations of porosity and GPL dispersion were both symmetric about the mid-height plane. Based on the established constitutive relations, the beam governing equation systems were derived within the framework of Timoshenko theory, and solved with the aid of the Ritz method and Newmark method. Consequently, the time history of beam mid-span deflections was estimated, in which the initial deformation status caused by self-weight was also embedded [5].

3 Results and Discussion

Based on aluminum composites, the parameters adopted in this study included $E_{Al} =$ 70 GPa, $\rho_{Al} = 2700$ kg/m^3, $v_{Al} = 0.34$, $w_{GPL} = 1.5$ μm, $l_{GPL} = 2.5$ μm, $t_{GPL} =$ 1.5 nm, $E_{GPL} = 1.01$ TPa, $\rho_{GPL} = 1062.5$ kg/m^3, $v_{GPL} = 0.186$. The beam section dimension was 0.1×0.1 m and the layer number was 14 [8]. The time length (2 s) was

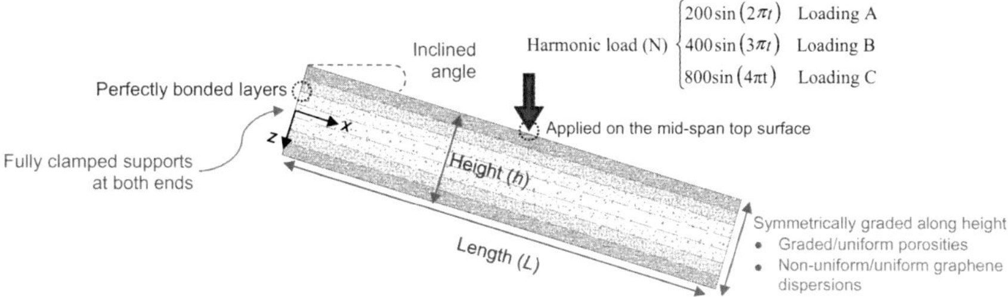

Fig. 1 Inclined functionally graded beam

the same for all calculation cases to obtain the maximum absolute value of the mid-span deflections. The beam boundary condition was taken as the clamped–clamped end supports. The present results were validated by being compared with ANSYS simulations (see Fig. 2). Figure 3 shows the mid-span deflection time histories of the examined beams under three harmonic loading scenarios. Results suggested that dispersing both internal pores and GPLs non-uniformly enhanced the inclined beam's stiffness due to reduced maximum deflections (~15%, ~18%, ~30%, as marked in Fig. 3a). It is also noticeable that the minimum deflections for the beams remained almost the same in loading case A but differed in loading cases B and C, because of the influence of gravity and the increased peak force in B and C.

For the purpose of simplification, the assessment given below is limited to responses subjected to harmonic loading case A. Figure 4a compares the harmonic vibration deflections of beams inclined at various angles. It is obvious that a larger inclined angle resulted in smaller deflections with lower harmonic force components along the height direction. The variations of porosity coefficient and GPL weight fraction are displayed in Figs. 4b, c. We can see that when the porosity coefficient increased from 0.2 to 0.6, the maximum deflection increased by ~10%. Meanwhile, an improved level of graphene weight fraction from 0.2% to 1.0% significantly

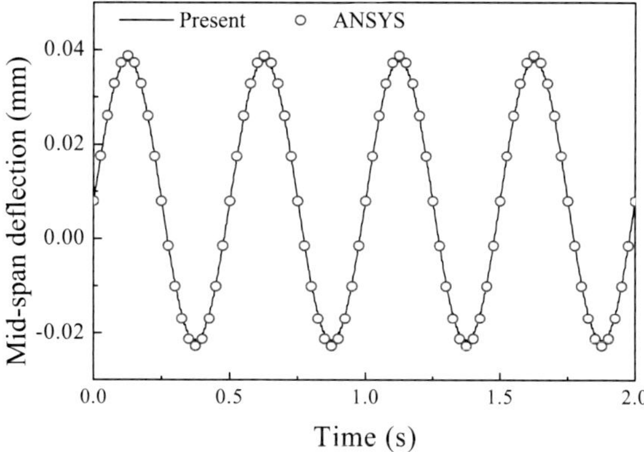

Fig. 2 Mid-span deflection time history of an inclined (45°) beam with graded porosity and non-uniform GPL dispersion under loading case C ($e_0 = 0.5$, $\Delta_{GPL} = 1.0\,wt.\%$, $L/h = 20$)

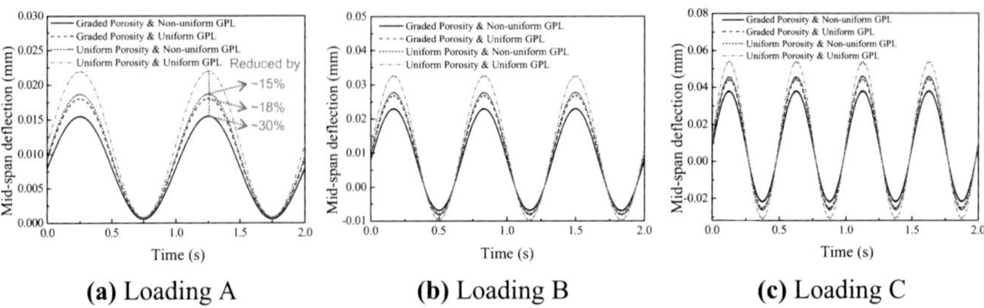

Fig. 3 a–c Mid-span deflection time histories of inclined functionally graded beams under harmonic loadings (inclined angle 45°, $e_0 = 0.5$, $\Delta_{GPL} = 1.0\,wt.\%$, $L\,/\,h = 20$)

stiffened the inclined beam of which the maximum mid-span deflection evidently decreased (~29%).

Table 1 details the influence of the slenderness ratio and inclined angle on the maximum mid-span deflections of inclined beams and still considers loading case A. Compared with two extreme cases (60° for $L/h = 20$; 0° for $L/h = 40$), a wide gap of 0.2476 mm was identified between their deflections.

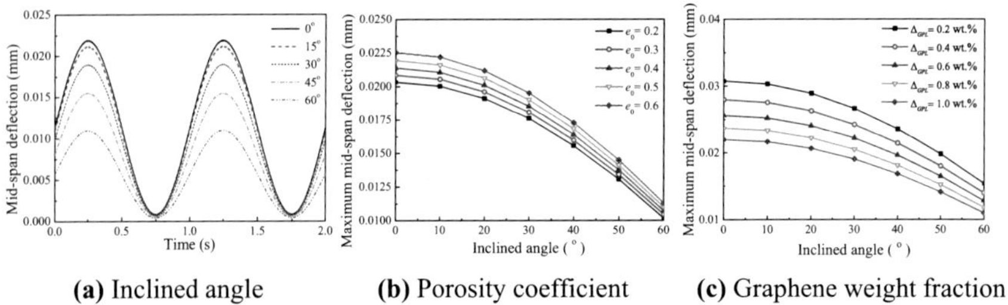

Fig. 4 a–c Mid-span deflections of inclined functionally graded beams under harmonic loading A (graded porosities & non-uniform graphene dispersions, $e_0 = 0.5$ and $\Delta_{GPL} = 1.0\,wt.\%$ for **a**, $\Delta_{GPL} = 1.0\,wt.\%$ for **b**, $e_0 = 0.5$ for **c**, $L/h = 20$)

Table 1 Maximum mid-span deflections (mm) of inclined beams under harmonic loading A (graded porosity & non-uniform graphene dispersion, $e_0 = 0.5$, $\Delta_{GPL} = 1.0\,wt.\%$)

Inclined angle	$L/h = 20$	$L/h = 30$	$L/h = 40$
0°	0.0219	0.0911	0.2586
20°	0.0206	0.0856	0.2430
40°	0.0168	0.0698	0.1981
60°	0.0110	0.0455	0.1293

4 Conclusions

We have discussed the influence of inclined angle, porosity, and graphene on beam harmonic vibration responses. We conclude that all three parameters are closely related to beam deflections under different harmonic loadings. The examined porous nanocomposites may be used to develop novel inclined structural components with reduced weight and enhanced stiffness subjected to periodic excitations.

Acknowledgements This research was supported by Australian Research Council (Chen: ARC DECRA DE220100876).

References

1. Yang D, Wang C, Pan W (2020) Further insights into moving load problem on inclined beam based on semi-analytical solution. Structures 26:247–256
2. Yang D, Wang C, Yau J (2020) Dynamic stability and response of inclined beams under moving mass and follower force. Int J Struct Stab Dyn. 2043004
3. Reddy RS, Panda S, Gupta A (2020) Nonlinear dynamics of an inclined FG pipe conveying pulsatile hot fluid. Int J Non-Linear Mech 118:103276
4. Nguyen DK, Tran TT, Pham VN, Le NAT (2021) Dynamic analysis of an inclined sandwich beam with bidirectional functionally graded face sheets under a moving mass. Eur J Mech A Solids 88:104276
5. Chen D, Yang J, Schneider J, Kitipornchai S, Zhang L (2022) Impact response of inclined self-weighted functionally graded porous beams reinforced by graphene platelets. Thin-Walled Struct 179:109501
6. Chen D, Yang J, Kitipornchai S (2017) Nonlinear vibration and postbuckling of functionally graded graphene reinforced porous nanocomposite beams. Compos Sci Technol 142:235–245
7. Yang J, Chen D, Kitipornchai S (2018) Buckling and free vibration analyses of functionally graded graphene reinforced porous nanocomposite plates based on Chebyshev-Ritz method. Compos Struct 193:281–294
8. Kitipornchai S, Chen D, Yang J (2017) Free vibration and elastic buckling of functionally graded porous beams reinforced by graphene platelets. Mater Des 116:656–665

Influence of Carbon Nanotubes on the Fracture Surface Characteristics of Cementitious Composites Under the Brazilian Split Test

Y. Gao, J. Xiang, Z. Yu, G. Han, and H. Jing

Abstract To better analyze the reinforcing mechanisms of carbon nanotubes (CNTs) in cementitious composites, the micromorphological characteristics of the fracture surface of cement-based specimens under the Brazilian split test were investigated. The results demonstrated that the addition of CNTs promoted nucleation and pore filling effects that optimized the pore structure and inhibited the development of microcracks in the cement matrix during loading. Due to the ultra-high specific surface area and bridging effects of CNTs, the formation of hydration products was promoted, internal microcracks and micropores were effectively reduced, and the regularity and integrity of the samples also improved. Therefore, under tensile loading, CNT-reinforced cementitious composites absorbed more energy, resulting in more complex stress paths and rougher fracture surfaces.

Keywords Cement · Fracture characteristics · Nano-reinforcement · Reinforcing mechanisms

1 Introduction

In the past decades, carbon nanomaterials, including carbon nanotubes (CNTs), graphene, graphene oxide and reduced graphene oxide, have attracted great attention as additives to reinforce the physical and mechanical properties of cementitious composites [1]. Due to their ultra-high specific surface area, superior mechanical

Y. Gao (✉)
School of Transportation and Civil Engineering, Nantong University, Nantong, China
e-mail: Y.Gao@ntu.edu.cn

J. Xiang · G. Han
Key Laboratory of Rock Mechanics and Geohazards of Zhejiang Province, Shaoxing University, Shaoxing, China

Z. Yu · H. Jing
State Key Laboratory for Geomechanics and Deep Underground Engineering, China University of Mining and Technology, Xuzhou, China

W. Duan et al. (eds.), *Nanotechnology in Construction for Circular Economy*, Lecture Notes in Civil Engineering 356,
https://doi.org/10.1007/978-981-99-3330-3_53

properties, and the densification effect on the microstructure of the cement, CNTs have been proved to be capable of enhancing the tensile strength [2] and durability [3] of cement-based materials with a minimal mixing ratio (0.01–0.05% by weight of cement) [4]. Moreover, the latest research suggests that carbon nanomaterial reinforcement is a cost-effective strategy for reducing the environmental impact of cement usage and maintenance, providing an effective way to reduce CO_2 emissions and energy consumption in the cement production and construction industries [5]. Hence, it is significant to carry out research on the development and application of carbon nanomaterial-modified cementitious composites and investigate the corresponding reinforcing mechanisms.

The micromorphological characteristics of the fracture surface of cementitious composites are closely related to the micromechanical damage evolution process of hardened cement-based materials. The micro-geometric features of the fracture surface record the irrecoverable deformation of the cementitious sample when it was broken, as well as micro-information from the initiation, propagation, penetration and nucleation of cracks to the final fracture and instability of the overall cement structure [6]. In scientific research and engineering practice, the relevant information recorded during the fracture damage process of cement materials can be obtained through the study of the micromorphology of the typical fracture surface of cement samples. Afterwards, mathematical statistics, induction and analysis of geometric features can be applied and then the mechanical mechanism of fracture could be reversed [7].

In the present study, we analyzed the reinforcing mechanisms of CNTs in cementitious composites by investigating the micromorphological features of the fracture surface of cement-based specimens under the Brazilian split test. First, we used 3D scanning technology to reconstruct the micromorphology of the cement-based specimens after the Brazil split test. Next, the micro-roughness characteristic parameters of the specimens were calculated and quantitatively analyzed. Considering the influential mechanism of micro-roughness on material fracture, a micro-dilatancy micro-element model of the fracture surface of the cement samples was constructed. The micro-dilation angle and the normal dilatation deformation of the CNT composite cement-based material after the Brazilian split test were calculated using the micro-roughness characteristic parameters. Finally, the influence of the CNTs on the fracture surfaces characteristics of cementitious composites under the Brazilian split test was discussed.

2 Methods

Ordinary Portland cement (PO. 32.5) was applied as the binder material and fly ash (FA), with a density of 2.4 g/cm^3, was used as a partial replacement for cement powder in the cementitious composites. Multi-wall carbon nanotubes (MWCNTs), fabricated via chemical vapor deposition, were selected as the nano-reinforcing material to enhance the composites. Polycarboxylic acid water reducer (PC), an ionic surfactant, non-toxic and harmless, containing both hydrophobic and hydrophilic groups in the molecules, was used to improve the dispersion of CNTs in the suspensions. The CNTs dispersion and cement-based material pouring process were consistent with our previous report [8]. Three groups, Ref-group, FA-group and CNT-group, were marked in this study. A 0.4 water-to-cement ratio was applied in all three groups. For the FA-group, 20 wt% FA was used as a substitute to reduce cement usage and enhance the workability of the paste to generate a cost-effective material. For the CNT-group, 0.08 wt% CNTs was mixed into the cement–FA-based slurry to further optimize the pore structure of the hardened matrix and enhance the mechanical properties of the cementitious composite.

Then, the 28-day-old standard cured disc samples with a diameter of 50 mm and a thickness of 25 mm were selected for the Brazilian split tests. The loading rate was 0.10 mm/min. In order to reduce the influence of errors in the test and improve accuracy, three samples were selected for each group for testing. Afterwards, a VR 3000 3D contour scanner with high precision was used to scan the fracture surface of the tested specimens. The manual stitching mode was selected for the scanning process. The maximum scanning size was 90 $\times$ 160 mm, and the highest resolution was 1 μm.

3 Results

3.1 Tensile Properties of the Cementitious Composites

The mean and standard deviation of the tensile strength of the cementitious composites are shown in Fig. 1. The tensile strength of the plain cement specimen was $\approx$3.6 MPa. After mixing in FA, the tensile strength of the cementitious specimens was significantly reduced to 3.17 MPa, $\approx$12% lower than that of the Ref-group. The deterioration in tensile strength was mainly due to the decreased cement powder content, which led to a decrease in the hydration degree of the composite. The hydration in the sample was uneven and incomplete, and there were more microcracks and micropores in the specimens. In contrast, the mean tensile strength of the CNT-group specimen was significantly improved to 3.77 MPa with only mixing 0.08wt% CNTs into cementitious composites, which was 4.6% and 18.9% higher than the Ref-group and FA-group specimens, respectively. The reinforcing rate of the tensile properties of cement-based materials after mixing CNTs was highly consistent with previous studies [9]. According to our previous report [8], by their nucleation and crack bridging effects, CNTs effectively enhance the microstructure of cementitious composites, thereby strengthening their mechanical properties.

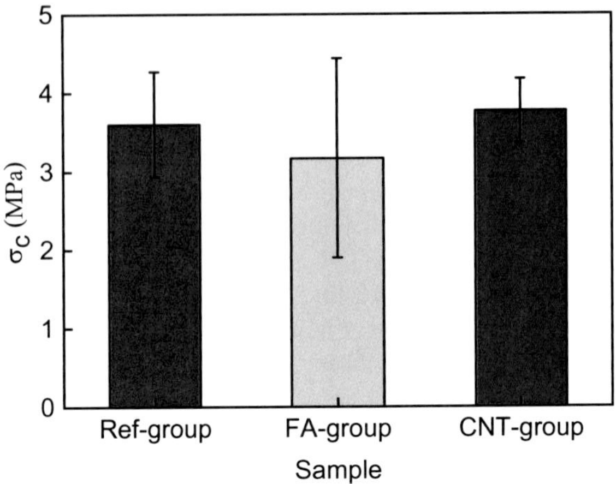

Fig. 1 Tensile strength of different groups of cementitious composite materials

3.2 Analysis of the Mesoscopic Asperities of the Fracture Surface

The fracture method describes the surface of the fracture through three aspects: shape, undulation and roughness [10]. Typically, macro-geometric features are described by shape and relief, and micro-geometric features are described by roughness. The micro-roughness is the degree of unevenness of the asperities attached to the macro-relief surface in a small-scale range, as shown in Fig. 2a. Previous studies have proposed several qualitative and quantitative methods for evaluating the microscopic roughness of rock fracture surfaces [7]. For example, Barton et al. [11] proposed evaluating the roughness of fracture surfaces by visual inspection, comparing and referencing 10 standard joint roughness coefficient curves. Based on their theory, many researchers have further explored and developed related statistical and fractal methods to more accurately characterize the roughness of fracture surfaces [12, 13].

In this study, the statistical method for the microscopic roughness of the fracture surface of cement-based materials was based on the method proposed by Belem et al. [14]. As demonstrated in Fig. 2b, the 3D fracture point cloud data were discretized into equally spaced points and gridded into a series of fine-scale planes. For each microscopic plane, the local inclination angle (αij) is defined as its angle with the horizontal plane, which is the angle between the normal vector n and z-axis. The αij and the height of each minutiae plane are counted to obtain the maximum, minimum, mean and standard deviation. For microstructure characterization, a 3D contour scanner was employed to image the typical fracture surface of the cementitious specimen after tensile damage to obtain the fracture surface point cloud data. The fracture surface point cloud image was cropped by Geomagic Wrap software, with a cropping size of 40 mm in length and 20 mm in width. The cropped 3D mesh data were exported by PolyWorks software and 3D reconstruction was performed using MATLAB. Representative 3D reconstructed images are exhibited in Fig. 4a–c.

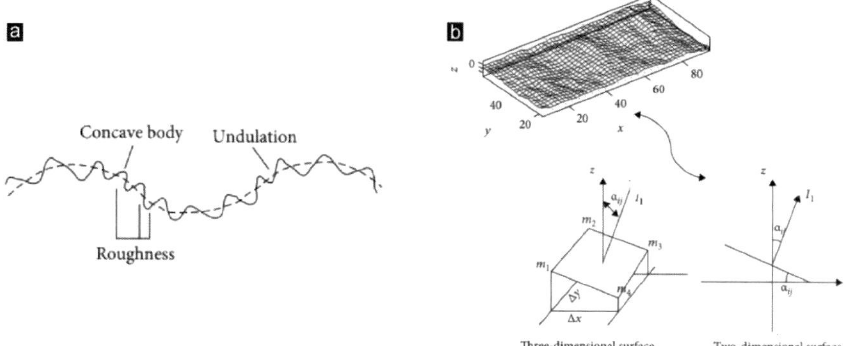

Fig. 2 Schematic of microscopic fracture surface: **a** microscopic roughness of fracture; **b** slope angle of reference surface in different dimensions

The fracture surface undulation height distribution for the three groups of cementitious specimens after the Brazilian split tests are shown in Fig. 3. It can be seen that the standard deviations of the undulation heights of the Ref-group, FA-group and CNT-group specimens are 0.83, 1.01 and 0.76 mm, respectively. The standard deviation of the undulation height of the FA-group specimens' fracture surface was ≈21.7% larger than that of the Ref-group, because the incorporation of FA decreased the hydration reaction rate and reduced the hydration products, resulting in more microcracks and micropores in the specimens. The fundamental cause of the failure of the sample was the original microcracks and micropores inside the sample gradually extending outward and deeper, causing the damage. Hence, the more microcracks and micropores inside the FA-group specimens caused a more complex and tortuous extension of the primary fracture. The undulation of the fracture surface was larger, which eventually led to a more significant standard deviation of the undulation height of the fracture surface of the FA-group than that of the Ref-group. With the addition of CNTs, the micropores inside the cementitious composites were optimized due to the nucleation and pore infilling effects [8], and porosity was also reduced. As a result, the extension path of the primary fracture developed from fewer micropores and microcracks became single and smooth, with fewer undulations on the fracture surface. Finally, the standard deviation of the undulation height of the fracture surface of the CNT-group specimens was smaller than that of the Ref- and FA-groups.

The measured results of the microscopic relief angle of the fracture surface are presented in Fig. 4. The standard deviations of the microscopic relief angles of the CNT-group and FA-group specimens were more extensive than those in the Ref-group, indicating that the plane slope changed more in space, and the fracture surface was correspondingly rougher. Cementitious composites continuously absorb energy during the loading process. For the FA-group specimens, after FA replaced part of the cement powder, the internal hydration reaction of the slurry was not uniform, resulting in nonuniform energy distribution inside the loading samples. The crack propagation path was more complicated when the unevenly distributed energy began

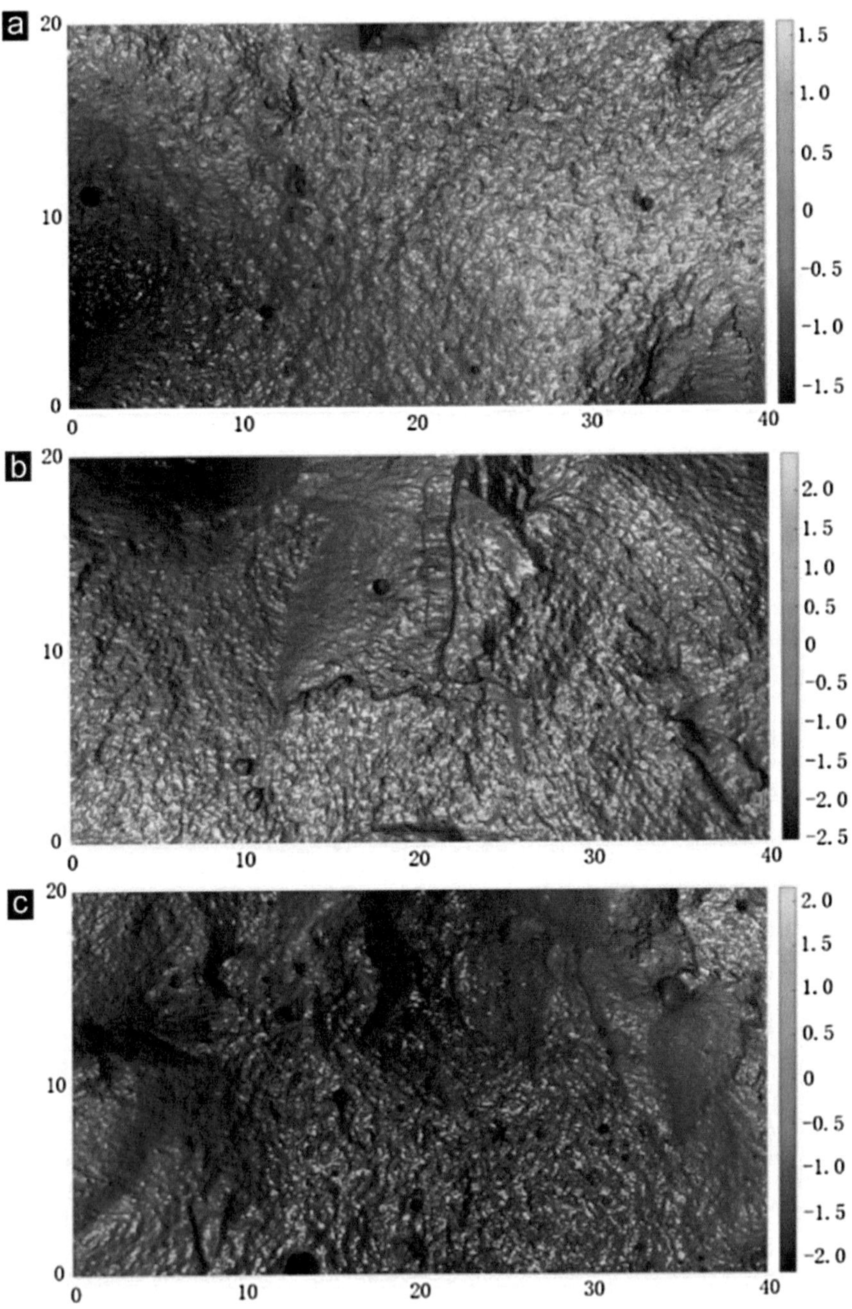

Fig. 3 Three-dimensional reconstruction images of fracture surface of typical cementitious materials under tensile damage: **a** Ref-group, **b** FA-group and **c** CNT-group

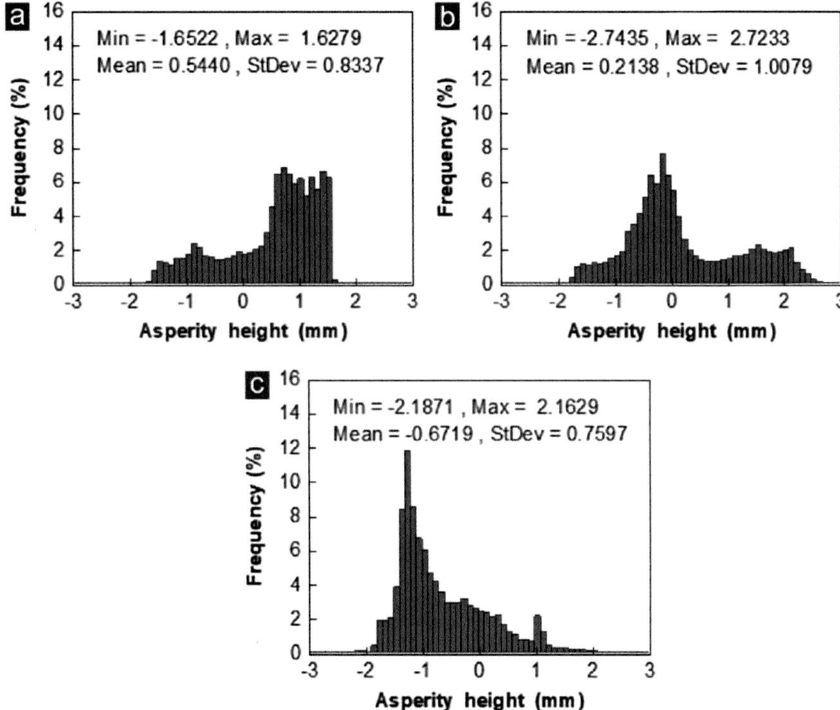

Fig. 4 Microscopic undulation height distribution of the fracture surface of cement-based samples under tensile failure condition: **a** Ref-group, **b** FA-group and **c** CNT-group

to be released into the stress concentration area at the tip of the microcrack. For CNT-group specimens, with the incorporation of CNTs, the nucleation effects promoted the hydration reaction, contributing to more and denser hydration products. The integrity of the specimens became higher and could absorb more energy during tensile loading. When the energy accumulated to a specific value, a large amount of energy was released instantaneously, and the stress was concentrated around the fracture surface, making the crack propagation path more complicated. In addition, the bridging and pull-out effects [9] of the CNTs effectively inhibited the development of cracks. More secondary microcracks were generated inside the sample, and the crack propagation path became more complicated, resulting in higher fracture surface roughness (Fig. 5).

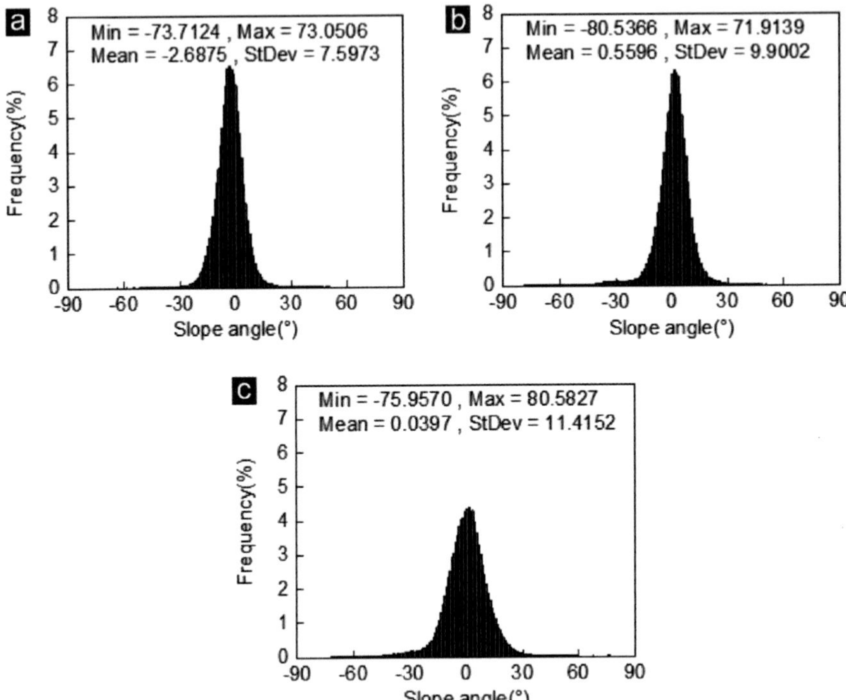

Fig. 5 Microscopic slope angle distribution of the fracture surface of cement-based samples under tensile failure condition: **a** Ref-group, **b** FA-group and **c** CNT-group

4 Discussion and Conclusions

In this work, CNTs were used as the additive in cementitious composites to assist FA in reducing cement usage and generating cost-effective, environmentally friendly, high-workability cement-based materials. The tensile properties of three groups of specimens were tested by the Brazilian split test. The fracture surfaces after tensile loading were scanned by a high-precision 3D scanning system. Based on fracture theory, the micro-enhancing mechanism of CNTs on cementitious composites was revealed.

Compared with plain cement, replacing the same cement mass in the slurry with 20 wt% FA resulted in a 12% deterioration in the tensile strength of the cement-based material. Nevertheless, adding only 0.08 wt% of CNTs into the cement–FA hybrid composites significantly increased the tensile strength by 18.9%. Compared with plain cement-based materials, the tensile strength of the hardened matrix was enhanced by 4.6%. This finding indicated that 0.08 wt% CNTs combined with 20 wt% FA could be a good substitute for cement in cementitious composites without affecting its mechanical properties.

The 3D scanning results of the fracture surfaces of the tested specimens demonstrated that the FA-group samples had a higher standard deviation of the macroscopic undulation height than the Ref-group at the macroscopic level. After mixing 20 wt% FA, more microcracks and micropores appeared in the cementitious matrix. As a result, the extension path of the primary fracture became tortuous and complicated in the FA-group samples. By contrast, due to the ultra-high specific surface area and bridging effects of CNTs, the formation of hydration products was promoted, internal microcracks and micropores were effectively reduced, and the regularity and integrity of the samples were also improved. Therefore, under tensile loading, the extension path of the final primary fracture was simple, and the standard deviation of the corresponding macro-fluctuation height was low.

At the microscopic level, energy was continuously absorbed during tensile loading of the specimens. Compared with the Ref-group specimens, the uneven hydration reaction inside the FA-group specimens made the energy distribution inside the sample uneven, resulting in a rougher fracture surface. The addition of CNTs promoted nucleation and had pore filling effects that optimized the pore structure and inhibited the development of microcracks in the cement matrix during loading. The CNT-group specimens could absorb more energy, resulting in more complex stress paths and rougher fracture surfaces. In conclusion, using 0.08 wt% CNTs combined with 20 wt% FA to replace the same cement mass in the slurry will reduce the microcracks and micropores in the hardened matrix, making the change rate of the macro-undulation height smaller. CNT-reinforced cementitious composites can absorb more energy during tensile loading. As the energy is released to the stress concentration zone at the tip of the microcrack, the stress path of the CNT-group specimens became more complicated, with a more extensive change rate of the microscopic inclination angle and rougher fracture surface.

Acknowledgements This study was supported by the Natural Science Foundation of the Jiangsu Higher Education Institutions of China under Grant (22KJB560010) and Nantong Basic Science Research Program of China under Grant (No. JC12022098).

Author Contributions The manuscript was written through contributions of all authors. All authors have given approval to the final version of the manuscript.

Declaration of Competing Interests The authors declare that they have no known competing financial interests or personal relationships that could have appeared to influence the work reported in this paper.

References

1. Ramezani M, Dehghani A, Sherif MM (2022) Carbon nanotube reinforced cementitious composites: a comprehensive review. Constr Build Mater 315:125100
2. Rocha VV, Ludvig P, Trindade ACC, de Andrade Silva F (2019) The influence of carbon nanotubes on the fracture energy, flexural and tensile behavior of cement based composites. Constr Build Mater 209: 1–8

3. Sarvandani MM, Mahdikhani M, Aghabarati H, Fatmehsari MH (2021) Effect of functionalized multi-walled carbon nanotubes on mechanical properties and durability of cement mortars. J Build Eng 41:102407
4. Reales OAM, Toledo Filho RD (2017) A review on the chemical, mechanical and microstructural characterization of carbon nanotubes-cement based composites. Constr Build Mater 154: 697–710
5. de Souza FB, Yao X, Gao W, Duan W (2022) Graphene opens pathways to a carbon-neutral cement industry. Sci Bull 67:5–8
6. Eberhardt E, Stead D, Stimpson B, Read R (1998) Identifying crack initiation and propagation thresholds in brittle rock. Can Geotech J 35(2):222–233
7. Shi X, Jing H, Chen W, Gao Y, Zhao Z (2021) Investigation on the creep failure mechanism of sandy mudstone based on micromesoscopic mechanics. In: Geofluids
8. Gao Y, Jing H, Zhou Z, Shi X, Li L, Fu G (2021) Roles of carbon nanotubes in reinforcing the interfacial transition zone and impermeability of concrete under different water-to-cement ratios. Constr Build Mater 272:121664
9. Gao Y, Jing HW, Chen SJ, Du MR, Chen WQ, Duan WH (2019) Influence of ultrasonication on the dispersion and enhancing effect of graphene oxide–carbon nanotube hybrid nanoreinforcement in cementitious composite. Compos B Eng 164:45–53
10. Yuan L, Zhou F, Li B, Gao J, Yang X, Cheng J, Wang J (2020) Experimental study on the effect of fracture surface morphology on plugging efficiency during temporary plugging and diverting fracturing. J Nat Gas Sci Eng 81:103459
11. Barton N, Choubey V (1977) The shear strength of rock joints in theory and practice. Rock Mech 10(1):1–54
12. Yong R, Ye J, Liang Q-F, Huang M, Du S-G (2018) Estimation of the joint roughness coefficient (JRC) of rock joints by vector similarity measures. Bull Eng Geol Env 77(2):735–749
13. Zhang G, Karakus M, Tang H, Ge Y, Zhang L (2014) A new method estimating the 2D joint roughness coefficient for discontinuity surfaces in rock masses. Int J Rock Mech Min Sci 72:191–198
14. Belem T, Homand-Etienne F, Souley M (2000) Quantitative parameters for rock joint surface roughness. Rock Mech Rock Eng 33(4):217–242

Effect of Carbon Nanotubes on the Acoustic Emission Characteristics of Cemented Rockfill

Z. Yu, H. Jing, Y. Gao, X. Wei, and A. Wang

Abstract The use of carbon nanotubes (CNTs) to reinforce cemented rockfill is attracting considerable interest due to the remarkable improvement in performance and the extremely low dose of the added nanomaterial. To reveal the enhancement mechanism of the CNTs on cemented rockfill, the acoustic emission (AE) characteristics of cemented rockfill specimens during the Brazilian split test were investigated. The results demonstrated that CNTs improved tensile strength by 17.2% and decreased the AE count. The nucleation and micropore-filling effects of the CNTs promoted the cement hydration reaction and formation of a denser structure, thereby improving resistance to loads. Meanwhile, finer pores avoid stress concentration, resulting in AE activity becoming more sparse. Finally, the AE b-value increased by 14.8%, which further indicated that the overall failure process was at a lower intensity.

Keywords Acoustic emissions · Carbon nanotubes · Cemented rockfill · Failure patterns

1 Introduction

Carbon nanomaterials, including carbon nanotubes (CNTs), graphene oxide and carbon nanosheets, have been widely recognized for their enhancement of cement-based composites [1, 2]. The strengthening mechanism of CNTs in cement-based composites has been the focus in recent decades [3, 4], and it is though physical properties such as extremely high mechanical strength and large specific surface

Z. Yu · H. Jing (✉) · X. Wei · A. Wang
State Key Laboratory for Geomechanics and Deep Underground Engineering, China University of Mining and Technology, Xuzhou, China
e-mail: hongwenjingcumt@126.com

Y. Gao
School of Transportation and Civil Engineering, Nantong University, Nantong, China

W. Duan et al. (eds.), *Nanotechnology in Construction for Circular Economy*, Lecture Notes in Civil Engineering 356,
https://doi.org/10.1007/978-981-99-3330-3_54

area that CNTs fill microscopic pores in cement matrix and promote the development of hydration products [5]. The resulting compact microstructure greatly optimizes the macroscopic mechanical properties of cement-based composites, including compressive strength, impermeability and corrosion resistance [6, 7]. At the same time, the addition of CNTs is potentially a way to reduce CO_2 emissions from cement production [8].

In construction and mining projects, cemented rockfill with higher mechanical properties is significant for structural stability and production safety. CNTs improve the mechanical strength of cemented rockfill materials [9], but due to the lack of sensitivity of current monitoring systems, the effect of CNTs on failure patterns during the loading process needs further research. Acoustic emission (AE) technology is a nondestructive testing method widely used in the study of the mechanical properties of rock materials [10]. During the failure process, the gradual development of microfractures will cause acoustic signals, which can be monitored by the AE system. The waveform of the AE signal reflects the microscopic failure pattern. For example, the RA (Rise time to Amplitude ratio) and AF (AE counts to Duration ratio) can be used as criteria for the type of failure [11]. Therefore, the AE technique is effective for studying the destruction mode of the cemented rockfill.

In this study, we add a very low content of CNTs with fly ash (FA) to partially replace cement in the cemented rockfill. The tensile failure strength of the specimens was measured by the Brazilian split test and simultaneously the AE system collected signals during the destruction process. Combing the AE activity, and the stress-–strain curve, the distribution and intensity of the failure events in each loading stage of the CNT-enhanced specimens were analyzed. The failure event accumulation of the specimen is discussed according to the peak and growth rate of the cumulative AE counts curve. The b-value in AE technology is calculated to characterize the severity of the overall failure process. The findings in this study promote the engineering application of CNTs and we discuss the enhancement mechanism of CNTs in cemented rockfill materials.

2 Methods

Waste coal gangue extracted from a coal mine in Shanxi province, China, was selected as the solid particles in cement rockfill. The composition of coal gangue does not react chemically with the cement. Ordinary Portland cement (PO. 42.5), conforming to the Chinese Standard GB-175-2007 [12], was the binder material. Multiwall CNTs (MWCNTs; manufactured by Nanjing XFNANO Materials Tec. Co., Nanjing City, Jiangsu Province, China) were added to the rockfill materials. Polycarboxylate superplasticizer was added to promote CNT dispersion in the suspension.

The preparation of CNT-enhanced cement rockfill samples mainly comprised the preparation of CNT–cement slurry and mixing of the slurry with solid particles. The process is detailed in our previous report [13]. A Ref-group and CNT-group were set in this study, and a 0.4 water-to-cement ratio was applied in both groups. In the

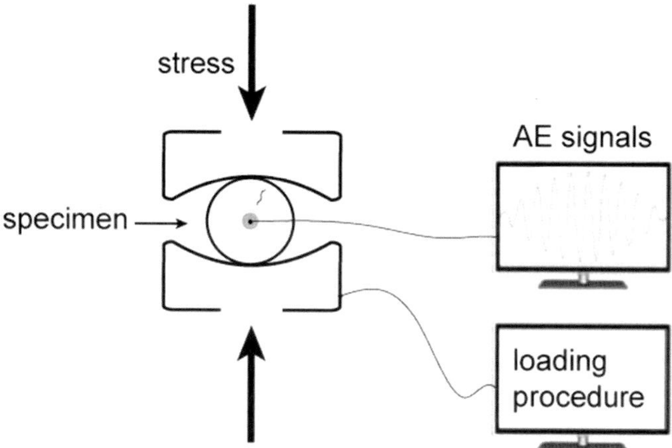

Fig. 1 Brazilian split test and acoustic emission (AE) signal monitoring

CNT-group mortar, there was 20% FA with 0.05 wt% CNTs to replace 20% cement. The other preparation processes of the two groups were consistent.

We chose 28-day-old standard-cured disc samples with a diameter of 50 mm and a thickness of 25 mm for the Brazilian split tests. The loading rate was 0.10 mm/min. To avoid the influence of errors in the test and improve accuracy, three samples were selected for each group. During the loading process, the AE signals were monitored by a Micro-II AE system (developed by the American Physical Acoustic Corporation), as shown in Fig. 1. The AE threshold and frequency were set at 30 dB and 10 Msps, respectively.

3 Results

3.1 Mechanical Properties of Cemented Rockfill

After curing for 28 days, the peak tensile strength of the specimens is shown in Fig. 2. The peak strength of the CNT-group was 3.4 MPa, an increase of 17.2% compared with the Ref-group, indicating that CNTs significantly enhanced the ability of the sample to resist tensile stress, which was consistent with previous studies [14]. Higher tensile strength can reduce the risk of sudden failure and improve engineering safety.

The increase in tensile strength was attributed to the strengthening effect of CNTs in the cement matrix. CNTs show a bridging effect between microfractures in the cement matrix [15]. When the sample is under tensile stress, the high tensile strength of CNTs can help the cement matrix resist external stress and avoid microcrack development [16]. As the sample's hydration reaction continued, the hydration products

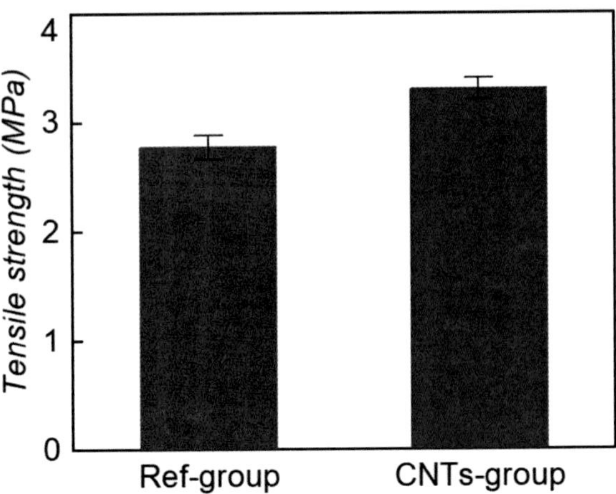

Fig. 2 Tensile strength of two groups of cemented rockfill specimens. CNTs, carbon nanotubes

became more closely connected with the CNTs and the resultant compact internal structure enabled the specimen to resist failure and deformation.

3.2 *Analysis of AE Activity During the Failure Process*

The stress–strain curve and real-time AE signals of the samples are shown in Fig. 3. As the stress gradually increased, the stress–strain curve divided into four failure stages. In the initial compaction stage and elastic deformation stage, there was almost no fracture inside the sample, accompanied by sparse AE signals. During the plastic deformation stage, cracks gradually developed and the AE activity gradually became denser. When peak strength was reached, the samples suddenly produced a lot of AE counts, corresponding to the development of major fractures. In the post-peak stage, cracks continue to expand.

With the addition of CNTs to the cemented rockfill, the AE activity became sparse and AE counts decreased significantly during the failure process. The cumulative

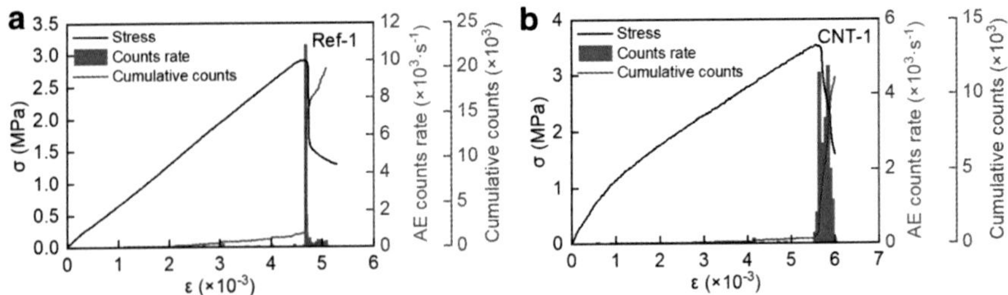

Fig. 3 Acoustic emission counts of the Ref-group (**a**) and CNT-group (**b**) during the Brazilian split test. CNTs, carbon nanotubes

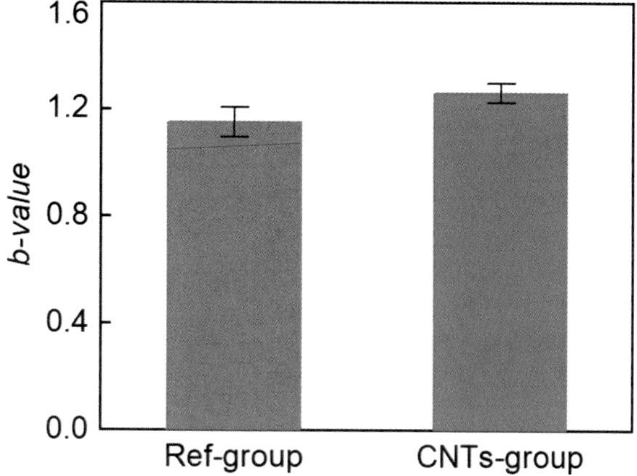

Fig. 4 The b-value of two groups of cemented rockfill specimens. CNTs, carbon nanotubes

counts of the Ref-group reached 20($\times 10^3$), much higher than in the CNT-group. At the peak strength point, the Ref-group showed a single and high-count AE event, whereas the CNT-group showed multiple peaks of AE activity. The reason for this phenomenon is the promotion of a dense microstructure by CNTs. The defect area is smaller, reducing the severity of failure, resulting in fewer counts and lower frequency of AE events, which leads to more gradual macro disruption.

The b-value reflects the ratio of the number of lower and higher grade AE events during the rock failure process, and can be calculated by the following equation [17]:

$$lgN_{(A/20)} = a - b \times (A/20) \tag{1}$$

where A is the amplitude of the AE event, $N_{(A/20)}$ is the cumulative frequency of AE events with an amplitude $\geq$ A, and b represents the b-value.

According to Eq. 1, the b-value of the CNT-group reached 1.27, increasing 14.8% more than the Ref-group (Fig. 4), which reflects the AE activity being mostly concentrated at lower energy levels during the failure of the sample. The reason behind this phenomenon is that CNTs promote the development of hydration products to reduce the size of micropores, further avoiding the high energy release of cracks during loading.

4 Discussion and Conclusions

In this study, CNTs and FA were added to a cemented rockfill mix to investigate the production of environmentally friendly and high-workability materials. The tensile properties of the three specimens in each group were tested by the Brazilian split test and AE signals were monitored by a PCI-2 system. The failure mode pattern of the

CNT-reinforced cement rockfill material was studied by the stress–strain curve and AE waveform.

In the Brazilian split experiment, the CNT-reinforced sample showed increased peak tensile strength by 17.2%. At the same time, the AE events became sparser and the count decreased, indicating that the intensity of destruction decreased. Finally, the addition of CNTs increased the b-value by 14.8%, which indicated that CNT-reinforced cemented rockfill material will have fewer high energy release destruction events, thereby making it a safer option.

Acknowledgements This study was supported by Yunlong Lake Laboratory of Deep Underground Science and Engineering Project (No. 104023002), National Natural Science Foundation of China (No. 52074259), Assistance Program for Future Outstanding Talents of China University of Mining and Technology (No. 2022WLJCRCZL050) and Postgraduate Research & Practice Innovation Program of Jiangsu Province (No. KYCX22_2580).

Author Contributions The manuscript was written through contributions of all authors. All authors have given approval to the final version of the manuscript.

Declaration of Competing Interest The authors declare that they have no known competing financial interests or personal relationships that could have appeared to influence the work reported in this paper.

References

1. Gao Y, Jing HW, Chen SJ, Du MR, Chen WQ, Duan WH (2019) Influence of ultrasonication on the dispersion and enhancing effect of graphene oxide–carbon nanotube hybrid nanoreinforcement in cementitious composite. Compos B 164:45–53
2. Yao X, Shamsaei E, Wang W, Zhang S, Sagoe-Crentsil K, Duan W (2020) Graphene-based modification on the interface in fibre reinforced cementitious composites for improving both strength and toughness. Carbon 170:493–502
3. Morsy M, Alsayed S, Aqel M (2011) Hybrid effect of carbon nanotube and nano-clay on physico-mechanical properties of cement mortar. Constr Build Mater 25(1):145–149
4. Li GY, Wang PM, Zhao X (2007) Pressure-sensitive properties and microstructure of carbon nanotube reinforced cement composites. Cem Concr Compos 29(5):377–382
5. Konsta-Gdoutos MS, Metaxa ZS, Shah SP (2010) Highly dispersed carbon nanotube reinforced cement based materials. Cem Concr Res 40(7):1052–1059
6. Liu C, Huang X, Wu Y-Y, Deng X, Zheng Z (2021) The effect of graphene oxide on the mechanical properties, impermeability and corrosion resistance of cement mortar containing mineral admixtures. Constr Build Mater 288:123059
7. Gao Y, Jing H, Zhou Z, Shi X, Li L, Fu G (2021) Roles of carbon nanotubes in reinforcing the interfacial transition zone and impermeability of concrete under different water-to-cement ratios. Constr Build Mater:272
8. de Souza FB, Yao X, Gao W, Duan W (2022) Graphene opens pathways to a carbon-neutral cement industry. Sci Bull 67:5–8
9. Wu J, Jing H, Gao Y, Meng Q, Yin Q, Du Y (2022) Effects of carbon nanotube dosage and aggregate size distribution on mechanical property and microstructure of cemented rockfill. Cem Concr Compos 127:104408

10. Zhao ZL, Jing HW, Wu JY, Shi XS, Gao Y, Yin Q (2020) Experimental investigation on damage characteristics and fracture behaviors of granite after high temperature exposure under different strain rates. Theor Appl Fract Mech 110:102823
11. Tragazikis IK, Dassios KG, Exarchos DA, Dalla PT, Matikas TE (2016) Acoustic emission investigation of the mechanical performance of carbon nanotube-modified cement-based mortars. Constr Build Mater 122:518–524
12. Standard C (2007) Common portland cement (text of document is in Chinese). GB 175. Standards Press of China, Beijing
13. Gao Y, Jing H, Yu Z, Li L, Wu J, Chen W (2022) Particle size distribution of aggregate effects on the reinforcing roles of carbon nanotubes in enhancing concrete ITZ. Constr Build Mater 327:126964
14. Siddique R, Mehta A (2014) Effect of carbon nanotubes on properties of cement mortars. Constr Build Mater 50:116–129
15. Tyson BM, Abu Al-Rub RK, Yazdanbakhsh A, Grasley Z (2011) Carbon nanotubes and carbon nanofibers for enhancing the mechanical properties of nanocomposite cementitious materials. J Mater Civ Eng 23(7):1028
16. Naqi A, Abbas N, Zahra N, Hussain A, Shabbir SQ (2019) Effect of multi-walled carbon nanotubes (MWCNTs) on the strength development of cementitious materials. J Mater Res Technol 8(1):1203–1211
17. Colombo IS, Main I, Forde M (2003) Assessing damage of reinforced concrete beam using "b-value" analysis of acoustic emission signals. J Mater Civ Eng 15(3):280–286

Effects of Graphene Oxide Content on the Reinforcing Efficiency of C–S–H Composites: A Molecular Dynamics Study

W. Chen, J. Xiang, Y. Gao, and Z. Zhang

Abstract Determining the graphene oxide (GO) content is the key to applying GO to reinforce the mechanical performance and durability of cementitious composites. However, most of the previous studies are conducted from the perspective of experiments and lack elaboration on the mechanism of the GO-reinforced cementitious composite under different GO content. Hence, we investigated the effect of GO content on the reinforcing efficiency of calcium–silicate–hydrate (C–S–H) to trade off the enhancement of GO in cementitious composites and the corresponding economic benefits. The results demonstrated that an appropriate number of GO nanosheets can reinforce the cementitious composite with simultaneous high enhancing efficiency and economic benefits. The microdamage evolution of GO/C–S–H composites and the GO reinforcing mechanisms are reported. Our findings deepen the understanding of the enhancing mechanisms of GO embedded in C–S–H nanocomposites and help to determine the suitable GO content in practical engineering.

Keywords Cementitious composites · Graphene oxide · Mixing content · Molecular dynamics simulations

---

W. Chen (✉)
Department of Mechanical, Aerospace and Civil Engineering, The University of Manchester, Manchester, UK
e-mail: weiqiang.chen@manchester.ac.uk

J. Xiang
Key Laboratory of Rock Mechanics and Geohazards of Zhejiang Province, Shaoxing University, Shaoxing, China

Y. Gao (✉)
School of Transportation and Civil Engineering, Nantong University, Nantong, China
e-mail: Y.Gao@ntu.edu.cn

Z. Zhang
Department of Chemical Engineering and Analytical Sciences, The University of Manchester, Manchester, UK

W. Duan et al. (eds.), *Nanotechnology in Construction for Circular Economy*, Lecture Notes in Civil Engineering 356,
https://doi.org/10.1007/978-981-99-3330-3_55

1 Introduction

By virtue of superior mechanical properties, ultra-high specific area and abundant oxygen-containing functional groups, graphene oxide (GO) exhibits outstanding performance in optimizing the microstructure [1], improving mechanical properties [2], enhancing permeability resistance [3], increasing durability [4], and improving the thermal and electrical conductivity [5] of cement-based materials through nucleation effects [6] and pore-infilling effects [7]. Furthermore, some latest studies indicate that graphene-based enhancement of cementitious composites could be a cost-effective strategy for decreasing CO_2 emissions and energy consumption in practical engineering [8].

The concentration of GO sheets significantly influences the mechanical performance of cement-based grouting materials when they are incorporated into the cementitious material. In the past study, researchers have primarily determined the proportion of GO with the help of experiments and experience [2]. There are few studies on the mechanism of composite enhancement by different ratios of GO. Hence, to better understand the regulatory effect of GO on calcium–silicate–hydrate (C–S–H) gel, it is necessary to understand the effect of different mixing concentrations of GO sheets on the mechanical properties of C–S–H gels. Therefore, in this study, we built a molecular dynamics (MD) model with different ratios of GO nanolayers incorporated into the C–S–H nanocomposite, and performed tensile tests to investigate the effects of the GO content on the reinforcing efficiency of C–S–H composites and the enhancing mechanism.

2 Methods

Following a previous study [9] and using the same model parameters of GO and C–S–H gel, GO/C–S–H nanocomposites were built in which one or two parallel GO nanolayers were embedded in the C–S–H matrix and the spatial orientation of the sheets was aligned with the x direction (Fig. 1). The whole system was firstly relaxed for 2 ns in the *NPT* ensemble at 300 K and 0 atmosphere to achieve equilibrium. Next, tensile deformation was applied to the nanocomposite with a stretching rate of 1 m/s. The *NPT* ensemble was applied in the y and z directions during the tensile deformation at 300 K and 0 atmosphere. The tensile process was completed when the tensile strain in the x direction reached 110%. The C–S–H gel was described by the CLAYFF force field [10] and GO was described by the consistent-valence forcefield [11], with a cutoff radius of 1.5 nm for both Lennard–Jones 12–6 potential and Coulomb electrostatic potential. The intermolecular force between the C–S–H gel and GO nanolayer was described by the Lennard–Jones 12–6 potential and Coulomb electrostatic potential, and the corresponding parameters were derived by

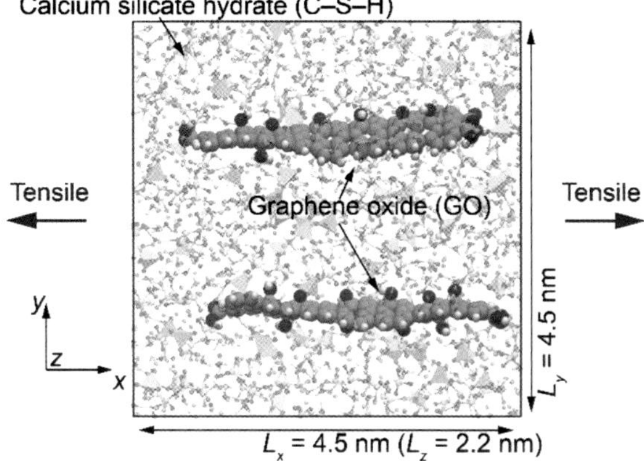

Fig. 1 Atomic model of GO/C–S–H nanocomposite with two GO nanolayers. The model with only one GO nanolayer can be found in Fig. 3a and is not shown here

the Lorentz–Berthelot combining rules. The time step was set as 0.5 fs and the long-range electrostatic interactions were resolved by the particle–particle–particle-mesh method.

3 Results and Discussion

3.1 Tensile Mechanical Properties of the GO/C–S–H Model

The ultimate tensile strength of the MD system did not change significantly after embedding different numbers of GO nanosheets, and the tensile stress–strain curves are presented in Fig. 2. Similar to the visible test results [2], when the MD model reached its peak strength, some damage was generated inside the C–S–H model, but the C–S–H model still had a certain amount of residual strength until its complete failure after further stretching. The monolayer GO/C–S–H model underwent damage at a tensile strain of 0.5. By contrast, the tensile ductility of the bilayer GO/C–S–H model was significantly enhanced to 0.9, with an enhancement ratio of 80%.

3.2 Failure Process of GO/C–S–H Models Under Stretching

In order to reveal the enhancement mechanism of GO incorporated into C–S–H, typical atomic model images (Fig. 3) during stretching were selected to display the microdamage evolution of the C–S–H models with different GO layers embedded. Figure 3 shows that the deformation of the GO/C–S–H model can be divided into the following stages: structural loosening stage, pore development stage, and failure stage.

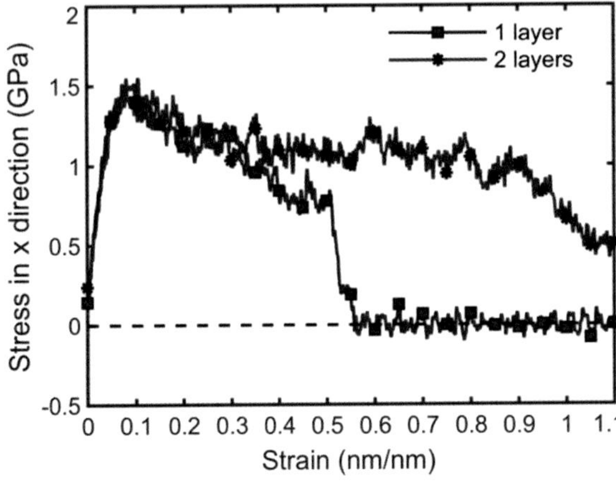

Fig. 2 Mechanical responses of GO/C–S–H models with different GO content under stretching

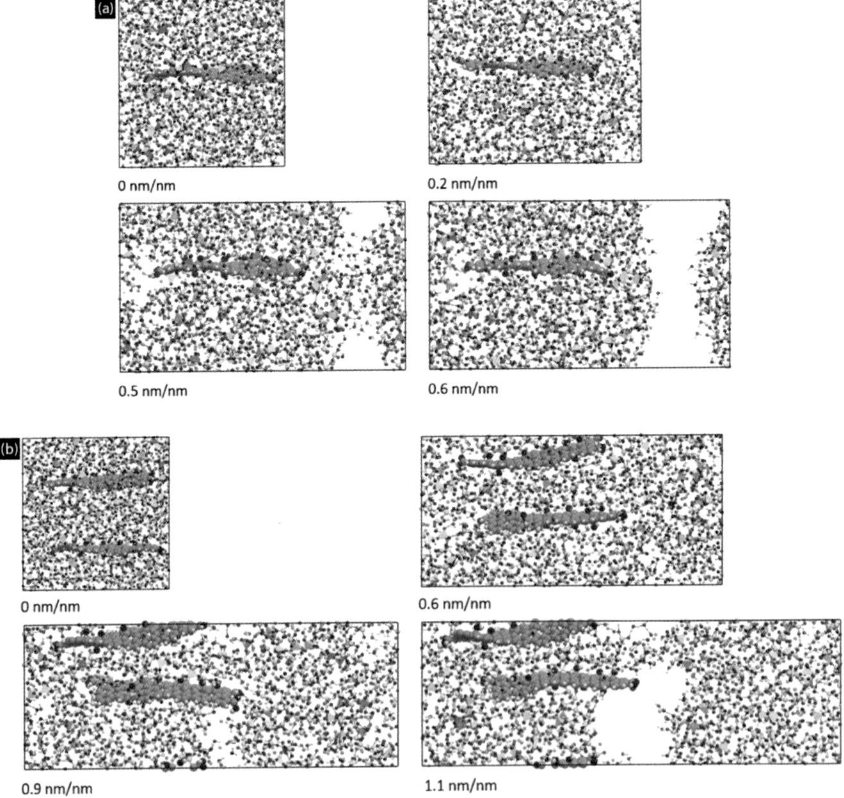

Fig. 3 Microdamage evolution of C–S–H models with different GO content: **a** single layer and **b** two layers

In the monolayer GO/C–S–H model, the C–S–H structure loosened as the tensile strain reached 0.2. As the tensile strain increased to 0.5, noticeable pores began to appear in the structure, but there was still a certain amount of bridging C–S–H, so the model was still a monolithic structure. As the tensile strain continually increased to 0.6, the C–S–H completely separated and fractured, consistent with Fig. 2. At the same time, the molecular structure of the bilayer GO/C–S–H model slowly started to be loosen. Pores gradually appeared and expanded when the strain increased from 0.6 to 1.1. At this time, the C–S–H gel still had residual strength and ability to resist tensile deformation. In addition, it was not hard to find that with increasing deformation, the molecular structure around the GO nanosheets became more closely arranged, especially the C–S–H molecules sandwiched between the two GO layers. It can be concluded that the ductility of the C–S–H was greatly improved with the addition of two GO nanolayers compared with the intercalation of a single layer of GO. Our results also proved that an appropriate amount of GO can effectively enhance the stretching mechanical performance of cement-based materials.

4 Conclusions

The effects of incorporating different numbers of layers of GO nanosheets on the tensile performance of C–S–H gels were studied by MD simulations, the micro-damage evolution of GO/S–H models in the tensile process were investigated, and the enhancement mechanism of GO on the tensile properties of hydrated calcium silicate composites was analyzed at the atomic scale. Our main conclusions are listed below.

(1) Incorporating an appropriate amount of GO was able to effectively improve the ductility of hydrated calcium silicate composites. MD simulations showed that the bilayer GO/C–S–H model increased the post-peak plastic strain by 80% compared with a monolayer GO/C–S–H model.
(2) The deformation of the GO/C–S–H model can be divided into structural loosening stage, pore development stage, and failure stage. Compared with the monolayer GO/S–H model, the strain values corresponding to the bilayer GO model entering the third stage were delayed.
(3) At the atomic scale, the structure of C–S–H around the GO nanosheets was more compact, showing a higher ability to resist tensile deformation, and also reflected the enhancement effect of an appropriate amount of GO on the mechanical properties of C–S–H composites.

Acknowledgements This study was supported by the Zhejiang Provincial Natural Science Foundation of China (No. LQ21E040003).

Author Contributions The manuscript was written through contributions of all authors. All authors have given approval to the final version of the manuscript.

Declaration of Competing Interest The authors declare that they have no known competing financial interests or personal relationships that could have appeared to influence the work reported in this paper.

References

1. Zhang C, Zhu Z, Wang W, Shao L, Wan Y, Huo W (2022) Mechanical properties and microstructure of expansive soil stabilized by graphene oxide modified cement
2. Pan Z, He L, Qiu L, Korayem AH, Li G, Zhu JW, Collins F, Li D, Duan WH, Wang MC (2015) Mechanical properties and microstructure of a graphene oxide–cement composite. Cement Concr Compos 58:140–147
3. Chen X, Zhang Y, Li S, Geng Y, Hou D (2021) Influence of a new type of graphene oxide/silane composite emulsion on the permeability resistance of damaged concrete. Coatings 11:208
4. Devi S, Khan R (2020) Effect of graphene oxide on mechanical and durability performance of concrete. J Build Eng 27:101007
5. Yang H, Monasterio M, Cui H, Han N (2017) Experimental study of the effects of graphene oxide on microstructure and properties of cement paste composite. Compos A Appl Sci Manuf 102:263–272
6. Li W, Li X, Chen SJ, Liu YM, Duan WH, Shah SP (2017) Effects of graphene oxide on early-age hydration and electrical resistivity of Portland cement paste. Constr Build Mater 136:506–514
7. Gao Y, Jing HW, Chen SJ, Du MR, Chen WQ, Duan WH (2019) Influence of ultrasonication on the dispersion and enhancing effect of graphene oxide–carbon nanotube hybrid nanoreinforcement in cementitious composite. Compos B Eng 164:45–53
8. de Souza FB, Yao X, Gao W, Duan W (2022) Graphene opens pathways to a carbon-neutral cement industry. Sci Bull 67:5–8
9. Gao Y, Jing H, Wu J, Fu G, Feng C, Chen W (2022) Molecular dynamics study on the influence of graphene oxide on the tensile behavior of calcium silicate hydrate composites. Mater Chem Phys:126881
10. Cygan RT, Liang J-J, Kalinichev AG (2004) Molecular models of hydroxide, oxyhydroxide, and clay phases and the development of a general force field. J Phys Chem B 108:1255–1266
11. Dauber-Osguthorpe P, Roberts VA, Osguthorpe DJ, Wolff J, Genest M, Hagler AT (1988) Structure and energetics of ligand binding to proteins: Escherichia coli dihydrofolate reductase-trimethoprim, a drug-receptor system. Proteins Struct Funct Bioinf 4:31–47

Graphene-Induced Nano- and Microscale Modification of Polymer Structures in Cement Composite Systems

Z. Naseem, K. Sagoe-Crentsil, and W. Duan

Abstract Redispersible polymers such as ethylene–vinyl acetate copolymer (EVA) have attracted attention in construction due to their enhanced flexural strength, adhesion, flexibility and resistance against water penetration. However, EVA particles cluster in a highly alkaline cementitious matrix and exhibit poor interaction with the cement matrix. The underlying mechanism of poor dispersibility of EVA is attributed to hydrophobic groups of polymers, a variation in the adsorption rate and molecular diffusion to the interface where they cluster together. This phenomenon can negatively affect the fresh properties of cement and produce a weak microstructure, adversely affecting the resulting composites' performance. This study highlights how graphene oxide (GO) nanomaterial alters the nano- and microscale structural characteristics of EVA to minimize the negative effects. Transmission electron microscopy (TEM) revealed that the GO sheets modify EVA's clustered nanostructure and disperse it through electrostatic and steric interactions. Furthermore, scanning electron microscopy (SEM) confirmed altered microscale structural characteristics (viz. surface features) by GO. The altered and enhanced material scale engineering performance, such as the compressive strength of the resulting cement composite, was notable.

Keywords Cement composites · Dispersion · Graphene oxide · Ethylene-vinyl acetate copolymer · Interactions

1 Introduction

Concrete is the most widely used construction material, but its quasi-brittle nature and low durability can affect its performance. Recently, polymer-modified concrete has become popular in response to the durability issue, because redispersible polymers

Z. Naseem (✉) · K. Sagoe-Crentsil · W. Duan
Department of Civil Engineering, Monash University, Clayton, VIC, Australia
e-mail: Zunaira.naseem@monash.edu

W. Duan et al. (eds.), *Nanotechnology in Construction for Circular Economy*,
Lecture Notes in Civil Engineering 356,
https://doi.org/10.1007/978-981-99-3330-3_56

such as ethylene–vinyl acetate (EVA) can modify and enhance flexural strength, adhesion, flexibility and resistance against water penetration [1, 2]. However, EVA shows poor interaction with the highly alkaline cement matrix, which adversely affects the material scale performance, such as the compressive strength of the resulting composite [3, 4].

Polymer additives can cluster in a highly alkaline cement composite due to their poor interaction with the cement matrix [5]. The resultant weak microstructure deteriorates engineering performance, such as the compressive strength of the cement composite [6, 7]. The underlying mechanism is the hydrophobic groups present in the macro-molecular long chain polymer [8]. Additionally, there is variation in the adsorption rate and molecular diffusion to the interface, where they cluster together and interact poorly with the cementitious environment [8, 9].

In this regard, a two-dimensional (2D) nanomaterial such as graphene oxide (GO) can potentially modify the nano- and microscale characteristics by its unique physical and chemical properties and larger surface area [10, 11]. In addition, outstanding mechanical properties of GO have been widely reported to enhance the compressive and tensile properties of resulting cement composites [12, 13]. In addition, the abundant oxygenated groups of GO can strongly interact with cement particles and modify their microstructure [14, 15]. As reported previously, adding a low dosage of 0.05% GO can significantly enhance the engineering performance of cement composites [16]. Muhammad et al. also reported that GO incorporation could alter the microstructure and enhance the transport properties of cement composites [17].

In the present work, we investigated the effect of 2D GO sheets on the nano- and microscale characteristics of the EVA polymer. Scanning electron microscopy (SEM) and transmission electron microscopy (TEM) were utilized to determine the effects of GO on the polymer's structural features. Engineering performance, such as compressive strength, was evaluated using the Instron 4204 50KN. Our experimental results revealed that the addition of GO altered the nano- and microscale characteristics of EVA, resulting in uniform dispersion of the EVA polymer and enhanced performance in the alkaline cement environment.

2 Methods

2.1 Nano- and Microscale Structural Characterization

Nanoscale structural characterization of samples was performed with TEM using a FEI Tecnai T20 instrument operated at 200 kV. Comprehensive nanoscale characterization of EVA polymer, as well as evaluation of the effect of the 2D GO sheets on the polymer, was performed in comparison with the reference EVA polymer. The microscale surface characteristics were investigated by SEM (FEI Nova NanoSEM 450 FEG SEM) under an accelerating voltage of 5 kV and the effect of incorporating GO was compared with reference EVA polymer.

2.2 *Specimen Preparation for Performance Evaluation in Alkaline Cement Matrix*

EVA polymer powder was first dispersed in water, followed by the addition of GO solution before ultrasonication for 10 min. Next, the solution was mixed with ordinary Portland cement (OPC) according to the procedures specified in ASTM C1738-11a to prepare cube-shaped specimens that were vibrated for the 30 s, covered with polyethylene sheets and demolded after 24 h, followed by the curing method. The engineering performance was specifically investigated as the compressive strengths of the cubic specimens. A universal loading machine, the Instron 4204 50KN, was used to test the cement specimens at 28 days of age. For each batch, a minimum of five specimens was tested and the average values were taken as the compressive strength.

3 Results and Discussion

3.1 *Modification of Nano- and Microstructural Characteristics by GO*

The nanostructure of the EVA polymer as examined by TEM is shown in Fig. 1. EVA polymer exhibited a clustered structure of polymer particles (Fig. 1a), which could be attributed to the rate of change of adsorption and diffusion, causing aggregation of the polymer particles. One of the critical reasons for their poor interaction with the highly alkaline cementitious environment is their aggregation, which deteriorates the material scale performance of the resultant cement composites. Remarkably, as presented in Fig. 1 (b, c), the incorporation of GO sheets disperses the polymer particles uniformly compared with the reference EVA polymer. The underlying reason could be the unique physical and chemical properties and larger surface area of GO, which alters the EVA polymer's nanostructure. In addition, effective electrostatic and steric interactions between the GO sheets and polymer molecules could hamper their aggregation [18].

Furthermore, SEM also showed the microscale characteristics, which confirmed the changes in the the nanoscale characteristics. Aggregated surface features were present in the reference EVA polymer (Fig. 2a). The addition of GO altered the microscale structure characteristics of the polymer (Fig. 2b), which was consistent with TEM results and confirmed the potential of GO to modify the EVA polymer characteristics at the nano- and microscale due to its unique physical and chemical properties and abundant functional groups on the basal plane that improved the interaction and material scale performance of the resultant cement composites [10].

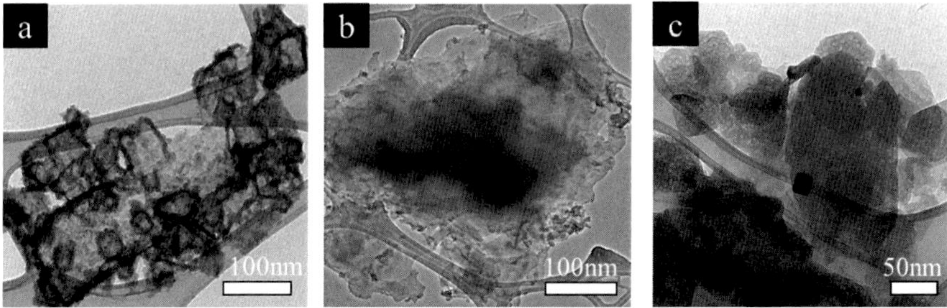

Fig. 1 Nanoscale structural characteristics by TEM. **a** EVA polymer shows clustering of molecules and aggregation, which causes poor interaction with the cement matrix. (**b**, **c**) GO dispersion of clustered polymer structure through electrostatic and steric interactions and alteration of the nanostructure

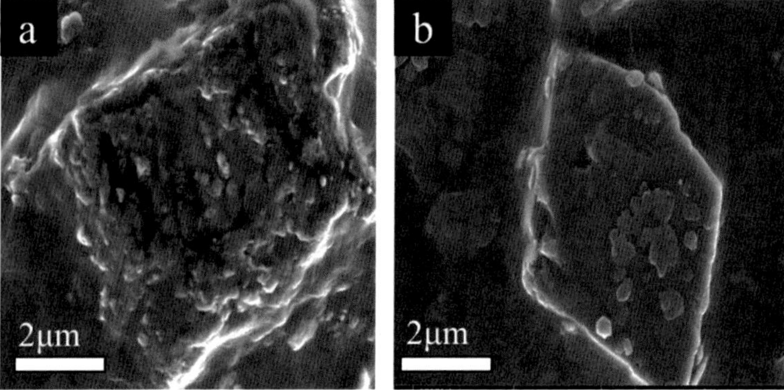

Fig. 2 Microscale surface features by SEM. **a** EVA polymer shows aggregated surface features at the microscale. **b** Altered aggregated surface features of the polymer with GO addition compared with the reference EVA polymer in (**a**)

3.2 Engineering Performance in the Alkaline Cement Environment

The material scale performance of the EVA polymer with and without GO in the highly alkaline cementitious environment was further assessed. Specifically, the compressive strength of the prepared cement composite specimens was investigated at 28 days of hydration age. As shown in Fig. 3, the EVA-incorporated cement composite (PMC) exhibited a lower compressive strength than OPC composites. The underlying reason is the clustered polymer structure and the poor interaction with the highly alkaline cement matrix, resulting in a deteriorated performance of the PMC. Remarkably, the addition of GO improved and enhanced the compressive strength by ~40% higher than the reference PMC samples. The underlying phenomena could be attributed to altered nano- and microscale characteristics of

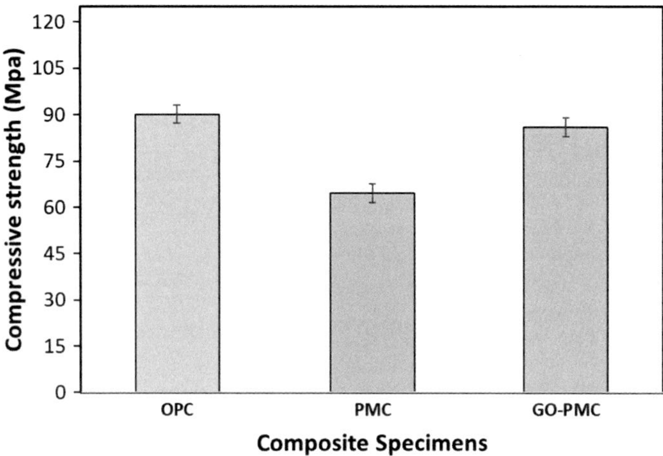

Fig. 3 Compressive strength of cement samples showing significantly enhanced compressive strength compared with the reference PMC sample

EVA by the GO sheets. In addition, the presence of GO can disperse the polymer particles through electrostatic and steric interactions and hamper their aggregation in the alkaline cementitious environment. As a result, significantly enhanced material scale performance was observed in the resultant cement composites (GO-PMC).

4 Conclusion

In this study, we used the nanomaterial GO as a novel approach to altering and improving the structural characteristics and the material scale performance of polymers in cementitious environments. TEM showed that GO altered the aggregated surface characteristics of EVA at the nanoscale, which was further confirmed by SEM displaying the modified surface features compared with the reference EVA polymer. The compressive strength of the composite specimens was quantified to signify their engineering performance at the material scale. Compared with the reference EVA composite, the GO–polymer-modified composite (GO-PMC) achieved ~ 40% higher compressive strength. This improvements in engineering performance is attributed to the altered nano- and microscale characteristics of the polymer by GO causing electrostatic and steric interactions and hampering particle aggregation in the cement matrix.

Acknowledgements The authors are grateful for the financial support of the Australian Research Council in conducting this study. They acknowledge the use of facilities within the Monash Center of Electron Microscopy.

References

1. Tarannum N, Pooja K, Khan R (2020) Preparation and applications of hydrophobic multicomponent based redispersible polymer powder: a review. Constr Build Mater 247:118579
2. Tran NP, Nguyen TN, Ngo TD (2022) The role of organic polymer modifiers in cementitious systems towards durable and resilient infrastructures: a systematic review. Constr Build Mater 360:129562
3. Wang R, Wang P-M (2011) Action of redispersible vinyl acetate and versatate copolymer powder in cement mortar. Constr Build Mater 25(11):4210–4214
4. Shi C, Zou X, Yang L, Wang P, Niu M (2020) Influence of humidity on the mechanical properties of polymer-modified cement-based repair materials. Constr Build Mater 261:119928
5. Zhong S, Chen Z (2002) Properties of latex blends and its modified cement mortars. Cem Concr Res 32(10):1515–1524
6. Pei M, Kim W, Hyung W, Ango AJ, Soh Y (2002) Effects of emulsifiers on properties of poly (styrene–butyl acrylate) latex-modified mortars. Cem Concr Res 32(6):837–841
7. Peng Y, Zhao G, Qi Y, Zeng Q (2020) In-situ assessment of the water-penetration resistance of polymer modified cement mortars by μ-XCT. SEM EDS Cement Concrete Compos 114:103821
8. Sun K, Wang S, Zeng L, Peng X (2019) Effect of styrene-butadiene rubber latex on the rheological behavior and pore structure of cement paste. Compos B Eng 163:282–289
9. Myers D, Myers D (2006) Surfactant science and technology
10. Montes-Navajas P, Asenjo NG, Santamaría R, Menendez R, Corma A, García H (2013) Surface area measurement of graphene oxide in aqueous solutions. Langmuir 29(44):13443–13448
11. Shamsaei E, de Souza FB, Yao X, Benhelal E, Akbari A, Duan W (2018) Graphene-based nanosheets for stronger and more durable concrete: A review. Constr Build Mater 183:642–660
12. Suk JW, Piner RD, An J, Ruoff RS (2010) Mechanical properties of monolayer graphene oxide. ACS Nano 4(11):6557–6564
13. Basquiroto de Souza F, Yao X, Lin J, Naseem Z, Tang ZQ, Hu Y, Gao W, Sagoe-Crentsil K, Duan W (2022) Effective strategies to realize high-performance graphene-reinforced cement composites. Constr Build Mater 324:126636
14. Gao W (2015) The chemistry of graphene oxide. Springer, Graphene oxide, pp 61–95
15. Lv S, Ma Y, Qiu C, Sun T, Liu J, Zhou Q (2013) Effect of graphene oxide nanosheets of microstructure and mechanical properties of cement composites. Constr Build Mater 49:121–127
16. Pan Z, He L, Qiu L, Korayem AH, Li G, Zhu JW, Collins F, Li D, Duan WH, Wang MC (2015) Mechanical properties and microstructure of a graphene oxide–cement composite. Cement Concr Compos 58:140–147
17. Mohammed A, Sanjayan JG, Duan W, Nazari A (2015) Incorporating graphene oxide in cement composites: a study of transport properties. Constr Build Mater 84:341–347
18. Ohama Y (1998) Polymer-based admixtures. Cement Concr Compos 20(2–3):189–212

Printed in the United States
by Baker & Taylor Publisher Services